Racing for the Bomb

Racing for the Bomb

General Leslie R. Groves,

The Manhattan Project's

Indispensable Man

ROBERT S. NORRIS

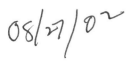

STEERFORTH PRESS
SOUTH ROYALTON, VERMONT

For information about permission to reproduce
selections from this book, write to:
Steerforth Press L.C., P.O. Box 70,
South Royalton, Vermont 05068

3 1172 04824 6540

Library of Congress Cataloging-in-Publication Data

Norris, Robert S. (Robert Stan)
Racing for the bomb : General Leslie R. Groves, the Manhattan Project's indispensable
man / Robert S. Norris.— 1st. ed.
p. cm.
Includes bibliographical references and index.
ISBN 1-58642-039-9 (alk. paper)
1. Groves, Leslie, R., 1896–1970. 2. Generals—United States—Biography. 3. Military
engineers—United States—History. I. Title.

UG128.G76 N67 2002
355.8'25119'0973—dc21   [B]   2001057629

FIRST EDITION

For Myriam

CONTENTS

PREFACE

Nearly sixty years after the end of World War II, the public's fascination with the Manhattan Project shows no sign of diminishing. The story of developing, testing, and dropping the atomic bomb is an exciting one, filled with the elements of great novels and high drama. The stakes at the time were huge, the political and moral issues profound, the legacy in the decades that followed lasting.

The most glaring omission in the vast literature about the Manhattan Project is the personal history of the key person who made it a success, Gen. Leslie R. Groves. Remarkably, there is no scholarly biography of him, of how he helped shape one of the great events of the twentieth century.[1] Most histories mention Groves, of course, usually having him appear on the scene in September 1942 to take charge. Sometimes passing reference is made to him having overseen the building of the Pentagon, or that he went to West Point and was an engineer, but not much else. His character remains indistinct and out of focus. While he is described as the chief administrator, he is not shown going about his task or depicted as the indispensable figure, integral to the success of the project. In overlooking Groves's role we lose valuable perspectives on many interesting debates that continue to surround the bomb. To place Groves back at the center of events, where he was and where he belongs, offers new insights into these important issues.

The purpose of this book is to provide a fuller picture of the life and career of Leslie Richard Groves, and his role in the beginnings of the atomic age. What did he do the first forty-six years of his life to prepare him for his most important assignment? Why did the leadership of the army pick him to manage the Manhattan Project, and how did he administer it? What influence did he have on the use of the bomb on Japan at the end of the war? Would the bomb have been ready in time for use without him? What did he do after the war?

Of equal import, and equally neglected, is the question of Groves's influence in shaping what is sometimes called the national security state. The phrase attempts to define the new governmental departments and agencies that emerged in the aftermath of World War II and evolved throughout the Cold War, and the procedures and practices by which they operated. These features include the widespread concern for security and secrecy, compartmentalization as an organizing scheme, "black" budgets, the reliance on intelligence and counterintelligence, and the interlocking relationship of government, industry, science, and the military. All of them first emerge during the Manhattan Project.[2]

At the center of the Cold War — the thing that made it different from all that came before — was the bomb. Though Groves was not alone in recognizing this, he did understand, long before it was used, how important this weapon was going to be in the new world to come.

I will argue that Groves was the indispensable person in the building of the atomic bomb and was the critical person in determining how, when, and where it was used on Japan. Without Groves's vision, drive, and administrative ability, it is highly unlikely that the atomic bomb would have been completed when it was. The Manhattan Project did not just happen. It was put together and run in a certain way: Groves's way. He is a classic case of an individual making a difference. Being in the right place at the right time is the secret of winning a place in history; rarely does a person arrive there by accident.

Though his perspective was always that of an engineer and an administrator, Groves's grasp of the pertinent scientific principles was more than enough to get the job done. He displayed a remarkable ability to choose among technical alternatives when it was hard to know which path would deliver results, and he did so quickly. He made many crucial decisions that, had they been the wrong ones, would have resulted in failure or delay. His judgment of whom to choose as his subordinates and whom to rely upon for counsel and advice was uncanny. To get such things right once or twice might be considered a matter of luck. To do so over and over speaks to a person having informed judgment and keen instincts. Of all the participants in the Manhattan Project, he and he alone was indispensable.

The way Groves has been depicted in the literature about the Manhattan Project is mostly a collage of inaccuracies, caricatures, or superficialities. Supposedly authoritative sources do not get even the most basic facts of his life correct. An example is the entry in Oxford University Press's *American National Biography*.[3] The author has Groves accompanying his father to Cuba and the Philippines, serving a three-month tour of occupation duty in France with the American Expeditionary Force immediately after World War I, having two daughters, winning the Nicaraguan Medal of Merit for his work there on a canal, spending six million dollars a month constructing military facilities on the eve of the war, and assuming control of the Manhattan Project (MED) on September 7, 1942. We also learn that by the spring of 1945 there was enough plutonium to create several weapons, that "a 400-pound device" nicknamed Little Boy demolished Hiroshima on August 6, and that four days later Nagasaki was bombed. Just about every "fact" cited here is wrong.[4]

In fact Groves was a larger-than-life figure, a person of iron will and imposing personality who knew how to get things done. He was ambitious, proud, and resolute. From the moment he was appointed to head the Manhattan Project, he was determined that the atomic bomb should be the instrument to end the war. For that to happen required enormous effort constructing and operating industrial plants of unprecedented size and function, with no moment lost.

He was brusque, sometimes to the point of rudeness, and cared little what others thought of him. He expected immediate, unquestioning compliance with his orders, and had little patience with abstract ideas, which might distract from the immediate job at hand. He reacted firmly and defensively to those he judged as threats to the attainment of his goals. His detractors often pointed to his corpulence as a sign of his unsuitability for the post he occupied. Exactly why this was the case is never explained, but it reinforced their negative image of him as arbitrary and ignorant. When he was overweight, which was most of the time, he did not look like what we expect a general to look like, but it had no effect on his ability.

The book is divided into seven parts and is largely chronological. The first three parts — ten chapters in all — cover Groves's life and career to the point of being chosen to head the Manhattan Project. Groves's roots are traced back to the mid–seventeenth century. The general's deep-seated patriotism derives in part from being an eighth-generation American, with special pride reserved for a few military forebears. Son of an army chaplain, Groves grew up on forts and posts all across America amid the military culture and traditions of the early twentieth century. This exposure, plus a highly competitive family situation, set him apart as a self-sufficient and driven young man. For reasons that we will see, his upbringing fostered self-reliance and independent judgment. From an early age he was iron willed and insistent on doing things his way. Self-assured and confident of what he wanted, he had decided by the time he was sixteen that the army would be his calling, and that it was essential to get into West Point. During his years there — abbreviated by the pressure of World War I — Groves

earned an excellent record, graduating fourth in his class and choosing the
elite Corps of Engineers as his branch.

Over the next twenty-three years, to mid-1942, Groves gained extensive experience participating in, and at times overseeing, a variety of construction and engineering projects. Along the way he met and worked
with scores of army officers, corporation executives, architects, engineers,
and contractors, many of whom he would later recruit to help him build
the bomb. It is crucially important to examine who these people are and
the relationships that Groves had with them. In this period we also see
Groves's personality shaped in fundamental ways by the special mores, traditions, and culture of the interwar army and Corps of Engineers.

Just before entering Engineer School in 1919, Groves and his classmates
went to Europe for a short summer study tour of the battlefields of France
and Germany, his developing engineer's mind soaking in everything he
saw. We also see him express his contempt and distrust of European culture and his belief in the superiority of everything American.

Groves married his longtime sweetheart and started a family. His career
advanced steadily as years of hard work increasingly drew the attention of
his superiors. A four-year tour in the chief's office in Washington put him
in a small circle of engineers that would have an enormous influence on
his career and perspective. Foremost among them was Col. Ernest Graves,
the person from whom he probably learned the most. Clearly marked for
advancement by senior officers, he was then sent to the Command and
General Staff School at Fort Leavenworth, Kansas, and later the War
College, completing the schooling expected of those who hold high positions in the army.

The busy three years prior to being chosen to head the Manhattan
Project started with an assignment to the General Staff, then to positions
with army construction during the mobilization period for World War II.
By the spring and summer of 1942 Groves was basically in charge of all
army construction in the United States, overseeing hundreds of projects
and expending huge sums of money. One of these many projects was to
build the Pentagon.

The core of Groves's life was the extraordinary experience, beginning
in the fall of 1942, of running the Manhattan Project. After the war, one
of his aides provided an acute characterization of the different roles that
he performed simultaneously: "General Groves planned the project, ran his
own construction, his own science, his own army, his own State
Department and his own Treasury Department."

New insights about the Manhattan Project emerge by sharply focusing on Groves's expanding responsibilities from 1942 to 1945. At the outset he was engineer and builder, charged with constructing the plants and factories that would make the atomic fuels — highly enriched uranium and plutonium. As the project accelerated, Groves came to oversee a vast security, intelligence, and counterintelligence operation with domestic and foreign branches. Through his final-decision-making power he was ultimately in charge of all scientific research and weapon design and kept close watch on the laboratory at Los Alamos. He was involved in many key high-level domestic policy issues, and in several international ones as well. By 1945, in addition to all else, he effectively became the operational commander of the bomber unit he established to drop the bomb, and was intimately involved in the planning, targeting, and timing of the missions. One is struck, in discovering all of his many activities, by just how much power Groves accrued. As a West Point classmate and friend later observed, "Groves was given as much power in that position as any officer ever has had."[5] A remarkable statement.

Basically Groves made nearly all of the significant decisions having to do with the bomb, from choosing the three key sites — Oak Ridge, Los Alamos, and Hanford — to selecting the large corporations to build and operate the atomic factories. He chose the key military officers who were his subordinates as well as such important figures as Robert Oppenheimer and William S. Parsons, the weaponeer aboard the *Enola Gay* in charge of dropping the bomb on Hiroshima. We are fortunate to have an extensive record of his daily activities while he ran the Manhattan Project, kept by his trusted secretary, Jean O'Leary. Groves's detailed appointment book notes thousands of telephone calls, the visitors to his office, the appointments he had outside the office, and even some of what was discussed in calls he made to Washington when he was traveling. It is a valuable contribution toward our understanding of how Groves managed the project day in and day out. The cast of characters is enormous. Familiarity with who they were and what roles they played in building the bomb is essential to understanding Groves's centrality.

Just as important as building the bomb are the complex and controversial questions surrounding its preparation, its use against Japan, and the immediate aftermath. Groves knew from the outset that the factor setting the pace for the entire project was the availability of the atomic fuels. When there was enough highly enriched uranium and/or plutonium, the use of a bomb would soon follow. In an extraordinary coincidence, there were sufficient amounts of each fuel ready at approximately the same time, and within a matter of days after the material was incorporated into the two types of bombs, they were used to destroy two Japanese cities.

This was essentially Groves's doing, through his role in the Target and Interim Committees and his supremacy in controlling the test at Trinity and the combat missions to Hiroshima and Nagasaki. In most of the vast literature about the decision to use the bomb and the bomb's role in ending the war, Groves has been slighted or ignored altogether. In fact, given his commanding position at the center of all the activities having to do with the bomb, he actually was the most influential person of all. It was primarily due to him that usable bombs were ready when they were and were employed immediately afterward. Groves alone knew all the details of the bomb and as a result controlled its testing, production, transport to the Pacific, and delivery on Japan. This of course was his job, and he performed it with a fervor and determination that few, if any, could have matched. He was the right man at the right place for the job. His superiors, civilian and military, were in perfect agreement that the bomb should be used when ready. As project head Groves had created a structure of such magnitude and momentum that it was, by mid-1945, unstoppable. By the summer of 1945 a decision not to drop the bomb would have been almost impossible. The "decision" to use the bomb was inherent in the decision made years before, to build it.

Groves's personal feelings about the bomb were not complicated: He never had any moral doubts, at the time or afterward, about using it. From the outset he believed that the bomb could be decisive in ending the war and saving American lives. As it turned out, with regard to the Pacific war it did both.

Groves never lacked ambition; nor was he modest about his accomplishments. From the outset of war he hoped that his efforts might contribute to winning it. In 1942 he was denied the opportunity to go to the theaters of war overseas. His actions make it clear that after a few moments of disappointment, he decided that building the bomb would be his achievement. Knowing it might end the war was one source of Groves's extraordinary determination and energy.

But Hiroshima and Nagasaki were not the end of Groves's job. When the bomb was no longer a secret, a host of new questions quickly emerged. Groves was instrumental in the preparation of much of the initial information released to the public about where and how the bomb was pro-

duced and the interesting personalities that were responsible. He did this
through overseeing the writing of the president's and secretary of war's
statements, press releases, and the Smyth Report, as well as allowing *New
York Times* science reporter William Laurence exclusive access to the
Manhattan Project.

Groves's final battles centered on the transfer of the Manhattan Project
to the civilian Atomic Energy Commission, his involvement in the questions of international control of atomic energy, and his position as chief of
the Armed Forces Special Weapons Project. Literally overnight Groves's
situation changed dramatically, and Groves had trouble adjusting to the
new rules of the postwar Pentagon and Washington politics. His often
high-handed style and treatment of people during the war eventually
caught up with him as several of the powerful enemies he made along the
way had their revenge. In the end he was forced into early retirement.

After he left the army in 1948, he worked for Remington Rand (later
Sperry Rand) and moved to Darien, Connecticut. He remained a public
figure throughout the 1950s, speaking and writing on issues of the day, but
from the sidelines. Much time was spent overseeing how his major accomplishment, the Manhattan Project, was being dealt with in histories, biographies, newspaper and magazine articles, films, and television. He wrote
his own account, *Now It Can Be Told,* which was published in 1962. He
kept up an extensive correspondence, served as president of West Point's
alumni organization, the Association of Graduates, and took charge of
building a retirement home for army widows. He retired from Sperry in
1961, and he and his wife moved back to Washington three years later. In
his final years he led a quiet life, occasionally visiting his son and daughter,
and enjoyed being a grandfather to his seven grandchildren.

General Groves left a deep imprint on the nuclear age that followed, one
not always appreciated. Along with his broad influences on domestic institutions and international relations, Groves's hand is more clearly seen in the
direct descendants of the Manhattan Engineer District itself. In many
respects the practices and culture of the Manhattan Project carried over to
the Atomic Energy Commission and its successors, and have lasted to this
day. Hanford, Oak Ridge, and the many other facilities in the nuclear complex always have been run on the government owned–contractor operated
(GOCO) basis that Groves first implemented. The university-laboratory relationship is intact, with the University of California running Los Alamos
and Livermore. The AEC from its inception was largely a nonaccountable

bureaucracy shielded from the public and most of Congress by layers of classification and secrecy, and the invocation of "national security." Arguments to build more and more bombs always took precedence over environmental, health, and safety issues, resulting in a troubling and costly legacy.

The Manhattan Project is also periodically held up as a model to emulate whenever the nation decides to undertake a large endeavor. It is worth asking what lessons from the Manhattan Project might be transferable to other large projects, and what was unique to the time.

Counterfactual arguments are popular in the literature about the bomb, and are sometimes useful exercises. One question worth pondering is what would have happened in the decades following World War II if the bomb had never been invented. Would there have been a major conventional war between the Soviet Union and the West? Probably so. As with all counterfactual arguments, however, we will never know. What we do know is that there was not a major war, and the bomb surely had something to do with this. The second half of the twentieth century remained a dangerous time, and the arms race threatened at times to spin out of control, but a major war between the superpowers was averted.

Two basic ways of attempting to live in a world with nuclear weapons confront us today, just as they did Groves. His way — and it is still the way of most — was the unilateral one. In a dangerous world a nation must depend upon itself, just as an individual has to. Trust and cooperation should not be relied upon given figures such as Hitler, Stalin, or their descendants.

The contrary path has its advocates, too. Two of the wiser men of the twentieth century saw early on that the bomb might bring a new type of order into the world. In 1943 Niels Bohr asked Robert Oppenheimer, "Is it big enough?" By this he meant, Was the bomb of such destructive power that it would make war impossible? Einstein's words convey the same thought: "It may intimidate the human race into bringing order into its international affairs, which, without the pressure of fear, it would not do." Order we do not have. But peace, of a kind, we do — or at least the absence of major wars between great powers following Hiroshima.

ACKNOWLEDGMENTS

It is sobering to realize that it took me longer to write this book than it took General Groves to build the bomb. I could argue that he had more people to help him, but when I began to count up all of those who assisted me, I may not be far behind.

I would like to thank Lt. Gen. Richard H. Groves, USA, retired, who has been extremely generous in sharing with me family letters, documents, and photographs. He has been patient in answering my endless questions and never once interfered with anything I wrote about his father or family.

I am fortunate to work for the Natural Resources Defense Council, an organization that encourages its staff to explore opportunities for professional growth. All of NRDC was supportive of my extracurricular adventure, but I would like particularly to thank John H. Adams, Frances Beinecke, and Judy Keefer in the New York office and Frederick A. O. Schwarz, chair of the board of trustees. The opportunity to take a sabbatical to finish the book was essential; I am grateful that NRDC has this wise policy. Closer to home my colleague Thomas B. Cochran has always been in my corner and was again this time, as were S. Jacob Scherr, Matthew G. McKinzie, and Christopher E. Paine.

From the day I started until the day I finished William M. Arkin, colleague, mentor, and friend, provided sound advice and wise counsel. Working with the staff of Steerforth Press was especially satisfying, particularly with Kristin Camp, Janet Jesso, Chip Fleischer, Helga Schmidt, and Stephanie Carter. My editor Thomas Powers read every word, held my hand, cracked the whip, and kept up morale. His suggestions and comments have improved the book considerably. Laura Jorstad's sharp red pencil smoothed the rough edges, Peter Holm produced an excellently designed book, and Patsy Fortney provided a thorough index. I would especially like to thank Gregg Herken and Priscilla J. McMillan, fellow laborers in the same vineyards, for their constant support and encouragement. Having colleagues read and critique draft chapters along the way is of special importance. I am grateful to those who read one or more of the chapters and suggested improvements: James L. Abrahamson, Steven Aftergood, William Arkin, William Burr, John Coster-Mullen, Gregg

Herken, Priscilla J. McMillan, Fred B. Rhodes, Coni O'Leary Watson, and Susan Watters.

The staffs of archives and libraries are essential to projects of this type. The National Archives II building in College Park, Maryland, is a pleasure to work in. Michael J. Kurtz provided encouragement, and John Taylor, Wil Mahoney, Tim Nenninger, and Marjorie Ciarlante helped me navigate through the vast Manhattan Project records and other holdings. The staff at the Library of Congress Manuscript Division was always helpful. The Washingtoniana Division of the District of Columbia Public Library helped with details about wartime Washington. At the U.S. Military Academy I would like to thank Alan C. Aimone, Special Collections and Archives Division, and Suzanne Christoff. Paula Kontowicz at Fort Leavenworth's Combined Arms Research Library helped with details about the Command and General Staff School in Groves's day. Martin K. Gordon, Office of History, U.S. Army Corps of Engineers, guided me through its important holdings. F. G. "Skip" Gosling, chief historian at the Department of Energy, responded to my requests. Los Alamos Archivist Roger A. Meade provided copies of documents and answered many queries. Alfred Goldberg, historian, Office of the Secretary of Defense, was a source of advice and kindly loaned me several photographs. Marjorie G. McNinch, reference archivist of the Hagley Museum and Library, made my visit to Wilmington to examine Du Pont records a productive one. The staff at the Niels Bohr Library of the American Institute of Physics Library were helpful, as were the staffs of the Darien Library in Connecticut and the Edgewater Library in Maryland. Much time was spent in the Pentagon Library, and I thank the professional staff for their help. Stanley Goldberg visited several libraries, and they deserve to be thanked as well. They include the Harry S. Truman Presidential Library, the Herbert Hoover Presidential Library, the Bancroft Library of the University of California, Berkeley, and the Public Record Office, Kew Gardens, England.

Lt. Gen. Ernest Graves Jr., USA, retired, was generous in providing photographs of his father and answered numerous questions. Robert F. Furman was a constant source of support and information. I would like to thank Barbara Vucanovich for loaning me photographs of her father, Thomas Farrell. Patricia Cox Owen was generous in loaning me photographs and other items. The correspondence and interviews I had with almost two dozen of the Sandia pioneers turned into a valuable resource and I would like to thank all of them.

A long list of people deserve thanks for help of one kind or another: Tom Allen, Sharon H. Bartholomew, Mara Benjamin, Barton J. Bernstein, Michael S. Binder, Kai Bird, Ellen Bradbury, Curtis Bristol and his discussion group, Joseph Cirincione, Avner Cohen, Hedy Dunn, Anne Fitzpatrick, Richard L. Garwin, Michele S. Gerber, Cary T. Grayson Jr., Chuck Hansen, Vilma R. Hunt, James Hershberg, William Lanouette, John Kennedy Ohl, Vincent Orange, George S. Pappas, Norman Polmar, Richard Rhodes, Edgar B. Smallwood, John Stevens, and Larry Suid. For her love of books and reading, which she passed on to me, I would like to thank my mother, Betty Norris.

My largest debt is to my wife, Myriam, for her patience and tolerance. I would like to dedicate the book to her.

R. S. N.
Edgewater, Maryland

A NOTE FROM THE AUTHOR

Abrief word is needed to explain how I became involved in writing this book. The idea of a biography of General Groves started with historian of science Stanley Goldberg, Ph.D. Dr. Goldberg was known for his efforts to demystify science and to explain issues of science and public policy. He was respected among his colleagues for the unusual breadth of his interests and the passion with which he communicated them to professional and lay audiences alike. He devoted many years of enthusiastic research to the subject of General Groves and the Manhattan Project. His industry was prodigious in visiting archives and libraries and accumulating copies of documents and records. He also interviewed many people who knew and worked with Groves, and he published a number of articles about the general and the Manhattan Project (see the notes and works frequently cited sections).

Stanley died suddenly in October 1996 before he had begun the writing of his book. His death was a great loss to all who knew him. Shortly afterward his family generously agreed to let me make use of the research material he had accumulated. This has been a valuable resource in my writing of the book, and the memory of his friendship sustained me during the long labor of recording Groves's life and achievement. For their generosity and support I would like to especially thank Susan Galloway Goldberg, Ruth Goldberg, David Goldberg, and Eve Goldberg.

It is clear that this book is different from the one that Stanley Goldberg might have written in its focus and conclusions. But knowing him as I did I am sure that he would have enjoyed arguing his views with me. All of this would have been done in the spirit of the intellectual exchange that he saw as a scholar's responsibility to better understand complex issues through debate and dialogue.

Racing for the Bomb

At the Top of His Game

There are few cities in the world as beautiful as Washington in springtime. The cherry trees around the Tidal Basin have thousands of relatives sprinkled throughout neighborhoods all over the city. Other fruit trees and dogwoods come into bloom as well, adorned by a rainbow of azaleas that prosper on the banks of the Potomac.[1] Examples of this bouquet in the spring of 1945 were prolific on Mount Saint Alban, specifically in the 3500 block of Thirty-sixth Street between Ordway and Porter in Northwest Washington. In the shadow of the National Cathedral, at number 3508, the alarm went off at 5:30 A.M., and Maj. Gen. Leslie Richard Groves awoke to prepare for another busy day.

The eight-room yellowish brown brick house built in 1924 was modest, joined like Siamese twins with its next-door neighbor in a style common in the Cleveland Park neighborhood. On the first floor were a living room, a dining room, a kitchen, and a back room facing a tiny yard in the rear. Upstairs Grace, the general's wife, slept in the front bedroom, the general occupied the middle bedroom, and their daughter had the bedroom in the back. When their son was home, he used the small fourth bedroom.

They had found the house and rented it in August 1939, for eighty-five dollars a month, soon after Dick finished his year at the Army War College.[2] From there he was assigned to a slot on the War Department's General Staff. The house was near a good school for their daughter; that was the principal reason why they chose it.

As they settled into their new home, the war in Europe was intensifying. Over the next two years, especially after Pearl Harbor, Dick's career accelerated rapidly. Grace did not know exactly what he did with the Army Corps of Engineers, and under his admonition, issued long ago, she never inquired. Now almost six years later in April 1945, this house was where they had lived the longest in their twenty-three years of marriage and nomadic army life.

Their son Richard was in his final year at West Point and would be graduating in less than six weeks. Sixteen-year-old Gwen was at home and would soon be getting up to go to the National Cathedral School for Girls, on Woodley Road just five blocks away. She was finishing her junior year and beginning to prepare for her College Board examinations. During the summer, she planned to take a typing class downtown at a secretarial school. Her father had always sought the best education for both of his children, even on an army officer's meager pay. Promoting individual excellence and enhancing intelligence were goals to be supported and fostered. It had been so in Dick's family growing up in the first two decades of the century, and now these values were being passed on to the next generation.

The general slept in a long, cotton marine corps T-shirt, a kind of makeshift nightshirt, a habit he had taken up long ago. He washed, shaved, and combed his thick black hair, now increasingly turning to an iron gray. He put on his uniform, noticing that the trousers were a bit tight. At just a shade under six feet, he weighed around 240 pounds. The diets he embarked upon never lasted long, undone in part by a ferocious sweet tooth and the enormous pressures of his job. Sugar gave him energy, and he needed lots of it to get the job done. He finished dressing and came downstairs to the kitchen. He never drank coffee, so breakfast was a quick drill. He fixed his own scrambled eggs, stirred with a fork as they cooked, while his wife and daughter continued to sleep upstairs. At about six-fifteen he emerged out the back door, walked across the concrete path through the small patch of grass that passed for a backyard to the garage, climbed into his 1942 apple-green Dodge four-door sedan, started it up, and maneuvered down the narrow alley to Thirty-sixth Street. He had been fortunate with this car. He had purchased it just days before Pearl Harbor. Very soon afterward, the automobile companies stopped producing passenger cars altogether, switching to tanks and military vehicles. Though now more than three years old, his Dodge was one of the newer cars in town and had less than ten thousand miles on it. Still, Dick was never one for material things. The car served only the utilitarian purpose of transportation, of getting him from point A to point B as quickly as possible. There were only so many hours in the day, and every one of them had to be used efficiently if this job was going to get done. All the better that the car blend in and not draw any attention to him or to what he was doing.

There were several possible routes he might follow to work. This morning he took a right onto Porter, then a right onto Thirty-fourth Street and a left onto Massachusetts Avenue, past the Naval Observatory and, just beyond it, the British Embassy, descending the long hill across the bridge that spans Rock Creek Park. Past Sheridan Circle, Groves took a right onto Twenty-first Street. He knew Twenty-first Street well. In the early 1930s he had walked almost its entire length with Colonel Ernest "Pot" Graves on their way to work in the Office of the Chief of Engineers in the Munitions Building on the Mall. He had learned a lot from Graves and was now putting it to good use in a job no one could have fathomed even a few years ago. Upon arriving on Virginia Avenue, he drove to the New War Building and parked in his space in the underground lot. The Corps of Engineers had moved into this brand-new building in mid-1941 from the Railroad Retirement Building across town. From those offices, Groves had overseen billions of dollars worth of army construction during the mobilization period and the first ten months of the war. One of his projects was the construction of the Pentagon, just across the river. Over the past two years, given the high level of secrecy of his real work, he presented himself in public as deputy chief, Construction Division, and later assistant to the chief of engineers, occasionally attending ceremonies to open a munitions plant here or a training base there.

Groves entered the New War Building through the art deco doorway and lobby of yellowish red travertine and black marble, past the security guard, and continued under the fifty-foot-wide mural, *America the Mighty,* taking one of the eight elevators to the fifth floor. After a short walk down the corridor, he unlocked the door at number 5120 and entered the outer office, where several secretaries and a few army officers would soon arrive. Their gray steel desks sat on the pale green carpet.[3] On them were telephones, typewriters, In and Out trays, pencils, paper clips, rubber bands, and huge ashtrays. The general had never smoked but many in his office did, especially his secretary, Jean O'Leary, Colonel Lansdale, his head of counterintelligence, and Gavin Hadden, an aide and engineer who had worked with him since his army construction days. On some of the desks were little racks from which hung rubber ink stamps. Some said SPECIAL DELIVERY or RUSH while others left the imprint CONFIDENTIAL, SECRET, or TOP SECRET on the flood of documents that entered and left this office. The windows had venetian blinds; there were a few leather chairs and a low, modern magazine table. Through the next door was the general's office, with two large desks, one for him and one for Mrs. O'Leary. There was an oak conference table, a leather sofa, and a very large safe, where highly secret documents, a pistol, ammunition, and ample supplies of caramels, peanuts, and chocolates were stashed.

The general's job was unique. Very few people knew what it was. Two and a half years earlier he had been chosen to oversee the building of an atomic bomb. Those few in the know had no idea then, or even now, if it would work or whether it would be ready in time to be used during the war. From the first hours he had been assigned to the job, he had given it every last bit of his energy. Indeed, he attacked most things he undertook that way, with drive, determination, and a will of steel. It was now all coming together. Tens of thousands of people were working on it, and soon almost two billion dollars would have been spent.[4] All of this activity had taken on a momentum of its own. Those to whom he reported and those who reported to him were all in this together when it came to their reputations and careers. Would it work, would it have a role in the war, would members of Congress undertake an investigation (when they found out about it), who would get the praise, or, if it failed, who would be blamed? His goal had always been for the bomb to have a role in ending the war. That is why every day counted, every hour, every minute. Generals in war commit young men to combat to fight and die, a terrible but necessary responsibility. If there was something available that could reduce the loss of American lives and end the war quickly, then it must be used. This new weapon might achieve that end, if it worked and was ready in time.

It was now clear, and had been for months, that the bomb would not be ready in time to be used against the Nazis. The newspaper headlines that week told of Germany on the verge of surrender. The Red Army had encircled, and was now entering, Berlin: elsewhere, U.S. Army units moving east and Soviet units moving west were about to meet at the Elbe. The front page of the *Washington Post* that morning had a picture of twenty-seven hundred dead "slave laborers" that American soldiers had discovered at the Nordhausen concentration camp. Italian patriots had captured Mussolini as he attempted to flee. It would not be long now. But the Pacific war ground on. The bomb could still be a factor in that theater, and the general was deeply involved in preparations for using it against Japan. The mechanism for its use had been set in motion long ago and was now inexorably moving at full speed, racing toward the finish.

To oversee an empire so vast from just a few offices and with such a small staff was the exception in Washington, especially in wartime. But that was the way Groves wanted it. In fact, he would have preferred an even smaller office. Later he said that he took as his model Gen. William T. Sherman, who, during his march to the sea, limited his headquarters to what would fit into an escort wagon.[5]

The Washington office almost qualified. A couple of dozen people, including secretaries, oversaw the activities and functions of tens of thousands in the field. A legal staff wrote the contracts with the contractors and vendors. His liaison officers procured anything that was needed at Oak Ridge, Hanford, or elsewhere, using an AAA priority rating if the situation required. One of Groves's first acts, within days of getting the job, was to make sure that he had the highest priority to obtain scarce materials.

Others in the office under the direction of Col. William A. Consodine were preparing for the day this secret program would become public knowledge. Groves had begun to plan for it on several fronts, and knew it would be a big story when it happened. Two weeks earlier, he had been to see editors at the *New York Times* to recruit their science writer, William Laurence, to cover the story. Just the day before, Mrs. O'Leary had spoken with the Princeton physics department chairman, Henry D. Smyth, who said he was working on what he hoped would be the final revision of his report describing the project.[6] In addition to informing the public, the always security-conscious Groves saw Dr. Smyth's report as defining the boundary lines between what could and could not be said by the participants after the bomb had been used.

The culture and practices of the Corps of Engineers had accustomed Groves to delegating authority and responsibility to trusted aides and subordinates in the field, and this is what he had done. He had worked with many of them on previous projects. Groves chose each of the principals, and he knew what he was looking for. Just a few months before, he had brought Brig. Gen. Thomas F. Farrell into the office to be his deputy. Secretary of War Henry L. Stimson had been concerned about what would happen to this two-billion-dollar project if Groves were killed in a plane crash or died suddenly. Groves had thus drawn up a list of about six officers, all from the Corps of Engineers, and decided upon his old colleague. Farrell was a few years older and a fine engineer. He had worked on the Panama Canal, served in France during World War I, and been an instructor at West Point, although he had not attended the academy. They had first met at Fort DuPont in Delaware. A decade later, during the mobilization, Farrell served as Groves's executive officer in the Construction Division. Most recently he had come from the China-Burma-India theater, where he had been building the Ledo Road.

Groves ruled his empire through a decentralized structure modeled after the corps, using the district engineer to oversee the area field offices. The Manhattan district engineer was Col. Kenneth D. Nichols, located at Oak Ridge, Tennessee, where the uranium for one type of bomb was being purified. By late April very good-quality uranium, enriched to more than 80 percent U-235, was being produced at a rate of three to four kilograms every two weeks and shipped to Los Alamos. Nichols and Groves had served together in Nicaragua fifteen years before. They knew each other well — sometimes a little too well from Nichols's perspective.

To oversee the vast operations at the Hanford Engineer Works in Washington State, Groves chose Col. Franklin Matthias. Matthias had helped Groves build the Pentagon, and now he was in charge of producing plutonium for the other type of bomb. He had plenty of help, of course: Tens of thousands were at the site, Nobel Prize scientists as advisers, and some of the best people that Du Pont could put on the job. Three "piles," as they were called, were now finally operating, the last of which went "critical" in February. The procedure was to extract the uranium-filled rods from the piles, let them "cool" in water, and then, in a chemical procedure, separate out the plutonium and ship it to Los Alamos for final processing. Groves ordered Du Pont and Matthias to speed up their alchemy. There were three ways to do so: shorten the time the rods were in the pile, shorten the time the rods were in the cooling ponds, and turn up the power level of the piles. Through unrelenting pressure over the past six months Groves had achieved all three; now as May approached, significant amounts of plutonium were about to be shipped to Los Alamos.

Also reporting to Nichols, and ultimately to Groves, were area offices in New York, Chicago, Boston, Berkeley, Detroit, Los Angeles, Montreal, and elsewhere. Groves visited them from time to time — whenever something was wrong or they needed to be driven to work faster. Much of his job involved dealing with problems that had not been resolved below. If things were running smoothly, only the most serious problems would rise to the top for decision. He spoke to Nichols almost daily and to the other area engineers as frequently as necessary to keep this far-flung empire operating at full speed.

Groves had an uncanny ability to pick qualified people. Usually in a matter of minutes he could tell if someone measured up or not. The best example was perhaps Robert Oppenheimer, the scientific director of the Los Alamos Laboratory, located on an isolated mesa in northern New Mexico, where the design, testing, and fabrication of the bombs were taking place. Against the advice of almost everyone, Groves chose the theoretical physicist to head the facility, seeing in him qualities that would allow them to reach their common goal. For both of them the bomb could be their route to immortality. The truth was that this odd couple saw in each other the fulfillment of their enormous ambitions. In the chain of command, Groves had Los Alamos report directly to him, not District Engineer Nichols.

Occasionally the general took time for a quick scan of the newspapers. He preferred the afternoon *Evening Star* and would try and read it this evening when he got home. But at hand was the morning *Washington Post,* shrunk to only twelve pages as a result of the wartime campaign to conserve paper. That late April morning Drew Pearson's "Washington Merry-Go-Round" column assessed how those who had once dealt with Roosevelt would now have to "learn how to get along with an unknown gentleman in the White House." Pearson went on to say that Gen. Brehon Somervell "will never become Chief of Staff or rise any higher in the Army." It was Somervell who had picked Groves for this secret job. There were some movies playing, but not the Marx brothers, his favorites. *God Is My Copilot,* with Raymond Massey cast as Gen. Claire L. Chennault, was at the Metropolitan on Tenth and F Streets. *For Whom the Bell Tolls* with Gary Cooper and Ingrid Bergman was at the Columbia on Twelfth and F Streets. But there was no time for movies. Maybe he could listen to the *Amos 'n' Andy* show at ten o'clock on WRC radio.

The baseball season had just gotten under way, and the eight teams in each league had played less than ten games. The Washington Senators beat the Boston Red Sox the day before 4 to 1 and would next play the Yankees in New York. There was a small article saying that Army would be playing the Hopkins Club in lacrosse the next day in Baltimore. His son Richard had been first-team All-American at the in home position last year and would be expected to score.[7] Busy as he was, Groves would take a few hours tomorrow afternoon and attend the game.

At around seven-thirty Jean O'Leary arrived in the office. She had come to Washington almost four years before from New York City. In 1940 her husband had died of leukemia, leaving her a widow at thirty, with a nine-year old daughter. She had applied for a government job and was called to Washington. Like thousands of other women who came to Washington to work during the mobilization period and then the war, she served in a secretarial pool. Groves, always on the lookout for capable assistants, went through many before choosing Mrs. O'Leary. They worked well together, and now two and half years into the Manhattan Project she was the only other person who knew almost as much about it as Groves himself. Mrs. O'Leary was more an executive officer than a secretary, making decisions herself when they had to be made. At times there were minor problems with some male officers who would not fully respect her or take orders from her. Groves had wanted her to join the WACS — the Woman's Army Corps — hoping that a uniform and rank might correct the problem, but it never happened.

The absolute epicenter of the empire was Groves's desk and, connecting him to those he served and to those who served him, his telephone. Being on the phone for hours a day was wearing out his elbows, and so he decided to use a headset with earphones and a microphone, similar to those used by switchboard operators. While the contraption kept his hands free, he might have startled an unsuspecting visitor who entered the inner office.

Every day that Jean O'Leary was in the office, she would listen to almost every phone call, transcribing the conversations using Pittman shorthand. She kept track of the general's busy schedule, which she recorded and then typed neatly in the appointment book. At about 8:00 A.M. the phone calls started in earnest, continuing for the next ten to twelve hours to and from the varied outposts of the general's suzerainty, which now stretched from the forward edge of the European battlefield to sites in a dozen states and islands in the Pacific Ocean.

Radiating out from the epicenter were circles upon circles of people at work in their scientific, industrial, military, and political orbits. All of the phone calls and the visits were to keep everyone in those revolving constellations in harmony and in motion.

In September 1942 his superiors had chosen Groves for this unusual job. There was much evidence in his past that he was the right man for the job. Twenty-five years of Efficiency Reports written by dozens of officers he served under, and who were in part responsible for his career, told a fairly consistent story.[8] In the 1920s and 1930s there was Colonel Peterson, Captain Kittrell, Major Schley, Lieutenant Colonel Sultan, Lieutenant Colonel Wilby, Colonel Sturdevant, Colonel Moore, and General DeWitt, among others. More recently Generals Gregory, Somervell, Robins, Styer, and Reybold assessed his performance in terms such as these:

"A hard driving officer who gets things done." "Very energetic, very capable and reliable." "A superior officer, forceful, intelligent, resourceful, loyal, not afraid to assume responsibility and possessing both initiative and a large capacity for work under trying conditions." "[W]orks rapidly, thoroughly, accurately and methodically." "[H]as the ability to turn out a tremendous amount of work and to see that undertakings are brought to a successful conclusion." "[A] good organizer, executive and administrator." "[N]ot easily diverted from his purpose." "[E]xcellent command of the English language." "Takes a definite stand, determined, holds to his convictions, sure of himself." "[A]mbitious" . . . "good judgement" . . . "cool even-tempered disposition" . . . "conscientious" . . . "keen mind" . . . "industrious" . . . "very resourceful in meeting new problems."

Of course, Efficiency Reports do not tell everything about a person. Their purpose is to identify qualities sought in military leaders. A fuller report assessing the total man would add that he saw much of the world in black-and-white terms, with gray a rarely used color. Endlessly judgmental, he believed that there were intelligent, competent, and qualified people — maybe 10 percent of the population — and then there were all the rest, which didn't much interest his pragmatic and instrumental mind. Humility was foreign to him. There was not a humble bone in his body, one of his aides later said of him. There were winners and losers, those who were strong and those who were weak. Tolerance, understanding, and compassion were in short supply when they stood in the way of getting the job done. Life was a pretty tough and harsh passage, and to make it through you had to be tough yourself. Whereas the outer trappings of Presbyterianism, such as churchgoing and the like, never took hold, the inner worldview was there. Deficiencies in people were, more often than not, their own fault. The better ones, the smarter ones, should be able to overcome their defects through fortitude, hard work, and effort.

Efficiency Reports also do not provide answers about what kind of husband or father a person is, how he plays tennis, or whether he has a sense of humor. By all accounts Dick was a good husband and father, a canny tennis player, with a sense of humor of the dry, sarcastic, and sardonic variety. Some photographs reveal a twinkle in his blue eyes and a smile about to unfold. With his children he would tease and cajole, sometimes a bit harshly to his son, but always out of love and what he thought was in their best interests. With others the sarcasm sometimes cut and left impressions of a stern taskmaster. Once, on a visit to the laboratories at Columbia University, the scientists, who knew he was coming, had put every test tube and beaker neatly in order. The place was spotless. When the general entered he surveyed the immaculate scene, ran his index finger across a table, and asked, "Where do you work?"[9]

Groves's preliminary job description had been to build the special plants where the materials for the new weapon were to be made. From working with private industry throughout the 1930s and during the mobilization of the early 1940s, Groves had built up an extensive list of dependable contacts. Beginning in the fall of 1942, he wasted no time pressing many of them into service. Du Pont, J. A. Jones Construction, Stone & Webster, Tennessee Eastman, Allis-Chalmers, Union Carbide, General Electric, Westinghouse, and Chrysler were some of the larger corporations and firms involved in this uncertain venture.

While that would have been quite enough to do, soon many other responsibilities were added to his portfolio. At Gen. George C. Marshall's behest he was put in charge of coordinating foreign intelligence about atomic matters, with the main focus on Germany. After all, it had been feared that the Nazis might be building a bomb. In the spring of 1945 it was now clear, and had been for at least four months, that they were not. Up until D-Day Groves had tried to gather every scrap of intelligence he could about what the Germans might be doing. For the Italian campaign and then after the Normandy invasion, Groves had formed a team of experts, code-named Alsos, to follow the Allied troops eastward across Europe targeting and capturing specific German scientists and locating and securing any uranium ore, equipment, and laboratories.

He asked that news about Maj. Robert R. Furman be sent to his office. He wanted to know what was happening in southern Germany on the Alsos mission's progress in rounding up the scientists and impounding their equipment. Furman was an engineering graduate from Princeton who had been with Groves since before the Pentagon project. Furman was not West Point, but he was intelligent and capable, and could keep a secret as well as the general.

One thing had led to another, and now Groves was deeply involved in military planning and even military operations. Last month he had felt it necessary to destroy a German plant in the Russian zone to prevent the Red Army from seizing it. He informed General Marshall, the chief of staff, who agreed. On March 15 more than six hundred B-17s from the Eighth Air Force got the job done in half an hour, dropping 1,684 tons of bombs on the target.[10]

Now, in late April, the military preparations for the testing and delivery of the bomb were proceeding at a fast pace. With the assistance of Gen. Henry Arnold, Groves had several dozen B-29 bombers modified to carry a five-ton bomb. He also directed that special crews be trained. Lt. Col. Paul W. Tibbets was put in charge of the 509th Composite Group, which was training in a desolate part of western Utah known as Wendover Field. To practice, the planes dropped bulbous-shaped test bombs, known as Pumpkins, at Sandy Beach, a naval air station on the eastern side of the Salton Sea in the California desert.

Only Tibbets and a few others at Wendover knew what their mission was to be, which was true of the thousands of workers at Oak Ridge and Hanford as well.[11] That knowledge would come only when it was over. Groves's bedrock tenet about security was to compartmentalize everything so that a person only knew what he or she needed to know to get the job done and nothing else. Everyone should "stick to their knitting." This way of working was even the rule in the small Washington office. Office mates sitting at adjoining desks did not know what the other was working on. The method had the added virtue of being an effective way to rule and manage: In effect, it was the secret of Groves's power.

Compartmentalization had a second meaning for Groves. It was the way his mind approached each problem and decision. Part of his success can be attributed to his ability to focus on the matter at hand, to give it full attention, and after its resolution to move on to the next problem or issue. He managed his time with great precision and efficiency. There were few lost minutes in a day. To remain informed on all the many parts of his responsibilities was an extraordinary effort.

With Wendover the rear training base, the forward operational base, from which the bombers would fly to Japan, was Tinian Island, code-named Destination, a tiny speck in the Marianas that had been captured in the summer of 1944. Groves had sent Navy Cmdr. Frederick L. Ashworth to the Pacific to find a base. Ashworth recommended Tinian, more specifically a spot next to North Field, and Groves approved. Tinian had been turned into a gigantic airfield, used by the Twentieth Bomber Command to bomb Japan relentlessly. On March 9 and 10 Maj. Gen. Curtis E. LeMay had directed a low-level incendiary attack on Tokyo that resulted in at least eighty thousand and possibly more than one hundred thousand deaths, destroying almost sixteen square miles of the city.[12]

Capt. William S. "Deak" Parsons of the U.S. Navy was in charge of final delivery of the bomb. Parsons was head of the Ordnance Division at Los Alamos and an associate director under Oppenheimer. After Oppenheimer,

Parsons was probably Groves's most important choice. In May 1943, after an interview of a few minutes, Groves was sure he was the right man for the job.

Just a few weeks ago the first personnel of the 509th arrived at Tinian. Other elements were scheduled to arrive in May and June, by ship and plane. At Groves's direction one of his most trusted aides, Col. Elmer E. Kirkpatrick, had been there for some time supervising a contingent of Navy SeaBees in building Quonset huts, tents, sheds, and mess halls next to North Field. There was also a unique creation that the hundreds of other bombers on Tinian would not need. It was a pit, dug into the ground, where the five-ton atomic bomb would be placed. Then the B-29 would be rolled over it, and the bomb would be raised into its modified bay.

Precisely where to test the bomb was a question that Groves had dealt with months ago. Various possibilities had been explored in the Southwest; the site finally chosen was a 432-square-mile section of the Alamogordo Bombing Range.[13] The site, nicknamed Trinity, offered several advantages: It was already on federal land, it was very desolate, and it was only a few hours from Los Alamos. Groves approved it in the late summer of 1944. Harvard physicist Kenneth T. Bainbridge was in charge of coordinating the test. Under Maj. Wilbur A. Stevens's direction, the base camp was completed by the end of the year. Now a wide variety of monitoring devices was being constructed to measure the effects of the weapon. The target date for the test was early July, ten weeks away.

It had been only fifteen days since President Roosevelt had died suddenly, and the nation was still absorbing the loss and adjusting to the new president. Roosevelt had not informed Vice Pres. Harry Truman about the bomb. As an influential senator with his own investigating committee, Truman had heard rumors, and at one stage even planned to look into them. Fortunately, Secretary Stimson intervened and persuaded him to defer. Through this kind of assistance, Groves had been successful in keeping Congress out of his hair. He had had enough trouble with Congress during the mobilization years of 1940–1942, when they had either demanded of him that camps or munitions plants be built in their districts, or criticized him for how much was spent in building the Pentagon. There was another time, in 1944, when certain congressmen were becoming too curious about the project. At this point Stimson, Marshall, and Vannevar Bush, Roosevelt's science adviser, went to Capitol

Hill to speak with the House leadership and ask for their trust and silence. There were no further problems. Now there was a plan to take a small congressional delegation to Oak Ridge in May to show them what they had bought with the hundreds of millions of dollars they had unknowingly appropriated.

It was important that the new president be properly informed of all of the consequences of the bomb. Groves had been working with Secretary Stimson's office to prepare a report. Two days ago, on Wednesday morning, they had gone to the White House to describe to the new president the scale and status of the program. Groves had prepared thoroughly for the briefing, writing an extensive memo detailing where matters stood. So as not to attract any attention, he did not arrive with Stimson but was taken in through a back door.

A little after eight o'clock the general left for an eight-thirty meeting in Gen. Lauris Norstad's office in the Pentagon. It was the first meeting of the Target Committee, and Groves had chosen the members, the best thing to do if one wishes to control an agenda.[14] There were General Farrell and Major Derry from his office, representatives from the Army Air Forces, and Los Alamos scientists to assist with technical advice. After briefing the group, Groves left General Farrell in charge and went to see George L. Harrison at nine-forty-five, then Harvey Bundy at ten-thirty. Harrison and Bundy were Stimson's aides who worked on the bomb project. General Groves was in touch with them sometimes several times a day.

After the meetings, he returned to his office for an eleven-thirty appointment with Dr. James Chadwick. Chadwick was the head of the British Mission to the Manhattan Project, all under Groves's control. A group of some two dozen British scientists was at Los Alamos, with others at Berkeley, Oak Ridge, New York, and Washington.[15] Groves was not particularly fond of the British. His Anglophobia was of long standing, inherited from his father. Whereas he had traveled tens of thousands of miles to every corner of the United States, he had very little experience abroad. Some impressions came from a three-month tour of the World War I European battlefields in the summer of 1919 soon after graduating from West Point. He found English "morals to be rather depraved," he had told his father. The only people worse, in his view, were the French. No, America was the best country in the world. Chadwick was an exception to the rule, and they got on well.[16]

At eleven-forty-five there was a phone call from Oppenheimer about different kinds of gauges. He and Groves spoke almost every day. In about ten days there was going to be a rehearsal for the Trinity test.[17] The scientists were going to detonate one hundred tons of TNT with a small amount of radioactive material added to check their instruments and calibrate their gauges. In this way the effects of the real test, soon to follow, could be measured more accurately.

General Groves also had a call that day from James B. Conant, president of Harvard University. Conant, along with Richard C. Tolman of the California Institute of Technology, advised Groves on various scientific matters. Conant was also a member of the Military Policy Committee, the group to which Groves theoretically reported. The other members were Vannevar Bush, Rear Admiral William Purnell, representing the navy, and Lt. Gen. Wilhelm D. Styer, Somervell's chief of staff. In practice Groves was given wide latitude in making major decisions; more often than not the committee would affirm his actions after he had taken them.

In the first few days after he was assigned the job, back in September and October 1942, Groves made huge commitments of resources and money. These huge commitments, by accident or design, left Groves's superiors little choice but to go along with everything he did. Groves never did things by halves. If they wanted this bomb built, then it would be necessary to pull out all the stops. One early projection had been that it would cost $133 million to build the bomb.[18] Groves spent that much in the first few weeks, and now more than ten times that amount was behind them with a two-billion-dollar price tag in sight. There were probably private moments when Bush, Conant, and the others became almost short of breath, stunned at the scale of what Groves was doing. Groves's decisive actions had the inevitable consequence of binding them all together. After the first few months there was no turning back. Groves had tried to see to it that there could be no future criticism; that every last effort had been made. He had given it his all, and now they were coming down the home stretch, charging for the finish line flat out.

At five-twenty-five the general took a call from K. T. Keller, president of the Chrysler Corporation. Just two years previously Groves, Nichols, and Percival C. "Dobie" Keith had traveled to Detroit to ask Chrysler to design and manufacture diffusers, the containers in which the barriers would be placed in the K-25 plant at Oak Ridge.[19] Keller was a no-nonsense executive, the kind Groves liked. He knew every aspect of his company. One of Groves's rules was to always try and do business with those who could make the binding decisions. When they had first seen Keller, the major problem that needed solving was how to plate the diffusers so that the uranium hexafluoride gas would not corrode them. Keller had put his people to work on it, and in six weeks they had a solution. Now, two years later, the diffusers had been produced, delivered, and installed.

The day was drawing to a close but the work was not yet done. Often his daughter Gwen would meet him at the office, and they would go out to dinner. Then he would return to do more work, and Gwen would do her homework in the outer office.[20] Mrs. Groves kept to a busy schedule, doing volunteer work downtown, chaperoning at the young officers' dances at the Soldiers and Sailors Canteen, and giving piano lessons. She also worked part time at the Garfinkel's store in Spring Valley, which brought in some extra money. She was quite the opposite of her husband. She was well read and could speak French, Spanish, Italian, and German, all fluently. Her special love was music; she sang and played the piano. By the same token, she had large gaps in her education and experiences. Like her husband she had been brought up in the army and raised on posts all over the country. But she had been taught by her father, Col. Richard H. Wilson (West Point, class of 1877), a frontier scholar who spoke many languages and knew about the world and history through his extensive reading. As a result, she had been shielded from society in odd ways. She had never attended a real school or taken an exam. This upbringing had produced a woman, a mother, a wife, and a friend whom everyone tried to protect, each in their own way. She was as fragile as her husband was tough. She was known as Boo or Buddha. She did not particularly like the name, and sometimes claimed to be embarrassed by it, but friends and close acquaintances called her by it.

Grace was in charge of the social life at 3508 Thirty-sixth Street, such as it was. She tried to improve her husband's social graces, but parties and dinners were not high on the general's list of favorite activities. Calvinistic currents ran deep, instilled long ago by his chaplain father, questioning the worth of such frivolous things. Out of spousal duty, more so than for enjoyment, the general would accompany his wife to plays, concerts, and operas, and stay awake at least part of the time. But lately he had been traveling so much and working so hard that he would often disappear for days on end, with Grace and Gwen not knowing where he was or when he would be back.

With all of these comings and goings at 3508, father, mother, and daughter rarely saw one another. To keep track of each other's schedules messages, orders, and directives were left on the hall table.

> Dear Mother,
> Mrs. O'Leary called. DNO won't be home tonight. I will be at Kate's. I got an A– on the French test, and Mlle Pellet says tres tres O.K. Bon pour moi!!

> Dear Gwen,
> I will be home for supper. Turn the oven to 300° at 4:30. DON'T FORGET. And pick up cookies at Taylor's. There are some bananas on top of the icebox in case you can't wait.

> Dear Gwen,
> Call Mrs. O'Leary when you get in. DNO wants to play tennis. If you go out to dinner with him, leave a note.[21]

Though much remained to be done the general put down his work, collected Gwen, and exited the office, locking the door behind him. They drove home quickly, up Twenty-third Street to Massachusetts Avenue, then right onto Thirty-fourth Street, right onto Ordway, and left down the alley and into the garage. It was after ten when they entered the house. As the general climbed into bed he thought of all the things that needed to be done tomorrow. That it was a Saturday made no difference; it was going to be another busy day.

PART ONE

Early Life and Education

· 1896–1918 ·

CHAPTER ONE

Family Heritage: The Groveses in America

Leslie Richard Groves Jr. was born in Albany, New York, on August 17, 1896, a member of the eighth generation of Groveses in America.[1] Dick's earliest American ancestor was Nicholas La Groves, born around 1645 possibly at or near La Rochelle, France. Nicholas was a Huguenot, a French Protestant, and to escape from religious persecution he moved to Jersey, the largest of the Channel Islands, located about fourteen miles off the Normandy coast. Originally he called himself Le Gros, a common name on the Isle of Jersey; later he Anglicized his surname to La Groves.[2]

Nicholas's primary reason for immigrating to America was to better his economic situation. Working out of Jersey as a seaman, he probably jumped ship in Salem, Massachusetts, though it is not clear precisely when he arrived. A group of Channel Islanders had settled in Salem, living on the south side and maintaining a degree of ethnic cohesion and exclusivity for a time. As early as 1660 they were known as good in business and had ongoing commercial relations with their native land. They imported indentured servants from Jersey and married English and French spouses almost equally.

La Groves's arrival must have been well before 1668, for by then he had prospered enough to sign a petition against taxes to the General Court of the Massachusetts Bay Colony. On May 16, 1671, he married Hannah Sallows of Salem, a woman whose grandparents had arrived in Massachusetts as early as 1624. Around this time Nicholas moved to Beverly, Massachusetts, on the coast northeast of Boston. He was a seaman at first, later a ship's captain who earned his living in trade with the West Indies.

Nicholas La Groves and Hannah Sallows had six children. The first three (Susanna, Nicolas, Hannah) took the surname La Groves; the fourth (Peter) La Grove; the fifth (John) Lagro; and the sixth (Freeborn) Lagroe. The family joined the Beverly First Church and rapidly became

Anglicized, adopting English customs. In the next generation, the prefix *La* was dropped, and from then the name was consistently *Groves*.[3]

The direct line from Nicholas to Leslie Richard Groves Jr. began with Nicholas's son Peter La Grove (1679–1755), though he was known throughout his life as Peter Groves. He, like his father, earned his living from the sea and was probably a ship's captain. In 1702 or 1703 Peter married Hannah Winter; a son, Nicholas, was born the following year in Beverly.

Nicholas Groves married Hannah Corning, also of Beverly, in December 1724 and, after having three children by 1730, moved to Brimfield, Massachusetts, west of Sturbridge, not far from the Connecticut border. There he was a practicing husbandman, landowner, and a prominent man of the town, serving at various times as assessor, selectman, and treasurer. In Brimfield four more children were born, the last of whom was a second Peter Groves, born on August 18, 1739. In July 1774, in response to the Intolerable Acts, Nicholas signed a covenant not to purchase any goods imported from Great Britain.

Peter Groves participated in the battle of Lake George during the French and Indian Wars and was a veteran of the American Revolution, thereby qualifying all of his descendants to become members of the Sons of the American Revolution or the Daughters of the American Revolution.[4] Peter married Lydia Lumbard, and they had nine children between 1768 and 1786. The eldest son, also named Peter, was born in October 1772.

Peter Groves Jr. — Dick's great-grandfather — was a farmer who lived and died in Brimfield, though the rest of the family moved to the Utica area around 1803. Peter married Jemima Allen around 1798, and operated the family farm after his parents and younger siblings moved to New York. Peter and Jemima had nine children between 1799 and 1818. Their seventh son, Allen Groves, was born on December 2, 1815, in Brimfield. After Peter died in 1840, Jemima lived until 1881, reaching the age of 107.[5]

In about 1850 Allen Groves — Dick's grandfather — moved to Chuckery, New York, near Clinton and Kirkland, in the vicinity of Utica. On March 31, 1851, Allen married Adeline Jane Beebe of Kirkland. They had four sons and a daughter. The fourth son, Leslie Richard Groves — Dick's father — was born on June 22, 1856, in Chuckery.[6] In April 1863, when Leslie was almost seven, the family moved to a dairy farm in Westmoreland, nine miles from downtown Utica. Allen died on June 6, 1878. Adeline died at age seventy-five, due to "paralysis, with complications and old age" and "nervous prostration" on October 10, 1900, when Dick was four.[7]

———

Dick's mother's roots in America were much more recent. Both sides of her family were Welsh, from the highlands of Caernarvonshire, where the principal business is raising sheep.[8] Her grandfather, Owen Griffith, and his family left Wales in search of a better life, sailing to New York City in 1795. Lured by the promise of good, cheap land in the Utica area, they sailed up the Hudson and Mohawk Rivers, first settling in Steuben and then moving to nearby Remsen in about 1798. After his first wife died, Owen married again and had at least two sons. The eldest — Dick's maternal grandfather, also named Owen — was a first-generation American born in Remsen on January 19, 1823.

Dick's maternal grandmother, Jane Owen, was born in Pen-y-Llys, in northern Wales, on March 2, 1827. She was nineteen years old when her family immigrated to the Utica area. Jane Owen and Owen Griffith were married on September 13, 1853.

In January 1856, with their infant son George Washington Griffith, they left upstate New York to join a Welsh community in Racine, Wisconsin, where Owen took a job as a clerk in a granary and later as a bookkeeper for a general store. His starting wage was $37.50 a month, which he described as a "fair salary."[9] A daughter, Frances, was born in 1856 but did not survive, and a second son, William Morton, was born in 1858.

When the War Between the States started, Owen Griffith had no intention of enlisting. But under pressure from friends and acquaintances — and, with the growing likelihood that he would be drafted — he signed up on August 9, 1862, with the Twenty-second Regiment, Wisconsin Volunteers. A few weeks later the Twenty-second Wisconsin was inducted into federal service and left for the front.

The regiment operated in Kentucky and Tennessee.[10] Owen felt strongly about the war and the issue of slavery, calling it:

> this most wicked rebellion . . . uncalled for . . . on the part of those Southern Slaveholders to overthrow the best government the world ever saw for the avowed purpose of establishing a government whose corner Stone was Slavery.[11]

Owen was captured by the Confederates at Brentwood, Tennessee, ten miles south of Nashville, on March 25, 1863. He was taken to Richmond, Virginia, and after a month in Libby prison was exchanged at Annapolis, Maryland, in May 1863. He eventually rejoined his regiment, but in April

1864 Captain Griffith resigned, driven out in the course of a bitter controversy involving a senior officer of the Twenty-second Wisconsin.[12] Whatever Owen's wartime difficulties and shortcomings, he was a revered figure within his family, and to successive generations. He became the origin of the family's military tradition; pictures of him in his Union uniform were cherished.

At the end of the Civil War, in April 1865, Owen Griffith and his family moved back to Utica. On October 13, 1865, Gwen Griffith — Dick's mother — was born, and eighteen months later on April 12, 1867, her sister, Jane, came into the world. For about five years, in partnership with his brother-in-law, William M. Owen, Owen Griffith was a dealer in flour and grain. Apparently there was a dispute in 1870 and the partnership dissolved. Owen and the family went to West Winfield, New York, where he operated a general store for nine years, but the depression that began in 1873 caused him to go bankrupt. Owen returned to Utica in 1879 and worked as a bookkeeper for his brother-in-law's firm, Wm. Owen & Co.[13] The years continued to be difficult. His wife, Jane, ran a boardinghouse at their home at 34 Cottage Street, Utica, where daughters Gwen and Jane did much of the work.

Owen's wife, Jane Owen Griffith, died on December 3, 1891, after accidentally falling down the stairs at their home and severing her spinal cord. Owen died on June 26, 1892, at age sixty-nine, after being taken ill during a visit to Albany to see his daughter Gwen, her husband Leslie, and their firstborn, Allen.

Dick's grandfather, Allen Groves, apparently prospered as a dairy farmer at Westmoreland. He sent his son Leslie Richard to Whitestown Seminary[14] and then, in 1877, to Hamilton College in Clinton, New York, three miles from Kirkland.[15] It was not, as were many nineteenth-century American colleges and universities, directly affiliated with a particular religion, though the mandatory chapel services were Presbyterian.[16] The expressed goals of a college education during this period were less preparation for a career or profession than "training the mind" and establishing the proper moral and religious precepts of a gentleman. Whether by inclination, by his upbringing, or by his Hamilton education, Leslie Groves became a devout Presbyterian, a person whose devotion to his God manifested itself in his daily comportment. Drinking, gambling, and swearing were strictly forbidden, and Sunday was for the Lord.

When it came to how he would earn a living and keep a family, Leslie

Groves exhibited a lifelong pattern of indecision, vacillation, and wanderlust. His Hamilton education, while sound in the classics, the landmarks of Western literature, and moral philosophy, prepared him for no particular calling.[17] As we shall see, he tried several vocations: teacher, lawyer, Presbyterian minister, and army chaplain.

Groves graduated in June 1881 at the age of twenty-five. He taught in country schools prior to and during his college years. The year following his graduation he was the principal of the Fort Covington Academy and Union Free School, in Fort Covington, New York, which meant, apparently, keeping all the school's administrative records in addition to teaching a full complement of classes.[18]

In August 1882 he entered the law office of Charles H. Searle, a prominent Utica lawyer, remaining there for two years; he was admitted to the bar in June 1884. Though he seems to have had some success as a lawyer,[19] that profession too did not suit him, for in the fall of 1886 he undertook a three-year course of study at the Auburn Theological Seminary, in Auburn, New York, to prepare for the Presbyterian ministry.[20] This new career path seemed more compatible with his moral convictions than the demands of the bar, where he had to deal with and defend people whose actions he detested. Leslie graduated from Auburn Seminary on May 9, 1889,[21] was ordained on June 18, and was installed as pastor of a small Presbyterian country church in McGrawville, New York, between Syracuse and Ithaca.

On April 29, 1890, Leslie Richard Groves married Gwen Griffith of Utica at the home of the bride's parents.[22] Groves's popularity among his parishioners, his status in the community, and the fact that his marriage was quite sudden and unexpected can be gleaned from the following poem, which was published in a small McGrawville paper soon after they were married:

> He was a young man, "smart's a whip,"
>   A minister was he,
>
> And all the maidens fair avowed
>   "He's nice as he can be."
>
> And so they looked their prettiest
>   And smiled their sweetest smile,
>
> And the minister he prayed and praught [sic]
>   and was thoughtful all the while.

> They sang their sweetest songs to him,
>   And asked him out to tea,
>
> Sometimes they asked his mamma, too,
>   So "very nice" was she.
>
> All of a sudden — hello, — swish,
>   And off to U-T-K [Utica]
>
> The young man went, my brethren,
>   And he wasn't long away.
>
> And when he did return — just hark,
>   He brought with him a wife;
>
> 'Twas the nicest, unexpected move
>   We've seen in all our life,
>
> And the prayer meeting room was full
>   As full as full could be,
>
> And he prayed for "all that came to pray
>   And for those who came to see."
>
> But "all is fair in love or war."
>   May heaven's blessings glide
>
> On Leslie Groves, our minister.
>   And Gwen, his bonny bride.[23]

As happened twice in the past, and would happen several times in the future, Leslie appears to have become dissatisfied with his lot. His salary as a minister of a congregation in a small upstate village was no doubt insufficient, especially when early in 1891 he and Gwen learned that they were expecting their first child. In July 1891 he accepted a transfer from McGrawville to Albany to become pastor of the Sixth Presbyterian Church, on Second Street below Lark, at an annual salary of fifteen hundred dollars. The Groveses' first child, Allen Morton, was born on November 12, 1891; a second son, Owen Griffith, soon followed on May 11, 1893.

On a salary of $125 a month the family's financial situation was difficult, and the demands of his parish proved a heavy burden.

> When I left the law practice and went to the theological seminary
> my plan and hope was to serve in a country church. I had seen the
> need of the country church and, having been brought up in the
> country, I wished to live and work where I had seen the need. But

circumstances had pushed me out of the country and into a city church and a city where life was hard. The church of working people, where a minister tries to hold his own with the churches that can afford a secretary and an assistant and all the machinery, has something to do. So after five years of two or three services and a Bible class in the Sunday School, Christian Endeavor meetings and prayer meetings with the sick and others to visit — no telephone — calls always on foot — my friends saw, and I knew, that my career was to be a short one, unless I changed my work and my residence.[24]

Ever restless and dissatisfied, Leslie felt it was time for a change. Among his summer parishioners in McGrawville was Daniel S. Lamont. Lamont was the secretary of war in President Cleveland's second cabinet from March 1893 to March 1897.[25] The Groveses and the Lamonts became good friends. Lamont was concerned with his friend's heavy workload and urged Leslie to become an army chaplain. Lamont painted for Groves an attractive picture of army life, suggesting "a quiet life in a frontier post."[26] At the start of his last year in office, 1896, Lamont sent Groves the forms required to apply for entrance into the military chaplain service and virtually ordered Groves to fill them out and submit them. Groves complied, even though his knowledge of what army life was really like was virtually nil. "It was a plunge into the unknown but I took it," he said.[27] In December 1896 he resigned as pastor of the Albany church and became a U.S. Army chaplain.[28]

> In total ignorance of the Army and all that pertained to it I went to Washington to learn something of what I had gotten into. My knowledge hitherto was built on having seen recruits coming out of stations in ill-fitting blue uniforms, a half hour's visit into Ft. Niagara and the army stories that pictured frontier post life as of a sort that one would not care to become a part of.[29]

Having the secretary of war and the secretary's wife as close friends had its advantages. Lamont sent Groves a list of possible postings and asked him to indicate what part of the country and what climate would best suit him, again painting a very attractive picture.

> I think I could assign you to Fort McPherson in the suburbs of Atlanta, possibly to San Francisco or Vancouver a delightful post near

Portland Oregon where the climate and surroundings are all that could be desired. Mrs. Lamont thinks you would like one of the Montana posts or a station in Arizona. We were so pleased with the climate and quiet of these latter that she would be quite willing and happy to exchange places with you. At the larger posts first mentioned the quarters are more modern. At *any* of them you will have a choice little society which I am sure you will enjoy. To your first station you have to pay your own traveling expenses and perhaps you will prefer that I should give you a brief tour of duty at some nearby place and then transfer you a month or two hence. Ordinarily a chaplain must undergo a physical examination but I think we will dispense with that in your case.[30]

Groves chose the position of post chaplain at Vancouver Barracks, Vancouver, Washington. Lamont wrote on Groves's behalf to the commanding general of the post, urging that the new chaplain be provided with a "good home."

But before beginning the trip west there was one matter that needed attention. Gwen was pregnant with their third child. Leslie requested that his entry into the army be delayed until after the birth of this baby. The request was granted, and so it was that Leslie Richard Groves Jr. — to be known by the family as Dick — entered the world on August 17, 1896, born in the manse of the Sixth Presbyterian Church.[31] A few months later the family undertook the arduous weeklong train trip across America to their new home and new life at Vancouver Barracks.

CHAPTER TWO

Growing Up in the Army

• *1897–1913* •

The Groves family arrived at Vancouver Barracks on February 1, 1897, and was provided with the "good home" that Lamont had arranged.[1] Writing much later, Chaplain Groves described the pleasures of his new life in the Pacific Northwest:

> One only needs to go from snow and cold of the East and land in Vancouver, Washington about February 7, as we did to be filled with the joy of living. . . . [The] East half of a double home at the East End of the line was assigned. All the East side windows looked out on Mt. Hood. . . . The grass was green, the days mostly fair with just enough rain to keep all fresh and beautiful. Never have I seen nature with so fair a face and for a year and two months the quiet life made good. With Mt. Hood and the other snow-capped peaks and all the accompanying scenery to delight the eyes; with strawberries, cherries, plums, and peaches to delight the taste; and balmy weather, new to those from the rough East, that blew hot and cold, changed without notice; just to be living was joy.[2]

But for all this, after about a year as post chaplain of Vancouver Barracks, serving the Fourteenth Infantry Regiment that was stationed there, and attending to his family, Groves was growing restless again. His army pay proved adequate to free him of the financial worries that had plagued him, and his duties as post chaplain were far lighter than the burden he bore as pastor of the Sixth Presbyterian Church in Albany. But he was seeking something else; something more, not yet clearly defined in his mind. He was sure, nonetheless, that it lay somewhere beyond the security of the established job, the idyll of the Northwest, and a settled family life that he was now enjoying. And the knowledge that he had not found it made him uneasy, unfulfilled. Suddenly, in the spring of 1898, an opportunity came for him to renew his quest.

Throughout the nineteenth century America had been troubled by conditions in Cuba, which occasionally flared up into direct confrontations between the United States and Spain.[3] During the 1890s, aggravated by grossly exaggerated yellow press reports of Spanish oppression on the island, differences between the two nations deepened into a major crisis. Then on February 15, 1898, the USS *Maine* was sunk while on a goodwill visit to Havana, and the crisis spun out of control.[4] Within ten weeks war was declared — on April 25 — and frantically the United States mobilized to fight what was generally thought to be a first-rate military power. Immediately, calls went out across the land for volunteers to fill the ranks of militia units, and most of the Regular Army's regiments were ordered to assemble in several southern coastal cities, such as Tampa, New Orleans, and Mobile, as well as other interior camps.[5]

On May 17 Chaplain Groves was ordered to join the Eighth Infantry Regiment in Tampa, Florida, part of the expeditionary force that would invade Cuba. The Eighth was a close-knit, proud unit with a long, distinguished record; its old-timers did not readily accept newcomers, like Groves, to their ranks. As the fates would have it, one of those old-timers was Capt. Richard H. Wilson, commanding Company A and the regiment's Second Battalion. In time, he would become Dick's father-in-law.[6]

Chaplain Groves departed Vancouver by train on May 19; traveled down the West Coast through Oregon and California, then east across Arizona, New Mexico, and Texas to New Orleans; finally arriving in Tampa on the morning of May 27. Every day he wrote to his wife, describing the varied scenery, experiences with his fellow passengers, how hot it was, how frugal he had been (spending less than ten dollars during the journey)[7] and expressing hope that she and the boys were getting along in his absence.

In Tampa Chaplain Groves continued his daily letters, giving his impressions of that busy and chaotic place. He said, "I would prefer to be with the 14th [Regiment], but I may be more needed here. Profanity seems the common speech. I have heard more of it today, I believe, than in the year at Vancouver."[8] "The soldier's language is simply hideous."[9] "The 71st New York just marched past us to a camp back of ours. A New York City crowd, I judge by their language."[10]

Precisely why Chaplain Groves chose to serve in Cuba is not altogether clear.[11] His letters to his wife and children, of which hundreds survive,[12] portray an internal conflict that is difficult for most modern Americans to understand.[13] His frequent allusions to his family are warm, caring, and loving — causing us to wonder why he so readily left them. The answer is probably found in his devout Presbyterian/Calvinistic beliefs. His most important relationship was to his God. Serving him by doing good works for all of mankind was the highest duty one could perform, coming ahead even of one's family and personal comfort.

Of course his family was important. We see in the letters to his three sons an absentee father's effort to preserve his traditional paternal role. A letter to middle son Owen,[14] then five, written shortly after Groves arrived at Tampa is typical:

> How is Father's little boy? I think he is well and happy. Do you and Dick have a nice time playing together out in the back yard? That is a nice back yard for little boys and chickens. . . . Mother wrote to me that Dick hurt his finger. I know little Owen was good to him then. I am very glad you are so kind to Dickey. If Father was home you would be kind to him too. Well sometime he will come home. And perhaps I will bring that pretty little box that you gave me with toothpicks in it. But the toothpicks will be all gone. You know Owen I live in a tent now and there are so many tents all around me, hundreds of them. There are a great many people here and none of them live in houses now. So many soldiers you never saw and so many officers. But they don't look like officers sometimes for they don't wear coats much of the time. Do you wear your officer caps now?[15]

The chaplain's grandson Richard lived with him for a while and observed the regimen that Groves established for his family. It was highly disciplined and focused on education and religious observance. The grandson's experience paralleled that of his father:

> Sunday at six in the morning you went to church. Come home after service. Read the Bible. Back to church at 11:00, then home to read Shakespeare, back to church at 5:00: a bowl of bread and milk upon return. Read [historian] Francis Parkman and then go to bed. Every Sunday that's all you did. No funny papers — they sat on the doorstep until Monday morning. That is the way he [Dick Groves] was raised.[16]

Chaplain Groves's life was, on Sundays and the rest of the week as well, largely a single-minded exploration of ways to serve his God. On his way to Cuba he told Gwen,

> As I get nearer the place of work, I feel more deeply the need of strength from above. Don't forget that part for a day. The Lord will take care of me, if I am of any use. I long to have Him use me for His glory.[17]

He saw his service in the army — especially with soldiers going off to war in Cuba — as a unique opportunity to serve his Master. He could bring the word of the Lord to the troops, provide formal services and religious guidance to those few who were already devout and earnest Christian soldiers, and offer salvation to the rest — the drinkers, swearers, gamblers, and hooligans who seemed to make up the majority of those he encountered.[18]

Today it appears almost to have been a convenient cover under which his restlessness and wanderlust could find an outlet — perhaps a manifestation of some inner dissatisfaction with the course of his life. Chaplain Groves and his son Dick, as we shall see, were never introspective about what drove them to do the things they did. Occasionally the chaplain's letters reveal varying degrees of dissatisfaction with himself as a man, as a parent, as a husband. But always, his greatest dissatisfaction is with his performance in serving his Master.

Some of his acquaintances considered his decisions to go to Cuba and later to the Philippines as the actions of an eccentric. His dogged determination to abandon his family for long periods of time (Cuba, it turned out, was only the beginning of his travels), to journey to distant lands, to live in the field under the less-than-ideal conditions of a bivouacked soldier, often in the line of fire, while his wife struggled at home to maintain the discipline of family life, seemed to many not only peculiar but irresponsible and selfish as well. But within the family there were no feelings of resentment toward him at all. Throughout his extended absences Gwen carried on as best she could, maintaining family discipline in a spirit that reflected his will and his values.

It wasn't easy for her. In 1898 a post chaplain received the pay of a first lieutenant, $125 per month or $1,500 a year.[19] Under normal conditions, while he was away, Groves kept for his own expenses one paycheck out of three, thus reducing the income for the family to about eighty-three dollars per month. Tithing reduced this to about seventy-five dollars, hardly a living wage for a family with three children. Chaplain Groves was extremely parsimonious, a trait he passed on to his son Dick. Army salaries left them little choice. And adding to Gwen's difficulties, when Groves joined the Eighth Infantry, he ceased to be post chaplain of Vancouver Barracks, leaving his family, in effect, squatters.[20]

It is quite evident that Groves had not bargained for the life of a combat soldier.[21] Yet his letters always showed him to be resolute and tough, never taking his eye off his duties as a chaplain. "There is no reason that I know of why I should have an easy, pleasant life. My Master did not and I want nothing better than the faith and courage to endure hardness as a good soldier."[22]

One letter to Gwen written shortly before he embarked for Cuba described life in the Tampa tent camp.

> We live in dirt and eat sand. I think we will be ready for the first installment of the rainy season. But they have some lumber and have begun putting in floors. That will help some. This morning I fixed up my tent a little putting up a frame in the end with nails and hooks to hang things on. I am wishing all the time that I were with the 14th [Regiment]. It is a long hard task to get to know the crowd and then I should like the association of the officers of the 14th. . . . The language of the soldiers is simply hideous. Of course they do not know who I am, many of them, and so are more free of speech. . . . We have no place for services in our own camp. I do not know how to arrange for one. The sun and heat are intolerable out of doors till evening. . . . We get absolutely no news here. The papers of the town are just nothing and know nothing.[23]

After twelve days of dealing with such adversities, Chaplain Groves finally boarded the transport ship *Seneca,* to which the Eighth was assigned. But once aboard, he went nowhere. Rumors of a Spanish fleet headed for Cuba caused the navy to delay the expedition for eight days — eight long days that seemed interminable to the soldiers packed tightly in the transport's hold.

Because of poor planning and a lack of ships there was room for only seventeen thousand of the twenty-five thousand men assembled in Tampa. Even so, horses, wagons, and hospital stores had to be left behind.[24] Living

conditions aboard ship were extremely unpleasant, and a barrage of profanity constantly assaulted the chaplain's ears.

From time to time Chaplain Groves reflected on just how effective he
was in his profession. He found:

> no sympathy with my work, at all. . . . This is a stiff job. . . . I escape
> from the profanity down to my stateroom, only the soldiers by the
> window are sometimes almost as bad as the officers. Oh, how I would
> like a decent little church! I don't feel half big enough for this work.
> I find myself wondering if I ever have done, or ever can do, any good
> in the world. I go out among the men for a little and talk with a few
> and come back and crawl into my bed and sigh.[25]

And he was "disgusted with the total lack of patriotism [he] find[s] in the
officers here . . . [and] the swearing beasts . . . opposed to the war . . . mocked
the sentiment in it."[26] His feelings were passed on to his son. Of the many
who reflected back on Gen. Leslie Groves's character, decades later, none
remembers him ever swearing, and they stressed his unremitting patriotism.

At last, on June 14, the flotilla of some thirty transport ships got under
way. Within a few days they were off the Cuban coast — and then they
waited. Groves complained repeatedly that no one knew where or when
they would land or what would happen when they did.

Then, on short notice, the Eighth Infantry Regiment led Shafter's Fifth
Corps ashore in Cuba on the morning of June 22, 1898, at Daiquiri, about
twelve miles southeast of Santiago.[27] Throughout the next week, the chaplain marched with the Eighth as it worked its way westward and north-
ward toward Santiago.[28] July 1, "The day that was to be the last of so many
men brought from the far ends of the earth to kill one another dawned
bright, hot and cloudless," wrote Captain Wilson.[29] The goal of the Second
Division that day was to seize the village of El Caney and cut off the garrison's escape route to Santiago. As the Eighth marched up the road, the
soldiers came under fire for the first time. Many years afterward Chaplain
Groves described his feelings:

> The first sounds of battle are startling. Following at the rear of our
> column, the bullets began falling all about us. The Regimental
> Surgeon said, "Let's get out of here. We are no use when we are
> dead." A wise word, proven good by experience. A non-combatant
> had best be out of the way when the guns are working.[30]

And in a letter to Gwen:

> The Hospital Corps followed the regiment. At once we were in the fire and fell out to one side. One of the Hospital Corps was wounded at the very first. We got into the dry bed of a stream and lay there till they stopped shooting our way. At once, the wounded were brought back where we were until we had about 30 with us. Then the doctors were called off to a big tree near the wounded and left me in charge of our place. Some were fearfully wounded. Many only slightly. It was horrible. My dread of war increased; I hope not to see any more.[31]
>
> In every contest, on the field of battle or in life, some are easily beaten. Even around the wounded, differences of men appear. A sergeant who, with the pioneer corps cutting barbed wire entanglements, had been shot through the shoulder so that he was one-handed, was helpful all day with the wounded — in contrast with others whose duty was to assist. The horrors of war are seen not at the front where men are shooting and being shot. There is the excitement of contest, but at the rear, where the mutilated are gathered, groans replace the cheers. The first shock of it is almost overwhelming. A feeling of helplessness — sickening at the sight and smell of blood, one blunders on and finds, after a little, what he can do to help.[32]

As he followed his regiment with the doctor and hospital corpsmen, the "road was strewn with the wreckage of battle. Men had been removed, but the bloated bodies of mules and the broken equipment told of the battle." Burial sometimes had to take place, and Chaplain Groves performed the service "without mourning family, without the usual formal church surroundings — is it less solemn? Rather more solemn. Those who stand by are conscious that their turn may come soon. The sound of death's messenger is in the air."[33]

Sickness began to overtake the regiment. The medical resources, inadequate to begin with, were no match for the yellow fever, malaria, typhoid, and dysentery that ravaged the troops. Later Chaplain Groves thought that he had caught yellow fever on July 2. Unmistakably a few days later he began to suffer from it. For him, the war was over. Throughout the next few weeks the fever left him weak and dimmed his perceptions. He was sent to a yellow fever hospital in Siboney, a small port west of Daiquiri. Along the way he also contracted malaria. On July 23 he boarded the hospital ship *Concho,* which sailed up the Atlantic seaboard the following week. On board he was able to write to Gwen for the first time in three weeks, informing her of his sickness. In the commotion he had lost all of his prayer books and told Gwen that he had to conduct from memory the shipboard funerals of other yellow fever victims.[34]

The ship arrived in New York City on July 31 and the chaplain was given a month's sick leave. Partly because of his weakened condition, he chose not to make the train trip across the country. Instead, he convalesced for two weeks in Millbrook, New York, near Poughkeepsie, as a guest of the Lamonts, gaining weight and getting stronger. After a visit to Albany and Utica he spent September and October on sick leave.[35]

Meanwhile, Gwen left the house and the two older boys in the charge of her younger sister, Jane, who was a member of the household. Accompanied by two-year-old Dick, Gwen Groves traveled east in mid-August to be with her husband, to spend time in Utica with family, and to accompany him back to Vancouver Barracks. They arrived back on November 4, 1898.

Even before he landed in Cuba, Chaplain Groves had become so discouraged by the behavior of the troops — the drinking, swearing, and gambling — that he thought he might resign from the army. In a letter to Gwen, while still in Tampa Bay, Groves wrote:

> Do you know it is a little hard to think of this war as the Lord's doing as I have been doing, when I meet the men who are in it. Certainly, there is nothing further from their purpose than serving the Lord. But they serve the Devil energetically. I suspect I shall be ready to resign when this war is over, or when I have had a few months' experience in it.[36]

But by the time he was shipped home six weeks later — with little encouragement from the few men who praised his efforts to do God's work — he had changed his mind and decided to remain in the army.

While Chaplain Groves was off saving the souls of the Eighth Infantry in Cuba, the Fourteenth Infantry had gone with the expeditionary force sent to the Philippines. Soon after returning to Vancouver Barracks, Chaplain Groves was ordered to join them. His enthusiasm about going is somewhat in doubt. A scandal of the day was the number of chaplains who found excuses not to go to the Philippines. But when Chaplain Groves's turn came and he was ordered to go, he made no excuses, he did his duty, and he went. There was strong objection to his going by the post commander, on the grounds that he had not yet fully recovered from the effects of yellow fever and malaria. But orders were orders and duty was duty.[37] And so began another odyssey for Groves — this one lasting almost two years — initially with the Fourteenth Regiment in the Philippines, and later in China.[38]

American interests in Asia in the 1890s centered on acquiring markets, winning concessions, establishing bases, and expanding missionary activity. It was a time of dramatic international intervention by the United States, and Chaplain Groves would be at several key interstices to bless the process.

The Spanish had ruled the Philippines since 1565. In late 1896 nationalist rebels, under the leadership of Emilio Aguinaldo, rose to try to oust them. A truce was arrived at in December 1897, with Aguinaldo given a sum of money and expelled to Hong Kong. At this juncture the United States saw an opportunity. Soon after the April declaration of war against Spain Commo. George Dewey won a decisive naval victory in Manila Bay on May 1, becoming an overnight hero. To take the city, and the island of Luzon, Dewey needed ground forces, and he sent a request to Washington for troops. While awaiting their arrival Dewey encouraged Filipino insurgents to besiege the city and permitted Aguinaldo to return; Aguinaldo in turn declared independence on June 12.

On July 25 Maj. Gen. Wesley Merritt arrived to assume command of VIII Corps, a U.S. force of 10,800 that had crossed the Pacific to conquer the Philippines.[39] Knowing they were beaten, the Spanish were willing to surrender to the Americans, but not to the Filipinos. After a prearranged sham assault on Manila the Spanish surrendered to the Americans on August 13.

The Treaty of Paris, signed on December 10, 1898, ceded the Philippines (and Puerto Rico and Guam) to the United States, much to the surprise and consternation of the Filipino nationalists. Unwilling to accept their new masters, Aquinaldo and his guerrilla rebels took up arms against the Americans. A two-year period of brutal counterinsurgency warfare ensued, with the movement eventually crushed, Aguinaldo captured in March 1901, rebels imprisoned in camps, and some even executed.[40] It was in the midst of this American suppression of the Filipino rebels that Chaplain Groves arrived to fortify the Christian resolve of the American soldier. If his letters are any indication, he never questioned what the United States was doing there — assuming, as did everyone else, that America's manifest destiny now spanned the Pacific to Asia.

Chaplain Groves sailed from San Francisco on December 6, 1899, aboard the transport ship *Sherman*. En route to Hawaii he wrote to Gwen and the boys, conducted services, taught Bible lessons, and led the singing of hymns. He reached Honolulu on December 13 after sailing twenty-one hundred miles, but he was not able to go ashore because "some Chinese had brought in the bubonic plague a few days ago."[41] And so on the next day the *Sherman* sailed for Guam, arriving on December 27; unfortunately there was no coal available, so it was off once again, this time a six-day sail, before finally dropping anchor in Manila Bay the morning of January 2, 1900.

At last on land Groves explored the town, marveling at the water buffalo and the little ponies, both pulling carts. He met his superiors and acquaintances from the Fourteenth, whom he had finally caught up with, and finagled his way into being assigned regimental chaplain of the Fourteenth Infantry in Manila. Half of the regiment's companies were at headquarters, and the other half scattered throughout the city.[42] Groves spent fifteen months in the Philippines — from January to July 1900 and from November 1900 to July 1901 — paying these companies regular visits. According to his commanding officer, Chaplain Groves

> worked with the men in a common-sense, tactful way and won their esteem. His influence was very great, and aided much in bringing the regiment up to a high state of discipline and efficiency.[43]

Groves seemed to enjoy his time in Manila. He rode horseback and eventually bought a handsome chestnut, Reddy; socialized with the other officers and expatriates; and interacted with his clergy brethren. He performed his religious duties,[44] set up a small library for the troops, and faithfully wrote to his family every day. There were some dangers and irritations, of course: The plague threatened, the "Insurrectos" might strike,[45] ravenous mosquitoes were ever present, the food was bad and the climate dreadful. And always there was self-doubt; as in Cuba, he questioned whether his efforts were really doing any good.[46]

Gwen apparently kept up her side of their correspondence, but unfortunately her letters to him did not survive, apparently destroyed or lost by

Chaplain Groves. We get hints of their content in his responses to her about what Allen, Owen, and Dick were doing in Vancouver. In almost every letter Chaplain Groves gave instructions to Gwen about how the boys should be brought up,[47] how much time they should spend on schoolwork, and how much time for play. He also spent a great deal of time mulling over the family finances, which were always in short supply.[48]

As spring wore on, he began to report that rumors were circulating to the effect that the Fourteenth might be sent off to deal with the growing crisis in China. He repeatedly told his wife that he was not interested: "There is no desire for adventure left in me. My job is good enough."[49] On July 8 rumor became reality: Chaplain Groves was informed that the Fourteenth was going to China, and that he was urged to go with them. With a new destination he was now eager to go, and after boarding the transport ship *Indiana* on July 14, they sailed the following afternoon, "to settle affairs there."[50] The affairs had to do with a group of Chinese known as the Boxers.

What we commonly call the Boxer Rebellion was a complex event, a full account of which lies far beyond the scope of this work, though given the influence it had on the chaplain and his youngest son, some of the events and details must be recounted.

Boxers was the name Westerners gave to members of a Chinese secret society — initially composed of peasants and later of the urban under-class[51] — which held that the performance of certain martial rites endowed them with supernatural powers. Emerging as a patriotic, nation-alistic force, the Boxers' distinguishing hallmarks were their anti-Christian and antiforeign sentiments.[52] In late 1899 and early 1900 dissident ele-ments of the Manchu court, in concert with some provincial governors, aligned themselves with the Boxers in a common policy to rid China of the "foreign devils" who were carving up China ("like a melon") into spheres of influence for their own benefit.[53]

By the spring of 1900 in Shansi and Inner Mongolia, and by June in Peking and Tientsin, Westerners and Chinese converts were being attacked, many fatally, with special targets the missionaries. Of equal con-cern were the threats to commerce from the ripped-up Western-owned railway tracks and the cut telegraph lines.[54] Every one of the affected gov-ernments reacted swiftly — not only to protect its own citizens and national interests in China but also to ensure that no other power would gain either military or commercial advantage over it in the process.

Of the several phases of the "Boxer War" Chaplain Groves participated in two: the march from Tientsin and the freeing of the foreign legation in Peking. In Peking most foreigners lived in a walled compound near the Forbidden City. On June 20, after the Chinese government issued an ultimatum for the foreigners to leave within twenty-four hours, imperial army forces, with some Boxer participation, laid siege to the legation area. The British, Russians, Germans, Japanese, Americans, and others inside put up a hasty defense.[55] After a month of fighting there was a brief truce as the court tried to reach a negotiated solution, but full-scale bombardment resumed in early August.[56] Urgent help was needed to lift the siege, and a multinational China Relief Expedition was quickly formed from military units already in China and from those in the region. As part of the American contribution, two battalions of the Fourteenth Infantry, including their chaplain, journeyed from the Philippines.[57]

Groves's transport ship sailed, in an odd coincidence, first to Nagasaki, arriving there Saturday morning, July 21. The chaplain was able to manage a quick tour of the city and the nearby countryside, admiring the terraced rice fields and breathing the cool, fresh air. His ship departed Sunday afternoon and sailed across the Yellow Sea, arriving near Taku on July 26. There he took a train to nearby Tientsin, where a savage battle had recently ended a siege, and troops were mustering for an assault on the diplomatic quarter in Peking.[58]

After a week of waiting for wagons, mules, horses, and supplies, they finally set off on August 4. Some twenty thousand troops (half of them Japanese) left Tientsin on a grueling eighty-mile march along the Pei Ho River toward Peking.[59] There was some fighting against imperial troops, with actions at Peitsang on August 5, and at Yangtsun on the 6th.[60] The Fourteenth Infantry suffered the most severe casualties. In four days, Groves recounted, the regiment sustained seven killed and fifty-six wounded, with fifteen of the wounded likely to die.[61] Many others were overcome by the oppressive heat, often over a hundred degrees.

Groves's devotion to his work is attested to by his commanding officer, Col. Aaron S. Daggett.

> During the campaign in China where so many men were prostrated by heat, the Chaplain was constantly with them, helping them in every possible way. His horse was more often ridden by soldiers than by himself. He would arrive at bivouac among the last and late at

night, and be ready to move early in the morning. No other Officer worked as hard. After the battle of Yang-tsun he was thoughtful enough to put the name of every soldier who was killed, in a bottle, to be buried with his body, for identification. These are only examples showing the manner in which he performed all duties."[62]

The relief expedition finally reached Peking. On the morning of August 14 the Fourteenth Infantry Regiment played a key role in lifting the siege, being the first foreign unit to enter the Outer City. Confronting the forty-foot wall near the Tung Pien Men gate, the Fourteenth's commanding officer, Colonel Daggett, asked for volunteers to scale the wall. Musician Calvin P. Titus, the chaplain's assistant and bugler of Company E — who would become a lifelong friend — volunteered and was able to accomplish the assignment while under heavy fire.[63] About one-half of E Company climbed the wall following Titus. Once over, they opened the gate, and many units forged through. They spent all day fighting to help free the legation. Resistance crumbled, and the fifty-five-day siege was lifted.[64]

The British role in the attack that day left a bitter aftertaste with the chaplain, one that would sour his son's attitudes toward them as well. Sir Claude MacDonald was head of the British legation. On August 8 he sent a message to the commander of British forces (and commanding general of allied forces), Gen. Sir Alfred Gaselee, who was approaching Peking, on how to best enter the legation quarter. The British selfishly decided to keep this information to themselves and not share it with their allies. These secret instructions enabled a British force of some seventy men to use the sluice gate (or water gate) of the canal to enter the legation quarter — to be welcomed by Sir Claude in immaculate tennis flannels — an hour or two before the Fourteenth Infantry arrived.[65]

Anglophobia ran deep in the Groves family, and the opportunity to watch the British close up in China did nothing to diminish it. For Groves this was how the British were. Everything they did was for selfish reasons, to gain an advantage for themselves. Whether trying to gain glory by being the first into the legation and jeopardizing the lives of their allies, or by using the knowledge they gained about atomic matters for their own purposes after the war — as his son would conclude — this was just the way they were.

After the siege was lifted they bivouacked at the Temple of Agriculture, a 275-acre cypress-covered compound in the southern section of Peking. The American forces, under the command of Major General Chaffee, did much to restore order and even provide programs of relief, justice, and public health.[66] Nonetheless, there was looting, and Groves strongly disapproved of it, especially that done by the British and the Russians.[67] As the only chaplain in Peking for all the American forces, he was kept busy with funerals, performing sixteen by August 19, of which twelve were of the Fourteenth Regiment.[68] He resumed Bible classes and services, visited missionaries and the sick and wounded at hospitals — sometimes on his horse Reddy, who made the trip — conducted funerals, toured the Forbidden City,[69] Coal Hill, the Temple of Heaven, and the Summer Palace, and wondered where they would be sent next. With the armies of several nations all in Peking he had the opportunity to compare and contrast, praise and critique. While there are things that the U.S. Army could learn from the others, "When it comes to getting there and fighting, our soldiers are out-of-sight of the whole crowd. Only the Japs can equal Americans and I hope we will never have to find out which is better."[70]

Finally, on October 21, Groves and approximately half the Americans of the expedition marched out of Peking with great ceremony. The Fourteenth got a twenty-one-gun salute, and General Chaffee expressed his appreciation for its efforts.[71] After an arduous journey back to the coast, marching and camping, the regiment boarded the transport ship *Warren* on November 3 and sailed for Nagasaki, arriving three days later. Then, despite rumors that it would go on to America, the Fourteenth headed back to Manila.

Soon after arriving Groves learned that his seventy-five-year-old mother, Adeline, had "gone home," a phrase he often used to describe dying and death.[72] He quickly resumed his routines, some of them astride Reddy, who had returned to Manila with him. These duties, with even a new one proposed — secretary of the Army Temperance Union — dragged on as he anticipated orders to leave and return home. But even after the rebels surrendered and "Augie" (Aguinaldo) was captured, fighting continued.[73] Finally, on May 15, he heard that they were leaving in July. On July 20 the transport ship *Sheridan* left Manila Bay bound for San Francisco, with stops in Nagasaki and Yokohama, arriving at the Presidio in mid-August.

By the time Groves returned to the United States after a twenty-one-month absence, Allen was almost ten, Owen eight, and Dick five. Dick was slow in learning to talk. According to legend his first word was *cheese,* at nearly five years old.[74] Gwen and her sister, Jane, who was, for all practical purposes, a member of the family, had overseen the children's day-to-day care and discipline. There is some evidence that in Leslie's absence the rigid rules that he insisted on were not followed to the letter. Gwen taught the boys card games; they were even allowed to play on Sundays.[75] But even from afar Groves exerted an influence on the boys. He wrote frequently, urging them to learn their lessons, to be strong, brave, and honest.[76]

According to Richard Groves, his father and the other Groves children were raised in a highly competitive environment. There was tremendous competitiveness within the family and between the family and the rest of the community. From an early age they were taught to excel. "If it is a game, you win it. If there is a class you stand number one, that was how they [Chaplain Groves's children] were brought up."[77]

These experiences were passed on, for, as we will see, when it came to raising his own children, Dick created the same kind of competitive environment. We see the zest for competition even in the letters that their father wrote the children from the Far East. Near his seventh birthday Owen received the following:

> I wonder if you can read this letter yourself. You must learn to read faster or Allen will be so far ahead of you that he will be through College before you get there. That would never do. You know you have money in the big bank in Albany to use when you go to College and bye and bye when Faddy [Groves consistently uses this word for "Daddy" in these letters] gets home you will earn more money to put in the bank. Dick has not so much as you and Allen but Dick will work hard some time when he is a big boy to get some more. So I think he will have enough. But the little boys will have to study for themselves to get to know enough to go to College when the time comes. I hope we will live farther east so that they can go to Hamilton College.[78]

About two weeks later, Owen was sent the following:

> That was a nice little letter you wrote to Faddy, only it made Faddy feel badly that he had not sent you any stamps before. But Allen is the bigger boy and the big boy gets ahead of the little in such things. When you are as old as Allen is now you will have more stamps than he has now. . . . I am going to ask an Englishman here to give me stamps for

my little boys and part of them will be for you. Not so many as for Allen though. He will save them more carefully because he can read what is on them. But you shall have some you nice little boy. . . . .[79]

Several months later Chaplain Groves wrote to Owen, from Nagasaki, on his way to China:

You will be a good boy while Faddie is away off in China. Get Allen to show you China on his map and the way we will go from Taku to Peking. He can tell you too how big a city Peking is. He will look it all up in the Encyclopedia. There I may find you some funny thing to send you. There are no monkeys there and I could not send a monkey in a letter any way. He would scratch out or starve before he got there. I think it was Dick who wanted a monkey sent to him. I shall have to tell him why I could not send him one. . . .[80]

These letters reveal much about the dynamics of the family, dynamics that were to have profound consequences for the life paths of the three sons. Allen, the oldest, was also the favorite and, as is often the case with first sons, the child who embodies the parents' grandest dreams for their progeny. Chaplain Groves's advice to Owen, that he consult Allen about China — where it is and what it is like — was not empty advice. Though Allen was only ten years old, he had already shown signs of being an exceptional child. He was not just smart, he was brilliant. He remembered everything he read, was versatile, and was able to quickly master whatever challenge was put before him, be it the playing of a musical instrument, the ability to draw decent illustrations, or any academic subject — classics, language arts, English composition, science and mathematics, or history. He also seemed to have been blessed with a warm and caring personality — someone who made friends easily and was always there to help, within the family or among his friends.

Though favored by his parents, the attention and the adulation that Allen received did not seem to generate any hostility among the other children. The reasons for this are evident in Chaplain Groves's letters to his wife and to Owen and Dick. Though Allen was clearly the "number one" son in the family, it was apparent to all that the parents loved Owen and Dick as well. Each of the children was of value in the parents' eyes, and the children knew it.

Being a military family meant being a family always on the move. Between August 1901, when he returned from the Far East, and the spring of 1913, when his wife, Gwen, died, Chaplain Groves moved, under orders, eight times, and the family six times.[81] This span of time would be the formative years of Dick's childhood and adolescence, from age five to completing high school and entering college.

Dick's first memories, recorded late in life, were of Vancouver Barracks, probably in 1900 and 1901, involving a close friend of the family, a retired medical officer named Lieutenant Colonel Wolverton, who lived in Vancouver. While stationed there he had become quite wealthy using the Homestead Act to acquire land. Wolverton arranged to take Gwen and the three boys on a camping trip for three or four days by wagon through beautiful countryside. A second memory was from the summer of 1900, when Gwen decided that a vacation was needed. She took them to Fort Canby, right on the ocean at the mouth of the Columbia River, not far from Astoria. Some vacant officers' quarters were made available to those stationed at Vancouver. The most excitement came as a result of a forest fire that had broken out near the post. As it approached, they were alerted to leave on a few minutes' notice, but the wind changed and they did not need to evacuate. A third early memory was of Wolverton's daughter Florence giving Dick a party on his fifth birthday. It was a small and simple affair, but Dick remembered that the refreshments were very good, and that they all paraded around the fenced-in yard carrying small American flags while somebody furnished the marching music by means of a comb and paper.

Prior to Chaplain Groves's return to the United States in August 1901, the Fourteenth Infantry — to which he was now permanently assigned — had been ordered to Fort Snelling, Minnesota, a stone fortress located halfway between St. Paul and Minneapolis at the confluence of the Mississippi and Minnesota Rivers. In July Gwen moved the family from Vancouver to Snelling. Then in September she traveled to Utica to join her husband, who was there to settle his mother's estate. Five-year-old Dick and his father visited the Lamonts' country estate, with its private bowling alley, and the family farm outside Utica, near Hamilton College. While in Millbrook they traveled to nearby West Point to visit the chaplain's former assistant in China, Calvin Titus, who was a cadet. Pres. Theodore Roosevelt had appointed Titus to the Military Academy, and the following June awarded him the Medal of Honor for his heroism in climbing the wall at Peking.

In November they returned to Fort Snelling and Dick started first grade. A memory Dick had of Fort Snelling was of seeing the Seventeenth Infantry marching out in formation, wearing blue flannel shirts with their bedrolls tied horseshoelike over their shoulders.

In the summer of 1902 the family moved to Fort Wayne,[82] Michigan, which was garrisoned by the Fourteenth's regimental headquarters and one battalion of the Fourteenth Infantry. Dick attended first grade. At Fort Wayne the fourth child, and only daughter, Gwen Griffith Groves, was born on August 7, 1902.

Soon after Gwen's birth the Fourteenth Infantry was ordered back to the Philippines. Chaplain Groves wanted to return with his regiment but was ordered by the War Department to exchange places with another chaplain who had not seen foreign service. Thus the family moved in early February 1903 to Fort Hancock, New Jersey, a coastal artillery post on Sandy Hook, protecting New York City, where they would spend two and a half years.

They lived in a large three-story house, at Number 7 on Officers Line, fronting on the bay, and had a cook and a maid. While there the boys obtained two years of formal schooling in the Fort Hancock area. Dick finished second and third grades in the post's one-room school, half a mile from the house. The school went only through fourth grade and had a total of thirty to forty children. Allen and Owen went by train to Atlantic Highlands, where Allen completed his first year of high school and Owen attended middle school.[83] Dick remembered his grammar school as well run, not least because of the strict discipline.

> It was customary for misbehaving children to receive a rather sharp whipping, either with a wooden switch, or with a ruler laid across the palm of their hands. It was also customary never to tell anyone at home that we had been punished for that merely meant that we would be reproved and, if it was bad enough, we might be punished again.[84]

Dick evidently misbehaved at least once, at home. One morning as punishment for some misdemeanor his mother put him in the small closet under the front staircase, with orders to stay until she told him to come out. As she went about her chores, she forgot where he was and he remained silent, as he had been told. When Chaplain Groves came home for lunch he called out for Dick, with no response. The parents became frightened — perhaps he had wandered to the beach. Search parties of soldiers were sent out. "At long last, the frantic mother remembered the closet and the small prisoner. The door was flung open. There he sat, in the dark, obeying orders to the end. It had been many hours."[85]

After feeling poorly for some time, Chaplain Groves discovered that he had tuberculosis, an affliction that he may have contracted while serving in China or the Philippines. His susceptibility likely had been heightened by the earlier bouts of malaria and yellow fever, which left him tired and weak, just the kind of condition that makes one vulnerable to tuberculosis. In January 1905 Chaplain Groves departed for the Army's Tuberculosis Hospital at Fort Bayard near Silver City, New Mexico. Once again Gwen was in charge of moving the family. Waiting until the end of the school year in June 1905, with the chaplain still recuperating in New Mexico, Gwen, Jane, and the children traveled by train (with passes for part of the trip supplied by Mr. Lamont) back to Vancouver Barracks to rejoin the Fourteenth Infantry Regiment.

Once again in familiar surroundings, they occupied half of a double house, with a large backyard. Nine-year-old Dick began fifth grade at the Vancouver school about a mile or so from the post, getting there and back in a mule-drawn wagon that served as the school bus. A few of his schoolbooks from this period have remained in the family: *The Boys' Book of Famous Rulers* by Lydia Hoyt Farmer, *Hero Tales from American History* by Henry Cabot Lodge and Theodore Roosevelt, and *Stories from English History* by Henry P. Warren.

In the spring of 1906 young Gwen, then three and a half, was diagnosed as having tuberculosis of the spine. Once again it fell to her mother to deal with a serious problem, and she was advised to go to Los Angeles to consult with the leading expert. Before she could leave, an epidemic of the measles hit the post; Dick, and a few days later Owen, came down with the disease. A quarantine period of three weeks delayed Gwen's departure, and fortunately also caused her to miss being in San Francisco during the earthquake and fire that occurred on April 18 and 19.[86]

Young Gwen would spend the next three or four years lying recumbent on a frame shaped to correct the curvature of her spine, being allowed to get up only to bathe. This difficult routine was an additional trial for her mother, who had to take care of the child's every need.

Three aspects of the Groveses' family life seem to have mitigated somewhat the disruptive effects that frequent moves, and the chaplain's comings and goings, might have had on the children's upbringing and education.

The first of these was Chaplain Groves's decision to establish residence in southern California. By the spring of 1906 his recovery from tuberculosis had progressed to a point where he was released from the hospital at Fort Bayard and granted a yearlong convalescent leave. After investigating several communities in southern California he decided that the Los Angeles area best suited the family's needs. He initially rented a house on the outskirts of Pasadena. There the boys were introduced to the ordeal of farmwork as pickers of walnuts, peaches, watermelons, prunes, apricots, and grapes, normally working ten hours a day to earn a dollar.

Just before Christmas 1906 Chaplain Groves bought a comfortable house on three-quarters of an acre, just north of the Pasadena city line, at 969 New York Avenue in Altadena, and the family moved in.[87] The purchase of a permanent residence must have come as a welcome relief to Gwen from the nomadic existence the family had experienced up until then.

Worried about his ability to support his family, Chaplain Groves had been trying for more than a year to be promoted to major — a rank for which he was eligible under a newly enacted law. He wrote to officers under whom he had served, and former patrons such as Lamont, to solicit their support. But the judge advocate general denied his application on the basis that there was no authority to promote him while he was undergoing treatment. Finally, after almost three years as a patient at Fort Bayard, with occasional sick leave, Chaplain Groves returned to active duty.

The choice of where he was to be posted, given the state of his health, was worked out: He requested Fort Apache, Arizona. In January 1908 he returned to duty, and on March 25 he was promoted major, receiving a welcome pay raise.

The rest of the family stayed behind in Altadena so that the boys could complete their school year, after which they reunited in Arizona. In May 1908 Gwen, young Gwen, her nurse, Dick, and their dog Othello made the arduous three-day journey to Fort Apache, first by train to Holbrook, Arizona, and then another ninety miles by a stagecoachlike Dougherty wagon, drawn by six mules. Established in 1870, the post had been the main base of military operations against the Coyotero Apache.[88] Chaplain Groves's quarters were a large one-story house with a private stable and a backyard with a small vegetable garden. Twelve-year-old Dick's responsibilities were to keep the garden, secure the chicken house at night from predatory animals, look after the cow and her calf — and help with the milking — and take care of the two horses, Dr. Gray and Zee. The family would spend the summers together in Arizona, and then

Gwen, Jane, and the children would return to California to begin a new school term.[89]

A second influence that tempered the effects of the constant changes on the children was Chaplain Groves's direct approach in shaping the moral and social development of each child. According to his grandson, Richard:

> Chaplain Groves believed in one-on-one contact with children during their formative years. In the spring of 1937, I became the object of his close scrutiny when he made a special trip to Deerfield [Academy, in western Massachusetts] where he spent a week with me, probing all the time my beliefs and my manners — an awesome experience for a thirteen year old, which I shall never forget. In like manner, in mid-1911, Chaplain Groves seems to have decided that his third son [Dick] needed his personal attention. Until this time, whatever interaction they had was always in the presence of others. Now they would be alone together for several months. And, like me, [Dick] came away deeply impressed. He had always respected his father from afar; after this encounter, they really knew each other — and they developed a mutual admiration, which endured through their lives.[90]

Allen and Owen had had their turns. Dick's came in the summer of 1911, just before his fifteenth birthday. As soon as school was out, he journeyed from Pasadena to spend the summer with his father at Fort Apache. Most days, while his father tended to his duties at the fort, Dick played tennis at the Indian Agency, often not leaving until after dark. He excelled at tennis and developed a passion for the game that continued throughout his life. Many years later, even after he had gained a great deal of weight, Dick was able to hold his own against much younger, more agile opponents. Even those who knew him well were surprised to find out just how good a tennis player he was.

Dick handled many of the responsibilities in the household: cooking meals, cleaning, filling the oil lamps, tending the vegetable garden and the chickens. Father and son spent many days together. They would pack a lunch and go off on horseback for daylong rides, beginning before dawn and not returning until dusk. Sometimes they would ride to the east fork of the White River and fish for speckled brook trout. Sometimes they would just find a comfortable place to sit and talk. In these direct ways Chaplain Groves passed on to his sons the essence of his own wisdom, values, and beliefs. This was the first of two extended periods of time that Dick would have with his father. At the end of the summer, Dick returned to Pasadena to begin tenth grade.

The third aspect of Groves family life, the influence of Gwen on the children, is probably the most important, but the most difficult to assess and describe. With the chaplain away for extended periods of time, in Cuba, the Far East, or Arizona recovering from tuberculosis, it was really Gwen who raised the children. Her influence seems to have been profound. We see from the hundreds of his letters that survive a portrait of an outwardly moral, upstanding clergyman trying to do what is best for his family, often from afar. His many changes in vocation, his wanderlust, and his long absences from his kin reveal a man and father who is indecisive and confused, though he is able to keep it hidden from everyone, even himself. Gwen must have seen these qualities, too, but could say little in protest. In response, she decided to be the formative influence on her children, giving them the essential daily direction and guidance they needed. To try to do this through letters from afar or sporadically in person was just not enough. From the limited written record that does survive, Gwen's strong character, stronger than her husband's, rings through.

In the fall of 1911 Allen began his junior year at Hamilton College. The money that had been put away for this purpose "in the big bank in Albany" was supplemented by scholarships from Hamilton. As an entering freshman in 1909, Allen had won the Fayerweather Entrance Scholarship and the Maynard Entrance Prize.[91]

Allen's career at Hamilton was one prize or award after another. In 1911 he won the sophomore essay prize on the subject "How Far Is the Newspaper Subsidized," and in 1912 he took the junior essay prize for "Nemesis in Greek Tragedy and in Shakespeare." But literature was not the only field in which he excelled. He had a photographic memory and total recall:

> He could recite the names and batting averages of every player in the major leagues; he could repeat the census figures of all the cities and larger towns of the United States — these were but *tours de force,* stunts which he set himself for fun. This faculty was of the utmost value to him in those linguistic studies in which he chiefly excelled. The facility with which he could master a page of difficult Latin or Greek on the way up the Hill to morning chapel was a mystery to

those who did not understand that the secret of his rapidity lay in his unusual power of concentration.[92]

He won essay prizes in three of his four years at Hamilton and prizes in American history and mathematics as well. He was valedictorian and Phi Beta Kappa. In high school he had studied for geometry exams by casually scanning the material as the streetcar carried him to school.[93] His high school essay, "On Moderate Drinking," took city, county, state, and national awards. Allen was an excellent basketball player, won second place in the tennis tournament in his senior year, and was a superb debater.[94]

In the fall of 1912 Owen (class of 1916) joined Allen at Hamilton, now beginning his senior year. Though Owen did not have Allen's brilliance or powers of recall, he was an excellent student nonetheless, majoring in English literature. Following in Allen's footsteps, Owen won the junior essay contest and was published frequently in the *Hamilton Literary Magazine,* eventually becoming its editor.[95]

Gwen Groves kept in close touch with her boys, writing often. In a series of letters that survive — from the summer of 1912 to her death, a year later — we see her firm pressure on them to excel at everything they did. To Allen, entering his senior year at Hamilton, and to Owen, just starting as a freshman, she inquired about how they were faring academically, socially, and athletically. She also informed them of Dick's behavior at home.

To Owen, who had just graduated from high school in June, she offered congratulations on his English honors:

> I am very glad indeed that you have second, if the judge wouldn't give you first. Second best in class is good. Next year we will try hard for first. . . . Your Father in all his letters tells me he is immensely pleased with you and all your work and that's no faint praise.
>
> Richard should be trimmed in tennis, in mind, manners and in everything. And A[llen] isn't going to be equal to the job. R[ichard] does play a smashing McLoughlin sort of tennis game. His return drives are awful sometimes. I am always afraid he will kill someone with them, and he reaches all over the court, and never tires, or hurries. Everyone likes him, which makes him unmanageable at home. I can only rule him by strategy."[96]

In October she told Owen of life at Fort William Henry Harrison. She and Chaplain Groves held a dinner for Colonel and Mrs. Wilson and three other couples. Richard and the Wilson daughter, who had apparently been feuding, now "speak as they pass by." Some political opinions were expressed, Gwen asking, "Don't you think Taft's chances are growing brighter every day. Richard assures me that Montana is for Taft."[97]

Over the next month in weekly letters she hoped Owen would make the Press Club, and perhaps the following year, the choir and glee club.[98] She advised him that a typewriter

> may help you to better marks in English. Your writing and your work is often carelessly done. . . . What is this about your German and why don't you put it out of your way? I thought your Pasadena credits were good in everything except Latin.[99]

She fussed and fretted that the two older boys might be injured in football:

> Dick is more interested in your football practice than in anything that has happened to either of you boys. Every few minutes he wants to know what I am going to do about [it]. . . . I noticed in the Hobart game Stone was injured and some boy had his nose spoiled. As I have often said you boys were a heap of trouble and expense to bring up, and I don't want any spoiled boys. Does the football management pay all damages? I personally hate the game. And you can't say I am no sport. Baseball and tennis are quite as good. Did you notice that Devore is out of the West Point game? One of the best tackles in the country. I never pick up the paper without reading of football injuries.[100]

And she lectured them on politics. After the Democratic landslide on November 5, she wrote:

> I am glad election is over tho it was distressing — and in it all the Roosevelt party is to blame [for Woodrow Wilson's victory]. Think of old Bill Sulger as Gov. of NY State. You keep away from so called pro-gressive influences. Let A[llen] take care of all the freak politics of the family.[101]

But always, she returns to the main point of the exercise: achieving excellence in a competitive environment:

Are you going to write for prize this year and on what subject? Do your best, I expect you to win.[102]

You must practice your signature. Make it larger, as a man of your importance in the world ought to.[103]

Now I wish you really would cultivate your handwriting. It is funny — so tiny you make your letters. Be a little more dashing.[104]

She commented on and promoted the competition between Allen and Owen:

I am sorry the typewriter was in such demand the last day that you did not beat it with yours to have it well done. After all I am disappointed in Allen. So many times I've written to him to look over your work, and at the last there seems to have been only blunders, disappointment, and misunderstandings. Never mind, we know you both and love you both, and it's the way many brothers have. We are tremendously pleased with your work just the same, and I think you'll win . . . next year you will have a typewriter and will be yourself an accomplished typist — and depending entirely upon yourself will be easier in many ways. So cheer up dear Owen. If this year you do not win, there are other years.[105]

In December 1911 Chaplain Groves was ordered to rejoin the Fourteenth Infantry, now garrisoned at Fort William Henry Harrison, about four miles outside Helena, Montana. Before reporting to his old regiment, he took a short leave in Pasadena, and when he left by train for Fort Harrison shortly after Christmas, Dick accompanied him.[106]

For a second time then, Chaplain Groves would focus his attention on his youngest son's moral and intellectual development. But Dick's year in Montana would also be significant on two other counts: He came under the influence of Lt. Edmund B. Gregory, and he met his future wife.

Lieutenant Gregory taught courses at the post for enlisted men interested in preparing for the entrance examination to West Point.[107]

Although [Dick] would have been too young to attend Gregory's classes at this time [he] discussed his future with the Lieutenant and seems to have formed some idea of how he would proceed.[108]

Dick was beginning to sort out his career and to take his education into his own hands, a process that would gather momentum and crystallize over the next two years. Gaining admission to West Point was the first big goal that he set for himself, and once he set it, he was determined to accomplish it.

When Dick discussed going to West Point with his parents, they advised against it, though they did not forbid it. His overburdened mother could hardly be expected to recommend the military life. His father, too, had been unprepared for the brutality of combat and the rigors of foreign duty, and he was frustrated by his inability to reach the large numbers of soldiers who were in his care, whose penchant for swearing, drinking, gambling, and fast women he found so abhorrent. Chaplain Groves would rather Dick follow his brothers to Hamilton College, but neither parent ever firmly insisted that he not go into the army.

The second important event during Dick's time in Montana was that he met a fourteen-year-old girl who would become his wife. Grace Wilson, known to family and friends as Boo, was the daughter of Col. Richard Hulbert Wilson. Colonel Wilson was born in Hillsdale, Michigan, on June 10, 1853, the son of a prominent attorney and later circuit judge.[109] From 1869 to 1871 he attended Hillsdale College, after which he entered West Point in July 1873. He ranked twenty-sixth (of seventy-six cadets) in the class of 1877.[110] For the next twenty-nine years he served with the Eighth Infantry Regiment pacifying Indian tribes at almost a dozen western frontier posts in New Mexico, Arizona, California, Nevada, Nebraska, and Wyoming, some very isolated.[111] On June 26, 1895, while detailed as agent[112] to the Arapahoe and Shoshone Indian tribes near Fort Washakie, Wyoming, Wilson, then forty-two, married twenty-one-year-old Grace Arents Chaffin of Cheyenne. Their daughter Grace Hulbert Wilson was born at Fort Washakie on May 25, 1897.[113]

As mentioned at the beginning of the chapter, Captain Wilson and Chaplain Groves served together in Cuba during the Spanish-American War with the Eighth Infantry Regiment. At that time Captain Wilson commanded the regiment's Second Battalion, landing at Daiquiri on July 22 with the main invasion forces. He served with distinction, showing initiative that later won him the Silver Star for gallantry in action. Like the chaplain, he kept a diary and came down with fever, although he refused to be evacuated.

Grace, as part of a military family, moved from post to post throughout her childhood, growing up in the army. The list is long and varied:

Huntsville, Alabama; Cheyenne, Wyoming (while Wilson was in Cuba the second time); Fort Logan, Denver, Colorado; Presidio of San Francisco; Fort St. Michael, territory of Alaska; Fort Slocum, New Rochelle, New York; Fort McKinley, Manila, Philippines (via Hawaii, Guam, and Nagasaki); Fort McKinley, San Juan, Puerto Rico; Fort Crook, Omaha, Nebraska; and Fort Logan A. Roots, Little Rock, Arkansas.[114] In June 1910 Wilson was promoted to colonel and appointed commanding officer of the Fourteenth Infantry Regiment, with headquarters at Fort William Henry Harrison.[115]

Dick's memoirs make no reference to the initial meeting with fourteen-year-old Grace, which occurred at the colonel's New Year's eggnog party. Nor does he supply any information about their ten-year courtship and subsequent marriage. This was in keeping with his reserve about private matters. Dick had a lifelong reticence about revealing himself, especially about such personal matters as his wife. Fortunately, Grace Wilson did record some memories of these events. About that first meeting she wrote:

> During the second winter [1911–1912], the Regimental Chaplain joined the Regiment. He had been ill and sent to Arizona for his health. He arrived at Ft. Harrison with only one member of his family, his youngest son Dick. The Chaplain alluded to him as "my little boy Dick", but he was anything but little. They were just in time for the New Years' Day eggnog party at the Colonel's house. The 14th Infantry was proud of its regimental punch bowl, a huge basin made of silver "looted" by the regiment during the Boxer fighting in China. . . . I'm sure that our Chaplain didn't approve of the refreshments. There must have been something less strong for him and his little boy, Dick, who was fifteen, to drink.[116]

Dick and Grace soon became fast friends. Later in the year Grace composed short poems about those at Fort Harrison, among them:

> This is Dick and this is fudge
> From it little Dick won't budge[117]

In just thirteen words Grace had fathomed two of Dick's most fundamental characteristics, his passion for chocolate and his stubbornness.

Grace Wilson also provides us with a perspective of life in Chaplain Groves's household:

[In the summer of 1912 the] Groves family lived next door to us —
brilliant Allen; plodding, silent Owen: "little boy" Dick; Tudy [the
family nickname for young Gwen]; Mrs. Groves; and the aunt,
"Dada" [pronounced *DAY-Dee,* the family name for Jane Griffith,
Gwen's sister]. The bandstand was located in front of the
Commanding Officer's quarters and the band concerts every Sunday
were, I'm sure, hard to take for the strict Presbyterians whose children could not even go out to play on Sundays. They stayed indoors,
reading religious books, while "Faust" and "Poet and Peasant" were
blaring outside.[118]

Shortly after New Year's Day, 1913, the Fourteenth Regiment moved
again, this time from Fort Harrison to Fort Lawton, on the outskirts of
Seattle, Washington.[119]

It was during this period that Allen began to show the first serious
symptoms of the congenital heart trouble that would cut short his life.
More than fifty years later, Owen explained the circumstances to his
brother Dick, who at the time had been in Pasadena. Owen began by
pointing out how competitive Allen had always been. "On the Sunday of
the last chapel of the semester," in January 1913, Allen and Owen had gone
to visit Uncle George Griffith, who lived near Hamilton. The trolley
bringing them back was late. This left them less than fifteen minutes to
run the mile and a half, with the last half up a rather steep hill. Allen made
it, even though once he was there he had to climb up a balcony to sneak
in. For what reason is not clear, since, according to Owen, there was "no
reason for our having to get to the chapel on time." Apparently, even this
challenge was not one that Allen could pass up. Owen, who described
himself as "having shorter legs and less interest" followed behind by several minutes. Examination week followed, and

> between sessions of cramming we played basketball and Allen not
> needing to cram, I am sure, was in the game frequently. The attack
> and diagnosis were during or just after the exams; and as you know
> he was excused from classes for several months and spent the time in
> bed or just sitting at Uncle George's reading through the *Encyclopedia
> Britannica.* . . . A strong will (of which Allen and you [LRG Jr.] had a
> great supply) is very important but in matters physical it should not
> always be obeyed.[120]

It is fairly clear that the brothers conspired to keep the seriousness and extent of Allen's illness from the family. About two weeks after Allen's attack, Gwen wrote to Owen:

> Allen's letter yesterday was very disturbing — and yet I was glad to have the suspense over. I knew you were keeping back something. Any details we know not yet. Simply A[llen]'s cheerful letter, telling us "No more athletics" — and "no organic trouble," "not to worry," and a statement to the effect that he was feeling well and would be careful.[121]

She urged Owen to take care of himself, counseled him that if his liver were "off" he should take "sodium phos.[phate] every day for a little" and to "give it up and come home" if he could find no respite from exhaustion. She begged him not to overwork.

> Go to a good Dr. and find out what's the matter with your "mind when it won't work." Get your sleep in, 9 hrs every night, and don't go into any athletics unless you are fit. . . . Why don't you draw out some of your savings [and use it for relaxation]? Why *haven't* you during this time of great distress just to have a little for the things you want awfully to do, and don't want to spend *Pa's* money for? You were lucky to get the scholarship. . . . Never mind about that essay — if you are bothered about the work. And keep A[llen] from that Bowdoin-Wesleyan debate — unless he is well. How much does he weigh? How much do you weigh?[122]

Three days later his mother asked Owen to

> . . . please take the best possible care of Owen Griffith Groves. Don't keep things back. We were perhaps unduly anxious about A[llen] when we did not know, only guessed. This is sad business but a mental strain is worse than muscular, isn't it — and that was what we feared at first. In what way did the warning come to him? Quite likely his nervousness was due to this heart strain. Before you go in to any athletics at all, will you be examined, please?[123]

In her letters to Allen we get an even clearer sense of how the boys had kept the full story of his illness from her. With obvious relief in her voice, she reflected back to Allen such phrases as, "You were in fine health," "walking as fast as usual — without hurry and feeling no ill effects."[124] In the meantime the chaplain was investigating the possibility of Allen returning to Seattle for the coming year, teaching high school and living at home.[125]

Gwen Groves's letters also reflect her concerns and misgivings about Dick. Dick never talked about his status in the family or his boyhood relationships to his parents or his siblings. Still, Gwen's letters to her children give us some insights into Dick's social and intellectual position within the family.

When Dick entered Queen Anne High School on January 24, 1913, to begin the second semester of his junior year, she wrote to Owen:

> Speaking of Richard, in Math he is way up, a shark I understand, in German and Latin he is good, he might be but if he will work. But his English is childish, his writing of English crude and careless. He won't read the text in Lit and insists upon it. "It is all because of Miss Frye." He needs another year in English at least before going to Hamilton. I don't want the family disgraced in H by Dick's foolish writing, and absurd contrariness, and contradictions, and he never could be made to appear in chapel.[126]

In mid-March:

> Dick is home [from school] this week and he is sad today because weather interferes with 8 hrs of tennis. So he is playing with stamps and teasing [his sister] Gwen. Boo [Wilson] is learning to play tennis, but she persists in wearing a tight skirt and won't or can't run at all. Richard [that is, Dick] is also playing chess with Capt. Harvey getting ready as you might suspect to trim Allen. He still talks about West Point and Annapolis, but will go to the University [of Washington] next year, and that may put some sense in his head. Who knows — He will have to make two credits this summer. But he thinks that's easy. Dick thinks he has a bean [a brain] — I sometimes wonder if he has.[127]

Allen's graduation from Hamilton was scheduled to take place in June 1913, and Chaplain Groves was busy making plans to attend. As was his custom, he planned to use the opportunity for more extensive travel. He arranged for a three-month leave from the regiment to begin about April 15 and planned to visit family and friends in upstate New York before arriving for Hamilton's commencement and attending a reunion of his own class.[128] As Gwen wrote to Allen:

> He [Chaplain Groves] is going to have, as Owen says, the time of his life. I am so glad he will be with you to share your honors which are many. I hope you will make for home as soon as you can and tell me all about it. It will be lonely without Pa — without a man in the house, for Richard [that is, Dick] don't count as you well know.[129]

It is not clear what his mother meant by this. Was it that Dick was too young (at seventeen) to be viewed (even temporarily) as the man of the house, or was it that he did not willingly contribute much to the day-to-day management of the house? Whatever she meant, there was no doubt about Gwen's feelings about his plans to attend West Point:

> Richard [that is, Dick] is full of plans for the University next year. Mr. McCameron says he will make Dick a Sergeant of a [ROTC] company at once. I suppose D[ick], with his military spirit and *bearing* will enjoy that. It is to be hoped that a year or two in the University will cause him to change his mind about West Point.[130]

As it turned out, Dick's military spirit and bearing were reliable barometers of his firm and unwavering commitment to pursue a military career. As fate would have it, however, his mother would never know. After a sudden, intense, and short illness, Gwen Griffith Groves died on Monday, June 30, 1913, at the age of forty-seven.

Many years later, almost as an afterthought to a description of his years at Fort Lawton, Dick described the circumstances of her death as follows:

> My mother died in June of 1913. She had been far from well, and had had considerable heart difficulty, which in the present day, could have been taken care of easily. The immediate cause of her death was apparently ptomaine poisoning. She went to a party in Seattle one afternoon and became sick that night. She died a couple of days later from various complications, including kidney and heart. Father was in the East at the time for Allen's graduation from college and Owen was also there finishing his Freshman year. At home there were only my aunt, my sister and I. . . . Because of the travel time involved,

> Father was not present at Mother's funeral. I do not recall whether
> Allen got back or not; I believe he had already started for the West
> while Father was visiting places in the East with Owen. [131]

Whether or not ptomaine poisoning was a factor is not clear, according to Dr. Robert M. Hardaway, the attending physician. He determined the cause of death was a "cerebral hemorrhage."[132] "Nephritis" was cited as a contributory cause, though he did not know how long Gwen had been afflicted with it. Dick was overcome with gratitude by the thoughtfulness of the regiment's officers who, among other things, insisted that he accept an offer of money to cover expenses until his father's return.

As Dick noted, Aunt Jane was present. She had only recently returned from California. In a letter to Owen, Jane said that they had thought of waiting for Chaplain Groves to return before having the funeral and burial but, "[I]t was better not to wait . . . although we did miss him so." We also learn from Jane that Allen did get back before his mother died. "He came when the attack was just becoming violent and I think felt that his coming had made her worse. It most certainly did not and we were thankful he came in time. She spoke with him Saturday evening for a few minutes." The doctor had given her heavy doses of morphine, "to quiet her." She dozed most of Saturday and Sunday, but sometime Sunday night she suffered a "hemorrhage of the brain." She was not conscious Monday; the end came quietly at 11 A.M.[133]

Jane described the funeral:

> Everyone has been so thoughtful and kind. . . . So many messages
> came in the back door from the cooks and non-com families. . . . The
> officers and ladies sent a beautiful blanket of sweet peas and lilies and
> it covered the white casket. . . . Capt. Ferguson took Tudy and me in
> the Col's carriage and the boys. . . . Everyone went down with us [to
> the cemetery]. . . . It was like one big family. . . . Dick has been a
> wonder of thoughtfulness and ability, and has tried to save all of us
> from anything harrowing.[134]

Photographs of Gwen Griffith Groves provide visual evidence of how her responsibilities and labors on behalf of the family had aged her over twenty-three years of marriage. Raising three active boys, taking care of a sick daughter, moving the family, and having her husband away for years at a time all took their toll, making her look older than her years. Grace's sister Mary described her as a "frail gray-haired woman with very blue eyes, fiercely loyal" to her husband, "and critical of locals who did not attend the Chaplain's Sunday services."[135] In the last photograph, taken shortly before her death, she looks tired and worn out, not overjoyed with her lot in life — some might even think her bitter. But she had fulfilled her marriage vows to love and obey her husband; she had faithfully executed his wishes in raising their children, and could be proud of what she had wrought. Dick for one was everlastingly grateful for her efforts. Although he rarely spoke of her, her portrait always hung alone on the wall above his dresser.

When Chaplain Groves returned to Fort Lawton from the East he found that his carefree youngest son had undergone a remarkable transformation. Dick, just about to turn seventeen, was focused and determined; he had mapped out clearly and with great assurance his life's trajectory. His immediate goal was to do everything necessary to get accepted as a cadet to West Point, in the shortest time possible. Nothing his father might say, about Hamilton, about family traditions, or about the drawbacks of an army career, could make any difference now.

CHAPTER THREE

Dick Defines His Future

• *Summer 1913–June 1916* •

During the spring of 1913, as Dick was finishing his junior year at Queen Anne High School, he took advantage of an opportunity to speed up achieving the first major goal of his life: gaining admission to West Point. He planned to enroll the following fall as a senior at Queen Anne, and, simultaneously, as a college freshman at the University of Washington, an opportunity available to any enterprising student. To help qualify for this arrangement Dick took extra courses over the summer, and got high marks.

Habits and patterns formed during childhood and adolescence were beginning to manifest themselves in adult behavior. Certain distinguishing personality traits of Leslie Richard Groves were becoming visible. His campaign to get into West Point was representative of how he would approach many problems, large and small, personal and professional. This strategy to achieve a goal, usually in a reduced period of time, would be a lifelong characteristic of Dick's. Perseverance would usually win out. If there were no alternatives, he would sometimes simply push and push until he achieved success. Brute force, wearing the problem down until it yielded, often worked. If there were really serious impediments, or failure appeared certain, new tactics would be tried. He would back off and try another approach, and another and another, until he got what he wanted. If none of this worked, he would redefine the problem and begin all over again.

As Dick went about this process, he was often judged to be impatient, brusque, intolerant, and irascible, all appropriate adjectives that were used by many to describe Groves during the Manhattan Project.

And yet while there is truth in all of them, they mask opposite traits as well. Groves was patient, focusing on a goal and trying strategy after strategy to accomplish it, until it was achieved. If something was in his way, he might have to knock it over, or he might have to go over, under, or around it. Often many would be bruised in the process. So it would be in his drive and determination to get accepted as a West Point cadet, and so it would be that the atomic bomb would be produced and used before the end of World War II.

In order to enter the University of Washington's College of Science on September 15, 1913, Dick had to complete a couple of prerequisite courses during the summer. He studied independently, cramming a year's worth of college Spanish and a semester's worth of government and civics into twelve weeks, getting high marks in both subjects. His performance in spoken Spanish was not as good as his written exam, as might be expected of a person studying the language on his own, but the instructor passed him.

The final hurdle he faced before attending the university was to convince his father. Chaplain Groves initially objected on the grounds that a major's salary was not sufficient to support a third child in college. How, with the two older boys in college — Allen at graduate school at Johns Hopkins (on a scholarship) and Owen a sophomore at Hamilton — could he afford to pay Dick's tuition? The family had been generous with the first two boys in supporting their education unswervingly, and Dick might have felt some bitterness about being slighted, but if he did it is not evident.[1] There was always deep respect for his father, and in this instance, he assured him that "I would limit my expenditures to what I had incurred in high school."[2]

At the university he had a busy course schedule, taking English composition, French, and advanced German. But to fit the German course into his schedule, he had to take a special session of it, for two hours on Saturday morning, organized for the benefit of local high school teachers.

> I was very much impressed by their lack of mental quickness and general intelligence. Towards the end of the term my instructor told me that I could not continue in that class as it was too easy for me and he felt that I wasn't really earning my four hours of credit. He also said that I was embarrassing the teachers in the class.[3]

Dick also took mathematics courses. The algebra class was taught by

> a Scotsman and decided eccentric named Bell. He had apparently wandered all over the world as a merchant seaman, and had engaged in many of the vices encountered in the worst of our seaport towns. It was said that he had been an opium user and had generally been down and out in various ports, such as Shanghai. By the time that I

met him, I believe he had reformed. His only eccentricity was a little fingernail, which was about an inch long.[4]

Dick also took a course in analytical geometry and calculus that had only four students: Dick and three girls. The instructor expected them to work every problem in the book. When Dick would complain to his father about this, his answer was, "You wouldn't want to let those girls beat you out."[5] The chaplain never passed up any opportunity to instill competition in his sons.

Dick also took military drill in ROTC at the university, but did not engage in any outside activities or join a fraternity. Instead, for recreation and exercise he played tennis and rode horseback along the peninsula where Fort Lawton was situated, sometimes with his father.

Dick's academic performance at Washington was not of the highest caliber,[6] but then he was taking a full course load at Queen Anne as well, with the goal one of acceleration rather than academic excellence. In this regard, unlike his brothers, when forced to choose between standing number one and finishing ahead of schedule (in an acceptable manner), he always chose to accelerate. This too would become a lifelong pattern of behavior. During the Manhattan Project his goal was to produce a usable bomb in time, rather than a better bomb too late.

He took a few regular courses at Queen Anne and arranged for some special ones during the year and a half he spent there.[7] He persuaded the Latin teacher to let him study the second semester of third-year Latin by himself, and to pass the course by examination. Overall his grades were slightly better than average, and in June 1914 he was awarded a high school diploma.[8]

During years 1914 through 1916 there were several routes for an aspiring young man to reach the U.S. Military Academy at West Point: congressional appointments, at-large presidential appointments, by being an honor graduate from certain "honor schools," or by being a soldier in the Regular Army.[9] Once he had been appointed, there was a second step in the process: The candidate had to pass a difficult academic examination and a physical.

The most common appointments by far were those made by congressmen and senators. At that time every member of the House of Representatives was entitled to have two cadets from his congressional district at the academy; each senator could have four from his state.[10] This meant in practice that a member of Congress could appoint on average a new candidate every other year. Whenever a vacancy occurred, the member nominated three candidates — one as the principal, one as a first alternate, and one as a second alternate. The nominees would undergo a rigorous entrance examination. If the principal was not acceptable, then the first alternate was considered, and so on until a candidate was found acceptable for admission. The presidential appointments, far fewer in number, were reserved mainly for the sons of army officers, soldiers, and others whose service to the nation made it difficult to obtain political patronage. Presidential appointments were based upon the results of an extremely rigorous mental examination. Regardless of the source of the appointment, when the candidate arrived at West Point he signed a form agreeing to serve in the army of the United States for eight years, took an oath of allegiance, and was admitted.

Dick made two efforts to get into West Point, the first in 1914, then again in 1916. His most obvious paths were to get either a congressional appointment or a presidential at-large nomination. Dick initiated the process in the month before he started his double academic year at Queen Anne and the university. On August 14, 1913, Chaplain Groves wrote to the acting secretary of war requesting an appointment for his son to the USMA.[11] The letter was forwarded to the president, who replied to the acting secretary of war designating an appointment at large, allowing him to compete to fill a vacancy. In further support of his application, on August 19, 1913, Maj. Gen. Thomas Henry Barry recommended Dick for appointment.[12] On February 21, 1914, the Hon. Charles Webster Bell, representative of the Ninth District of California, nominated Dick as first or second alternate for one of his appointments. On March 17, 1914, Dick took the examination competing for a presidential appointment, but did not achieve a high enough score.[13] Apparently, Mr. Bell's principal nominee accepted, so Dick wound up with nothing.

Dick was not discouraged by this initial failure to gain entry to West Point. If at first you don't succeed, try again. What was needed was an alternative plan, and he had one. He would go to Boston and become a full-time student at the Massachusetts Institute of Technology (then generally known as Boston Tech), the best undergraduate engineering school in the country. He believed that taking courses in science and engineering, and even getting a degree, would give him an advantage over his competitors the next time he got an appointment and took the examination for West Point.

Dick passed the entrance examination, and with everything in place, he set out in September 1914 on a seven-day, trans-Canadian train trip to the East Coast. His tourist-class accommodation on the Canadian Pacific saved him some money, but it was not a comfortable trip. The weather was extremely hot — more than one hundred degrees in Manitoba and Ontario — and the cars were crowded. "The cars were designed for immigrants and had a cooking stove in the compartment next to the men's washroom." War had broken out in Europe the month before, and Englishmen were trying to get back home to join the armed forces. They filled every space.

In the fall of 1914 MIT was located on Boylston Street, near Copley Square in Boston's Back Bay section.[14] After consulting a hotel directory during the train ride, he decided to stay at the Copley Plaza on Copley Square, the hotel closest to MIT. Upon taking the subway the motorman, sizing up Dick's limited circumstances, advised him to stay at the much cheaper Copley Square, just up the street, "where all the ball teams stay."

The next day Dick registered at MIT and looked for a place to live. Using a list of people who had registered rooms with the institute, Dick, on the tightest of budgets, chose the cheapest he could find, a room on West Rutland Square for $2.50 a week. It was far from satisfactory; the room was large, but its only source of light was from open gas jets overhead.

The neighborhood was "in the center of a most disreputable district." He had hardly reached his room when he was accosted by his next-door neighbor, a woman "who was most friendly indeed." Dick spotted "her at once as a professional prostitute." Upon questioning her Dick learned that she was not alone in the surroundings of West Rutland Square. It was time to move. The following day he found a rooming house on St. Botolph Street for three dollars a week, about three-quarters of a mile from campus. His new apartment, a fourth-floor walk-up, was in the house of a skilled leather worker and his wife.

> St. Botolph Street did not enjoy a savory reputation either but it was
> far better than West Rutland Square. It was known as a street where
> some female residents were of easy virtue but were discreet about it.
> Personally, I never saw any signs of them, but all Bostonians of that

era were inclined to raise their eyebrows slightly when its name was mentioned.[15]

Dick lived the life of the poor student, allotting himself $3.50 a week for meals. The routine he followed on weekdays was: no breakfast, a ten-cent lunch — two very thin hot dogs and two slices of bread with margarine — and a thirty-five-cent dinner, of poor food, at the student dining hall. He spent whatever funds remained to feed himself on the weekends.[16]

It is not clear just how much Dick's father contributed to his expenses at MIT. Surviving records of Dick's expenses show that he lived quite frugally.[17] Chaplain Groves contributed $104.65 toward the year that Dick spent at the University of Washington and another $1,165.00 for his stay at MIT between 1914 and 1916. A few years later Dick paid his father back; the money he returned was used to help with his sister's college tuition.

That fall in Boston an extraordinary sports story was unfolding, and Dick took a day off to enjoy it. On the Columbus Day holiday Dick decided to go to the third game of the 1914 World Series. He got up at four in the morning and walked the mile and a half to Fenway Park. He was about twenty-fifth in line and stood there for five hours waiting to buy a ticket. At around nine o'clock a man offered Dick ten dollars for his place in line, but Dick refused. Later, when he reported this to his father, he was reproved for his lack of intelligence. Not being much of a baseball fan, the chaplain would have taken the money. Finally Dick bought his ticket and went into the park, taking an excellent seat about twenty feet in back of first base, in the third or fourth row.

The Boston Braves had been in last place in the National League on July 18, then proceeded to win thirty-four of their remaining forty-four games, clinching the pennant, and ending up 10½ games ahead of the second-place New York Giants. Nevertheless, going into the World Series they remained underdogs against Connie Mack's formidable — but over-confident — Philadelphia Athletics, series champions in 1910, 1911, and 1913. The series opened on Friday, October 9, at Philadelphia's Shibe Park, and the Braves won the first two games. After traveling back to Boston on Sunday, game three was played at Fenway on Monday, Columbus Day, with Dick a close observer.[18] It was an excellent contest, tied 4-4 until the bottom of the twelfth. Braves catcher Hank Gowdy led off with a double against Athletics pitcher Joe Bush. Les Mann replaced Gowdy as runner. After an intentional walk, Herbie Moran bunted and Bush threw wildly past third base, allowing Mann to scamper home with the winning run. The next day the "Miracle" Braves swept the series, four straight.

For a break from his studies Dick played tennis at the YMCA courts and later at Jarvis field at Harvard. There was also an occasional movie, but never "two of the favorite spots of Tech students — the Old Howard burlesque theatre, and a rather famous saloon which was reputed to have a wonderful free lunch."[19]

The civil engineering program at MIT took four years.[20] Dick felt that because of the year he had spent at the University of Washington, he should get the MIT degree in three years, but the authorities did not agree with him. And there was another impediment to Dick's realizing his goal, as the catalog made clear:

> In order to enter any of the Engineering or allied Courses in the second year, it is essential that applicants have preparation in Analytic Geometry and the elements of the Calculus, and highly desirable that they be familiar with Mechanical Drawing and Descriptive Geometry. A single year of college work will rarely suffice for admission to the second year.[21]

He was given credit for the French, solid geometry, German I, and physics courses he took at Washington; still, he lacked the first-year courses in descriptive geometry, mechanical drawing, freehand drawing, and chemistry, which now had to be taken in addition to the normal course load. It was a heavy burden indeed. As he said,

> Our schedule was supposed to require 48 hours of class and study, but actually this meant about 60 or 70 hours, depending on the intelligence of the student. In order to make up my work I was taking about a 20 percent overload.[22]

Yet despite all his effort, this was not sufficient in MIT's opinion to assign Dick full sophomore status. In the 1914–1915 catalog, he was listed as a second-year "unclassified" civil engineering student.[23]

Dick was not deterred and pushed ahead to hold to his timetable. When he entered MIT Dick was promised that he would be allowed to take a six-week version of the introductory chemistry course during summer school, but as he was about to register for the course, the chemistry instructor informed Dick that he was not eligible because he had taken no chemistry in high school. When a conversation with the head of the chemistry department did not resolve the matter, Dick went to the head of the civil engineering department, who convinced the department head that the promise should be kept. When Dick went back to register, the instructor told him he was making a mistake and he would see to it that Dick would not pass. Typically, Dick ignored the threat and went ahead with the course — six hours of laboratory and two hours of class every day, for the two three-week terms. Fortunately the instructor did not carry out his threat, although Dick got only a low pass for each term, not the best marks to support his case for obtaining a degree in three years.

Next on the relentless schedule Dick had set for himself was the six-week civil engineering summer program in surveying, held on the shores of Gardner's Lake at East Machias, Maine, near the coast and the Canadian border. In the camp the students lived two to a tent. The work was all "practical surveying," some of it done in the dining hall, which was used as a classroom in rainy weather.[24] Dick received grades of P (passed), for railroad fieldwork, surveying plane, and summer reading, merely average.[25] His grades for the spring term of 1915 had been only average as well[26] — again not the strongest evidence that he should graduate in three years instead of four.

On August 10, 1915, while Dick was in Maine, his fifty-nine-year-old father took a second wife: Jane Griffith, Gwen's younger sister, who was then forty-eight. Jane was no stranger to the children: She had spent most of the past twenty-five years living with the Groves family and had played a significant role in the children's upbringing, especially little Gwen's after her mother died. Her transition from aunt to stepmother was remarkably uneventful, made without any apparent difficulty. They all had been close before this marriage, and they remained close after it.

When Dick returned from Maine for the fall term he found new housing, on Newbury Street — a more convenient room costing fifty cents a week more, but only a block or two from the institute. Coincidentally there was another student on campus that year who would play a central role in Dick's life. Vannevar Bush was admitted to the Ph.D. program in the Department of Electrical Engineering in November 1915. Five months later he submitted his thesis and got his degree.[27]

Meanwhile Dick's campaign for full junior status was not progressing very well. In the 1915–1916 MIT catalog Dick was designated "u 3" — that is, unclassified third year.[28] The jury was still out, and his course work for that term did little to improve his chances.[29] It was time to redefine the problem and try anew.

At the beginning of 1916 a major turning point in Dick's life occurred. On Saturday, January 8, his older brother Allen, who had never fully recovered from his heart attack, died, at the age of twenty-four, of pneumonia at Walter Reed Hospital in Washington, D.C.[30]

Dick came down from Boston to represent the family at the funeral service, which was held Monday afternoon at Washington's First Congregational Church on Tenth and G Streets, Northwest, with burial at Arlington Cemetery. While Dick never described his feelings about the death of his brother, he elected, when the time came, to be buried beside him.[31]

While he was in Washington Dick got his second chance to compete for a presidential appointment to West Point. This fortuitous turn of events caused him to interrupt his studies at MIT. He could not afford to fail the exam again, so for seven weeks, between late January and mid-March, Dick crammed at the Columbia Preparatory (Shadman's) School, which specialized in preparing students for service academy exams.[32] While living at 1519 Rhode Island Avenue, Northwest, he studied geography, U.S. history, modern European history, algebra, plane geometry, English grammar, and English and American literature, the subject areas in which he would be examined.

On Tuesday, March 21, Dick began the three-day written examination for admission to West Point. He took it at Fort Banks, Massachusetts, in Boston Harbor, one of about two dozen army posts across the country where the exam was given. On the first day he took a four-hour exam covering European history since 1453, along with U.S. history; on the second, a four-hour algebra exam in the morning, and a three-hour geography exam in the afternoon; and on the third, a four-hour geometry exam in the morning, and a four-hour exam, covering English grammar, English composition, and English literature, in the afternoon.[33]

While awaiting word on the outcome he returned to MIT to resume his studies, cramming in one more course, applied mechanics.[34] That spring he heard that he had passed the exam — and in May he passed the physical exam as well. The next stop was West Point.

CHAPTER FOUR

West Point

• *June 1916–November 1918* •

roves entered the U.S. Military Academy at West Point, New York, on June 15, 1916. He later recalled that:

> Entering West Point fulfilled my greatest ambition. I had been brought up in the army, and in the main had lived on army posts all my life. I was deeply impressed with the character and outstanding devotion of the officers I knew. I had also found the enlisted men to be good solid Americans and, in general far superior to men of equal education in civil life. I was imbued with the idea that the West Point graduates were normally the best officers and on the whole enjoyed higher respect from the enlisted men.[1]

During the next two and a half arduous years Dick was unremitting in his efforts to make the most of this opportunity and to fulfill his life's ambition of becoming an officer in the Regular Army.[2] In contrast to his rather mediocre performances at the University of Washington and MIT, he excelled in academics at West Point, was active in sports, and held the rank of cadet first sergeant when his time was cut short by early graduation. Ranking fourth in the General Order of Merit of his class, he had his pick of branches and chose a commission in the Corps of Engineers.

All this lay ahead of him when he climbed the hill from the railroad station to the Plain on that June day in 1916. Going to West Point was his decision, something he wanted intensely. Though his father was in the army he did not push Dick to follow him, and in fact he preferred that his son choose another career.[3] Groves said that his mother, before her death in 1913, tried in discreet ways to influence him against an army career. It was to no avail.

West Point is a special place with a long line of heroes and many traditions that weigh heavily on every class that enters. The cadet is constantly

exposed to the past through the names of the buildings and the monuments that recall the important military figures.[4] There is the Gen.
Thaddeus Kościuszko monument on the northeastern edge of the Plain,
the open, unifying space that has been utilized as a fort, cadet summer
camp, cavalry and artillery drill ground, and parade ground. There is the
statue to Gen. John Sedgwick (class of 1837) — once on the northwestern
end of the Plain, but relocated to Trophy Point. According to academy
legend, spinning the spurs on Sedgwick's statue while in full-dress uniform at midnight ensures the cadet good luck and a passing grade on the
final exam.[5] Battle Monument, a commanding shaft of polished granite
with the figure of Fame (or Victory) atop, commemorates Regular Army
officers and soldiers who died to preserve the Union and occupies one of
the most scenic views overlooking the Hudson.

Buried in the cemetery are several of the nation's military heroes, amid
many regular officers. Colonel Sylvanus Thayer, superintendent from 1817
to 1833 and "Father of the Military Academy," is buried there, as is General
of the Army Winfield Scott, who fought in the War of 1812 and was a general in the Mexican War and the Civil War.

In the early years of the twentieth century West Point provided a place
where excellence was encouraged and, through competition, emerged in
those who ranked at the top of their class.[6] Not everyone could be first, of
course. Hard work, self-discipline, and determination were qualities that
could set one apart. Rank was supremely important; it is how the army is
organized and operates. Psychologically, if one was in the top 10, or better
yet the top 5, percent of one's class, an extra measure of elitism became part
of one's curriculum vitae, and perhaps of one's personality as well. Those
who took it seriously were more likely to become imbued with an inde-
finable sense about who they were that lasted a lifetime. They knew where
they ranked in their class as well as they knew their names, and who might
have been ahead of them. As they proceeded through their careers and on
into retirement, a copy of the *Register of Graduates and Former Cadets of the
USMA* was never far away. Just as they were marked and graded continu-
ously while at the academy, they would continue to judge, assess, and critique one another to their dying day. This was the competitive environment
that Groves found himself in. He took to it with enthusiasm.

More so in the first half of the twentieth century than in the second, the
mores and culture of West Point bound people together,[7] setting family priorities and practices. Sons of West Point graduates were expected to attend
the Military Academy. West Point classes were small enough that a cadet

would probably know most of his classmates, and even many in the classes just ahead or just behind.[8] Graduates would often marry young women whose fathers were graduates themselves. The Regular Army officer corps was inbred and in many ways isolated from the society at large.

Like every new cadet, Groves wanted to realize the promise of West Point. In a contemporary account a former instructor describes what a cadet feels when he arrives.

> To the candidate [his arrival at West Point] conjures a vision of all that he hopes to be. The honor of being a cadet, the privilege of wearing the uniform, the immense possibilities of physical and mental achievement, the soul-satisfying fear of an ambition about to be realized, the glamour of military life, and, it must be admitted, a secret feeling of righteous superiority over his boy friends at home — all these thoughts crowd his imagination so that for once he sees frozen the vague ideal that he always has had of himself.[9]

No sooner did one class graduate in early June than a new plebe or fourth class arrived a few days later, trudging up the hill from the train station on the western bank of the Hudson River.[10] Dick was one of the 322 cadets of the class of 1920 to enter that June and July of 1916, the largest up until then.[11] Two days before Dick arrived President Wilson had delivered a half-hour address to the graduates of the class of 1916 and given them their diplomas, among them the third-ranking cadet, Wilhelm D. Styer, someone we will meet again shortly.[12]

From all Groves had heard at Forts Harrison and Lawton, he knew that his first three weeks as a cadet in "Beast Barracks," the indoctrination into West Point ways, would be the most demanding experience of his young life. And indeed it was, with early reveille, long hours of physical training, and military drill. In 1916 this rite of passage also included a lot of hazing by upperclassmen on the Beast detail.[13] The severe treatment of plebes had its purpose. As a contemporary of Groves said,

> The standards of West Point are entirely unlike those of any other institution. To preserve those standards unchanged it is considered necessary that a young man entering the Academy be subjected to the severest discipline, that even his personality be more or less suppressed in order to give the spirit of West Point time to get hold of

him, to allow him to adapt himself to the ideals of the Corps, and to keep those ideals from being, through him perverted.[14]

For Dick, Beast Barracks was followed by two months in summer camp with F Company at the northwest corner of the Plain, named that year Camp John P. Story. At the time summer camp was supposed to more or less replicate actual field duty, though any actual resemblance was only coincidental. Cadets lived in tents, drilled and exercised, heard lectures on military topics, attended evening concerts, hops, or balls, and rested up for the academic regimen that was soon to come.

At the end of August Groves moved into barracks to begin the first academic term, which ran from September 1 to December 23, followed by the exam period from December 26 through December 31. His plebe roommate (and a tent mate at camp) was George Bryan Conrad.[15] Like every plebe, Dick was issued a copy of *Bugle Notes,* a pocket-sized handbook known as the Plebe Bible and containing useful information about the rites and rituals, schedules and procedures of West Point. There was also a section titled "English as She Is Spoke at the U.S.M.A."[16] The plebe was required to learn a new vocabulary that helped describe his new life. In the first year he was "braced" many times, no doubt "skinned" as well by a "quill," for offenses listed in the "black book." But if he "hived" (understood) or "speced" (memorized without understanding) his lessons, he might "max."

The cadets lived in either South Barracks, built in 1848 but improved in 1907, or North Barracks, completed in 1908, in the Elizabethan Gothic style that is predominant at West Point. Each room was identical, fourteen feet by twenty-two feet, with the same furniture for its two cadets: two iron cots, two plain wooden tables, two wooden chairs, two steel clothes presses, and two washbowls. Bathing required a trip to the "sinks" (lavatories) in the basement. Rugs on the floor and pictures on the wall were prohibited. A mirror hung flanked by notice of the hours of instruction and the schedule of the cadet. Failing to have it correctly posted was an offense. There were also hooks on the wall to hang clothing, in the proper order: the first hook for the raincoat, the second for the overcoat, the third for the dress coat, and so on. Shoes were aligned toes-out alongside the bed, sized from high to low. During the day mattresses were folded, with bedding folded and piled on top, ready for any inspection that might occur.

Almost every minute of the day was prescribed. The cadet was up at 6:00 A.M. with breakfast at 6:30. Normally there were two classes from

8:00 A.M. to 12:35 P.M. Forty minutes for dinner were followed by another class until 3:45 P.M. Then it was drills, parade, athletics, and supper at 6:25 P.M. Call to quarters followed, with a study period from 7:30 to 9:30, Taps at 10:00, and lights-out.[17]

A cadet was issued an extensive wardrobe of uniforms of different types and combinations, depending upon the season, weather, or function.[18] Gray, a color adopted in 1814, predominated but was supplemented by white. Cadets knew the proper uniform for each occasion by looking out their barracks room window to see which signal flag flew in the court-yard. Full-dress gray over white trousers would be worn for chapel. Dress gray was worn to class. For riding, gray kersey-reinforced breeches were worn with canvas leggings. For dismounted drills, when leggings were worn, flannel breeches were prescribed. There was a short overcoat, a long coat, and a raincoat. Several pairs of gloves, several kinds of hats or caps, leggings, shirts, shoes, belts, collars, cuffs, and "Comfortables" (slippers) rounded out the wardrobe. To distinguish classes, a thin stripe of mohair was put on the sleeve of the dress coat and full-dress coat, above the stan-dard wide black band. A plebe's sleeve was bare, a third classman had a single stripe, a second classman two stripes, and a first classman three stripes. Chevrons designated rank for sergeants and corporals.

Cadets were not allowed to have money. Before admission, a deposit would have to be made with the treasurer to cover the cost of the uni-forms, about $160 when Dick was a plebe. His pay was $600 per year, but that too was deposited, and all costs incurred were drawn against it. The cost of meals — $29.19 per month in 1918 — books, laundry, blankets, sheets, pillowcases, mattresses, candy, tobacco, and furloughs kept the cadet constantly on the verge of debt, or in it.

During meals there were ordered procedures that had no variation. A bugle, drum, or fife call summoned the cadets to formation to be marched to Grant Hall, the cadet mess. Cadets marched everywhere, at all times. If even two cadets were going from the chapel to the library, they marched in formation. Once in the mess the cadets arranged themselves in proper rank, standing at attention behind their chairs. After the order "Take seats," the meal began. Usually a first classman, known as the table commandant, was in charge of each table, and he sat at the head. Second and third classmen came next, positioning themselves down the table until at the foot sat three or four lowly plebes, who performed the arduous duties of serving their superiors. One plebe, known as the water corporal, poured the water and milk; another, the meat corporal, carved the "bone"; and a third, the coffee corporal, poured the coffee. The gunner supervised the supply of food brought from the kitchen by civilian waiters. If plebes were in short supply, they might have to double up on their duties. Busy with their chores, the plebes ate in spasms when they had a free minute. Meals were not leisurely affairs. Including announcements from the officials about various matters, the meal was over in half an hour. At its conclusion the cadets were called to attention and dismissed, to make their way back to their rooms.

During plebe year many drop out for one reason or another. In Dick's class seven were gone in the first three weeks, not making it past Beast Barracks. "[T]hree score" departed as a result of academics, and a few were lost to the Code of Honor, the strict rules that demand the reporting of prohibited behavior.[19] The academy's Code of Honor — "A cadet does not lie, cheat or steal, nor tolerate those who do" — stood as the keystone in fulfilling West Point's mission.[20] The code is about trust. If officers in the Regular Army could not trust one another, in peace but especially in war, then the army itself would be undermined.

In light of the endemic cynicism of today it is difficult to recapture the tenets and beliefs that permeated West Point eighty years ago. West Point graduates believed in the values taught there and used them as a moral compass throughout their lives. Groves was a prime example; he lived by the code and expected others to as well.

The curriculum of Groves's day, as well as the nature and purpose of the West Point education, was still deeply rooted in nineteenth-century traditions, though some reforms were occurring.[21] A single course of study helped forge a unified and homogeneous officer corps. Mental discipline and character building were the essential ways to prepare young men to be army officers.[22]

Cadet Groves's first-year courses included mathematics, English, history, practical military engineering (surveying), and drill regulations (infantry and artillery).[23] The technical curriculum was heavily weighted in the first two years toward mathematics, consuming about 45 percent of the course work. Dick had already taken several math courses at the University of Washington and MIT, and partially as a result did well in math and surveying, placing fifth and second, respectively. Overall at the end of his plebe year he ranked twenty-third and was one of the twenty-six distinguished cadets of the fourth class entitled to wear gold stars on his collar.[24]

The first cadet "hop" or dance was on October 21, 1916. That same day army swamped Trinity 53–0. Elmer Oliphant, "a Hoosier lad," scored four touchdowns, three on spectacular runs of ninety-six, fifty, and sixty-five yards respectively. Army played Navy on November 25, 1916, at the Polo Grounds in New York, winning 15–7 and ending an unbeaten season.[25] Upon their return on Sunday six hundred cadets pulled the team up the hill from the ferry landing in an ancient stagecoach, and then cheered them. The line coach, Ernest "Pot" Graves, whom we shall soon meet again, served that year with American troops in Mexico. The Hundredth Night entertainment was a musical comedy, *The Devil to Pay,* on March 3, 1917. The lower classes, enlisted men, and families attended the after-noon performance and the upper classes, officers, and invited guests the evening.

The reelection of President Woodrow Wilson in November 1916 required a second inaugural, and on Sunday, March 4, 1917, the Corps of Cadets, seven hundred strong, with the band and fourteen horses from the cavalry detachment went to Washington on special trains to march in the parade down Pennsylvania Avenue the next day.[26] After the parade, on a bitterly cold day, a tea dance was held at Rauscher's on Connecticut Avenue, after which the cadets marched back through the darkened streets to board their trains in the yards north of Union Station.[27] It was one of the few times that Dick left West Point. A month later, on April 6, the United States entered World War I. At West Point orders were given to graduate the class of 1917 immediately.

During Groves's yearling, or third-class, summer from June to August 1917 he was in I Company and then a corporal in G Company, and lived with Arthur Pulsifer, who would end up 198th in his class.[28] A new superin-tendent was appointed in June. Col. Samuel E. Tillman (class of 1869), a former chemistry professor who had retired in 1911, replaced Col. John Biddle.

For the academic year that began in September 1917 and ended in June 1918, Groves took mathematics, French, drawing, military hygiene, and drill regulations (infantry).[29] Dick's grades improved, and by the end of his yearling year Cadet Corporal Groves ranked second out of a class of 232 and was one of fifteen distinguished cadets.[30] He kept close track of how he was doing, calculating how to advance in class rank.

After the turn of the twentieth century, a new emphasis emerged on the physical conditioning of the cadet.[31] In 1905 President Theodore

Roosevelt ordered all cadets to attend gymnasium daily. By 1910 they used the new gymnasium for boxing, fencing, bowling, and squash. The following year an indoor swimming pool was completed, in a building three stories high and lighted by skylights.

Physical exercise also came in the courses that exposed the cadet to the other branches of the army besides engineering: the cavalry, field artillery, and infantry. Equestrian exercises were practiced in the riding hall, with its vast curved roof.

Exercise also came in daily drills and parade. In order to drill in formation and appear visually as a coherent whole, the corps was "sized." This meant to line up the classes by height and assign them to companies based on how tall they were. The number of companies would depend on class size, but would begin with A and might run to I, as it did for Dick's class.[32] The tallest cadets would be in Companies A and I, the next tallest in Companies B and H, and so on, until the class was sized. This presented an illusion to the eye when the corps assembled in line on the parade ground — as it was intended to — for it appeared as though everyone was the same height. Cadets in A and I Companies were known as Flankers. Shorter cadets in the middle companies were known as Runts. While the primary purpose was for the corps to appear uniform when drilling, the cadets used it as a means of social differentiation, leading to some real animosity and contempt among them. Flankers held themselves superior to the lowly Runts, who were considered officious, aggressive, and hypermilitary. For his plebe year from September 1916 to May 1917 Dick was in H Company, at five feet ten and one-half inches — tall in his day, a Flanker.

There was also an extensive athletic program, with intramural and some intercollegiate competition. Dick favored the physical-contact sports, football and wrestling. In his plebe year he played on the Cullum junior varsity football squad, and in his yearling year he played on the varsity football team. The coach that year was Geoffrey Keyes, whose only loss in eight games was to Notre Dame. The schedule did not include playing archrival Navy, the contest being canceled for the duration of the war. Dick did not play in two-thirds of the games and thus did not earn an A letter.[33] The star of the team was Elmer "Ollie" Oliphant, a year ahead of Groves, an All-American fullback and co-captain. On November 10, 1917, Army played Carlisle and beat them 28–0. Groves was second-string center behind his roommate Pulsifer and played in the game.[34] In the winter months there was wrestling. In his plebe year during the indoor meet he came in second in the light heavyweight class. As a third classman, he was on the wrestling team as well.[35]

To his classmates Dick was known as Greasy, a reference to his thick and oily hair. He kept somewhat apart. While his roommates and hall mates were devoting significant amounts of time in the evenings to gossiping, cards, or horseplay, Dick would study or write letters home, all the while seriously pontificating to the others about not wasting their opportunity to excel. Dick thrived in the competitive atmosphere, having been well prepared for it at home. He was respected for his ambition and his dedication; it was clear to several of his classmates, even then, that he was destined for leadership. For all that, he had few close friends, and others generally kept their distance.

More than twenty-five years later his roommate Arthur Pulsifer wrote to Groves soon after the end of World War II:

> Dear Leslie:
> Just back from Okinawa and I find the papers full of you. Congratulations on your war success old man. You certainly refuted my peoples' ideas as to your progress in the Army. I can just see old Williamson's and G B's eye's [sic] popping out as they read the paper about you getting a permanent star this early in life.

Later in the letter he asked Groves for his organization's shoulder patch, saying that he had "patches from most of the commanders from IKE on down and yours certainly now ranks up with them."[36]

West Point was afflicted with influenza and quarantined in 1918; in February a widespread measles outbreak caused classes to be canceled for ten days.[37] The news about the European war, the battles being fought, the ships being sunk, some of the older classmen getting ready to go, gave them much to think about.

During second-class summer from June 1918 to August 1918 Groves was a Flanker in I Company as a cadet lieutenant. Moving to second-class barracks in September, he was first sergeant of A Company and roomed with Pulsifer again.

In a two-month second-class academic year, from September 1 to October 31, 1918, Groves took classes in natural and experimental philosophy (now called physics); chemistry; mineralogy and geology;

drawing; and drill regulations (cavalry and artillery). He ranked first during that period.[38]

The curriculum at the time offered no choices. Many plebes entered West Point after already having taken four years of college elsewhere. Groves himself had had three years. No matter what the previous schooling, every cadet took every subject, every cadet had to be proficient in every subject to advance, and every cadet had to recite every day and was graded on his answers. There were no exceptions.

The cadets were marched to class to the East Academic or West Academic (now Pershing Barracks) Building, by the section marcher. After the cadets were in the classroom, the section marcher closed the door, faced the instructor, saluted, and reported either that all were present or some were absent. In Dick's day there were probably a dozen or so cadets per section. After the order "Take seats," the instructor asked if there were any questions about the day's assignment. If there were this might take up possibly half of the class time. Afterward the instructor would say, "Take boards," and each cadet would command a blackboard, write his name in chalk in the corner, and listen for instructions. If it were a math class, for example, the boards were marked odd and even. Cadets at the odd blackboards would work on two or three problems over the next thirty or forty minutes while their neighbors at the even boards would work on two or three different ones. Whereas the honor system was operative every minute of the day, the odd-even system helped prevent an inadvertent observation that might arise due to peripheral vision. With the order "Cease work," the cadet stopped where he was. The instructor would then give the pointer to a cadet, who would have to explain how he solved the problem. For every hour-and-twenty-five-minute recitation period a cadet was expected to spend twice that in preparation. After the class was dismissed, the instructor assessed each of the boards and recorded a grade. At the end of the week all the marks were compiled into the weekly report ("tenth sheets") and posted on Saturday on bulletin boards in the sally ports of the barracks, for all to see a cadet's progress, or lack thereof. Once a month the cadets were "resectioned," with the most capable moving up to higher sections and those less so moving down. Each month the grades were mailed to the parents to keep them informed. A few of Groves's monthly reports home survive. This was a highly competitive environment where tenths of a point could mean a great deal in class ranking. It was the kind of atmosphere that Dick was used to, and he thrived in it.

The professors and instructors gave grades using a scheme laid down by Sylvanus Thayer, superintendent from 1817 to 1833. Perfect was scored 3.0; passing was 2.0. In reality there were ten passing grades between 2 and 3 — 2.1, 2.2, 2.3, and so on — and the grade competition was fought out there ("fighting for tenths"). The system was slightly skewed, with instructors of the top sections more likely to continue to give higher grades to those cadets, often for identical work, than did his counterparts in the middle or bottom sections.[39]

From the daily recitations to weekly compilations, to the monthly, semester, year, and finally entire four years, a composite grade, to the second decimal point, was calculated. This score — along with demerits — determined the General Order of Merit, or class rank, and would follow the cadet to the end of his days (and even beyond), largely determining what sort of career he would have. In Groves's day a cadet in the top 10 or 15 percent likely would have chosen the engineer branch, as did Robert E. Lee (second in the class of 1829). If he was in the middle or below it might be the infantry, like Ulysses S. Grant (class of 1843).

The commandant of cadets is responsible for the military instruction and discipline of the corps. In Dick's time Commandant Guy V. Henry, and the officers in the tactics department, prescribed the order of the barracks, the mess hall, and just about every other detail of cadet life. For breaches of regulations cadets were reported — "skinned" in the parlance. With so many rules there were ample opportunities for infractions, from being late for class to inattention and missing a button on one's uniform. The cadet had to submit a written explanation of the offense and, depending on the severity, might have to undergo punishment. It might be a "punishment tour," one hour of walking an assigned pattern, on Wednesday or Saturday afternoon in the Area. It might mean confinement to one's room, loss of furlough or deprivation of privileges, a reprimand, or, in the most serious cases, suspension or dismissal. Conduct was weighed into overall class standing, and when dealing with tenths of a point demerits could make the difference in where one eventually ranked.

Dick's demerits were never significant in severity or in number.[40] As a plebe he ranked 109th in conduct. Like most plebes he was awarded the most demerits in the first six months.[41] The purpose of this constant attention to detail was to see how well cadets coped with stress and pressure under the constant surveillance. At the time there were seven classes of offenses. A Class 1 offense was the most severe; playing at cards or games of chance, for example, could receive eleven to twenty demerits. Least severe were Classes 5, 6, and 7, which received three, two, and one demerit, respectively. Almost all of Dick's demerits were 7s and 6s, with an occasional 5. His worst day was October 14, 1917, when, under the watchful eye of tactical officer Captain Kelly, Dick was awarded six demerits for three offenses at inspection: a dirty rifle, a soiled collar, and a tarnished breastplate.

Whereas Dick's academic performance at the University of Washington and MIT had been mediocre to average, at West Point he was an excellent student. Combining his entire record Dick ranked fourth in the class of 1920, graduating, as did everyone else, a second lieutenant on November 1, 1918.[42] He achieved a percentage of 93.2377 through earning a score of 1,005.41 points out of a possible 1,078.33.[43] Graduating first in his class was Cadet David W. Griffiths, who earned 94.55 percent.[44] Graduating last in his class was Cadet Wheeler, the class "goat" with 70.35 percent. Given his steady climb in academic rank in the two and a quarter years he spent at West Point, Dick might well have ended up number one in his class, had time not run out.[45]

Because of his high standing Dick could choose his branch. He opted for the prestigious Corps of Engineers, the almost automatic choice of the best cadets, decade after decade. West Point was, after all, founded as an engineering school. Beginning with the superintendent, the most prestigious vocational choice was engineering. Indeed, leadership of the academy was confined to the Corps of Engineers until 1866, when Congress passed a law allowing those from other branches of the army to serve as superintendents. Nevertheless, of the thirty who served from 1802 to 1918, twenty were from the Corps of Engineers.

That particular year there were a large number of spaces available in the corps, so Dick had a lot of company. Twenty-nine out of the top thirty cadets, and forty-four out of the top fifty, chose the Corps of Engineers. Twenty-five percent of the entire class — sixty-two cadets — chose the corps. The choices for most of the remaining second lieutenants were infantry (fifty),[46] field artillery (thirty-two), coast artillery (twenty-nine), and cavalry (twenty-two). One other option was available, air service, and Eugene Luther Vidal (ranked seventy-second) and a few of his classmates took it.[47]

The pressure of the European war, under way since 1914, was slow to influence the conservative academy.[48] Occasionally the war would intrude in the form of visitors. Gen. Joseph Joffre, hero of the Marne, visited West Point in May 1917. The following May the Blue Devils, French combat veterans, visited. The West Point Band played "The Marseillaise" and nine hundred cadets marched as one for them.[49]

The initial response at West Point to the Great War in Europe was that no change was needed at all in the plodding, tradition-bound crucible of the incipient American officer class. Until 1917 the academy refused to acknowledge the Great War in tactics or academics, adhering to the curriculum that had been established prior to the American Civil War. In military instruction, the momentous battles of the great captains who had graduated from West Point were the only objects of study, unencumbered by such modernities as the Gatling gun or trench warfare. Academy officials pointed to the success of their graduates as reason to resist change.[50]

Finally, after the United States formally entered the war on April 6, 1917, the War Department, badly in need of officers, accelerated the graduation dates of West Point cadets. The classes just before Dick's were graduated, and the officers headed off to war. The class of 1915 would turn out to be an extraordinary one, though no one knew it at the time. Dwight David Eisenhower, Omar N. Bradley — 36 percent of the 164 graduates attained the rank of brigadier general or higher. Many of them gained some experience in World War I.

Eventually the six wartime classes that entered between 1913 and 1918 would all graduate in a thirty-eight-month period from April 1917 to June 1920, leaving the Military Academy practically empty.[51] This rush to produce West Point officers through accelerated graduation was somewhat controversial. Granted, large numbers of officers were needed to lead the huge numbers of men going off to war. A partial "solution" was to give commissions to thousands of civilians, so-called emergency officers..[52] In 1916 there were 5,175 officers in the army. The following year there were 34,224; in 1918, 130,485.[53] This largely unqualified group was filling hundreds of officer slots, slots that should have been going to West Point graduates. If it kept up, there would not be any spaces for West Point officers.[54]

On the morning of October 2 a rumor began spreading that the classes of 1920 and 1921 would graduate on November 1.[55] At dinner that evening an announcement confirming the rumor was made to the whole corps in the main mess hall. It was a moment of high drama. Immediately after the order had been read, the acting commandant held up his hand for silence and said, "Now let's see what sort of discipline you have . . . I want everyone to keep silent for five minutes." After the excruciatingly long period was over, they were told to make as much noise as they wanted; what followed was pure pandemonium.

A hectic four weeks followed, "a nightmare of lectures and drill. . . . busy morning, noon and night . . . a mad attempt to give us a year's work in three months."[56] The hurriedly arranged graduation ceremony took place on the parade ground in the presence of about three thousand people. Representing the War Department was Assistant Secretary Crowell; representing Gen. Peyton March, chief of staff, was Maj. Gen. Henry Jervey. In a short address Crowell told the 511 new second lieutenants that it was the government's intention to have them in France within four months.[57]

Groves continued to maintain contact with Grace Wilson, now living with her family in Amherst, Massachusetts. They saw each other twice during his West Point years. Grace's father, Colonel Wilson, had retired on June 10, 1917, but was placed back on active duty and detailed in September 1917 as a professor of military science and tactics at Massachusetts Agricultural College, Amherst, and later at Amherst College, where he served until August 1919.[58]

Dick visited Grace in Amherst during his Christmas leave of 1917; she was also his date ("drag") at West Point for the "Furlough Hop" weekend earlier that spring. Many years later, in her written reminiscences, Grace admitted that she had set her cap for Dick the first time she had met him in their teen years at Fort William Henry Harrison.

The newly minted officers assigned to the Corps of Engineers were ordered to report to the commanding general at Camp A. A. Humphreys, Accotink, Virginia, on December 2, 1918, for a course of instruction.

PART TWO

An Engineer in the Peacetime Army

· 1919–1930 ·

CHAPTER FIVE

Caught Behind the "Hump"

• *December 1918–June 1921* •

Dick graduated from West Point on November 1, 1918, ten days before the armistice that ended World War I, starting his career at a difficult time. With the war over, and the demobilization of the national army that soon followed, there was great turmoil in the Regular Army officer corps. One problem was how to integrate the large numbers of wartime officers into the peacetime army. Because most of them were ahead of the wartime West Point classes, this created a situation that would affect the Regular Army for the next two decades.

The operative description for this uneven distribution of army officers was the "Hump." Given the strict seniority system, cohorts of West Point graduates crawled from rank to rank during most of the interwar period. The Hump had a profound influence on the careers of thousands of Regular Army officers, leaving a grim prospect for advancement without war. For example, members of the class of November 1918 remained approximately fourteen to sixteen years in the grade of first lieutenant.[1]

On the other hand, now that the war was over some of the gaps in an officer's education, which had resulted from the accelerated West Point program, could be filled. Ever since the founding of the academy the highest-ranking West Point graduates had always chosen the Corps of Engineers. The corps was seen as an elite branch, with slightly better chances of promotion due to its own rules. For Dick, near the top of his class, there was really no choice. He was going to be an engineer. After a monthlong leave, spent in part with his uncle William M. Griffith in Queens, New York, Dick reported to Engineer School on December 1.[2]

Camp Humphreys had just recently become the home of the Corps of Engineers.[3] In the spring of 1914 a parcel of about fifteen hundred acres acquired from the District of Columbia was surveyed, and the ruins of Lord Fairfax's house — Belvoir — uncovered. The land was used for field training and exercises until America entered the war in Europe in April

1917. Maj. Gen. William M. Black, chief of engineers,[4] foresaw that large numbers of men and quantities of materials would be needed for the war effort, and decided that Belvoir would be an excellent place for training engineer units. Adjoining portions of land — eventually more than six thousand acres — were acquired, and roads, barracks, buildings, a five-mile railroad spur, parade grounds, and training areas were built. The military base that emerged was named Camp Andrew A. Humphreys.

No sooner had all of this been accomplished than the war ended, and the question arose of what to do with the camp. General Black recommended that all engineer activities be concentrated at Camp Humphreys. The Engineer School was based nearby at Washington Barracks in Southwest (now Fort Lesley J. McNair), having moved from Willets Point, New York (later named Fort Totten), in 1901.[5] The Engineer School at Camp Humphreys, as conceived by Black, was to be a superior graduate school producing professional engineers who could better address the army's unique needs.

Initially Black had doubts that his high standards could be achieved, given the sudden influx of undereducated officers from the abbreviated West Point classes that was now forced upon him. Black made his feelings known to Colonel Tillman, superintendent of the Military Academy.

> As you probably know, I protested with the greatest vigor possible against the assignment to the Corps of Engineers of the sixty-two young men that passed only thru two years of the four-year course at West Point. For reasons best known to higher authority, my protest was disregarded and these young men were commissioned officers of Engineers when they were not Engineers and had not received the training requisite for an Engineer.[6]

After a first week devoted to physical training, bayonet exercises, and pontoon drill,[7] the academic courses got under way for Dick and his classmates. To complete the education that had been cut short by the war, General Black saw to it that a special set of engineering courses were taken by the graduates of the classes of June and November 1918. The regimen was strict. Dick took advanced mechanics and structures, physics, chemistry, and drawing.[8] Dick was particularly impressed with one of his instructors, Professor Moore from the University of Washington, one of the civilians who had been brought in for the war effort. He felt that Moore's course in advanced mechanics was the equivalent of a year of graduate work, and he was grateful for learning the newest theories in the field.

Dick got good grades, ranked at the top in the engineering classes, and received an excellent Efficiency Report. The assistant commandant, Virgil L. Peterson, wrote for the period December 2, 1918, to September 1, 1919:

> This officer has done the best work in the Engineer School, of the men who graduated in his class at the Military Academy. . . . Lieut. Groves has a very keen mind and should make an excellent officer for any duty pertaining to the Corps of Engineers.[9]

Groves was now launched on a career path. The assumption of the day was that to graduate from West Point meant a lifetime commitment. At West Point and now at the Engineer School Groves's hard work and self-discipline were paying off. What did he see himself doing; how would he make his mark?

The army had traditions, West Point had traditions, and the profession of military engineering had its traditions as well. The conduct of warfare and engineering have been entwined for thousands of years. Fortresses, great walls, bridges, pontoons, battering rams, and catapults were among the ancient accomplishments. With each change in technology and other inventions and improvements, military engineers adapted their skills to the new challenges. Notable military engineers from antiquity through the Renaissance and into the modern period are famous in their own right, and knowledge of them is part of the engineer's education and identity.

In the United States military engineers were responsible for exploring, surveying, developing, and improving the expanding nation from colonial days to the present.[10] In 1802 Pres. Thomas Jefferson established the U.S. Military Academy at West Point as an engineering school, the first and only one until 1849. Its first graduate, Joseph G. Swift, would be chief of engineers from 1812 to 1818. The Corps of Engineers' involvement in planning and executing large public works is an almost uniquely American phenomenon. Its civil works role began with passage of the General Survey Act on April 30, 1824, which authorized the president to use army engineers to survey road and canal routes "of national importance, in a commercial or military point of view."[11] West Point would produce hundreds of engineers who would oversee the construction of the nation's canals, railroads, coastal fortifications, river and harbor works, public buildings, and monuments.[12]

Army engineers contributed to the planning and building of the nation's capital, Washington, D.C., from the beginning. After the Civil War the corps's responsibilities were broadened significantly when Congress established the Office of Public Buildings and Grounds in 1867 and placed it under the Corps of Engineers. In 1878 Congress permanently replaced Washington's elected government with a three-man commission. An army engineer holding the title of engineer commissioner was responsible for the city's physical plant (and would remain so until 1967).[13]

There was also an illustrious line of military engineers, figures crucial to winning battles and wars. Montgomery C. Meigs (class of 1836) was quartermaster general during the Civil War.[14] Andrew A. Humphreys (class of 1831) had been General Grant's chief engineer and would later serve as chief of engineers from 1866 to 1879. Some projects built by engineers were so significant and so monumental that their names would ever be associated with them. The accomplishments of army engineer George Washington Goethals (1858–1928, class of 1880) in building the Panama Canal from 1907 to 1914, for instance, brought him lasting fame.

At the end of the first part of the special course in June 1919, the students of the Engineer School were ordered to France for a tour of the European battlefields and supporting installations built by the Army Corps of Engineers. After visiting Uncle Will and Aunt Jo in Queens, and seeing his brother Owen — who had returned from serving with the army in Italy two months earlier — Groves sailed for France on June 21 to begin his adventure. This almost twelve-week trip is well documented in Dick's numerous letters home, and elsewhere.[15] It was the only time in his life when he tried to express his feelings in a comprehensive way. His descriptions of the things he saw show a keen eye for military matters, and his opinions about the mores of the French and British show the development of some strongly held social and political views.

The converted German merchant ship *Cap Finisterre* left from Hoboken, New Jersey, with some 120 engineer students and instructors from Camp Humphreys and instructors from West Point. After an uneventful ten-day passage, the ship arrived in Brest, France, on July 1. Dick spent a night in the Hotel de Paris and then went to Camp Pontanezen, a U.S. Army base, where he was assigned to a tent. Groves called Brest "a rotten town" with much ill feeling exhibited toward Americans, who were charged double for a bottle of red wine and were also "running up prices on prostitutes."[16] He noted in his diary the rampant venereal disease problem.

From July 3 to July 9 Dick visited Rennes ("[v]ery dull town"), St. Malo, and Dinard with two of his classmates, after which they went back to Brest and embarked on a monthlong tour aboard a sixteen-car Red Cross hospital train visiting supply depots, ports, battle sites, and hospitals in France, Belgium, and Germany.

The first stop was St. Nazaire, where the U.S. First Division had landed in 1917. An enormous engineer depot had been established, much of which still remained. "There does not seem too much to the town of St. Nazaire. Thumbs down."

On Saturday, July 12, Groves arrived in Bordeaux, "a real city, more on the style of an American town."[17] While there were some good aspects to the city — tramways, real stores, "although prices were a little high," restaurants with wine included — Groves claimed there were forty thousand registered prostitutes and eighty thousand others, in a city of half a million. He does not inform us of how he calculated this ratio — high, he would probably have to admit if pressed, even by French standards. He was once accosted on the street by a woman who grabbed his arm in broad daylight at seven in the evening; he could not get rid of her for about five blocks.

The students visited Bassenes, the American docks eight miles south of Bordeaux on the Gironde River, and Groves made his analysis. No ship drawing more than eighteen feet could come up the channel. There was room for three ships at the French docks and twelve at the American. The warehouses, nine miles east, were made out of wooden beams and corrugated sheet iron, satisfactory for temporary work, Groves concluded. Unfortunately, St. Supplice is "at present a venereal camp."

These kinds of notations in his diary and letters are typical of his analytical approach. He continually assessed how well military installations achieved their purpose, or how they might do more with less. On the docks at Bassenes, for example, were cranes to unload the ships, "by linking two 10 ton ones together it was possible to lift 30 tons. There is also an enormous refrigerator plant there. After things were really organized, meat was boned before shipping saving about 1/2 in cu. ft. and 1/3 in weight."[18] At Gievres "it was found that frozen beef could be shipped from ports to here without ice in airtight cars. Result big ice plant here was not used." His eye and mind were always focused on ways to save time and space, on how to solve problems, on how to get the job done. He was attentive to quantities, distances, time, space, area, and volume; he frequently broke down a task into its component parts to examine whether it was efficient or not. How did they feed eight thousand men in an hour?

How could large numbers of men bathe and have their clothing deloused and washed? Was it done efficiently, or might young Lt. Dick Groves recommend a way to do it better?

By July 16 they were in Verneuil, a motor transport base near Dijon. Groves learned lessons at every stop. Machines needing twelve work hours or more were shipped to Verneuil usually by rail from the front. Three experts inspected the machine and decided whether it should be repaired or torn down for parts. There were forty-three different makes of cars, necessitating a huge number of spare parts. Of the trucks in use White, GMC, Quad, and Pierce-Arrow were among the best. The base at Verneuil was built under the direction of Colonel Hegaman, "who was the pioneer in truck transportation in the army." German prisoners were utilized for everything except handling food, and those still there "seem a very superior lot. I think that we are going to have trouble someday if we go on with our present propaganda about the inferiority of the Hun."[19]

Groves comments repeatedly on the difficulties between the American soldiers still in France and the French. "Two Frenchmen were killed in Brest the night before we arrived and 7 more in Bordeaux about 2 months ago. All by Americans during riots."[20] Nevertheless Dick holds out the possibility that instead of returning to the States to become an instructor he "would like to stay over here for a year or two."[21]

In Is-sur-Tille, an advance supply depot of the AEF First and Second Armies, Groves comments on the huge bakery capacity — 860,000 pounds a day, almost all handled by machinery, with bread reaching the troops in ten days. An engineer must take note of everything, from logistics to the feeding of troops.

On July 18 they visited a sector of the battlefield, halfway between St. Mihiel and Apremont, held by the Germans for years, with its concrete bunkers, trenches, and tunnels. Groves was often infuriated at the seemingly lackadaisical attitudes of the French and other Europeans in getting the job done. "The villages are all knocked to pieces but, if the people were awake, could be rebuilt very rapidly. However, at their speed, it will take 20 years." In his diary especially he is repeatedly critical of the Europeans.

> The more I see of the Europeans, the more I favor isolation for America. The Germans are too brutal and the rest of them are too wooden and indifferent. Where the French got their reputation for politeness is a mystery. I wish we would come off of our pro-English, pro-French, pro-Wop propaganda and realize what we are up against.

I wouldn't mind war for a reasonable cause, but simply because Europeans are too selfish to live in peace together is not sufficient cause for us.[22]

They visited Verdun, where a million had died. They surveyed the battlefield and the forts and reconstructed what happened. "On our way to Rheims we saw the grave of Quentin Roosevelt." They proceeded on to Belgium and to Germany. To his father he wrote from Coblenz on August 7,

I honestly believe that Germany will come back to her former power and that as soon as she can see a way to keep us out of it, she will start another war.

Also to his father:

The more I see over here, the more I see the necessity of a strong national policy and the giving-up of this policy of supporting the rest of the world. It would be different if the world had the same ideals as we have, or even appreciated what we are doing for them, but such is not the case and we are merely being played for suckers, to put it rather crudely. I certainly hope the Senate will have the necessary guts to stand off the sentimentalists, of which we seem to have a superabundance.

The group finally reached Paris and was granted a two-week leave to visit England, Scotland, and Ireland. Dick went to London and told his father on August 19,

Here I am and must confess I don't like the English. I find their morals to be rather depraved, to say the least, and the whole nation seems to be ready for a general bust in some direction.

Back in Paris, the last stop before going home, Dick did some sightseeing and shopping:[23]

I spent one afternoon in the Louvre trying to find something worthwhile gazing upon, but had to give it up as a bad job. Someday they may wake up and arrange things so that they can be found. As it is, things seem to have been thrown together instead of arranged. I finally succeeded in finding Mona Lisa and found that she was as ugly as I had anticipated. I can't understand why anyone ever thought the lady worth stealing. . . . Most of the time I wandered around the city, visiting many stores and incurring the wrath of many proprietors by telling them what robbers they were.[24]

This was the only period in his life in which Dick kept a narrative diary. Taken together with the lengthy and descriptive letters he wrote to his father, stepmother, sister, and brother, we get a definite picture of the man who was emerging.

Groves comes through clearly as having a sharp eye for the relationship between the design of military installations and the locale and purposes to which they will be put. This is reflected in his comments and critiques of the strengths and weaknesses of the various sites and installations he visited.

Groves's comments on Europe and Europeans reveal him to be politically conservative.

I certainly hope that the League of Nations is turned down, as I have a feeling that we would be about the only nation that would dream of keeping it. England and France have already shown about their national honor, as their moral standard is so low. The chief difference is that France is openly immoral while England makes some attempt to keep it hidden.[25]

Socially he exhibited a kind of puritanical rectitude, partially reflective of his father's religious outlook and upbringing, and partially reflective of the ideas of the day. He frowned on drinking and casual relationships with women. This combination of political conservatism and puritanical outlook made him not merely patriotic but indeed a chauvinist, firmly convinced of the soundness and superiority of the American system.

His sensitivity to the immorality in the military camps, and to the prevalence of venereal disease, was a prevailing view among many Americans of the day. The entry of the United States into World War I set off a determined effort by American reformers to protect its new soldiers from the ravages of venereal disease and from the other sins — gambling, alcohol use, and profanity — associated with the military.[26] The image of the military camp — and its immediate surroundings — as a place of immorality was deeply ingrained in the American mind. The memory of the frontier post, the conditions of the camps during the Spanish-American War (which Chaplain Groves knew firsthand), and the recent stories about American troops on the Mexican border only strengthened these beliefs. With America's entry into the war, President Wilson and his secretary of war Newton Diehl Baker were determined to do something about it. Driven by the progressive vision behind so much political and social activity at the time, they called for a crusade to "redeem barbarous Europe" abroad and to raise America to a high moral pinnacle at home.

French culture disgusted young Groves. His diary is filled with detailed, lurid accounts of having to defend himself against the importuning of the French prostitutes who, if his description were to be taken literally, thronged every street corner of every French village, town, and city.[27]

He was equally unkind to the British, though his focus shifted from streetwalkers to the corruption and hypocrisy of the British, regardless of class. Groves was a near-perfect expression of the Anglophobia that was widespread at the time.[28]

From the Revolutionary War to World War II strong currents of anti-British nationalism were ever present among large numbers of Americans. The British were distrusted, feared, and disliked, felt to be oppressors and rivals to an expanding America that was defining its identity. Since World War II and the crafting of the "special relationship" by Roosevelt and Churchill, these attitudes have virtually disappeared, but they were once very pronounced. In the early decades of the twentieth century the historical enmity was still alive. Chaplain Leslie Groves felt it deeply and passed it on to his children. This fervent moralistic patriotism clearly shaped young Dick, who saw himself as an American through and through. These attitudes would later be evident in his treatment of British scientists and diplomats during the war.

His diary also reveals him to harbor some anti-Semitic and racist attitudes, quite typical at the time. His attitudes toward Jews might be considered simply a part of inherited cultural baggage. When he became head of the Manhattan Project, the fact that a person was Jewish was not important. Getting the job done was. While he despised Leo Szilard, he had an excellent relationship with Robert Oppenheimer, I. I. Rabi, and Edward Teller, to name a few. Groves, like many of his day, was firmly convinced that blacks were slothful, dumb, and inferior. On this he never changed his mind.

On August 31 the navy transport ship USS *Kroonland* left Brest, docking at Hoboken on September 10 and unloading its now more worldly young engineers.

Groves returned to Engineer School to complete the special engineering course, studying electrical and mechanical engineering and civil engineering, among other subjects, during the academic year from September 1919 to June 1920. He continued to get good grades, ranking sixth out of thirty-two in the final standings.[29] On May 1, 1920, he was promoted to first lieutenant (Regular Army), and on June 16 he graduated from the special course. The secretary of war, Newton D. Baker, spoke, but Groves and his classmates — sitting in the open with the temperature more than a hundred degrees — remembered little of what he said.

On June 19, 1920, First Lieutenant Groves was assigned to the Seventh Engineers, stationed then at Camp Gordon, Georgia. A day after arriving he went to Camp Benning, near Columbus, Georgia, and joined Company B of the Seventh. Camp Benning was in the process of establishing the Infantry School to provide for the professional development of career infantrymen, just as Camp Humphreys was doing for engineers.

Groves's major duties at Camp Benning included commanding Company D of the Seventh Engineers; constructing floored and walled tents, mess halls, and latrines for the Twenty-ninth Infantry; repairing a highway bridge; and, as camp engineer, constructing a narrow-gauge railway. He made many lasting acquaintances there, among them Nathan F. Twining and Matthew B. Ridgway. His marks on his Efficiency Report were almost all "superior" with a few "above average," but he was awarded only an "average" for "Tact."[30]

On February 10, 1921, after eight months in Georgia, Dick returned to Camp Humphreys for the four-month Engineer Basic Officers' Course, entirely on military subjects designed to teach the duties and responsibilities of a company officer and staff officer of engineers. Dick got good grades, was judged "a promising young officer well above average," and stood seventh in a class of forty-four.[31] Upon graduation in June he was assigned to the Sixth Engineers, Camp Lewis, Washington.

CHAPTER SIX

Married with Children

• July 1921–July 1931 •

For the next ten years, from 1921 to 1931, Dick moved from one assignment to another with the Army Corps of Engineers, learning his craft and gaining experience in the civil programs associated largely with rivers and harbors. Throughout this period, his efficiency ratings by his commanding officers consistently gave him high marks for his understanding of engineering principles, for his ability to command the respect of his men, and for his administrative and organizational skills. With equal consistency he was rated somewhat lower when it came to exhibiting tact or maintaining a military bearing. During this period he would also get married and have two children.

After a month with the Sixth Engineers at Camp Lewis, Dick was sent in early August to Fort Worden on the Olympic Peninsula to prepare a topographical map to be used for the coastal defenses of Puget Sound. The posting to Fort Worden was convenient to Seattle, allowing him to pursue his courtship of Grace Wilson — and not a moment too soon.[1]

Dick and Boo had of course first met at Fort Harrison in 1911. In 1913 Col. Richard Wilson moved, with the Fourteenth Infantry Regiment, to Fort Lawton, Washington, where he continued as commander. During the Mexican border crisis Colonel Wilson and the Fourteenth went to Arizona in May 1916, returning to Vancouver Barracks a year later when the crisis passed. Colonel Wilson retired in June 1917 at age sixty-four, precisely on his birthday, after forty-four years of active service.

While he was in Arizona his wife and his two daughters, Grace and Mary, went east, first staying about eight months with her aunt and uncle in Chicago, and then, during the summer of 1917, going to Newport, Rhode Island, where young Grace studied music.

Three months after retiring Colonel Wilson was recalled to active duty to serve as a professor in Massachusetts, an appointment that lasted two years. It was during this time that Grace visited Dick at West Point, at least once and perhaps twice, and Dick came for a visit to Amherst. For several months in the winter of 1918–1919 Grace went to Boston and took cooking classes at Pratt Academy.

When World War I ended Colonel Wilson was returned to retired status. With his academic responsibilities completed, Colonel Wilson and the family moved to Seattle in August 1919 and found a large frame house at 1028 Fifteenth Avenue North, across a busy street from Volunteer Park. Grace regretted leaving Massachusetts, where, after her "pathologically shielded existence," she had begun to meet some nice boys.[2] After Boston, she felt Seattle was "crude and dull," and while Dick wrote once in a while, the Amherst boys wrote much more often. In Seattle Grace took a two-year course to become a kindergarten teacher, eventually graduating at the head of her class.

Grace had a much more active social life than her future husband. On a blind date aboard a ship at the Bremerton Navy Yard she met Rockwell J. "Rocky" Townsend, a recent Naval Academy graduate. Overwhelmed a bit by the uniform and the romance of it all, she found herself engaged to him after only three or four weeks. They stayed engaged for a year and a half. During this time Rocky served aboard a destroyer and was at sea for long periods, but he wrote to her and sent presents.

> But one fine day, while Rocky was cruising some far-away ocean our doorbell rang — there stood DNO. . . . He appeared frequently after that; Rocky returned; DNO still hung around.[3]

In late September 1921 Dick was ordered to the Presidio of San Francisco. After Dick departed for San Francisco, Grace broke off the engagement with Rocky, apparently breaking his heart.[4]

Later that year Dick telegraphed, and then telephoned, from the Presidio to propose marriage to Grace. She accepted. On February 7 Dick was granted a twenty-five-day leave of absence and departed immediately for Seattle. The wedding took place on February 10, 1922, at St. Clement's Episcopal Church. As Grace described it:

> [W]e were married on a dark rainy afternoon in February, in a small church, no guests, but at least I had a bouquet of orchids and lilies of the valley. We spent a rainy, cold week in Vancouver and Victoria [on their honeymoon]; and [then went] to the Presidio in San Francisco.[5]

Late in her life Grace Groves recalled that on the first day of their marriage, Dick told her that she was never to ask him any details about the nature of his work. She went on to say that he never talked about his work at home, and, as he had demanded, she never asked. Grace listed the following reasons why she never tried to know about her husband's work:

> One reason was my father's saying that women should remember they had no "rank," though their husbands did. Another reason — his work actually didn't interest me; I was busy with my own. Principal reason — very early in our marriage DNO said, 'You cannot possibly *help* me, but you *can* hurt me.'"[6]

It was an arrangement that she accepted, though occasionally she found it acutely embarrassing. As we shall see, her first inkling of what her husband worked on during the war came with the announcement over the radio of the atomic bombing of Hiroshima.

At the Presidio the newlyweds lived in his two rooms in the bachelor officers' quarters. There was no kitchen so they had to take some of their meals at the officers' mess, but they often went into the city to wonderful French or Italian restaurants, which abounded there. Later they moved into a large house in the East Cantonment that did have a kitchen. The coal-burning army range was soon removed and replaced with an ancient gas stove. As Grace recounted it, one day she turned on the oven, went to get something, and then returned to light the oven. There was an explosion that badly burned her face. Luckily Dick was in the house and a doctor lived next door, so she was attended to promptly. For some time Grace remained in bed, swathed in bandages. Dick got her an orchid corsage and told her he would always love her, even without a face.

During the summer of 1922 the engineer training companies went into the field for a month and Grace traveled home to Seattle.[7] Upon her return she found Dick wearing a mustache. For a short while Grace complained about it, to no avail. Dick would wear a mustache for the rest of his life.

At the end of August Dick was ordered to the Third Engineers at Schofield Barracks, Honolulu, territory of Hawaii. On October 28 he and Grace sailed from San Francisco aboard the army transport *General Thomas*. Unlike the time she had traveled with her father on the upper decks, they had a tiny cabin, far below, across from the galley, and Grace was seasick the entire trip.

For the next two and a half years — from November 1922 through late June 1925 — Dick was commander of Company F of the Second Battalion, Third Engineers. Groves's commanding officer was Edward H. Schulz, first in the class of 1895.[8] This was his first troop duty since the Spanish-American War, and he was considered by the other officers to be something of a joke.

Their quarters were a large house in the Engineer Loop, next to the General's Loop. It had two bathrooms upstairs, a coal-burning stove, and a small cabin in the back for the help. Dick bought a Ford automobile, borrowing the money in a rare act of profligacy. Dick was careful with money throughout his life, never having much of it until some modest sums came in after he left the army and worked for Remington Rand. He pledged that he would never do that again, and didn't.

In 1923 the Engineers had finished last in rifle marksmanship among the Regular Army service teams (Infantry, Cavalry, Coast Artillery), as well as the Navy and the Marine Corps, the perennial winner. Rising to a new challenge, Dick decided that he could help the cause. He practiced and became an expert in rifle, pistol, and automatic rifle marksmanship. His superiors noted this new skill, and he was given responsibility for all marksmanship training within the regiment. His efforts may have done some good: the Engineers won the 1924 competition, with members of the Third Engineers well represented on the twenty-one-man squad.[9]

Groves also coached the baseball and football teams and helped advance the careers of promising young soldiers like Bert Muse, an Arkansas boy. Muse had enlisted as a private in December 1922 at age eighteen and was sent to Schofield Barracks, rising quickly to first sergeant of Company F, Third Engineers, the youngest first sergeant in the army. His drive and determination got his commander's attention, and Groves helped him get into West Point, where he graduated in the class of 1929. Tragically, his career as an aviator ended in a fatal air crash in April 1930.[10]

On July 10, 1923, Dick and Grace's first child, Richard Hulbert Groves (named for Grace's father at Dick's suggestion), was born at Tripler General Hospital, Fort Shafter, Honolulu, at a formidable nine pounds, fourteen ounces. Family tradition would later have it that he was the second heaviest baby delivered at Tripler up until then.[11] His father often said that even as a baby his son's hands seemed bigger than his own, which were surprisingly small, almost dainty. With such attributes his father predicted a career in the ring as a prizefighter.

There was time for tennis, of course. Dick became the post singles champion and, with his partner, Lt. Max Taylor (the future chairman of the Joint Chiefs of Staff), doubles champion as well. The tennis partners met their match, though, when they were quickly beaten by the visiting Japanese Davis Cup team that was passing through Hawaii. In an effort to save money so he could get married Taylor would occasionally eat dinner with the Groveses. Several of Dick's classmates and contemporaries in Hawaii would become lifelong friends.

Groves also had construction duties, and on April 12, 1924, he received his first official commendation, for work on the Kahuku-Pupukea Trail. Under his direction 11,500 feet of mountain trail, an average of 6 feet in width, were completed in thirty-seven working days.[12]

On June 8, 1925, Groves got a new assignment — assistant to the district engineer, Galveston, Texas — and was granted almost four months' leave and travel time before taking up his duties. Grace and baby Richard[13] sailed ahead by commercial liner[14] and went to Seattle to stay with her parents. On June 27 Dick sailed from Hawaii by army transport and eventually caught up with his family. They stayed in Seattle for several weeks, and in the time they spent together Dick came to better know and appreciate his father-in-law. In August the family traveled to Pasadena to see Chaplain Groves and Jane; from there they continued on to Texas, where Dick reported for duty on November 3. Grace recalled how dreary it all looked: the low and flat landscape, unpainted houses, unpaved streets, no trees, and general air of desolation.[15]

After a stay in the hotel, they found a tiny bungalow to rent, at 702 Avenue H, that they could afford on Dick's salary, amid neighbors who were predominantly black.

Groves was assistant to the Galveston district engineer, Maj. Julian L. Schley. Later Schley would be governor of the Panama Canal Zone and, from 1937 to 1941, chief of engineers. Schley thought highly of Groves, considered him his best officer, and wrote favorable Efficiency Reports.[16] The work Groves did in Texas was his first real opportunity at major engineering projects. One of the jobs was opening the channel at Port Isabel, which later became the entrance from the Gulf of Mexico to the ports of Brownsville and Harlingen. During his two years in Galveston Groves was away from home almost half the time, leaving his wife and young son alone.

An early memory of young Richard is a visit, with his mother, to Chicago to see her aunt Bossie and uncle Duncan Hines, the famous food connoisseur.[17] Other memories include swimming with his father, building sand castles on the beach, and playing with a group that could have come straight out of an Our Gang movie.

Groves's other duties were dredging operations in Galveston Bay, in part connected to improving access to the Houston Ship Channel by widening and deepening the channel to accommodate larger ships. The official history mentions Groves, and an incident that reveals his two sides:

> While attached to Galveston, Leslie R. Groves served a tour of duty on the Harrisburg quarterboat, for a duration considered "too long" by the other men aboard the vessel. This man, who in 1942 was pegged to direct the development of an atomic bomb, was unsurpassed at getting the job done, but he lacked those qualities that would have endeared him to his fellow workers. One day he was out with a crew working in the bay when the weather became very rough. The captain of the vessel decided it would be wise to return to shore, but Groves disagreed and ordered him to keep on going. As the weather continued to worsen, the captain asserted that as long as they were afloat he was in command and that once they were safely ashore, Groves might exercise his authority. Whether Groves was more influenced by this line of reasoning or by the crew member who stood ready to throw him overboard remains questionable, but he did acquiesce.[18]

In mid-September Groves received orders to proceed to Fort DuPont, Delaware; on October 15 he became commanding officer, Company D, First Engineers, Fort DuPont. With no quarters available, the family moved into a drafty spare suite at the bachelor officers' quarters. There was no kitchen, so they took most of their meals at the First Engineers' mess.

While Grace may have hoped that the family would now be together, it was not to be the case. Less than a month after arriving Dick left to lead a detachment to Fort Ethan Allen in Vermont. Floods were raging in New England, with Vermont suffering the heaviest loss of life and property.[19] The military assisted in many ways. Groves and his troops undertook the construction of a pontoon bridge across the Winooski River to service the town of Burlington, which had been cut off from the town of Winooski by the rising waters. At noon on December 1, 1927, the pontoon bridge was swamped and most of it swept away downstream.[20]

The commanding officer of Fort Ethan Allen, Col. Frank E. Hopkins, charged Groves with negligence and requested that a competent engineer replace him. Hopkins appointed an officer to investigate the accident and to fix responsibility for it. In his report of December 20, Major J. F. Stevens found:

> That First Lieutenant L. R. Groves, Jr., Corps of Engineers, the officer in charge of the bridge was negligent in failing to be present and take personal charge of the bridge when it became endangered by the rising waters and before the bridge or any part thereof had become flooded.[21]

Despite the difficulties he was having with the local military authorities, Dick organized the procurement of the materials necessary for a new pontoon bridge, which was in place by December 15. This was a good example of how well Groves could perform under extreme pressure; there would be many more.

Soon after this incident Dick took a ten-day leave beginning on January 10, 1928, to help settle his family at Fort DuPont. But Dick's difficulties in Vermont were not over. On the afternoon of February 2, at about four-thirty, a catastrophe occurred.[22] Dick and several of his men were seriously injured when a block of TNT — intended to clear ice from the river, so as not to endanger the pontoon bridge — exploded prematurely in the hand of 1st Sgt. Steward A. Littlefield. It was a cold and windy day. It was difficult to light the fuse — cut to about a three-minute length — and then sometimes difficult to tell if it had caught or not. The fuse caps and the TNT were World War I stocks. Some had deteriorated so badly that they were discarded. Groves was only six feet away when the TNT exploded, blowing off Littlefield's entire lower arm, tearing through his clothing and covering Groves with blood. Groves saw at once that his first sergeant was not going to live, but got him to the hospital where the following morning he died without regaining consciousness.[23] Dick was luckier, though he was hospitalized for more than five weeks with injuries to his left eye, cuts and bruises on his face and hands, and some loss of hearing in his left ear.[24] A piece of bone from Sergeant Littlefield penetrated his left hand at the forefinger knuckle. His heavy clothing had saved him, and the visor of his cap prevented a direct hit to his eye. An official investigation of the accident found no evidence of negligence or misconduct on the part of Groves or any of his men.

Though Colonel Hopkins endorsed these findings, he also immediately requested that a "suitable Engineer Officer" replace Groves. His efforts to have Groves transferred continued through February and March with letters to the chief of staff and to the commanding general of First Corps Area. Upon being released from the hospital Groves was granted a month's sick leave, from March 10 to April 9. Vermont was the low point of his career, and it shook him profoundly. On March 11 he left Burlington for Fort DuPont and did not return.

Inevitably Colonel Hopkins got the last word with a scathing Efficiency Report:

> [I would] object to have this officer serve under [my] command in peace and in war, in his present or higher grade — would not want him under any consideration on a similar job. . . . He is not practical and lacks good judgment. . . should not be given any command similar to the one held here. . . . This officer is a failure in his present position. Lieut. Groves was sent to this post with a pontoon bridge following the flood. He showed lack of good judgment from the day he arrived, and the bridge was lost over the dam. Later one man was killed and he and others injured while working under his direction. . . . many times he was called up and advised and assisted but could do no better.[25]

Higher authorities never supported the December finding. The chief of engineers, Maj. Gen. Edgar Jadwin, interceded, and though the letter is missing, he apparently chastised Groves's superiors for trying to attribute to Groves errors of judgment that had clearly been their own responsibility.[26]

Later during the Manhattan Project certain scientists, without knowing the details or the context of the incident, spread rumors about it in their cultural war with Groves. According to them, not only was he ignorant of the science and physics of the atom, but he had also killed a man in the past through negligence and incompetence.[27]

Upon Groves's return to Fort DuPont he was made headquarters and service company commander, and also served as regimental supply officer. The family also moved, into one half of a brand-new brick duplex. Over the next year and a half Groves gradually regained his health, worked diligently performing routine garrison duties, and continued his work on

A1. Young Leslie Richard Groves Jr., about 4 years old, Vancouver Barracks, Vancouver, Washington.

A2. Father, Leslie Richard Groves Senior (1856–1939), c. 1890 at the time of his marriage.

A3. Mother, Gwen Griffith Groves (1865–1913), c. 1890 at the time of her marriage.

A4. The four Groves children, clockwise from top, Allen, Leslie, Gwen, and Owen. Taken c. 1905 at Fort Hancock.

A5. Cadet Groves in 1918 from West Point yearbook, *The Howitzer,* wearing the stars of a distinguished cadet — those with an overall grade of at least 90 percent.

A6. Grace Hulbert Wilson, the future Mrs. Groves, at the Wilson summer cottage on Puget Sound in 1921. Dick and Grace were married in February 1922.

A7. Baby Grace Wilson being held by Chief Washakie (then about 100 years old), at Fort Washakie on the Shoshone Indian Reservation, Wyoming, 1897. Standing at right are her parents, Captain Richard Hulbert Wilson and Grace Arents Chaffin Wilson.

A8. First Lieutenant Groves served with the 3d Engineers at Schofield Barracks, Honolulu, Hawaii, from late 1922 to mid-1925.

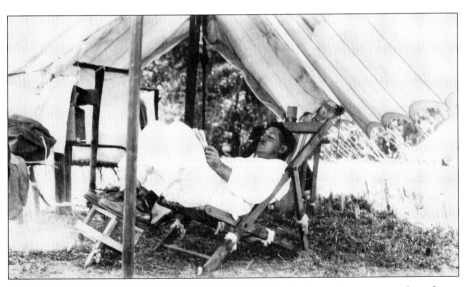

A9. Reading in his tent in the Nicaraguan jungle. Groves served as commander of A Company of the 1st Engineers from October 1929 to July 1931 surveying possible interoceanic canal routes.

A10. Ernest "Pot" Graves (1880–1953). The frontispiece from his book *The Line Man's Bible: A Football Textbook of Detailed Instruction,* published in 1921. Later Colonel Graves would be an important influence on Groves especially during his time in Washington in the Chief of Engineers' office, 1931–1935.

A11. Family reunion at Chaplain Groves's home in Altadena, California, summer 1935. Standing from left: Marion, Gwen, Owen, Richard, Leslie Senior, Dick, and Grace. Kneeling: Gwen and Charlotte.

marksmanship training, though he could no longer fire the rifle as before because of the wound to his left hand in the Vermont accident. On October 25, 1928, Grace gave birth in Wilmington to their second child, a daughter named Gwen, for Dick's mother.

There was also time to be a father to his son, a son who sometimes got into trouble and was subjected to corporal punishment. For a time five-year-old Richard roamed the post with a gang of small boys, slashing screens, throwing rocks through glass windows, and, on a particularly memorable occasion, smashed some brand-new toilet bowls that had been left outside the building where they were to be installed.[28] Each day Groves would ask his son why he did these things, and other infractions, never whether he had done them. Many years later Richard learned why his father had framed the inquiry in this way. He said he believed that to ask a child if he has done something for which he will be punished can only turn him into a liar.

After discussing the offenses they were tallied up; three belts for being rude to Mama, one belt for not coming when called. Credit was allowed for good behavior. At the end the account was settled using a web belt delivered on his son's bare legs. Reflecting back, he believed his father's system was just and taught that there is a price for everything.[29]

From Fort DuPont the family traveled extensively in their 1927 Chevrolet coupe, once to West Point to visit Lt. Max Taylor, then an instructor in French, and often to Baltimore, Philadelphia, and nearby Wilmington. Among Wilmington's attractions was the Reynolds candy store. Groves was a frequent visitor to purchase a particularly scrumptious chocolate-covered peanut cluster.

Groves's Efficiency Reports were excellent. In report after report — from April 1928 to October 1929 — he was judged "An able and competent officer who has achieved notable results in the performance of his varied duties" . . . "A superior officer" . . . "a very able officer with considerable initiative" . . . "a very efficient officer."[30]

In early 1929 Dick heard that an expedition to survey an interoceanic canal through Nicaragua was being proposed, and that a provisional battalion of army engineers would contain a company from the First Engineers (the Corps of Engineers' oldest) at Fort DuPont. He made it known to his superiors that he wanted to be in that battalion. As a result, he was named head of Company A, the designated company, of the Twenty-ninth Engineer Provisional Battalion, under command of Maj.

Charles P. Gross. Maj. Daniel I. Sultan directed the survey.[31] The congressional joint resolution of March 2, 1929, authorizing the survey also established the Interoceanic Canal Board.

The family was to be separated again. Grace, six-year old Richard, and not-quite-one-year-old Gwen went by train to Seattle to live with her parents, stopping on the way to visit Aunt Bossie and Uncle Duncan in Chicago. Habits and personalities at the Wilson household on Fifteenth Street were somewhat unusual. Grace's sister Mary occupied a large corner room with her dozen Pekingese dogs, constantly yapping. Two uniformed Filipino boys, Nemesio Pascual and Fulgencio, did the chores. Grandmother Wilson stayed up late and rarely came downstairs before midday. With his elderly grandfather Richard would go on walks through Volunteer Park. Sometimes the older man returned exhausted and — usually at the dinner table — fainted dead away, winding up on the floor. While the Filipino boys carried him off to bed, Grandmother Wilson blamed young Richard for being so inconsiderate of her husband's frail health.

Young Richard enrolled in the local school. His mother's previous efforts at home education, plus poor learning habits, required that he start in the first grade. More important to his early education was the influence of his learned grandfather, who patiently taught his grandson to read and write.[32]

Dick sailed with his company from Brooklyn for Nicaragua aboard the army transport *Chateau Thierry,* on October 15, 1929. The voyage, as Groves remembered it, took six days to Panama and two through the canal, arriving at Corinto, the chief Pacific port, on October 24, a few days before the stock market crash. Dick spent his time aboard ship studying Spanish. They anchored in the harbor of Corinto and were taken ashore in small boats. Groves and his company, along with detachments from some other companies of the Battalion, stayed in Corinto to wait for, and then unload, the army freighter transport *Kenowis,* which carried much of the battalion's equipment. Corinto, according to Groves, "was at that time one of the toughest seaports in the world."[33] His men, fresh from the United States, where Prohibition was in place, were unfamiliar with foreign customs and not used to either the tropical climate or the women. There could have been problems, but luckily Lieutenant Groves was there to keep them in line.

I had no trouble with my own men from the First Engineers who, with a few exceptions among the privates, were thoroughly familiar with my reputation for discipline.[34]

As Sultan later recounted, the expedition was "no holiday jaunt." The annual rainfall averaged some 255 inches, so the troops were never dry. There were swamps and dense vegetation, insects by the millions, mosquitoes, fleas, ticks, scorpions, alligators, poisonous snakes, even sharks.[35] There had been continuing political problems with factions fighting one another in an ongoing civil war. The American marines had been in Nicaragua from 1912 to 1925. After they left, fighting broke out again. The marines returned, and President Coolidge sent his personal representative, Col. Henry L. Stimson, in April 1927; he arranged an armistice and an election.[36] Under American supervision — assisted by the presence of five thousand heavily armed marines — an election was held in 1928, and Gen. José M. Moncada, leader of the army at the time of the cease-fire, was elected president. The marines also trained and provided sergeants for the Guardia Nacional as a police force to keep order. Nevertheless, the engineers of the Battalion remained armed at all times, watchful for the rebel forces of Augusto Sandino, the only one of Moncada's generals who did not accept the cease-fire.[37]

Visions of a canal across Nicaragua had been around for nearly a century. Prior to 1900 five major studies and surveys were made. The survey of 1852 had fixed the logical route. A later survey in 1901, in which Robert E. Peary — later to explore the North Pole — participated, further mapped many of the jungle rivers in the valley of the lower San Juan. Though Panama had been chosen earlier, and a canal built, some thought there was need for another to accommodate larger vessels and increased traffic.

As conceived in the late 1920s, an interoceanic Nicaraguan ship canal would carry more tons of traffic per year than the Panama Canal. Though it would be three and a half times longer — at 173 miles — it would take less time for most ships to transit, saving the shippers money. The canal would take about ten years to build and cost an estimated $722 million.

The survey headquarters was in the old Monastery of San Francisco in Granada, a city of some twenty-two thousand and the main port on Lake Nicaragua. Groves's company — A Company of the First Engineers — of approximately one hundred men, with equipment and supplies, departed on December 10 from Granada on the antique lake steamer *Victoria*,[38] disembarking at San Jorge, the port for the old city of Rivas, at about five in the afternoon. Groves planned an all-night march to avoid the heat. On horseback, he set forth with thirty carts each pulled by four oxen, slowly lumbering through the night to a campsite near Brito, and then on to the small port village of San Juan del Sur the next afternoon. From there over the next five months A Company would gather topographical, geological, and hydrological data along the proposed line of the canal, along the Costa Rican border, from Fort San Carlos to El Castillo, a forty-mile stretch, and survey a possible railway route between these same two places. At the end of May 1930 Company A returned to Granada, and from there sent detachments to survey Tipitapa and the Tipitapa River. Beginning in mid-August the company began boring, using diamond drills, to investigate proposed lock sites, and continued this work for many months.

While in Nicaragua Groves began to systematically read a series of important and influential books to try to fill the gaps in his education. At his request his father-in-law had chosen for him one hundred best books. The list survives and attests to Colonel Wilson's erudition, as well as to Dick's determination to expand his intellectual horizons. The list was broken up into groups, with the history and literature of the West the central focus: Greece, Rome, the age of Christianity, the Renaissance, Europe in the sixteenth, seventeenth, eighteenth, and nineteenth centuries, Asia, Africa, and America. Colonel Wilson starred twenty-seven books that he recommended for purchase, at an estimated cost of ninety dollars. To buy all of the books, he thought, would cost approximately three hundred dollars. He calculated the time required to read them at four years, at a rate of seven hours per week, ten hours per volume.

It would be a more than adequate guide to follow today, covering the ancient and modern classics, plays, political philosophy, novels, poetry, and histories. And so in his tent in the jungle Dick began his reading, perhaps starting with Thucydides, *The Peloponnesian War* ("esp Bk II, Ch 6," noted his father-in-law), H. G. Wells, *The Outlines of History,* or James Frazer, *The Golden Bough.* Long after leaving Nicaragua Dick would keep at it. His son Richard remembers his father reading Gibbon and Shakespeare in Washington, years later.

With the field survey basically completed by the summer of 1930, Grace and the children joined Dick for the second year while he worked on preparing the battalion's report. The family arrived at Corinto harbor aboard the army transport *Somme* in August 1930, and Dick went out to the ship to greet them. After unloading Mama's piano and the rest of their belongings, they went by launch to the quay at Corinto. Unfortunately, they missed the train back to Granada and had to spend the night in a hotel. As was customary in the tropical climate, the walls of the rooms did not extend to the ceiling, to allow for the circulation of air. Mrs. Groves was not particularly pleased with her accommodations, especially when she saw a large rat run along the top of the partition.

The next day they took the train to Granada. The car was dirty and crowded, and the air inside stifling. The woven-cane seats were hard and uncomfortable; there was nothing to eat. The hot and sweaty newcomers directed their complaints at the head of the family, who sat silently with no good response. Finally by late afternoon they reached Granada, where Dick had rented half of a house that covered an entire city block. Its walls were made of whitewashed adobe and its roof of orange-red tile. There were no windows, but the rooms were spacious and airy. Everything was arranged around two large interior courtyards or patios, one for the family and one for the servants. Dick had hired four to help run the household — Carmen, the maid and overseer, Manuela, the cook, Modesta, the laundress, and Valentina (or Tina), the nursemaid — all for twenty dollars a month. Only Tina lived in the house and took care of two-year-old Gwen, who soon started to talk — in Spanish.

The day after arriving Groves took his son to a private school, known as El Colégio Particulár de Varones. There they met Don Salvador de Barbereno, the principal who ran the school, and young Richard was placed in a classroom with about a dozen Nicaraguan boys. Since English was not spoken, he quickly became fluent in Spanish. As it turned out, the school was an excellent one, and its course of instruction rigorous. It was arranged in four grades, though not comparable to the American system. Completing all four was somewhat equivalent to an American high school education. Because of the initial language problem, seven-year-old Richard was placed in the lowest grade, a situation that bothered his father. To regain what he thought was the proper place for his son, Groves hired a tutor and supplemented his formal schooling by assigning readings and drilling him in arithmetic.[39] Education was all-important. No effort must be spared to take advantage of every opportunity.

There was a lively social life for the engineers in Granada. Families with teenage daughters such as Majors Gross or Sultan were ever alert to the overtures of Nicaraguan Romeos. The goings-on of some of the young American lieutenants set tongues wagging in the small community. Still, there was time for tennis, and swimming at a fine beach on Lake Nicaragua.

Shortly before the engineer battalion was scheduled to leave the country, a devastating earthquake struck Managua, the capital and a city of sixty thousand inhabitants, at 10:19 A.M. on Tuesday, March 31, 1931.[40] At Granada, twenty-six miles to the southeast, the tremors were felt, but there was no damage. When the engineers learned how severe the quake had been in the capital, six officers, including Groves, and twenty-eight enlisted men were dispatched to join with American marines to assist, arriving by train in the city at 3:30 P.M.

The engineer troops took over the job of fighting the fires that had engulfed a fifteen-block area in the heart of the city. Their battle was a perilous one. There was little water because the mains had ruptured, and a continuous wind, at times quite brisk, caused the fire to spread. Recurring tremors brought down already weakened buildings, but by Thursday morning the engineers had confined the fire, though it continued to burn and smolder for more than a week. The earthquake and the fire caused the loss of about one thousand lives and fifteen million dollars in property damage.

With the main danger past a group of engineers, led by Groves, turned to restoring the water supply to the city. For their efforts, the American ambassador, Matthew E. Hanna, singled out Sultan and Groves for special commendation in a telegram to the secretary of state on April 15. The Nicaraguan government was appreciative as well. Groves and two other officers were awarded the Nicaraguan Presidential Medal of Merit.[41] President Moncada's citation (in Spanish) describes what Groves did, and in the last sentence identifies some of the qualities that we shall see at work again.

> Lieutenant L. R. Groves operating under the orders of Lieutenant Colonel Dan I. Sultan, Corps of Engineers, rendered effective and almost continuous service in directing efforts to control the conflagration during several days following the Managua earthquake of March 31, 1931. Subsequently Lieutenant Groves was placed in direct charge of the hazardous and exceptionally difficult task of the Managua water system, located within and at the base of a precipitous volcanic crater. This task he successfully completed in the face of recurrent landslides, which constituted a continuous menace to life. Lieutenant Groves, by his technical skill, cheerful and courageous leadership, and unfailing energy contributed outstanding service in the work of relief following the earthquake.[42]

On July 16, 1931, the Groves family sailed from Corinto on the *Chateau Thierry,* sister ship of the *Somme.* At the entrance to the Panama Canal, Colonel Schley, now serving as lieutenant governor of the Canal Zone, came aboard and took Lieutenant and Mrs. Groves for a day of sight-seeing and shopping. They purchased a set of Wedgwood china and that evening went to a nightclub with their old friend from Galveston. The voyage resumed, and after several days of heavy seas and foul weather the ship arrived in New York Harbor on the morning of July 25, docking in Brooklyn. By taxi the family went to the Hotel McAlpine just in time for lunch. The next day Groves helped get his soldiers on their way back to Fort DuPont and got the Wedgwood china through customs, leaving time for a little sight-seeing. Dick and his son went to the top of the newly opened Empire State Building, where Richard practiced spitting over the side when no one was looking. The following day the family separated. Mrs. Groves and the children took the train to Greenfield, Massachusetts, to spend the month of August on a farm in Buckland, with old friends Gracie and Nellie Williams. Dick returned to Fort DuPont, and from there went to Washington to report to the Office of the Chief of Engineers and to prepare to receive his family.

PART THREE

Getting on the Fast Track

• 1931–1942 •

CHAPTER SEVEN

Learning the Ropes

• July 1931–June 1935 •

Groves's appointment to the Office of the Chief of Engineers was no doubt made in recognition of his hard work and solid accomplishments in the preceding eleven years. He was marked for advancement, and it was time for him to gain experience in the functioning of the War Department and the official Washington. In the next four years he worked and associated with men who would have a profound and lasting effect on the rest of his career.

At the end of the summer Dick traveled to the Buckland farm in a shiny new 1931 red and black Chevrolet roadster to retrieve his family, returning them to a capital caught in the grips of the Great Depression. The situation in the army was equally dire.[1] The Corps of Engineers was marginally better off. As the secretary of war reported for fiscal year 1932, "The Corps of Engineers, in the course of its nonmilitary activities, has engaged in the greatest program of productive public works in its history, with resultant stimulation of employment and industry."[2]

The Corps of Engineers in the 1930s was a small part of a small army. The number of commissioned officers in the entire army averaged about 12,000 in the interwar period, with a total strength of 135,000.[3] The portion of officers that were in the Corps of Engineers was in the 575 to 675 range, about 5 percent. In the fall of 1931, when Dick arrived, the corps had one major general, one brigadier general, twenty-two colonels, thirty lieutenant colonels, ninety-eight majors, 107 captains, 206 first lieutenants (of which Groves was one), and 110 second lieutenants.[4]

The chief of engineers from 1929 to 1933 was Maj. Gen. Lytle Brown. His office was then located in the Munitions Building, a temporary structure built during World War I on the Mall between Nineteenth and Twenty-first Streets. The office Brown ran was close knit, hardworking,

and highly motivated with probably only about eighteen to twenty officers assigned.[5] A few sentences from an address Brown presented to the
annual dinner of the American Engineering Council suggest what was
expected of the engineer officer:

> No man who has large responsibilities is free from executive duties
> no matter whether he be engineer or builder. If he directs others, he
> performs executive duties. To do these duties well, he must have nat-
> ural qualities beyond the ordinary. The knowledge of what these
> qualities are may tend to call them out of slumber and summon the
> will to strive for their development.
>
> The executive qualities are the same as those of a commander.
> They are such as make possible rapid and accurate decisions, and con-
> stancy in abiding by them. These qualities are quick intelligence,
> sound judgment, energy, a strong will, and far, far above all else, a
> great moral courage. These qualities are greatly aided by a broad gen-
> eral knowledge of men and affairs.[6]

Dick had found an apartment for his family in Cathedral Mansions,
Central, at 3000 Connecticut Avenue directly across from the entrance to
the National Zoo.[7] The two-bedroom apartment was directly over
Napoleon's Bakery. The schooling of his children was always a first priority. Gwen was only three and not old enough to go to school. Richard,
after attending schools in Delaware, Seattle, and Granada, entered grade 3B
in the James F. Oyster Elementary School, six blocks away on Calvert
Street. Throughout their schooling, from kindergarten to postgraduate,
their father looked over their shoulders and prodded, coaxed, bribed, consulted, cajoled, and insulted them to do better. He also, from time to time,
made known to school authorities and teachers what kind of education
he thought proper for his children. Taking a page from his own education,
he negotiated with the Oyster School to have his son enter grade 3B, skipping grade 3A.

Grace kept up her musical interests. She sang for pay with several choirs
and at a local radio station, earning enough to provide the family with
things that otherwise would not have been affordable. With her earnings
supplementing Dick's meager army salary they could have a maid: Martha
in the Cathedral Mansions, then Inez at Calvert Street. The maid took

Gwen to the zoo during the day and baby-sat in the evenings when Lieutenant and Mrs. Groves went out. The pay was seven dollars, for a workweek of five full days and half days on Thursdays and Sundays. If they were out late, Groves would have to drive Martha home to Southwest, near the War College, as her mother was afraid to have her go by cab.[8] At that time Washington was mostly white, with the "colored" sections in Southwest and on U Street in Northwest.

During this period Dick and Grace became close friends with Edmund and Verna Gregory. Lieutenant Colonel Gregory worked in the quartermaster general's office; later, in 1933, he would go to Governors Island, New York. It was First Lieutenant Gregory who had met young Dick at Fort Harrison more than twenty years earlier and steered him toward West Point. Though Gregory was fourteen years older than Dick, the two became lifelong friends. The wives too became close, with Verna initially bringing Grace into the French club during this first Washington stay. Eventually the two families would host Thanksgiving and Christmas dinners for each other.

Groves's first assignment on reporting to the chief's office was to put Lieutenant Colonel Sultan's report on the Nicaragua survey into final form and to prepare his report for the printer. In these tasks he worked closely with Col. Ernest "Pot" Graves, for several years resident member of the Mississippi Commission in the Office of the Chief of Engineers. By example and counsel Graves was to have a greater influence on Groves than any other mentor.[9]

Pot Graves was a legend in his own time: captain of the West Point football team; second in the General Order of Merit in the class of 1905; twice head football coach at West Point; and at the behest of Pres. Teddy Roosevelt detailed as line coach under Percy Haughton at Harvard.[10]

He served in Cuba; supervised construction of the U.S. fortifications on Corregidor in the Philippines; was district engineer in Vicksburg; built a 250-mile road into Mexico during the Punitive Expedition against Pancho Villa in 1916; went with Pershing to France and was in charge of all railway, road, depot, camp, and airfield construction in the Intermediate Section between the coast and the combat zone.[11]

Retired for deafness in March 1921, Graves returned to active duty in April 1927 at the request of the chief of engineers, Maj. Gen. Edgar Jadwin.[12] Graves became Jadwin's right-hand man in preparing a new flood-control plan for the Mississippi River and obtaining passage of the Flood Control Act of 1928 following the disastrous 1927 flood. The act gave the corps a central role in flood control. Pot was adept at working with Congress, getting members to support corps projects. He was someone who knew how Washington worked. By the time Lieutenant Groves began working with him in 1931, Graves was, in Grace's words, "the famous near-deity of the Corps."[13]

The association between Graves and Groves extended well beyond the normal relationship of a superior and subordinate. Almost every day, rain or shine, Dick would leave his apartment,[14] walk down Connecticut Avenue, cross the Taft Bridge, and stop at the front door of the Graveses' apartment building at 1835 Phelps Place, Northwest, to pick up the colonel. From there they would make their way down Twenty-first Street to their offices in the Munitions Building next to the Reflecting Pool on the Mall. These daily walks to and from work gave them time to talk. For Dick, Graves was a figure of great eminence, someone to learn from, by his example and by his words. Their conversations must have been a bit strange to behold. Because he was quite deaf, one had to shout at the colonel and repeat oneself constantly. Groves's deep voice usually got through, and Graves's responses were normally brief and laconic.

One of the best sources for understanding Pot Graves's approach to life — with some advice about football as well — is captured in a little book he wrote, *The Line Man's Bible: A Football Textbook of Detailed Instruction*.[15] Football can teach a great deal and be an opportunity to learn many of life's lessons. *The Line Man's Bible* is filled with advice that helps produce winning teams, but has wider applicability off the gridiron as well.

> . . . the game of football places squarely before the young American the problem of dealing with human nature, of handling men, of overcoming the keenest competition, both mental and physical. In football he has to bring into play and justify the mental training so necessary to success in all vocations.
>
> In addition this game teaches a moral. It teaches a player to stay within the law. No matter how desperate the physical and mental competition, a player must exercise self-control, and he soon learns that infractions of rules work to his disadvantage and sometimes even to his disgrace. In the game as now regulated, he grows to respect the rules and the judge.

Graves's advice was to focus on the fundamentals and keep things simple. "Napoleon's maxim of war, that simplicity is a prime requisite, applies throughout the game of football." The number of plays a team uses in a season should be limited to about fifteen; the players must be in top physical and mental condition, so the coach can "weld them into a machine that will work like a clock."

The line is the backbone of the team. "The best line wins the game." All young men are about equal in the inherent courage they possess. If some do not have it on the field, they have not been properly coached. "The feeling of domination over the enemy and the seeing of his opponents doubtful, uncertain, ducking and bewildered is sufficient reward for any line man."

> . . . perfecting line play is essentially mental. Study yourself in detail on and off the field and solve your own problem. The coaches can help but they cannot do it.

Reading the *Bible* returns us to a time seventy-five years ago when normal-sized young men played a rough–and–tumble game based on mental and physical training and toughness. Graves offered useful techniques, complete with pictures, for effective defensive line play: notably the Thompson Straight Arm Divide, the Bloomer Side Step, the Burr Slice, the Knee Grab, and the Crawfish Under.

> In short, this system develops a man. It does not train an animal. The kind of playing is no get rich quick, get something for nothing scheme that will relieve the men of the hardest kind of work both mental and physical. Nothing but vicious, fierce, desperate play will win from a worthy opponent. The methods outlined are simply guides as how to apply such play.

Much of what Graves passed on to Groves had a fundamental influence in shaping his character. Certain qualities that were respected were essential if one was to be an effective engineer. The core standards by which men were measured, according to Graves, were brains, willpower, and integrity. *Brains,* broadly defined, meant a composite of intelligence and mental discipline, the ability to analyze problems and people, all with the purpose of devising courses of action that could lead to results.[16] Willpower was needed to get the job done. To get the job done effectively you had to delegate, get your decisions down to the lowest levels in a clearly understood manner. Normally, Graves was opposed to written instructions (as we shall see, so was Groves). Orders should be simple and direct enough that they could be conveyed orally. If you were smart enough, you could keep it all in your head, no matter how large the project.[17]

What many admired about Graves was that he quickly got to the heart of the matter, whether it was how to build a road or how to tackle a quarterback. Give people a clear idea of what you expect from them and then give them the freedom to do it. This approach, he felt, made the American soldier the best in the world.

These attributes were infused by a strong sense of personal integrity and ethical behavior. Graves had spent several years after his first retirement in 1921 in private business, but was uncomfortable with the lack of ethics he encountered there. It was one of the factors that led him back into the corps.[18] Engineers dealt with private contractors every day, and there were ample opportunities for what would later be called conflicts of interest. But among this group there was never a hint of scandal or impropriety. One Christmas during this period a contractor sent a nice card table set to Groves. It was immediately sent back.

As political operatives in Washington, members of the Corps of Engineers were many steps ahead of their army brethren in the infantry, cavalry, and artillery branches. Because of their long experience with civil works, engineers were much more savvy when it came to dealing with Congress. The engineers were forever testifying before one committee or another, informing the members of past performance and future plans. They also met with the members individually, to discuss projects that might be needed in their districts or states, and how much money would be required to accomplish them. Every district engineer knew his congressmen and senators and they knew him, and between them they put *pork barrel* into our political vocabulary.[19] Pot Graves was quite adept on Capitol Hill, always looking out for the best interests of the corps. He was particularly influential in drafting legislation for committees, whose small staffs appreciated the assistance.

Lieutenant Groves was not Graves's only protégé.[20] His down-to-earth, practical approach to engineering and to administrative problems inspired and influenced, among others, Brehon B. Somervell, whom we shall soon meet as Groves's boss and principal advocate. Others in the circle whom Groves met, worked with, or came to know were John C. H. Lee, Thomas B. Larkin, Philip B. Fleming, Raymond A. Wheeler, Glen E. Edgerton, Max C. Tyler, and John J. Kingman.[21] This group was slightly older than

Groves, but part of a senior officer's professional responsibility was to sponsor and promote the careers of promising junior officers, for the betterment of the entire army.[22]

The Corps of Engineers benefited enormously from many of Roosevelt's Works Progress Administration programs. Several in this circle oversaw some very large projects during the 1930s. When it came time for the war mobilization, a cadre of experienced engineers was ready to take it on. Somervell, Fleming, Lee, and Larkin are good examples. Throughout the 1930s the slightly younger Groves came to embody the lessons and example of his mentors and colleagues. When his time came he was ready.[23]

The basic culture and ethos of the corps was well in place long before Pot Graves, of course. The Corps of Engineers have always been somewhat separate from the rest of the army, semi-independent, never really integrated with the other branches. The engineers perceived themselves, and were perceived by others, as special and apart, the elite branch with the best and the brightest West Point had to offer. Even the buttons on their uniforms were different from those of other officers, inscribed with the motto of the corps — ESSAYONS — proclaiming their separateness and informing the other branches of an elite within an elite.[24]

In general officers of the Corps of Engineers, in Washington or in the field, were not as isolated from the civilian society as were those of the other army branches or services. Much of the time engineer officers dressed in civilian clothes and lived in residential neighborhoods. Most noncorps officers lived at army posts, wore uniforms, and did not interact very much with the civilian community.

Their evolution within the army in part explains the derivation of the ethos; Pot Graves only fostered and amplified it. After it was instilled into an engineer, by whatever means, it resulted in confidence, strength, and a high degree of professionalism, "natural qualities beyond the ordinary," to use Lytle Brown's special phrase. At the same time it also bred arrogance, egotism, and condescension toward others. Those on the receiving end of these latter traits might be resentful, and some occasionally were. This group of engineers of the interwar period displayed all of the positive and negative attributes, to one degree or another, with Groves perhaps the prime example.

After completing the Nicaragua survey report, Groves was assigned to the Supply Section of the Military Division of OCE, and was the officer in charge of procurement.[25] At the time the major responsibilities of the

Corps of Engineers were civil works, mainly rivers and harbors. The Construction Division of the quartermaster general did almost all domestic military construction. The small amount of military work done by the corps had to do with coastal defense against ships and aircraft. At a time when military aviation was evolving and expanding, planners had to consider the potential role of airplanes in a future conflict. To better defend against attacking aircraft the army needed good searchlights, and in the spring of 1932 Groves became involved in their development.

The Supply Section of the Military Division of OCE was responsible for all research and development that related to its procurement efforts. The chief of the Supply Section was Lt. Col. Francis B. Wilby, who would soon go on to be chief of the Military Division.[26] The development of a metal mirror for army searchlights was given high priority. The best searchlight mirrors of the day were made of a fine quality of glass, ground and polished to a parabolic form and backed by a silver-reflecting surface.[27] During World War I, and for a time afterward, there was only one domestic manufacturer of the glass mirrors, and, given the technical sophistication of the work, the firm produced only one mirror per week. An added difficulty was that the primary source of the optical glass required for such mirrors was in Germany. Metal mirrors, though difficult to manufacture, had qualities that made them far superior to glass ones. They would not shatter if struck by a projectile, could be produced in quantity, were more economical, and, most important, were more effective than glass searchlights.[28]

Beginning in the spring of 1932 Groves's main responsibility was developing the LOCATOR program. The purpose of LOCATOR was to allow searchlights to lock on to and track incoming planes so as to effectively shoot at them.[29] Over the next three years Groves consulted with contractors, engineers, and scientists at various laboratories and factories. For example, he went to New York at the end of July 1932 to inspect Bart Laboratories and confer with Mr. Blasius Bart, a Swiss-born inventor who had developed a special electrolytic process that could be used for plating metal mirrors. Mr. Bart is a good example of a contact made in this period whom Groves would call upon to solve a technical problem during the Manhattan Project.

A dozen years later, in 1944, Groves would need a method for coating the approximately 575 miles of special piping for the K-25 gaseous diffusion plant at Oak Ridge. The corrosive effect of the uranium hexafluoride was the problem. Nickel was the answer, but how to implement it was the question. To make the pipes from solid nickel was impossible; they had to be plated on the inside. Blasius Bart's son, Sieg, now running the laboratory, and his chemist Millard Loucks came up with an original method to electroplate the pipes.[30]

Groves worked every Saturday morning,[31] leaving Saturday afternoons free — often to go to football or baseball games with his son. A high point of this period was attendance at two games of the 1933 World Series between the Washington Senators and the New York Giants. As Richard recounted it:

> We had seats in the bleachers for the first one we saw, and thought they were too far from the action; so, DNO splurged and bought reserved seats for the next game. When we got to our places, we found that one of them was occupied by a pillar supporting the stadium's upper deck. DNO complained vociferously and, when the argument finally subsided, we were put in the Griffith's (the Senators owners) box, right behind first base, to watch the game, along with all the other celebrities.[32]

Army football games, of course, have special pride of place. Efforts were made to see as many as possible, with the game against Navy something akin to a Crusade for any son of West Point. Groves and his son managed to go to three out of four during this period.

Sundays had their own routines. The first Sunday of every month was given over to making calls, a social ritual of army officers for families to pay respects to one another. Lieutenant and Mrs. Groves, with Gwen and Richard in the backseat, would drive to the homes of their friends and colleagues. A calling card would be left if they were not at home or a short visit was paid if they were, with the kids left in the car. Other Sundays were for journeys to nearby Civil War battlefields — Manassas, Antietam, Fredericksburg, Chancellorsville, and Gettysburg.

In the much smaller Washington of the day, army officers and their wives would regularly be invited to the White House to attend a musical performance or a reception. Among Groves's papers are invitations by President and Mrs. Hoover and President and Mrs. Roosevelt to such events. There were also courtesy calls to be paid at 1600 Pennsylvania Avenue. "Mrs. Roosevelt will be glad to receive Lieut and Mrs Groves on Friday afternoon January twelfth [1934] from five to six o'clock" read one invitation delivered to the Calvert Street home.

Within the family General Groves was called DNO (originally spelled — and pronounced — *Deno,* and later shortened to *DNO*) by his wife and children. The derivation of the name is not known with certainty. Two theories by family members are proposed. According to his daughter Gwen it was derived from the diminutive of the Italian word for "father" or "daddy." While the actual Italian words are *babbo* and *babboino,* it became *papadino* or just *DNO.* A competing theory offered by his son Richard suggests that the name comes from a contemporary Italian aviator named Dino Grande, a contemporary of Lindbergh, who was frequently in the newspapers and bore a physical resemblance to Groves.

After some investigation I would like to offer a third theory. I have not been able to find any Italian aviator named Dino Grande. What I did find was the following. In mid-November 1931 the Italian foreign minister, Dino Grandi, visited Washington. Dick and the family had moved to Washington just a few months before. Grandi and his wife's every move was followed during their time in Washington, and later while they visited New York and Philadelphia. Throughout the tour American anti-Fascists, Communists, and Socialists denounced and demonstrated against him as the representative of Mussolini's policies. In Washington Grandi stayed at the Mayflower Hotel, had discussions with President Hoover, Secretary of State Henry L. Stimson, and Secretary of the Treasury Andrew Mellon. Upon arrival in the United States the original plan was to have Col. Charles A. Lindbergh fly Grandi from New York to Washington in a Pan-American seaplane. The flight was canceled due to fog, but Lindbergh paid a visit to the Fascist diplomat on November 19, before a state dinner at the White House that evening. The two-week visit could hardly have had more press coverage. Upon leaving Grandi said he was taking three thousand clippings with him back to Italy. The many pictures that were published in the newspapers show a resemblance to Dick that might have been noted by his wife. Perhaps this is when the family started to use the nickname DNO.

During this period Groves did a great deal of traveling. His travel records show that for the second half of 1933, all of 1934, and the first half of 1935 he traveled several times a month, between Washington and New York, Cleveland, Detroit, Pittsburgh, and Schenectady, and several other cities.[33] Many of the trips were to General Electric Company offices and laboratories. The success of Groves's efforts in this regard can be judged from a letter to the chief of engineers from the manager of General Electric's Federal and Marine Department:

It is our confirmed opinion that the Mobile Type Anti-aircraft Searchlight Unit has been brought to a degree of perfection never before attained, largely through the efforts of Captain Groves. This has been due in no small measure to his efforts in improving the standards of design, his close cooperation with the manufacturer and his insistence upon careful inspection of the equipment during production.

The LOCATOR project involved the always-difficult problem of engaging in research and development work in a manner that assures the best results for the expenditures appropriated. We believe that he was very successful in enlisting the cooperation of those associated with the project in the Army and we know that it was particularly so with respect to the Engineers in our Company.[34]

This was Dick's first intense experience in dealing with contractors, engineers, and scientists. He made many contacts and filed away a list of people whom he would call upon later from General Electric, Westinghouse, Chrysler, General Motors, and many smaller companies. He had worked intensely with corporate America in the solution of a technical problem. It was good training for solving the much larger problems to come.

The hard work was paying off. In March 1935 Groves, recently promoted to captain,[35] learned that he had been selected to attend the Command and General Staff School at Fort Leavenworth, Kansas, another sign that he was being prepared for leadership positions.

CHAPTER EIGHT

Finishing Schools

• *June 1935–July 1939* •

Fort Leavenworth and the Command and General Staff College have a long history and a special role within the army.[1] Fort Leavenworth sits on a high bluff on the west bank overlooking the Missouri River, twenty-five miles north of Kansas City. Col. Henry Leavenworth established the post in 1827. The frontier outpost protected wagon trains heading west along the Santa Fe Trail and later the Oregon Trail and sought to maintain peace with the Plains Indians. Commanding General of the Army William T. Sherman established the first army school there in 1881, called the School of Application for Infantry and Cavalry. After several name changes, Elihu Root's reforms of 1903, and the creation of a general staff, by the 1930s the Command and General Staff School had become an essential conduit to high command and staff positions in the army.

Before beginning his classes at Fort Leavenworth Dick, Grace, and the children used the opportunity for a vacation. Piled into the 1931 Chevrolet roadster, now repainted apple green, with a trunk strapped onto the luggage rack over the rear bumper, the four spent two months touring historic sites, stopping at national parks, visiting relatives, and returning to familiar places.

Road travel sixty-five years ago was far different from today, though not necessarily inferior. Driving slowly on two-lane roads offered the chance to experience the diversity and vastness of America, envisioned as one of the intents, and clearly one of the results, of the family's journey. Travelers of the day stayed in cabins, boardinghouses, and "tourist courts," a forerunner of the motel.

Richard turned twelve that summer. Decades later, details of "The Trip" remained firmly etched in his mind.[2] With a well-orchestrated plan devised by the patriarch of the family, the courageous band headed westward. In the first few days they passed through Maryland's Cumberland

Mountains, on across Ohio to Chicago for a short visit, and some deli-cious meals, with Aunt Bossie and Uncle Duncan. The few days there were spent visiting the Planetarium, the recently held World's Fair, the Field Museum, and the stockyards.

After Chicago there were stops in Ames, Iowa; Kearney, Nebraska; and Cheyenne, Wyoming. On the road the backseat of the car was a perpetual combat zone with Richard and his six-year-old sister Gwen pinching, poking, and generally harassing one another constantly. The parents in the front seat tried to bring order, through admonitions or diversions, such as playing word or geography games or counting white horses in the fields.

In Wyoming the family visited the Shoshone Reservation and the rem-nants of Fort Washakie, where Mrs. Groves had been born. They found Mr. Roberts, the Episcopalian priest who had baptized her forty years ear-lier.[3] Next came Jackson Hole and Yellowstone Park, where they camped in a tent facing a lake. From there it was on to Salt Lake City, then across the Bonneville Salt Flats to Reno, Nevada, and Yosemite Park.

Central to the summer's master plan was a family reunion at Chaplain Groves's Altadena home. Dick's brother Owen, his wife Marion, and their nine-year-old daughter Charlotte came all the way from Garden City, Long Island. There were sharp differences of opinion between the two families on topics ranging from politics to child rearing. Dick and Grace were disap-proving of their niece's bad manners. Little Charlotte never said "please," "thank you," or "you're welcome;" what was worse, she called her parents by their first names. Richard would have encountered a serious session with the belt if he ever uttered such words or failed to be proper and polite.

Marion was the family "Socialist," by virtue of her voting for Norman Thomas and propounding a range of views and opinions that caused Groves to bristle. Owen was the epitome of the university professor, his nose always in a book and his head in the clouds — or at least his brother sometimes viewed him that way.

Chaplain Groves had some extended time with his grandson Richard, to pass on, as grandfathers often do, the kind of advice and knowledge that shapes young lives and lasts a lifetime.

The sights and attractions of Los Angeles, San Diego, Tijuana, and Aguascalientes, Mexico, were sampled and experienced. There was a full day at the San Diego World's Fair. The strange denizens of Ripley's Believe It or Not made a lasting impression on Richard. One day he and his father climbed Mount Wilson, and on another they went "deep-sea fishing" with Owen and Chaplain Groves off the Santa Monica coast.

As the summer finally came to an end, it was time to resume the journey to their ultimate destination, Fort Leavenworth. A southern route was chosen, first through the stifling Mojave Desert with a stop at the not-yet-completed Hoover Dam in Nevada.[4] That night they stopped at a primitive tourist camp in a ramshackle shantytown named Las Vegas. Next was the drive across the Painted Desert to the Northern Rim of the Grand Canyon, where it was a tent again for quarters. The South Rim was reached after a drive through Zion National Park and Bryce Canyon.

Groves took his family to revisit scenes of his youth in Arizona, passing through Flagstaff, Holbrook, and Fort Apache. Then it was north to the final park on the tour, Mesa Verde, and on to Albuquerque, Santa Fe, and Taos. Finally there was a sustained push until Fort Leavenworth's clock tower came into view. The Trip was over.

Upon arriving at Fort Leavenworth they moved into quarters at 109-D Meade Avenue, formerly used as a machine shop. The six-room apartment was spacious, though hardly elegant — with a few leftover pipes and drive shafts — but it did include a study for Dick, as was required for all student officers. The year before, his old tennis partner Maxwell Taylor had occupied the same apartment. As Captain Groves returned to the classroom, Richard enrolled in the post school's 9th grade — having just skipped grade 7A and the entire 8th grade — and Gwen, about to turn seven, started 2nd grade.[5] Grace kept busy with numerous clubs, organizations, and her position as the accompanist of the music group.

In the fall of 1935 there were two classes at the C&GSS, one of 122 officers who had arrived in 1934 and were pursuing their second year of course work, and Groves's class, of equal size, taking the just reinstituted one-year course.[6] Including Groves there were eight officers from the Corps of Engineers, all of whom had ranked near the top of their West Point classes.[7] The other army branches, proportionately larger than the corps, sent more students.[8] The selection process at the time began with the Personnel Division (G-1) of the War Department General Staff, which established class size, quotas from each branch or service, standards of age, rank, years of service, and efficiency ratings. Then the chiefs of the branches and services and the adjutant general's office, in consultation with G-1 and the C&GSS commandant, would make up the final class list.[9]

Dick's year began on September 3, 1935, with opening exercises, the address by the director, and the class picture. The following day classes began in earnest, and, much like West Point, almost every hour was filled.

The pressure was intense; a few dropped out. There were also some nervous breakdowns and even an occasional suicide.[10] Groves felt an additional pressure. For the first time his weight problem became a serious professional cause for concern, and his superiors told him to do something about it, or else. It had been a personal problem long before. In family letters it is remarked that when he went to Hawaii he weighed more than two hundred pounds. He lost twenty pounds, but later gained some of them back. His father also noted, "He tries to go light on the most fattening things but eats so much of the others that fat flows to him like water downhill."[11]

The faculty members who taught the one-year class of 1936 used five basic means of instruction: lectures, map exercises, map maneuvers, map problems, and conferences.[12] Some of the lectures Groves listened to included "Historical Development of Infantry, Cavalry, and Field Artillery"; "Principles of the Defensive and the Offensive"; "Essentials of Command"; and "Development of the War Departments Mobilization Plan." Map exercises were illustrative problems: drills neither graded nor turned in for faculty critique, which instead required students to produce a brief oral or written solution to a tactical problem. Groves's class participated in sixty-two map exercises during the school year, many taking an entire day. Groves's class also had half a dozen map maneuvers — two-sided, tactical problems in which large portions of the class assumed the roles of command and staff officers to play out a competitive scenario controlled by faculty umpires.[13] Groves's class had fifty-four map problems derived from the yearlong war game, written tests completed in a limited time and graded competitively. That year the yearlong map problem was set at Gettysburg, and in Dick's study a map of that area was tacked up high on the walls and circled the room. Finally, there were conferences, or group discussions that reviewed and reinforced lessons learned. They were usually conducted in a Socratic question–and–answer format; Groves's class had 158 sessions over the school year. There were also tactical rides, terrain exercises, and equitation. The course of instruction from September to June totaled 1,340½ hours.

Instruction embodied, among other things: combat orders, field engineering, leadership and psychology, military history, equitation, methods of training, strategy, tactics, planning, and troop leading. The theory of instruction was based upon applicatory learning. The students learned generally by applying military principles to the solution of tactical exercises and map maneuvers. Some of the problems were two-sided war games in which students vied as commanders and staffs of opposing forces. Some of the exercises were mounted terrain exercises where tactical and logistical problems were actually solved on the ground.[14]

To prepare the aspiring officers for their future leadership positions, the students began the year studying brigade-level functions and operations for a period of two and a half months. The second period, three months long, proceeded to the division level, where they became knowledgeable of the range of General Staff positions, from chief of staff to the Personnel (G-1), Intelligence (G-2), Operations and Training (G-3), and Supply (G-4) Divisions. The final four-month period repeated the process of General Staff procedure at the corps level.

> Leavenworth instructional methodology taught a systematic, logical process of solving tactical problems. By presenting students with an increasingly complex series of such problems, to be solved individually and in groups, the course sought to develop powers of judgment, analysis, and decision making.[15]

Classic military writings were required reading; great battles were analyzed. Hannibal's defeat of the Romans at Cannae in the Second Punic War was treated extensively. *Cannae* was also the generic term used by the German strategist Gen. Alfred von Schlieffen to describe any battle of annihilation, such as Austerlitz, Jena, and Sedan.[16] The Leavenworth students used von Schlieffen's book, titled *Cannae,* as a text to study the different battles.[17] The strategies of the American Civil War and of World War I were also studied. There were several conferences analyzing the August 1914 German defeat of the Russian Second Army at the battle of Tannenburg. There was some attention to contemporary geopolitics. On June 2 and 5 the instructor led conferences on Japan's conquest of Manchukuo (Manchuria) in 1931. In March 1936 German troops marched into and occupied the Rhineland, an aggressive move that caught everyone's attention. Concerns about a bellicose Germany in Europe and an aggressive Japan in Asia were leading some to predict that war might be inevitable.

At the end of the year they were ranked, and on June 19, 1936, the one-year class graduated.[18] While it is difficult to assess precisely what Groves took away from his year at C&GSS, we can see some influences that it may have had in his later responsibilities with army construction and the Manhattan Project. Groves was thirty-nine years old and already had an approach to problem solving that was fortified at Leavenworth. What

C&GSS probably did for Groves was to further discipline an already disciplined mind. The students were taught to be decisive, to finish an action,
and to do nothing piecemeal. Leavenworth taught and sought uniformity
and regularity, not innovation and originality. In this regard an engineer's
experience would be quite similar to those in the infantry or cavalry.
C&GSS made its critical contribution by giving

> the Army a common body of staff procedure, a common tactical lan-
> guage, and an accepted methodology for solving tactical problems. A
> graduate of the Command and General Staff School could function
> effectively on any Army staff, anywhere in the world — a priceless
> element of commonality in a largely citizen-soldier war Army.[19]

While no doubt there was much to be learned that some found valuable in their later careers, the army schools functioned, then and now, as
way stations to higher command and staff positions.[20] The army tracks its
future leadership from West Point on, keeping tabs on positive efficiency
ratings, which leads to the C&GSS, to the War College, and to the General
Staff. The professional schools also functioned in the socialization of the
officer corps, allowing an ever-more elite cohort to come to know one
another better.

For the greater number of officers who became generals during the
war, this staff schooling was indeed the single most important part of their
preparation for command. Studies of World War II leaders demonstrate
that Leavenworth was a common experience for virtually all of the corps
and division commanders in both theaters.[21]

In late March Groves learned what his next assignment would be after
graduation in June. The initial notification in February ordered him to
report for duty to be district engineer in Providence, Rhode Island. With
Richard about to attend Deerfield Academy, that suited everyone just fine
— especially Grace, who longed for the richer cultural life and less oppressive climate back east. Disappointment soon followed, however. New
orders directed that Groves was to go to the Missouri River Division in
Kansas City as assistant to the division engineer.[22] As Dick and the family
left Leavenworth in late June 1936 to take up rivers and harbors duties, the
German war machine was gaining momentum, having recently disregarded
the Treaty of Versailles and the Locarno Treaty of 1925 by marching into the
Rhineland. The world was beginning to take serious notice.

In Kansas City Dick found a spacious and comfortable house to rent at 507 East Forty-seventh Street, not far from Country Club Plaza. The plaza was the first shopping complex of its kind in America, according to local authorities.[23] The house, directly across the street from the William Rockhill Nelson Art Gallery, was conveniently located near the Rockhill Tennis Club, which Dick joined immediately upon arrival. Richard spent that brutally hot summer practicing and improving his game. Often he played with his father, and occasionally teamed with him against some older mixed-doubles opponents. Other days that summer were spent learning to type, having his teeth straightened, and preparing a proper prep school wardrobe. As the start of school approached, it was decided that Grace would drive Richard to western Massachusetts in the new Chevrolet. In mid-September they set out, traveling via Chicago and staying with the Hineses for several days. Then it was on to Detroit to catch the Great Lakes steamer to cross Lake Erie to Buffalo. After another day and a half they arrived at Deerfield Academy.

The family brought with them a young black girl, Edith, the last of their maids at Leavenworth. While at Leavenworth Edith had attended social functions with members of the Tenth Cavalry, one of the army's four "colored" regiments at that time. Apparently more than dancing took place, since shortly after arriving in Kansas City, Edith announced that she was "a little bit pregnant." It fell to Mrs. Groves to drive her to the clinic for checkups and, when the time came, to take her to the hospital, where she delivered twins.

Groves's office was in the heart of downtown at the Postal Telegraph Building at Eighth and Delaware Avenue.[24] The two main concerns of the Missouri River Division at the time were maintenance of navigation along a portion of the Missouri River from Sioux City, Iowa, to St. Louis, and the construction of the Fort Peck Dam near Wheeler, Montana. Because Groves had been away from troops longer than allowed by regulation, he was given an additional duty for several months to instruct and inspect reserve units in the Seventh Corps Area throughout Kansas and Missouri. The division engineer at the time was Col. Richard C. Moore, who had intervened to have Groves come to Kansas City instead of Providence and would play an important role in Groves's career three years later.[25]

The Corps of Engineers ran many WPA projects. It designed and supervised construction, hired and fired vast labor forces, and administered millions of dollars. Such projects as the Quoddy, Dennison, Bonneville, and Fort Peck Dams provided invaluable experience for officers who would soon have enormous administrative and organizational responsibilities during World War II.

The Fort Peck Dam project was authorized in October 1933; work began ten days later.[26] The Missouri River was closed off on June 24, 1937, but the dam was not totally finished until 1940.[27] As the largest hydraulically filled dam in the world, it is a monumental engineering feat, testifying to the American spirit of accomplishment against adversity. In damming up the Missouri River, Fort Peck Lake was created, 130 miles long, 16 miles wide at its widest, and 220 feet deep at its deepest. Built in the depths of the depression to provide jobs, flood control, irrigation, and navigation, the dam remains a symbol of New Deal determinism and the Corps of Engineers' attitude that nature could be tamed to its will.[28]

Boomtowns sprang up for the fifteen thousand Fort Peck workers, hired by the corps and paid in silver dollars. The most famous was Wheeler (named after Montana's senior senator, Burton K. Wheeler, a Democrat and backer of New Deal programs), a cross between Wild West Dodge City and Klondike gold rush Dawson, with jobs as the lure. Barkeepers, gamblers, bootleggers, and prostitutes inevitably followed in their train. The first issue of *Life* magazine, dated November 23, 1936, had a photograph on the cover titled *Fort Peck Dam* taken by Margaret Bourke-White. The first article in the new magazine had more than a dozen additional photos by Bourke-White showing "at close range the labors and diversions" of the inhabitants at "Mr. Roosevelt's new wild west."[29]

The Fort Peck project had Groves working with District Engineer Thomas Larkin, one of Pot Graves's protégés, and later Larkin's successor, Maj. Clark Kittrell, a friend of Groves's from West Point and Hawaii.[30] Several West Point classmates worked at Fort Peck, among them his close friend Claude "Chorp" Chorpening. Chorpening would later have this to say about Groves:

> I could see tremendous ability and drive, as well as a keen intellect. I felt that if any member of my class would achieve great fame — notoriety — he would probably be the one. When he decided to get something done, his will was like steel. And he had no patience whatsoever with inefficiency or lack of drive on the part of his subordinates. Groves became well known in the Engineers as a very capable officer. During WWII, Leslie Groves became involved in the devel-

opment and testing of the atomic bomb. *Groves was given as much power in that position as any officer ever has had.*"[31]

Another young engineer at the time had an opportunity to observe Groves and later made some telling comments about his character. William Whipple had several relatives who had gone to West Point. He graduated third in the class of 1930, and, after studying as a Rhodes scholar at Oxford, served with the corps.[32] While at the Omaha district office he had contact with Groves, whom he remembered as "large, tough and very intelligent"; although "only a captain, no one took this man lightly. I was careful to have no trouble whatsoever with him. When you looked at Captain Groves, a little alarm bell rang 'Caution' in your brain."[33] He did not resort "to the usual ploys and swaggering to magnify his own importance. He gave the impression of a man of great latent power, who was biding his time. He was not rude; but neither did he go out of his way to be friendly. He was obviously highly intelligent. His subsequent career did not astonish me.[34]

In late February 1938 Dick learned that he had been selected to attend the Army War College, to begin in September. At the end of August 1938 the family arrived back in Washington. After a few days in a hotel, Dick rented a house at 3132 Oliver Street, Northwest, in the Chevy Chase section of the District of Columbia, and the family moved in.

The secretary of war and the assistant secretary reported that plans were being laid for wartime procurement, for the mobilization of industry in the event of a national emergency. But for now, Groves and his classmates would be relieved of such immediate concerns.

The War College is the final step in the education of those who are expected to hold the highest positions in the army.[35] Like West Point and Fort Leavenworth, the War College has its own lineage and traditions; to graduate from it puts one among the select few of whom substantial things are expected.

Under the reforms of Elihu Root the Army War College was established under General Order Number 155 of November 27, 1901, as an element of the General Staff.[36] In December 1902 the Army War College moved into a brownstone at 22 Jackson Place, Northwest, facing Lafayette Square, where academic work began in 1904. Of the twenty officers among that first group of students were Maj. George W. Goethals and Capt. John J. Pershing.

In May 1903 Root approved plans for a magnificent building at the tip of what was then called Fort Humphreys or Washington Barracks, and what is today Fort Leslie J. McNair. The chief of engineers had engaged McKim, Mead and White, the renowned architectural firm, to design it; the War College moved in June 1907.

The requirements to get into the War College changed throughout the 1930s, but by Groves's time a candidate had to have above average Efficiency Reports and to have graduated from the C&GSS. Similar to the C&GSS process, the class of approximately ninety Regular Army officers was selected by the chiefs of the arms and services branches to proportionally represent the different parts of the army. There was also a 1928 policy stating that at least half of the War College selectees be available upon graduation for assignment to the War Department General Staff. And a 1929 decree (known as the "Manchu law") mandated that no officer could serve more than four consecutive years in the District of Columbia. The purpose of this was to inhibit political connections and prevent War Department cliques in the German mode. The student must also be listed on the General Staff eligibility list.

Captain Groves, at age forty-two, was one of the eighty-six Regular Army officers selected for the Army War College course.[37] As at C&GSS, the other branches — proportionately larger than the corps — sent more students. Infantry and field artillery dominated with more than half the students, while the cavalry, coast artillery corps, and air corps sent another 30 percent of the total. Groves was one of five engineers — two lieutenant colonels, a major, and two captains.[38] Groves already knew some of his classmates and would get to know others whom he would soon work with again.[39]

The course work was comprehensive and intensive. Classes were held from 8:45 A.M. to noon, and from 1:30 to 4:15 P.M., Monday through Saturday, with Wednesday and Saturday afternoons excepted. The divisions of the War Department General Staff were covered (the so-called G courses), with about a month for each from September to January. Then came analytical studies, in which the strategy, tactics, and logistics of the past conduct of war were studied for lessons for the future. The preparation of staff memoranda on subjects assigned by the faculty was emphasized in the G courses and analytical studies. Next on the schedule was the command course; here, field operations of the army, group of armies, theater of operations, and general headquarters were studied and analyzed. The preparation for war course examined the organization and functions of the War Plans Division of the General Staff; at the end of the school year they studied a command post exercise. Included within the courses were twice-weekly discussions of current world affairs.

The method of instruction was different from that at the Command and General Staff School. The applicatory method and participation in collective problem-solving exercises were used.

The role of the faculty was to devise and pose the problem and then to monitor and guide only as necessary the committees' research, analysis, and presentation of solutions. Lectures, whether presented by faculty members or by authoritative guests, were introductory or supplementary to the problem solving.[40]

The so-called committee method broke down branch parochialism and blurred the distinction between faculty and student. Cooperation in getting a task done was encouraged over competitive individualism, though this was not always achieved. Competition to get into the War College was intense, and it remained so among the more competitive students.

It is on this point that the War College philosophy deviated most markedly from both lower-level military education and from civilian graduate- and undergraduate-level education. It deviated for sound reasons. These officer students were expected to provide the leadership of the army in those most serious of endeavors, the preparation for and conduct of war. They were not necessarily expected to think alike, but their thinking was expected to be guided by mission rather than ambition.[41] Clear exposition in both oral and written presentation were special goals.

One of the lecturers in 1939 was Douglas Southall Freeman, who had won the 1935 Pulitzer Prize in biography for his four-volume work on Robert E. Lee. From January 1936 to February 1940 Freeman gave a series of nine lectures to the students of the Army War College on Lee's leadership, morale in the Army of Northern Virginia, and the objectives of the Union and Confederate armies. The winter lectures focused on Lee as a leader, and Groves was in attendance to hear Freeman on February 2, 1939.[42] His presentation made a lasting impression on Groves, who several years later wrote him a letter:

> I have never forgotten your giving your opinion at the Army War College in the Spring [sic] of 1939, as to why Lee selected Ewell to succeed Jackson. You then branched out on Lee's selection of his lieutenants in general. This discussion had a profound influence on my selection of subordinates to the most important positions in the

atomic bomb development. It encouraged me to select men much younger in years and with much less experience and reputation instead of deferring to normal practices. One of my two most important and successful selections was made against the strongest advice of all my most senior and valued advisors.[43]

The week before Freeman's lecture — on Thursday, January 26, to be exact — as Groves went to class an event was taking place across town at George Washington University that would eventually change his life, and everyone else's as well. At a physics conference Niels Bohr, just recently arrived from Copenhagen, and Enrico Fermi reported to their colleagues about the recent experiments of Otto Hahn, Fritz Strassemann, Otto Frisch, and Lise Meitner, and their discoveries of a strange new phenomenon called fission.[44]

In mid-April, as the academic year was nearing an end, Groves was chosen to serve on the War Department General Staff. The commandant of the Army War College, Maj. Gen. John L. DeWitt, wrote Groves's Efficiency Report for the period September 6, 1938, to June 30, 1939, and noted his academic rating:

Cheerful, even disposition, sense of humor, serious. Good team worker. Works rapidly, thoroughly, accurately, methodically. Open minded, appreciates views of others. Takes a definite stand, determined, holds to his convictions, sure of himself. Broad minded, views problems from all angles. Original independent thinker, produces practical ideas, active imagination. Prompt with good judgment, exhibits firmness.... Proficient in theoretical training for: High command and WDGS [War Department General Staff] duty. Graduate full course Army War College. Academic rating: Excellent. General personal rating: Excellent. Recommended for general staff duty with: Div, Corps, Army, GHQ or Corps Area. Recommended for duty with G-1, G-2, G-3, G-4 or WPD [War Plans Division] WDGS. Suited for civilian contacts and duty with civilian components. No special aptitude shown.[45]

CHAPTER NINE

Final Rehearsal

• July 1939–Summer 1942 •

After the conclusion of his academic year at the Army War College, Captain Groves reported for duty on July 1, 1939, to the War Department General Staff as a General Staff officer in G-3 Division, Mobilization Branch. Appointment to the General Staff was considered a significant vote of confidence. The General Staff was limited by law to a very small number of officers.[1] Promotions for Groves had been slow; he had spent fourteen years as a first lieutenant and five years as a captain, but was now well positioned to advance his career. The next three years would put him in a position that no one could have imagined. But before that happened, he would spend these three years meeting and working with numbers of military officers and civilians who would play crucial roles in the Manhattan Project.

Before taking up his duties for the General Staff, Groves was sent on a mission to Nicaragua. In the fiscal year 1940 appropriations bill Congress had ordered that a survey be made for a barge canal and highway across Nicaragua. Given his past experience, the chief of engineers requested that Groves be part of the survey team. On August 9, 1939, Groves, along with three other officers — Col. Charles P. Gross, Col. Dr. Paul R. Hawley, and Capt. Thomas H. Stanley — who had served in the 1929–1931 survey, and two civilian engineers, departed for Panama aboard one of the ships of the Grace Line, with a brief stop in Port-au-Prince, Haiti.

In Panama, in order to get an aerial view of the canal, Groves took his first airplane flight. Later he recalled,

> In order to get a good view, I sat in the bomber's plexiglass compartment at the front of the plane. No one gave me any instructions on how to put on my parachute and, as I recall, the passageway was too narrow for anyone who wore a chute to get down into the nose.

The flight only took about fifteen minutes, and I spent most of that time in trying to figure out how to put on my parachute. . . . What good the chute would have done me I never could figure out unless I had gone through the bomb exit release.[2]

A day or two later another flight took them to Nicaragua, where they set up headquarters in a Managua hotel. The fact was that the trip was really unnecessary. The earlier survey, combined with additional hydrological and seismological data gathered throughout the 1930s, would have been enough to write the report without having left the United States. But it was felt that it would be more diplomatic to go to Nicaragua and be seen conducting the survey. With little real work to do, Groves made day trips to various sections of the country that he had not been to before, and concentrated on losing weight. Daily walks of six to eight miles plus a strict diet under Dr. Hawley's supervision resulted in a loss of about twenty-five pounds during his three-month stay.

Groves returned to Washington on November 6, 1939, to learn that his father was dying. He was granted leave and traveled to Pasadena in time to be with him for a short time before he died on November 21. For the past ten years his younger sister Gwen had been living with, and caring for, Chaplain Groves. Owen came west as well, and the three children arranged the funeral and sorted through their father's papers, saving those of interest, such as the letters to the family from Cuba, the Philippines, and China.

Six years younger than Dick, Gwen, like the rest of the family, had moved from post to post until finally settling in Pasadena with her step-mother, Aunt Jane, in 1916. Gwen attended Pasadena High School and then went to Mills College in Oakland, the fourth child to receive a higher education. After college, she returned to Altadena and lived at home. Following Jane's death of pneumonia following an operation on March 15, 1929, Gwen took care of her father until his death ten years later.

The back problems that Gwen had suffered as an infant would plague her throughout her life. Much of the time she wore a brace, which tortured her. Twice in the 1930s she underwent surgery on her spine to try to correct the problem. It was thought that her tuberculosis of the spine was caused by her being dropped at a young age.

In a liberating move after her father's death, Gwen, at age thirty-seven, finally got married, to the son of a navy lieutenant commander and a Stanford graduate who lived a few doors down the street at 993 New York

Avenue. Spencer Dodge Brown and Gwen Griffith Groves were wed on March 28, 1940.[3] After a honeymoon in Hawaii, they moved into the Brown family home, where Spencer's invalid mother resided. In a time when it was normal for a son or daughter to take care of an ailing parent, Spencer, like Gwen, had been left to care for his mother. Mrs. Brown lived with them until she died in the 1950s. Gwen and Spencer continued to live in the house until she died of a brain tumor on March 27, 1964.

After the funeral on November 24, and helping to settle his father's affairs, Groves returned to Washington and took up his duties with the Operations and Training (G–3) Division of the War Department General Staff.[4] He walked to and from his office in the historic Winder Building, a pre-Civil-War-vintage brick building on Seventeenth Street, across from the Department of State building (formerly the State, War and Navy Building and now known as the Old Executive Office Building). While he was in Nicaragua war had broken out in Europe with Germany's attack on Poland on September 1, triggering British and French entry two days later.

Throughout the 1930s American opinion varied over what to do about Hitler. One popular response was isolationism, a deep-seated, almost instinctive approach by many Americans to issues in Europe or elsewhere. These views found expression in various pieces of neutrality legislation, which sought to insulate America from the world. Others believed that rather than limiting our influence in Europe's affairs after World War I, we should have been more involved. Hitler's rise and vengeful expansionist policies had their roots in unresolved issues from World War I and the Peace Treaty of Versailles. America's refusal to join the League of Nations further contributed to the conditions that now had brought about another war.

There was near unanimity among the American public, however, following the outbreak of hostilities in Europe: Germany was to blame, Britain and France should win, and the United States should stay out.[5] American armed forces were woefully unprepared, with total military personnel numbering some 335,000, including 190,000 in the army, 125,000 in the navy, and 19,000 in the marines. The army total included about twenty-two thousand in the fledgling air corps.[6] Most of the equipment was World War I vintage; navy ships and air corps aircraft were few. Initially President Roosevelt's approach was to urge Congress to overturn the neutrality laws so that the United States could help Britain and France defeat Germany. FDR also foresaw that eventually the United States might have to enter the war, but only after adequate preparation.[7]

At this time the responsibility for military construction still lay with the quartermaster general. The Army Corps of Engineers customarily took over construction during wartime, but only within theaters of operation abroad to support combat units.[8] For years there had been intense political infighting within the army over this less-than-satisfactory organizational scheme.[9] The engineers, other factions within the army and War Department, and certain members of Congress wanted army construction shifted to the corps. To them the quartermaster corps was a supply organization, pure and simple, whose primary mission was to clothe, equip, and feed the army. Its methods were slow, and it was overcentralized.[10] Logically, the engineers believed, military construction, even in peacetime and especially in the continental United States, should be done by the corps. The quartermasters had their own supporters, who wanted construction left where it was. A third group argued that a separate civilian-run construction division be created, out of the hands of both, as had been done during World War I. After World War I, a compromise of sorts had been reached and construction returned to the quartermaster corps, but dissatisfaction persisted and the issue remained contentious.

Throughout the 1920s and 1930s the corps went about building civilian projects, large and small, from the Bonneville and Fort Peck Dams to levees, dikes, jetties, locks, reservoirs, and channel dredging. But now war had broken out in Europe, the United States was mobilizing, and the engineers were in a double predicament. Their civil budget was shrinking, and the quartermasters were getting the military work. The future of the corps was at stake, and it was time to do something about the persistent problem. Groves would soon find himself in the middle of the fray.

After the initial German invasion to the east in September 1939, the Nazis did not immediately strike to the west, as had been feared. The period from the fall of 1939 to the spring of 1940 is known as the "phony war." This came to an end with German attacks on Denmark and Norway in April; Belgium, the Netherlands, and Luxembourg in May; and France in June. Italy entered the war on the Axis side on June 11, and Japan joined its Fascist brethren in September by signing the Tripartite Pact. With the Continent occupied Britain was now imperiled and could be the next victim. War was drawing closer to the United States, and the pace of mobilization began to increase.

While everything needed to be done, from recruitment to training and procuring equipment, much depended on construction as a controlling factor to produce soldiers ready for battle.[11] First and foremost, especially after institution of the draft in September 1940,[12] new troops had to be housed as they began their training. Large tracts of land were purchased, leased, or seized. Proposed sites were logically laid out. In each case decisions had to be made quickly about the placement of the barracks, mess halls, hospitals, and other buildings, as well as locating the infrastructure of water, gas, electricity, sewers, roads, and railroads. The army's construction budget was starting to increase at a rapid pace: from less than $10 million a month in July 1940 to what would become a staggering $720 million a month (under Groves's direction) in July 1942.[13]

On April 1, 1940, Maj. Gen. Edmund B. Gregory became the quartermaster general. A month earlier Col. Charles D. Hartman had become chief of the Construction Division. Hartman had been a branch head in the World War I version of the Construction Division. In his new position Hartman recruited some of his old colleagues and drew upon the practices and procedures from that era. But this was shaping up as a new and different war, and during the summer of 1940, as Hartman reorganized the division, it became evident to many that he was overwhelmed by the enormous expansion he was expected to carry out. Promotion to brigadier general on August 1 did not improve the situation.

> General Hartman was in a precarious position. Time was short. Winter with its poor construction weather loomed ahead. Unsuitable sites, inadequate engineering data, and uncertain markets were but some of the factors that threatened delay. If induction dates were to be met, the Construction Division would have to do a job of unusual difficulty with unprecedented speed. But speed meant money. Building funds were insufficient to pay for the program even if rigid economy were practiced. To complete the camps on schedule and to keep within the available funds was impossible. But that was Hartman's assignment . . .[14]

Supervision of the entire mobilization effort, including construction, was of course the responsibility of the Secretary of War Henry L. Stimson, Gen. George C. Marshall and the General Staff, and Robert P. Patterson, who had assumed the position of assistant secretary of war in late July.[15]

For the first six months of 1940 Captain Groves was assigned to work on mobilization planning in G-3.[16] The responsibility of G-3 was to recruit and train draftees, National Guardsmen and Regulars, and to work in concert with G-4 to find the best sites for the expansion of current posts and the building of new ones. Though a junior officer, Groves was the only engineer in the Mobilization Branch, and his expertise gave him more influence than might otherwise be expected. With the battlefield between the quartermasters and the engineers prepared — over who would be responsible for construction — the first skirmishes began.

The General Staff advised Quartermaster General Gregory to "pick a good officer to check on the progress of construction."[17] Gregory, of course, had known Groves for years, and they were good friends, as were their wives. The two families had been in Washington from 1931 to 1933, and now that they had both returned, they resumed sharing holiday dinners, occasionally traveling together to the Gregorys' farm, on Maryland's eastern shore. Much earlier as a young lieutenant at Fort William Henry Harrison, Montana, almost thirty years before, Gregory had filled young Dick's head with thoughts of attending West Point and becoming an army officer, and had coached him for the exam. Now Gregory turned to Groves for help and advice on how to proceed with this enormous effort, especially the problems that confronted Hartman.

According to Groves, it was only because of the great pressure applied by his old friend that he finally consented to allow Gregory to ask General Marshall to detail him from the engineers to the quartermasters as the general's special assistant.[18] There is no question that Gregory and Groves must have realized that they were treading in dangerous political waters. The formal request by Gregory to the adjutant general for Groves's services was made on July 8, 1940. On July 9 he was promoted to major, after almost twenty-two years of service. Interestingly, in this request Gregory stated that "Captain Groves has expressed a desire for this detail."[19]

The battle between the quartermasters and the engineers was now fully joined. On July 11 Gen. Julian S. Schley, chief of engineers[20] and Groves's onetime boss at Galveston, replied to the adjutant general that in view of the reduced budget for civil programs, the engineers were "now in an exceptionally favorable position to undertake immediately and carry to quick completion any amount of emergency construction which may be desired of the War Department." Schley went on to say that the current workload of the corps was such that there was no excess officer strength within the corps, and the "detail of officers of Engineers to other branches of the service cannot be concurred in."[21] Of course these arguments could not both be true, but in maneuver warfare it is best to have as many options available as possible.

Given all of his other responsibilities these territorial squabbles were of little interest to General Marshall, chief of staff. An immediate problem confronting the War Department was how to expedite the construction of the facilities necessary to allow for rapid mobilization, to provide enough shelter for the impending draft call-up to proceed. If Gregory needed an assistant, then so be it. On July 22, 1940, Marshall confirmed the orders, immediately detailing Groves to the Office of the Quartermaster General.[22]

Groves was in a delicate situation. On the one hand, his apparent willingness, even eagerness, to serve under Gregory marked him within the engineer bureaucracy as a possible traitor. At the least he was to be viewed with suspicion and not fully trusted. On the other hand, to those in the quartermaster general's office, with whom he now had to work, he could not be trusted at all. Not only was he an outsider, but he came to them from the camp of their longtime rival for domination over army construction. Furthermore, his assignment was to make field inspections of construction progress and to report his findings and recommendations directly to General Gregory.

Major Groves reported for duty as Gregory's special assistant on July 24. With the army construction program in full swing, Groves often worked seven days a week, and eighteen-hour days were not unusual. Much of his time was spent either on the road, traveling from construction site to construction site, or on the telephone checking on the progress from his office in the Railroad Retirement Building, at Fourth and D Street, Southwest, or from home. Groves crisscrossed the country, from Camp Jackson, South Carolina, to Fort Custer, Michigan, Westover Field, Massachusetts, Camp Ord, California, and a dozen other places in between. For each inspection he would write up a short memo for Gregory, with a copy to Hartman.[23] They were always written in language that avoided direct criticism of Hartman or his office.

To Groves, it appeared that Hartman did not know how to delegate authority effectively. He relied too much on the advice of the civilian contractors with whom he had worked during World War I and afterward. While they might have represented the best in construction then, many had not remained competitive with the more competent firms that emerged twenty years later.

During the summer and fall of 1940, unbeknownst to Groves, General

Gregory tried to convince Hartman that Groves should become his deputy and move into his office. Three times Hartman agreed, only to change his mind. On November 11 Gregory informed Groves that Hartman had finally acceded to Groves becoming his deputy, and that he wanted him to accept. Gregory warned Groves that there was a great deal of anxiety within the construction bureaucracy concerning his presence. Many viewed it as an attempt by the engineers to take over, which was not too far from the mark. Gregory told Groves that in order to give him the authority he required in such a situation, he should be promoted. Groves modestly responded that he was amenable to whatever Gregory thought appropriate, but — seeing an opportunity — if he was to do the job, he would have to be a colonel. That afternoon Gregory saw Marshall, recommending the new arrangement and the promotion. General Marshall accepted Gregory's recommendation, and on November 13 Groves was promoted to colonel.[24]

Groves had just been promoted to major in July, and regular promotions were awarded on strict seniority grounds based on time in grade. Now, just four months after becoming a major, he had catapulted to the rank of colonel, skipping over lieutenant colonel altogether. In the process Groves, a member of the original West Point class of 1920 (graduated November 1918), had leaped ahead of most of the class of 1912, and all of the intervening ones.[25] Groves was a man on the move, maneuvering adeptly through the bureaucratic labyrinths of Washington.[26]

As a complex social institution, the military naturally breeds its own unique culture, mentality, and worldview, which are often difficult for outsiders to fathom. At the core are deep-seated loyalties to one's service and to one's command. At an even deeper level there are ties to one's colleagues based on shared experiences, at the academy, C&GSC, War College, or, most intensely, under fire on the battlefield. Recognition of, and respect for, rank is also one of the most basic features of military life. There are reasons for wearing medals, ribbons, and devices on your chest and bars, eagles, and stars on your shoulders. It is an instant visual curriculum vitae. It shows others who you are and what you have done and how much respect you deserve. No matter how fine a colonel may be, you are not a general, and everyone knows it. All of this, coupled with the normal jostling for career advancement, makes for a special kind of politics within the military, different from civilian practices, but similar in the jealousies, pettiness, and strategies that it engenders. The aspiring officer must not be too ambitious

— or too timid. These facts of life are rarely talked about in public, and the scholarly literature on the topic is virtually nonexistent.

Military officers are constantly assessing one another among themselves, criticizing actions they think they could have done better. Much of the purpose of the military, after all, is to produce leaders, to fight and win wars, if need be, or to prepare for them and be ready when they come. It is not that different in many other fields, but in the military — with clearly defined rankings, and often with well-defined results on the battlefield — the activity becomes omnipresent.

To advance one must be perceptive to the larger forces, internal and external, that influence bureaucracies. So it was with Groves's entanglement in the construction battles of 1940 and 1941. He was becoming known as someone who could get things done. By the end of 1940 he was recognizable, known to the chief of staff, the deputy chief of staff, the G-3 and G-4, the quartermaster general, the chief of engineers, and even to the American public at large. The December 30 issue of *Life* magazine included Groves among ninety leaders in the mobilization effort, with their pictures.

Within the quartermaster general's office the atmosphere was so charged that on the day Groves's promotion to colonel was to be announced, he arranged not to be there. So while Groves was inspecting the construction at Camp Blanding, Florida, on November 13, Hartman held a meeting with his officers to announce the promotion and the change in organization. He told the group that anyone who felt he could not work within the new structure could be transferred to other duties without penalty. Though no one requested transfer, it was clear that many were not pleased with this turn of events.

The situation was becoming more and more untenable, and something had to be done. Pressure from within the War Department and from outside was pushing for the creation of an independent civilian-run office of construction. Unless some action was taken soon, construction might be taken out of the army's hands altogether. A crucial meeting was held on November 19 in Marshall's office. Among those present were Brig. Gen. Richard C. Moore, deputy chief of staff Brig. Gen. Eugene Reybold, and Michael J. Madigan, an engineer from New York and now special assistant to Undersecretary of War Robert P. Patterson, who would later play a role in Groves's career.[27] Significantly, no one from the quartermaster corps was invited. The group decided that Hartman had to go. On November 25

Gregory presented Hartman with a list of complaints about the Construction Division and told him to take corrective action. In early December Gregory demanded that Hartman decentralize and set up regional offices on the model of the Corps of Engineers. When Hartman refused Gregory took immediate steps to replace him.[28]

In earlier discussions between Gregory and Groves, the name of Brehon Burke Somervell had come up as a potential replacement for Hartman.[29] Groves had first known Somervell when he served in the Office of the Chief of Engineers in the 1931–1935 period. Somervell was another protégé of Col. Ernest Graves, Groves's mentor. In 1935 Somervell had wanted Groves to be one of the assistants to work with him in Florida, but Groves, about to go to Command and General Staff School, declined.[30]

Somervell had had an interesting and unusual career as an army engineer.[31] He graduated from West Point 6th of 107 cadets in the class of 1914 and was commissioned a second lieutenant in the Corps of Engineers. He served under General Pershing with forces on the Mexican border pursuing Pancho Villa in 1916. During World War I he supervised the training of more than eleven hundred officers and men of the Fifteenth Engineers in the United States, then performed construction jobs in France and served in occupied Germany after the armistice, earning the Distinguished Service Cross (the second highest decoration for bravery) and the Distinguished Service Medal.[32] Upon returning to the United States he was assigned to the Office of the Chief of Engineers on three separate tours of duty, interspersed with service in several districts. From 1926 to 1930 he was district engineer, Washington district. He also attended the Engineer School in 1921, the Command and General Staff School in 1923, and the Army War College in 1926.

In 1935, while serving as district engineer in Ocala, Florida, to initiate work on the controversial Trans-Florida Ship Canal,[33] Somervell met Harry L. Hopkins, FDR's confidant. The two became close friends, and in 1936 Hopkins chose Somervell to be Works Progress Administration (WPA) administrator for New York City.[34] In this capacity over the next four years he acquired a reputation as an extraordinarily accomplished administrator and a clever politician. In November 1940 Somervell's tour of duty with the WPA in New York City came to an end.[35] He approached Brig. Gen. Moore, requesting a choice position, and also approached President Roosevelt. All that emerged by the end of the month was an assignment by General Schley to be executive officer of the new Engineer

Training Center at Camp Leonard Wood, not the most exciting prospect.

Unbeknownst to Somervell, however, high officials in the War Department were examining him carefully to see if he might be Hartman's replacement. While waiting for the Camp Leonard Wood position to be finalized, he was placed in the Office of the Inspector General at the specific request of Secretary of War Stimson, who wanted "to get a look at him without Somervell being aware of this."[36] Madigan agreed that Hartman must go and was impressed with Somervell. Over lunch Madigan told Somervell about the Construction Division. Somervell said he would "love" the job.[37] A consensus quickly formed around Somervell; on December 11 he took over.

Lieutenant Colonel Somervell entered the job in a strong position, with Stimson's and Marshall's support and a direct line to the White House through Hopkins.[38] He reorganized the division by reducing the eleven branches to five (the Engineering, Real Estate, Administrative, Accounts, and Operations Branches) and adding two new sections (the Control and Public Relations Sections).[39] He also decentralized operations by establishing nine zones, reflecting the nine corps areas, each headed by an officer who was given responsibility for everything that went on in his zone.

> [Somervell] favored a type of setup known as line and staff and characterized by a high degree of decentralization, a minimum number of bosses, and a sharp distinction between those who gave orders and those who advised.[40]

He relieved dozens of Hartman's quartermasters and brought in scores of engineers, further consolidating the coup and embittering many.[41] His deputy was his old friend Lt. Col. Wilhelm D. Styer, who also served as his executive officer. Lt. Col. Edmund H. Leavey was brought in from the New York City WPA to head the Engineering Branch. Prominent civilians were recruited for full-time jobs or as consultants. Groves was appointed chief of the Operations Branch.[42]

Groves had a high regard for Somervell. Somervell struck many as arrogant and self-confident, traits that were attributed to Groves as well. Groves surely saw a part of himself in Somervell and could learn from him. They shared a driving ambition and a highly competitive nature that expressed itself in hard work and tough assignments. Somervell had a short temper and did not suffer fools easily nor tolerate mediocre performance.[43] Physically he looked the part of an army officer, a courtly well-bred southerner with an impeccably tailored uniform, a close-cropped mustache, and gray hair. He attended to congressmen, giving special concern to committee chairmen and ranking members.[44] There were qualities of Somervell's that Groves did not have, or want.[45] Somervell could be, when the occasion demanded, extraordinarily charming and urbane, something rarely said about Groves.

Somervell was a master of public relations.[46] One of his first appointments upon taking over the Construction Division was George S. Holmes, to head the new Public Relations Section. Holmes was soon issuing a flood of press releases, and newspapers and magazines eventually began to pay attention with favorable coverage. In the field at the project level, public relations officers issued their own information, involved the local officials, and gave tours. Somervell attended many ceremonies to break ground, open camps, and give awards. He spoke at many dinners, banquets, and functions — especially those involving the construction industry, where he extolled their contribution. None of this was lost on Groves, as we shall see. One of the few criticisms he had of Somervell was this apparent lusting after publicity. On the other hand, at a certain point in his running of the secret Manhattan Project Groves began to prepare for the time when the story of how the bomb was built would become public. By that time his own public relations skills were quite highly developed.

By February of 1941, with a new staff and a new rank of brigadier general, the reorganized Quartermaster Construction Division, under Somervell, was running fairly smoothly. Groves, now chief of the Operations Branch and at the center of the effort to complete the camps, and other projects, recruited a fresh team of his own,[47] many of whom he would call upon shortly to build the bomb. Groves favored West Point graduates and filled his ranks with them, whether quartermasters or engineers, though anyone demonstrating intelligence, initiative, and trustworthiness could qualify. Lt. Col. Thomas F. Farrell, chief engineer of the New York Department of Public Works, was recalled to be Groves's executive officer. In early 1945 Groves would recruit Farrell again, this time to be his deputy and carry on the project if something should happen to him. Capt. Clarence Renshaw, who would soon help him build the Pentagon, joined, as did Elmer E. Kirkpatrick, who would undertake several assignments for Groves in the coming years.

The harsh winter of 1940–1941, strikes, shortages, union rules, and the myriad other difficulties that Groves faced cast doubt on whether the camps would be ready by the deadlines. Soldiers from National Guard divisions were entering the army in great numbers: 100,000 in January, 150,000 in February, 200,000 in March. By April the army was a million strong.[48] Somervell used all of his skills to get more funds out of Congress, to play for time, and to keep everything moving so that by late April he could announce, "The new Army is housed."[49]

Among Groves's most important responsibilities as chief of the Operations Branch — as well as after July 1, 1941, when he was promoted to assistant chief of the Construction Division — was to create a government-owned, contractor-operated (GOCO) munitions industry in record time.[50] Not only did soldiers need to be housed in camps and cantonments, but they needed to be armed as well. When Hartman left in December 1940 there was a modest program under way. Over the next year with Somervell and Groves in charge, the Construction Division, in concert with the Ordnance Department and the Chemical Warfare Service, was responsible for designing and building an enormous range of plants to produce tanks, armor plate, TNT, small arms, rifles, shells, smokeless powder, ammonia, toluol, charcoal-whetlerite, and depots to store ammunition prior to its shipment overseas.

As with the camps and cantonments, Groves's involvement in the munitions works and plants was extensive.[51] In the process he came to know many contractors whom he would soon call upon again. Most significantly, Du Pont — "the colossus of American explosives and propellant production" — was involved with construction or operation of ordnance works in half a dozen states.[52] Other firms involved in munitions work overseen by Groves were: Stone & Webster building the Kankakee Ordnance Works in Joliet, Illinois, the H. K. Ferguson Company building the Gulf Ordnance Works in Mississippi, and the M. W. Kellogg Company with the Dixie Ordnance Works in Sterlington, Louisiana. The Chrysler Corporation and the Monsanto Chemical Company were also involved in munitions work with Groves. As we will shortly see, all were recruited by Groves to play major roles in building the bomb.

Groves had to contend with the priorities and allocation system, and in the process became aware of how important it was to completing projects. The first Army Navy Munitions Board (ANMB) priority classification dated from the summer of 1940. Under this system AA was the absolute top rating, reserved for emergencies. High-octane gasoline, synthetic rubber, and naval vessels were among the programs with the highest priority. The basic nonemergency ratings went from A-1 to A-10, with gradations within each, such as A-1-a, A-1-b, and all the way to A-1-j.[53] The higher the rating, the more likely it was that a program would be able to obtain materials that had been placed on the list of allocated items. Through 1941 and 1942 an increasing number of materials were added to the allocation list, creating "rating inflation." An engineer might get some help with a one-time rating for a single item, but the best situation was to have blanket authority for the entire program. Many of Groves's military construction programs were not highly rated — airfields, for instance, were A-1-e, and cantonments A-1-j.[54] On the other hand, munitions plants were rated A-1-a, which qualified them to obtain the huge amounts of steel they required.

For Groves and his engineers the materials problem was never ending. Not just steel, but rubber, tin, aluminum, nickel, chromium, copper, zinc, lead, iron, cadmium, magnesium, mercury, vanadium, hemp, and sisal, to cite a partial list, were always needed, and always in short supply. Central to what it meant to be an engineer was to be innovative in using the means at hand to get the job done, especially so if it were a combat situation. On the battlefield materials are not always available. On the home front, too, the pursuit of materials was constant. One either needed to know how to obtain them, use less of what one had on hand, or find substitutes. These would be lessons Groves learned well long before he came to building the bomb.

By reading the daily entries in Groves's appointment book for this period, we begin to see how this experience contributed to his being chosen to run the Manhattan Project. Colonel Groves made and received as many as two dozen local and long-distance telephone calls a day. They were to or from his subordinates at posts, camps, munitions plants, ports, hospitals, and other facilities all across the country. The calls would be about problems large and small, from plumbing to labor rates, sewer lines, housing for workmen, fences, roads, runways, concrete mixers, and the thousand other details and aspects of military construction. Groves was always pushing to keep projects on schedule or, if they fell behind, to correct the problems and get them back on track.[55]

Sometimes this required a trip to the job site to break the bottleneck. One of his aides later provided a description of Groves's visit to the Lake Ontario Ordnance Plant, which had fallen behind schedule.[56] Groves got off the train in Buffalo and drove to the project, spending the morning going over the physical aspects of the job. At noon he met with the contractor's representative and the on-site engineer staff, and listened to their conflicting accounts of the problems. Afterward Groves took the representative aside and talked to him for more than an hour, showing concern and understanding but trying to instill in him confidence and purpose. Finally the connection was made. The representative brought out the progress reports, explained the precise details of why he was not obtaining the materials he needed and why things were not working. Finally, with the problems correctly identified, realistic solutions could be attempted. Groves promised assistance and directed that modified reporting methods be implemented so that he could keep better progress of the project from Washington. At the end of the day Groves took the night train back to Washington and was in the office the next morning.

His days also were filled with answering inquiries from senators and congressmen or their staffs, who asked about army construction projects in their states and districts. Was he planning to build an army base in Texas? How close to completion was the Denver Small Arms Plant? Would Groves be interested in using a local construction company?[57] On January 7, 1941, Sen. Harry Truman called Colonel Groves to inquire about the status of the military hospital that was being built in Independence, Missouri. The following day Groves called back with his reply.[58] After all, with members of Congress appropriating the money, they had to be listened to and served.

Groves spent a portion of his time in meetings with civilian engineers, architects, lawyers, contractors, and a wide assortment of businessmen who wanted to obtain contracts from the army. He also had to prepare testimony to give before congressional committees or subcommittees. He met with government officials from other bureaus, departments, or offices of the federal government. He seemed to keep track of it all, from sewage disposal at Fort Huachuca to labor rates at the Kankakee Ordnance Plant and housing for the "colored" troops at Fort Warren.

There are some clues to his administrative methods among his papers at the National Archives. In 1940 and 1941 he kept small pocket notebooks and filled them with notes to himself about costs, dates, and deadlines of his many projects, with reminders. He kept a card catalog with ratings of how architect-engineers and contractors performed on their work of camp construction. The ratings were: "I — Highly desirable to seek for a new job. II — Desirable to employ on a new job. III — Acceptable to employ on a new job. IV — Should not be given a new job. V — Should not be given a new job under any circumstances." The firm of Alvord, Burdick & Howson of Chicago got a V rating for its work on Fort Leonard Wood, with the following comment: "Should not be given a new job under any circumstances because they have proved unsatisfactory and inadequate. They have proved neither pleasant to deal with nor cooperative and have demonstrated their inability to keep ahead of construction." On the other hand, there were firms that got good ratings. Doyle-Russell & Wise of Richmond, Virginia, got a II rating as contractor on Camp Lee, with its work "conducted with due regard to regard to economy and the contractor has proved himself to be capable, reliable, energetic and cooperative." The J. A. Jones Construction Company of Charlotte, North Carolina, got a I rating for Camp Shelby, Mississippi. "The contractor has splendid organization whose attitude was alert and energetic. This firm is particularly desirable for future work where time is a vital element."[59]

Through these experiences Groves acquired an intimate knowledge of major contractors and subcontractors across the country. He was ruthless in making sure that they toed the line and lived up to their commitments. He was given a free hand by Somervell and organized his office to his liking. He knew what he wanted and thought he knew how to get it. He kept contact with all the projects under his control by keeping the chain of command to a minimum, avoiding the creation of a large staff or a complicated bureaucracy. He carefully gathered around him a small group of officers on whom he could rely to carry out his wishes to the letter. If a project began to experience trouble because the contractor or his army subordinates were not performing properly, he would quickly arrange for a personal visit. Typically, these visits lasted less than eight hours; more often than not, by the time he was ready to leave Groves had gotten to the heart of the problem and taken steps toward a solution.[60] Working in this way, Groves was signing construction contracts at the rate of almost $600 million dollars a month.[61] Despite the vastness of the operations he was up to date on the progress of each and every one of these projects.

In carrying out these responsibilities from 1940 to 1942 Groves met hundreds and hundreds of people, in many walks of life, with a variety of specialties. After being chosen to head the Manhattan Project, as we shall see, he would recruit many of these same people to help him build the bomb. Similarly, several of the military facilities he built during this period would have a role in the Manhattan Project, and/or in the Cold War that followed.[62]

Besides heading the Manhattan Project, the other thing for which Leslie Groves is usually remembered is the building of the Pentagon.[63] The Pentagon is one of the three or four most recognizable buildings in Washington, after the White House and the Capitol. It symbolizes, probably more than any other building, the vast transformation that the federal government and the military went through from 1940 to 1945. Since 1945 it has come to represent America's active role in the world. While Groves may be correctly remembered for overseeing its construction, it was Somervell who envisioned putting the entire War Department under one roof, was the driving force in clearing the bureaucratic hurdles, got the funds from Congress, and designed it; it was Clarence Renshaw who actually built it.

In July 1941 the offices of the War Department were scattered throughout seventeen buildings in the District of Columbia, and additional ones in Virginia, for a workforce of twenty-four thousand. The largest was the eight-wing Munitions Building, an eyesore occupying the Mall between Constitution Avenue and the Reflecting Pool from Nineteenth to Twenty-first Streets, Northwest. It was built during World War I as an emergency structure, and was still there. Next to it to the east from Seventeenth to Nineteenth Streets, also on the Mall, was the Navy Building, a similar factorylike structure of the same period.[64] A few blocks away was the New War Department Building, at Twenty-first and Virginia, just opened in June but already inadequate for the War Department's growing needs.

The number of War Department employees assigned to the metropolitan area was increasing at the rate of a thousand per month and was expected to reach thirty thousand by the beginning of 1942. Many of the current offices were small, some providing on average only forty-five square feet per person. There was also a demand for more storage space for records. Of the 2.8 million square feet then occupied by the War Department, almost one-quarter was devoted to record storage. This was projected to increase as well. Something had to be done.

On July 14 President Roosevelt transmitted to Congress for its consideration a proposal for new office buildings for the War Department. On Thursday, July 17, the Deficiency Subcommittee of the House Appropriations Committee heard testimony from the commissioner of public buildings, W. E. Reynolds, and the assistant chief of staff, G-4, Brig. Gen. Eugene Reybold. The subcommittee was not satisfied with the president's proposal; Chairman Clifton A. Woodrum of Virginia requested that the War Department investigate the feasibility of using a site just across the Potomac in Arlington.

This recommendation had been prepared in advance by Brigadier General Somervell, who had already decided that the best site was a tract of land east of the main gate to Arlington National Cemetery facing the Potomac River on the former Experimental Farm, transferred by the Department of Agriculture to the War Department in November 1940.[65] Earlier he had spoken to Brig. Gen. Moore about this site. Moore urged Somervell to speak with Congressman Woodrum, who would likely be amenable to the idea. Somervell lost no time in acting upon Woodrum's recommendation. On the evening of July 17 he summoned his top engineers and architects. These included Lt. Col. Hugh J. Casey, chief of the Design Section,[66] Col. Leslie R. Groves, Col. Edmund H. Leavey, and G. Edwin Bergstrom, his chief consulting architect and former president of the American Institute of Architects.[67]

Somervell's instructions were blunt. He told them that by nine o'clock Monday morning, July 21, he wanted the basic plans for a four-story office building of five million square feet that would house forty thousand persons on a site along the south bank of the Potomac about three-quarters of a mile below the Arlington Farms tract on the site of the old Washington-Hoover Airport.[68] As they began to investigate their project, Brigadier General Reybold, the G-4 — but soon to become chief of engineers — proposed moving the site to the north and the west away from the floodplain of the Potomac. Over the next month the building's size, shape, and location changed several times as Capitol Hill, Frederic A. Delano (chairman of the National Capital Park and Planning Commission, and FDR's uncle), Gilmore D. Clarke (chairman of the District of Columbia Commission on Fine Arts), Arlington County officials, the press, President Roosevelt, and others weighed in with their views. The site eventually chosen, after firm intervention by FDR, was originally intended for a quartermaster depot, just to the southeast of Arlington Farms.[69]

Somervell stubbornly wanted to stay with the Arlington Farm site to the north. When the bill finally reached his desk for signature on August 25, FDR reserved the right to make the final choice on the location. He announced it would be at the depot site, and that the building would be half the size of Somervell's. On August 29 FDR summoned Clarke, Budget Director Harold D. Smith, and Somervell to the White House to discuss the building. This was followed by a tour of the site by car. With Somervell,

Clarke, and Roosevelt in the backseat and Smith and Roosevelt's dog Fala in the front seat, they drove across the Potomac. Somervell kept insisting that the Pentagon be located just south of Memorial Bridge east of the cemetery facing the river. As recounted by Clarke,

> the President was getting increasingly annoyed by General Somervell's insistence about this, and when we got on the 14th St. bridge, the President leaned in front of me and said, "My dear General, I'm still the Commander-in-Chief of the Army."
>
> As they approached the depot site, Roosevelt had the car stop and, pointing, said "we're going to put the building over there, aren't we?" To emphasize that the matter was now closed, he said to Somervell, "Did you hear that General? We're going to locate the War Department building over there."[70]

While this would have ended the matter for most people, it did not for Somervell. He may have lost the battle over location, but he was determined not to lose the one over size. As the official army historians delicately put it, "Pulling down a curtain of secrecy over the project, Somervell followed an independent course."[71] Somervell broke ground at the depot site — Roosevelt's choice — on September 11, but with plans for a building nearly as large as his original proposal, one significantly larger than what FDR had directed.[72]

Somervell chose a contractor and a constructing quartermaster. His choice for general contractor was John McShain, Inc., of Philadelphia. McShain had built the Jefferson Memorial, the Naval Medical Center in Bethesda, the just completed National Airport, and the New War Department Building in Foggy Bottom.[73] On the day of the selection, July 25, 1941, Colonel Groves's appointment book shows that Mr. McShain and Mr. J. Paul Hauck, the contractor project manager, came to the office to discuss the "New building in Arlington."[74] Over the next year they would meet many times. Somervell also chose the Doyle-Russell & Wise Contracting Company of Richmond, Virginia. The cost-plus-fixed-fee contract was awarded in September, with an estimated cost of thirty-one million dollars. The building of reinforced concrete, faced with Indiana limestone, would consist of five concentric pentagons[75] occupying thirty-four acres with four million square feet of floor space and parking for approximately eight thousand cars. To direct the work Somervell named Capt. Clarence Renshaw.[76] Renshaw reported to Groves, who had overall supervision. The chief architect was Bergstrom, who recruited David J. Witmer as his assistant and, later, his successor.[77] Specialists were brought in to design and plan highways and transportation, sewage treatment plants, landscaping, heating, air-conditioning, electricity, and plumbing. The complete range of tradesmen was put to work carrying out those plans.[78]

On October 10 Somervell took his plans to the White House, presenting FDR with a fait accompli. The building was already a month under way, and there was little the president could do. After Pearl Harbor an even faster construction schedule, and other changes to the building, increased costs by $14.2 million. The pace was relentless: Three shifts of workers labored around the clock. At the peak of construction the total number of workers on site reached fifteen thousand. The building was contructed in five sections; the first occupants moved into section A on April 30, 1942, and they kept moving in at a rate of a thousand per week over the next seven and a half months. On July 20 Somervell directed the addition of a fifth floor to the three interior rings to match the innermost and outermost rings, gaining another 356,000 square feet of office space. Secretary of War Stimson and Chief of Staff Marshall moved into their adjacent offices on the third floor of E ring[79] on November 14, 1942. The entire building, with approximately thirty-five thousand occupants, was completed by January 15, 1943, sixteen months after breaking ground.[80]

Everything about the building, then and now, is gargantuan.[81] As finally completed, the building had a total gross floor area of 6.24 million square feet, with 3.64 million square feet of usable office space. The building covers twenty-nine acres, including a five-acre courtyard in the center, known later in the nuclear era as Ground Zero. The Pentagon has three times the floor space of the Empire State Building, seventeen miles of corridors, 7,748 windows, 2,306 toilets in 275 rest rooms,[82] 546 drinking fountains, and fifty-three thousand light outlets. Each of its five sides is 921 feet long, making the outer perimeter almost a mile in circumference. The seventy-one-foot high building rests on 41,492 concrete piles, ranging from twenty-seven feet to forty-five feet in length. A total of 435,000 cubic yards of concrete was poured.[83] Thirty miles of new highway were constructed, together with twenty-one bridges and overpasses. Two parking lots covering sixty-one and a half acres have a total capacity of more than eight thousand cars.

Groves's name, one of only a dozen, was chiseled into a rectangular limestone plaque on the left side of the Mall Entrance commemorating those responsible for the building.[84] As we shall see, after his retirement from the army and for the rest of his life, Groves closely monitored what was written about him and his accomplishments. If during his reading of a newspaper or magazine article, or a book, he found a mistake or differed with an interpretation, he would contact the author and point it out.

Groves was ever alert to charges by Congress, the press, or anyone, that public monies had been squandered on any of his projects. This was true with the Pentagon as well as with the Manhattan Project. During 1943 he had to defend against attacks by Rep. Albert J. Engel, who had charged the corps, and Groves in particular, with gross, wasteful mismanagement with regard to Pentagon construction. This required Groves to devote many hours to gathering data and writing reports, and still more hours preparing for and testifying before the House Appropriations Committee.

In a 1968 article in *The Journal of the Armed Forces* an author stated that the Pentagon building cost eighty-three million dollars.[85] Groves fired back a letter informing him that the figure was an error. Groves said he did not know how the author had arrived at it, but it "must have come from one of the wild speeches made by Congressman Engel of Michigan."[86] A better estimate would be less than fifty million dollars. Of course, including the roads, parking areas, and additions by the tenants, the total cost was higher. Stockstill replied admitting the error, and a revised version of Groves's letter was published in the June 15 issue.

Despite the significant amount of work being accomplished by the quartermasters, the long-standing issue of military construction remained unresolved.[87] Over the summer of 1941 Undersecretary Patterson and his special assistant Madigan studied the problem and concluded that the engineers should be in charge. In point of fact an engineer now headed the Construction Division, most of the staff were engineers, and it was organized in the decentralized style of the engineers. The only anomaly was that the division was still within the quartermaster corps.

Madigan submitted a twenty-page report on August 15, detailing why all War Department construction should be consolidated in the Corps of Engineers. Patterson immediately approved it and got Stimson's agreement the following day. A few days later Madigan went to see General Marshall and convinced him as well. All of this went on without Gregory's counsel or even knowledge. After Patterson got the president's approval on August 29, a bill was introduced in the Senate and the

House, followed by hearings held in September and October. The measure passed and was signed by the president on December 1, 1941. In the interim, on October 1, General Reybold had become chief of engineers, succeeding Schley. Somervell, who had badly wanted the job, oversaw some of the early phases of consolidation, but on November 25 became assistant chief of staff, G-4, a temporary station on the way to a more powerful position he would assume a few months later as commanding general, Services of Supply (soon after renamed Army Service Forces, or ASF).

The attack on Pearl Harbor on December 7 galvanized the nation into war. The Munitions Program of June 30, 1940, had called for a four-million-man army by the spring of 1942, a target that was now within reach. But that goal had been overtaken by the Victory Program of September 11, 1941, which called for a nine-million-man army by June 1944.[88] The job facing the engineers was, as General Reybold termed it, "colossal." The official historians described the undertaking as

> . . . truly gigantic, dwarfing those previous endeavors, the building of the Panama Canal and the emergency construction programs of 1917–18 and 1940–41. In urgency, complexity, and difficulty, as in size, it surpassed anything of the sort the world had ever seen. The speed demanded, the sums of money involved, the number and variety of projects, the requirements for manpower, materials, and equipment, and the problems of management and organization were unparalleled. So formidable was the enterprise that some questioned whether it was possible.[89]

On December 16, 1941, a week after Pearl Harbor, the Construction Division was formally transferred to OCE. The chief of the Construction Division was Maj. Gen. Thomas M. Robins, a respected senior officer, with Groves as chief of the growing Operations Branch.

By the time the merger was nearly accomplished, a major reorganization of the War Department led to further changes.[90] On March 9, 1942, the army formed three overall commands — Army Ground Forces (AGF), under Lt. Gen. Lesley J. McNair; Army Air Forces (AAF), under Lt. Gen. Henry H. Arnold; and Army Service Forces (ASF), under Somervell, now promoted to three-star rank, at the age of forty-nine.[91] The Corps of Engineers became one of the operating divisions of ASF, along with the other technical services.[92] New positions and promotions resulted as personnel were shifted around. General Styer became chief of staff, ASF, and Groves, at Somervell's insistence, was appointed deputy chief of construction, OCE, on March 3, 1942. Col. Frederick S. Strong Jr. replaced Groves as chief, Operations Branch. Strong was almost another Groves. Graduating first in the class of 1910, he resigned from the army as a lieutenant colonel after service in Europe in August 1919 and was in real estate, land development, and investments until he was recalled to active duty in June 1941. He served in various positions before becoming chief of the Operations Branch for the period December 1941 to April 1943. Groves said of Strong that he

> had been thoroughly indoctrinated with my viewpoint that no delay was excusable and that most delays were caused by slowness in decision, not only on the site, but in the higher echelons. Strong was a brilliant man. He stood Number One in his class and had lost none of his intellectual keenness. His criticism of other people's work was always extremely sharp.[93]

On April 7 the Construction Division moved from the Railroad Retirement Building to the New War Department at Twenty-first and Virginia, in Foggy Bottom, where the Office of the Chief of Engineers was located.[94]

Just as before in CQM Groves exerted strong control over his projects in the field, by phone and, if necessary, in person. Groves had new orders for camps, depots, air bases, staging areas, hospitals, bomber plants, munitions plants, and ordnance igloos, as well as directives to expand much of what had just been built. To meet these new demands he adopted some novel methods. For example, there were only a few companies qualified to build munitions plants; recruiting new ones might entail risks. How could those already overburdened take on more work? Groves suggested a "master design and procurement" contract, whereby one company would supply drawings, provide services, and purchase equipment for a group of projects. In early 1942 Groves contracted with Du Pont to build three TNT plants using this method. Getting higher priority ratings and paying for overtime could also save days or weeks.

Groves's daily challenges were monumental and unrelenting. Throughout the early months of 1942 they increased as the number of projects rose into the hundreds; the amounts of money were staggering.[95] In July 1942 the peak was reached with a million men at work and a monthly value of $720 million ($14 billion in 2001 dollars), a figure larger than the total for all military projects from 1920 through 1938. For the year more than two thousand jobs were completed at a cost of almost five billion dollars.[96]

The priorities for Groves were how to save time, make people work harder, make them work longer, overcome bottlenecks, and meet deadlines. Speed was of the essence: The United States was at war and anything less than a full effort could jeopardize the soldiers — and the nation.[97]

Groves's meteoric rise and sojourn in the quartermaster corps still rankled many engineers. Many viewed him as opportunistic and overly ambitious. His ruthless manner got the job done, but many were bruised along the way.[98] The traditional administrative style for the engineers was decentralization to division and district engineers. While Groves continued to follow the traditional corps methods for the most part, he also exerted firm control from Washington to get things done on time — which, not surprisingly, brought complaints. There was also great dissatisfaction within the corps with Somervell, and Groves was seen as one of his protégés. Somervell's ruthlessness and publicity seeking rubbed many the wrong way.[99]

In addition to all else beginning in the spring of 1942, Groves was charged with acquiring the property and building the relocation centers for Japanese Americans living on the West Coast subject to evacuation. He had nothing to do with the policy issues, which were the responsibility of President Roosevelt, Assistant Secretary of War John J. McCloy,[100] Attorney General Francis Biddle, Director of the War Relocation Authority Milton Eisenhower, and others in Washington. For Groves, building the centers interfered with his other work, and he was of the opinion that the issue would come back to haunt the nation. As he told the Southwestern Division engineer, "On this Jap thing . . . there is nothing that is going to cost us more embarrassment than that."[101]

The Corps of Engineers acquired the property through its division engineers, working with Commanding General John L. DeWitt, Western Defense Command and Fourth Army.[102] The large tracts of federal land — located at a safe distance from "strategic installations," at the army's request — were largely inhospitable and situated mainly in the deserts of Arizona, California, Idaho, and Wyoming. The design of the buildings emphasized economy and speed of construction. After the specific plans were prepared at the district engineer's office, in the district where the center was to be built, they were submitted to the Western Defense Command. After approval, the district engineers awarded contracts to private builders.

Initially there were ten relocation centers. The first evacuees began to arrive near the end of May 1942; by the end of the year there were 106,770 people at the centers.[103] In a trip to San Francisco, from June 3 to June 10, Groves worked out a schedule to complete nine more. During the trip Colonel Groves and Lieutenant General DeWitt signed a detailed agreement on June 8 that set uniform standards for the construction of future camps.[104] In a typical center designed for ten thousand evacuees there were thirty-six housing blocks, each with twelve barracks buildings, a mess hall, and a combination H-shaped building that had toilet and bath facilities for men and women. Also specified were hospital facilities, schools, churches, theaters, and stores. Buildings for the nonevacuee staff and the military police were listed. Adequate utilities must be provided for water supply (one hundred gallons per capita per day), sewage disposal (seventy-five gallons per capita per day), electric power (two thousand KVA per ten thousand population), and lighting.

To save money — and, more important, to save time and materials — the corps used existing buildings to meet its goals, including schools, warehouses, racetracks, fairgrounds, and a large number of hotels and resorts. During the summer and fall of 1942 the corps took possession, through lease or outright purchase, of hundreds of hotels. The Breakers at Palm Beach and the Greenbrier at White Sulphur Springs became hospitals, the Stevens in Chicago (at three thousand rooms the biggest hotel in the world) were barracks for new trainees, and the golf resort at Pinehurst, North Carolina, became an air force station.

Several division engineers approached General Reybold, chief of engineers,[105] urging that Groves be relieved of his position as deputy. Reybold refused. They then approached Somervell's chief of staff, General Styer. Styer informed Groves that he was persona non grata with a number of the engineers. Styer wondered whether it would not be better for all concerned if Groves were moved to another position. First he offered to switch Groves with Lucius Clay, who was then assistant to Gen. Donald H. Connolly, administrator of the Civil Aeronautics Authority. When Groves refused this offer, Styer suggested appointing him engineer for the

Pacific Division in San Francisco. It was clear that Styer, who by this time was a friend of Groves (and a contemporary and friend of his critics as well), agreed with the latter and told Groves it was in his own best interest that he get as far away from Washington as possible.

That is where matters stood in June 1942, when Somervell recommended Groves for the Distinguished Service Medal,[106] and when Groves first heard of an unusual, highly secret project that Styer had just turned over to the Syracuse District engineer, Col. James C. Marshall.

CHAPTER TEN

Fateful Decisions

Much of the story of the atomic bomb has to do with decisions, who made them and why, and what impact they had on the course of events. Histories of the Manhattan Project usually depict the decision to build the atomic bomb as one reached by consensus, but much evidence suggests that it was Vannevar Bush, largely alone, who successfully orchestrated the decision to build the bomb in the spring of 1941.[1] Furthermore, it was Bush who, literally on the eve of Pearl Harbor, decided that the army must be brought in to run it. These two decisions would set the stage for the selection of Col. Leslie R. Groves to build the bomb.

Vannevar Bush was born in Everett, Massachusetts, in 1890, the son of a Universalist clergyman.[2] He received his bachelor and master of science degrees from Tufts College in 1913 and three years later had his doctorate of engineering from MIT and Harvard, under a joint program between the two schools at the time. Bush went on to combine several careers, as a university professor and dean, as an inventor and founder of several companies, and as an adviser to the federal government. A major turning point in his life occurred in 1939 when he came to Washington as president of the Carnegie Institution of Washington, a leading center for scientific research since its founding in 1909. In addition Bush served, in effect, as President Roosevelt's science adviser. On June 12, 1940, Bush proposed that the president establish the National Defense Research Committee (NDRC), a government organization to direct and coordinate the nation's mobilization effort in the field of science and weapon development. President Roosevelt approved the plan that day, appointing Bush to run the new committee two days later. Bush chose the other members, held some preliminary meetings, and the NDRC officially came into existence on June 27, 1940.[3] Included as part of the president's directive was a provision that the NDRC take responsibility for the Committee on Uranium, which until then had overseen a modest program of fission research.

The U.S. government had been slow to take interest after the announcement of the discovery of fission in January 1939. Research efforts over the following two years were small scale, uncoordinated, and decentralized. Most of the funding was limited to about a dozen universities and a few government bureaus or departments.[4] Scientists differed about the levels of funding and the kinds of resources that should go into fission research. There was great uncertainty about whether the atom could be harnessed as a source of energy, much less used for a bomb. No one knew how much an attempt to build one might cost or how long it might take.

In 1940 Bush himself was not enthusiastic and was initially skeptical. Many other worthy projects needed support — radar, submarine detection, the proximity fuse — but Bush thought it wise to at least explore the possibilities of fission; others might be doing the same.

By March 1941 the Uranium Committee, under the leadership of Dr. Lyman J. Briggs, director of the National Bureau of Standards, had little to show for itself. Critics began to mobilize around Ernest O. Lawrence of the University of California, Berkeley, and Arthur H. Compton of the University of Chicago to urge a greater governmental effort.

In time-honored Washington fashion Bush decided to form a committee to write a report assessing where things stood on the question of fission and to make recommendations. On April 19, 1941, Bush requested Frank B. Jewett, president of the National Academy of Sciences, to appoint a committee to consider "possible military aspects of atomic fission."[5] Four weeks later the committee reported back to Bush, recommending that "during the next six months a strongly intensified effort should be spent on this problem," though it was "unlikely that the use of nuclear fission can become of military importance within less than two years."[6]

While Bush was initially satisfied with this conclusion, his deputy, James Bryant Conant, president of Harvard University, was not. Conant had entered Harvard as a brilliant freshman in 1910 and by 1916 had received a doctorate in chemistry. After a stint in the army's gas warfare program during World War I, he returned to the Harvard faculty and became president in 1933.[7] When the war broke out in Europe, Conant involved himself with efforts in Washington to mobilize American science; Bush soon recruited him to be his deputy at the NDRC.

Conant criticized the report, and Bush went back to Jewett for a reassessment. A second report, dated July 11, was transmitted to Conant, now chairman of the NDRC, but the conclusions were still vague and generally pessimistic.

The reports left Bush between two groups — one led by Lawrence and Compton, who wanted the government to move faster with a bigger commitment, and the other by Jewett and the National Academy Committee, who remained skeptical and urged moderation. This impasse left the choice to Bush, who had become more enthusiastic about the prospects of atomic energy over the previous year.

Throughout this period, as the war in Europe intensified, Bush had been preparing and implementing overall governmental science policy. When the Office of Scientific Research and Development (OSRD) was established by the president in June, Bush became director of OSRD, and Conant became chairman of NDRC, absorbed by the new organization.[8] The following month the Uranium Committee, under the NDRC, was reorganized with a more active membership.[9]

Over the next four months a feeling grew that war was drawing nearer and, ominously, that the Germans might already be engaged in fission research. Word of British studies was beginning to reach key American scientists and officials. While the American reports of May and July 1941 had been inconclusive about the prospects for a bomb, a British technical committee concluded on July 15 "that release of atomic energy on a large scale is possible and that conditions can be chosen which would make it a very powerful weapon of war."[10]

Bush learned of the so-called MAUD Report while it was in draft form and received an official copy in October 1941.[11] Visits that summer and fall by British and American scientists to each other's countries, and the private conversations and communication that resulted, began to forge a consensus accepting the conclusions of the MAUD Report.

Several key decisions were taken after Bush met with Vice Pres. Henry Wallace in July and the president in October. Bush informed Roosevelt of the MAUD Report, and complete interchange with the British was endorsed. A Top Policy Group was formed consisting of the president and vice president, Secretary of War Henry L. Stimson, Chief of Staff Gen. George C. Marshall, Bush, and Conant.

With this as a backdrop a third report of the National Academy committee, dated November 6, 1941, "radiated a more martial spirit than the first two," to use Conant's words, and recommended "urgent development" of the program. In a prescient sentence the report concluded, "The possibility must be seriously considered that within a few years the use of bombs such as described here, or something similar using uranium fission, may determine military superiority."[12]

Over lunch at the Cosmos Club on Saturday December 6 — the day before the Japanese surprise attack on Pearl Harbor — Bush discussed with Conant and Compton his idea of asking General Marshall to assign an officer to look after bomb research and development. Ten days later Bush formally recommended to the Top Policy Group that the army should take over.[13] On the same day he wrote to Conant:

> There was no objection expressed to this point of view. I also said I felt at the present time there ought to be one officer, of fine technical qualifications, assigned to become utterly familiar with this whole matter.[14]

The primary reason for bringing in the army was to hide the expenditures for the project within the corps's massive budget. Throughout the war the budget for the Manhattan Project was principally concealed in line items of the Corps of Engineers' budget labeled "Procurement of New Materials" or "Expediting Production." Bush did not hide his reason for concealing the budget: He did not want to have to defend the project to the Congress. This would result in unacceptable delay, and it would undermine the extensive secrecy in which he felt the project had to be cloaked.

On March 9, 1942, Bush provided a status report to FDR stating that by summer, "the whole matter should be turned over to the War department," and that "Present opinion indicates that successful use is possible, and that this would be very important and might be determining in the war effort. It is also true that if the enemy arrived at results first it would be an exceedingly serious matter."[15]

Two days later the president responded to Bush saying he had "no objection to turning over future progress to the War department on condition that you yourself are certain that the War department has made adequate provision for absolute secrecy."[16] The fateful decision of transferring the project from civilian to military control had been made.

It is clear that in the first few months of 1942 General Marshall and the army did not give the project the highest priority. In a February 1942 memo to Stimson's assistant, Harvey Bundy, Marshall suggested that Col. Wilhelm Delp Styer, executive officer for the chief of the Construction Branch, might be the best man for the job. For one thing, Marshall wrote, the job would be only "part time."[17]

In March 1942 he suggested that Bush discuss the turnover with Lieutenant General Somervell, who in turn assigned Brigadier General Styer to administer the transition and serve as liaison.[18] It is clear from General Styer's memoranda at the time that he believed most of the research and development had been completed. This was completely wrong. Bush knew the money was running out from sources at his disposal and much more was going to be needed. On June 13 the Top Policy Group approved Bush's plan to bring in the army, and a few days later Roosevelt did as well.

The choice to head the new project was Col. James C. Marshall, Syracuse District Engineer, an accomplished administrator with an excellent record.[19]

Groves does not tell us precisely when or how he learned of the secret project, but there is evidence suggesting that he became aware of it on the day that Colonel Marshall was chosen. Marshall recorded in his diary that on Thursday, June 18, 1942, after learning from General Styer that he had been selected, he returned to OCE and told McCoach, Robins, and Groves that he had been instructed to form a new district.[20] The next day Marshall spent several hours in Bush's office learning further details, reading various letters, and receiving from Styer the authority to carry out the project.

> The Army's mission included building both pilot and full-scale plants for producing fissionable materials to be used in the manufacture of atomic bombs, letting contracts for these plants and others to be under OSRD direction, and extensive site selection, acquisition, and development.[21]

Marshall informed Styer that he wanted Lt. Col. Kenneth D. Nichols as his assistant[22] and would need office space in the New War Building. Afterward Marshall reported back to Groves and to General Robins on all that had occurred.

Groves and Robins, according to Marshall's diary, were not enthusiastic about this new project, and were skeptical of the supposed carte blanche authority he was given. They had seen other cases where such authority was given only to be considerably modified at a later date. For now, Marshall's project only interfered with their own. His personnel demands would take people away from their ongoing projects; also, he needed valuable office space somewhere in the New War Building, eventually landing in Room 6223. During the last week of June Marshall familiarized himself with the project and set up an office in Lower Manhattan on the eighth and ninth floors of 270 Broadway, headquarters of the North Atlantic Division of the Corps of Engineers.[23]

The secret project needed a name. On June 26 Somervell, Styer, and Reybold came up with *Laboratory for the Development of Substitute Materials,* or *DSM.* When Groves learned of the proposed code name he thought it would only arouse curiosity. In August he proposed *Manhattan,* following the custom of naming districts for the city in which they are located. Most important, it revealed nothing about the nature of the work. And so on August 13 General Reybold issued a general order establishing

> a new engineer district, without territorial limits, to be known as the Manhattan District, . . . with headquarters at New York, N.Y., to supervise projects assigned to it by the Chief of Engineers."[24]

A meeting of the S-1 Executive Committee of the OSRD was held at the Carnegie Institution on the morning of June 25, with Marshall, Nichols, and Styer in attendance. Among the topics covered were where to put the plants, which construction firms should build them, and how to obtain the critical materials and equipment. Though Groves would not take charge of the project for another three months, in September he influenced some of these early decisions, such as advising Marshall to use Stone & Webster, a Boston construction firm that he had used for his army projects.[25] Groves was much involved in drawing up a letter of intent, seeing to the transfer of money to the Corps of Engineers, and getting approval to purchase land for a site in Tennessee.

The major challenge facing the project was how to enrich uranium and, as a related task, how to produce plutonium. The goal of enrichment was to separate one isotope of uranium from another.[26] Uranium as found in nature consists of approximately 0.711 percent of the fissile isotope uranium-235, the remainder almost entirely the nonfissile uranium-238. While many heavy atomic nuclei are capable of being fissioned, only a fraction of these are fissile, which means fissionable by slow neutrons as well as fast neutrons. From a practical point of view, a fission weapon using uranium would have to be enriched in the isotope 235 to a level well above 50 percent.[27]

Plutonium splits as easily as U-235 but is found naturally only in trace amounts and must be made artificially. Discovered in 1941 by a group at Lawrence's laboratory in Berkeley, the new element — named plutonium for the outermost planet — could also serve as a potential ingredient for a bomb, if enough of it could be produced.

From early July to early September Marshall and Nichols visited the research laboratories where work was being done on U-235 and plutonium, talking with the scientists and making arrangements to support their research.[28] At the same time they were trying to decide where all of this work should be conducted — in one place or in several? Should pilot plants be built first to see which was most promising before going ahead with a major industrial effort?

Throughout this period, although busy with dozens of projects of his own, Groves was carefully observing how Marshall was going about organizing and managing the project. Groves had asked Marshall for a weekly report to be shown only to him, General Robins, and General Reybold, detailing the progress. Though Groves never publicly criticized Marshall, then or later, it is obvious from the actions he took in late September that he had decided that Marshall's way was not the right way to get things done.[29] Groves urged Marshall to act on the reports and studies that recommended a fifty-six-square-mile tract near Knoxville, Tennessee, for the electromagnetic and diffusion plants. The power supply was ample with the Tennessee Valley Authority nearby. Still Marshall held back, waiting "for further developments in the research for Dr. Compton's method."[30]

Marshall saw the need to get blanket priority ratings from the War Production Board and the Army Navy Munitions Board but never fought hard enough to obtain the highest ones. On September 2 he wrote in the joint diary,

> I showed Colonel Groves the two progress reports covering the project to date. He commented to the effect that a decision should soon be made with regard to the type of plant to be built — that the Army has to begin stepping in on OSRD. Groves would like us to make a site selection, get plans in shape, and start building with all speed. In this way he believes we can better justify an improved priority rating.

And a week later he again noted Groves's impatience:

> Groves appears to be most anxious for us to expedite final choice of a site by the S-1 committee and that we actually begin to build something at the main site.[31]

While Generals Reybold and Robins were satisfied with Colonel Marshall's progress, Bush and Conant were not. They had a better sense of what was required for success; Marshall was moving too slowly, and was too indecisive a personality.[32] Both of them — but especially Bush — had bet on being able to build the new weapon in time to be effective in this war. Without the knowledge of Congress, they had already committed significant sums of money to the effort.

As fall approached, Bush proposed to Stimson a reorganization plan whereby a small Military Policy Committee be formed to supervise the growing project. According to Bush's timetable, after formation of the committee, an army officer would be chosen to carry out its policies in a more vigorous manner. Initially Bush wanted Somervell but that was unreasonable, given his position as commanding general of Army Service Forces and his three-star rank. Bush's next choice was Styer. Styer was not inclined, and in any event, Somervell was unwilling to give up his loyal and competent chief of staff. Unbeknownst to Bush, however, Somervell and Styer had a candidate in mind.

Bush recognized he was facing a dilemma. On the one hand, he wanted the army to provide firm direction over the project and to fund it. On the other hand, this meant that he might lose control of it. Bush was aware of Somervell's formidable bureaucratic and political skills. Even if Somervell did not personally direct the project, as Army Service Forces chief he still might control it. Bush's fears were well grounded, but he was worried about the wrong person.

Determined to try to keep control, Bush met with Somervell and Styer in mid-September to discuss the makeup of the Military Policy Committee.[33] Bush would be chairman, with Conant as his alternate, and there would be army and navy members. Forming the committee would permit him to say that the scientists still had a voice in atomic policy, even while the army was in control.[34]

For most people living in Washington, September 17, 1942, was not particularly special. The weather that Thursday was typically hot and humid, with thundershowers expected later in the afternoon. The exquisite days of Washington's long fall were still a way off.

Eight months after Pearl Harbor, the city had adapted to being at war. The *Evening Star*'s and *Washington Post*'s headlines that week centered on the U.S. Marines' gritty repulsion of Japanese counterattacks on Guadalcanal and the Solomon Islands; on the belated announcement of the sinking, in

June, of the aircraft carrier *Yorktown;* and on the courageously defiant Soviet defense of Stalingrad, then in its fourth week. The country was girding itself to face fuel oil rationing for the coming winter. William M. Jeffers, head of the Union Pacific Railroad, had just been named Rubber Czar. The New York Yankees had already sewn up the American League pennant; in the National League the St. Louis Cardinals held on to a slim two-game lead over the Brooklyn Dodgers. Washington was preparing itself for the first citywide, door-to-door collection of scrap metal.

Regardless of the weather or the headlines, Col. Leslie R. Groves was looking forward to this day. It seemed as if he was finally going to get out of Washington and get into the war. The past two years had been busy ones, and he was weary of the constant strain of being in charge of all domestic army construction. It was not easy keeping track of it all, always demanding that contractors meet schedules, making sure they conformed to army regulations, rooting out the incompetent or unscrupulous ones, and choosing able engineer officers. Groves knew that he had made himself unpopular, but his prime concern was that the jobs be done on time, and done right.

Groves was a professional soldier, and to move from a desk in Washington to the battlefield to fight in this war was the best way he knew to express it. Serving overseas in an attractive assignment was also the high and wide road to rapid promotion, and Groves, while patriotic, was also very ambitious. He knew as well that his days as deputy chief of construction were numbered. Domestic construction had reached its peak in July 1942 and was now declining; the focus was shifting overseas. The next place where engineers would be needed would be in the theaters of war, with combat units. A few days before, General Eisenhower had taken command as chief of the Allied Expeditionary Force, with headquarters in London, for the upcoming northwestern African campaign, code-named Torch.

It was also no secret that many of his colleagues in the Corps of Engineers were hostile to Groves because of what they considered his high-handed, forceful style and his unwillingness to respect the traditional lines of authority or seniority. Styer had repeatedly informed him that he was not the most popular engineer in the corps. Many considered him an unprincipled opportunist who had manipulated his rise through, or rather over, the ranks.

Lastly, he saw field duty as a respite. It would remove him from the turbulence of pork-barrel politics that was so much a part of army construction — and the Washington maneuvering that went with it. Typical of

Groves's ironic sense of humor, he later recalled, "I was hoping to get to a war theater so I could find a little peace."[35]

As Groves later told it, in mid-September he "was offered an extremely attractive assignment overseas."[36] All he now required was release from his current post. For that, he would need the approval of General Somervell — now head of Army Service Forces, the umbrella organization that included the Corps of Engineers — and General Styer. He was about to find out that they had other plans for him.

Whatever other assignment may have been in the works, it was withdrawn, for on September 16, after the meeting with Bush but without consulting him further, Somervell and Styer decided that Col. Leslie R. Groves would be the ideal person to replace Colonel Marshall to head the Manhattan Project.[37] As Styer later said, he

> had a very high regard for General Groves' engineering, administrative and organizing ability; his capacity for work and the fearlessness with which he tackled difficult jobs and the drive and determination with which he pursued them to successful conclusion.[38]

For Styer this solved several problems. The prospect that he (Styer) would have to take direct responsibility for the project was now moot. Styer and his boss were somewhat skeptical of this top-secret, "longhair" stuff about making a bomb based on atomic energy. Maybe so, but probably not in this war. Styer could also satisfy any disgruntled engineers by removing Groves from his position as deputy chief of construction. Moreover, all of this would be accomplished without the appearance of penalizing, or in any way disciplining, Groves. Lastly, and perhaps most important, Somervell and Styer no doubt honestly believed that if anyone had a chance to bring off this huge gamble, it was Groves. He did have a track record of getting things done, and they could honestly say that even if it failed — which it very well might —they had provided their best man for the job. If it failed there would be consequences; blame to go around, reputations sullied, careers ended. But, they could argue, they had given it their best shot.

This fear of failure would be an ever-present shadow and stimulus until the test in mid-July 1945 and the bomb's use in early August. Everyone associated with the project, but especially Groves, felt it deeply. As he later said, "The President has selected me to carry the ball, which is another way of saying that I am to be the Goat if it doesn't work." Or even worse,

"If our gadget proves to be a dud, I and all of the principal Army officers of the project . . . will spend the rest of our lives so far back in a Fort Leavenworth dungeon that they'll have to pipe sunlight in to us."[39]

In addition to his own career he carried the reputation of the Corps of Engineers on his shoulders — as General Robins not too delicately reminded him:

> I hate to see you get this assignment, because if you fail in it, it will destroy you. I would be sorry to see that. But it would be still worse if it destroyed the Corps of Engineers. That would really make me sad.[40]

Groves's schedule was light on the Thursday morning in September when he received the appointment that changed his life. His only commitment that morning was to testify at ten o'clock — along with the newly appointed chief of engineers, Maj. Gen. Eugene Reybold, who was to take over officially on October 1 — on a military housing bill before the House Military Affairs Committee in Room 1310 of the New House Office Building.[41]

At 10:30 A.M., as Groves left the hearing room, he met Somervell in the corridor. Groves asked him if he would object to him (Groves) being relieved of his construction duties. Somervell's response surprised and dismayed Groves, and would also change his life forever:

"The secretary of war has selected you for a very important assignment, and the president has approved the selection."

"Where?" Groves asked.

"Washington," was the reply.

"I don't want to stay in Washington."

"If you do this job right," Somervell said, "it will win the war."

Groves suddenly realized what Somervell was referring to, and his spirits fell.

"Oh, that thing . . ."

Somervell went on, "You can do it if it can be done. See Styer and he will give you the details."[42]

Groves now understood the reason for yesterday's telephone call from Marshall's deputy, Nichols. Nichols had called to tell him that General Styer wanted to see Groves and him in his Pentagon office at noon on Thursday. When Groves had asked what it was about, Nichols pointed out, rather tartly, that he was not in the habit of interrogating generals in that way.[43]

So now, rather than going overseas and getting out of Washington, it seemed he was being pushed aside, doomed to spend the duration of the war working with a bunch of prima donna scientists, on a project filled with enormous uncertainty.[44] He was not pleased, but, he knew what it meant to be a good soldier, and Groves considered himself a very good soldier. Still, this was a bitter pill to swallow.

Just before noon Nichols picked Groves up outside his office at the New War Department Building for the short drive across the Potomac to the Pentagon. Groves was grimly quiet. Nichols had known Groves since they had both participated in the 1929 corps expedition to survey Nicaragua for a canal. Since June he had been with Marshall working on the project, and the two had seen one another from time to time. For Nichols, his future, at the moment, was a bit unclear itself. His boss, in effect, had just been demoted by now having to report to Groves. He knew better than to try to make idle conversation. Groves was not Nichols's favorite person.[45]

At the noon meeting Styer listed the reasons why they were unhappy with Marshall, what needed to be done, and why they had chosen Groves. Uncharacteristically, Groves burst out saying he wouldn't take it, as he had been promised an overseas position.

Styer ordered Nichols out of the room for a few minutes[46] and then made it clear to Groves that his selection to head the atomic bomb project was cast in stone. The secretary of war had asked for him, and the president had already approved the appointment. Styer was fond of Groves and in the past had protected him from the ire of those in the corps who wanted to see Groves removed from his post as deputy chief of construction. Now Groves reproached Styer "for letting me get hooked into this."[47] Styer did his best to mollify Groves. Realizing he was trapped with no way out, Groves made the most of it. He pointed out that he would have to be promoted to brigadier general; otherwise the scientists with whom he would be working would have no respect for him. After the actual promotion, he would formally take over. It was exactly this kind of quick-witted decisiveness that made Groves the man for the job. Styer concurred and promised a promotion within the week.[48]

Several other matters were discussed and decided upon. Groves would continue to oversee the completion of the Pentagon. He could best deal with any problems on Capitol Hill, and to suddenly disappear would arouse suspicion. The long-standing priorities problem was raised. Styer gave Nichols a letter from Bush on the topic, instructing them to return it to him later in the day with an explanation of how they would go about obtaining higher priorities. Styer told Nichols to draft a letter for Groves to use with Donald M. Nelson, chairman of the War Production Board, signed by Gen. George Marshall assigning an AA-1 rating to the Manhattan Project. At Styer's behest Nichols called Bush's office and made an appointment for three o'clock that afternoon.

Styer assured Groves that he would be virtually independent — with guidance and final approval from the about-to-be-established Military Policy Committee — and that he would have the full support of the War Department. As they left the office, Styer urged Groves to look on the bright side.

> The basic research and development [for the atomic bomb] are done. You just have to take the rough designs, put them into final shape, build some plants and organize an operating force and your job will be finished and the war will be over.[49]

Though Groves was still no doubt angry, he never let emotions intrude into his work habits, and upon leaving Styer's office he was all business. He asked Nichols to give him a rundown on the project and to review for him the progress that had been made since June. What he heard did not please him, for it confirmed what he had suspected since he had first heard about the atom bomb project: It was long on theory and short on practicality. The rosy account Styer had just given him had nothing to do with reality.

Upon returning to his office, Groves drafted the memorandum to the chief of engineers, to be signed by General Somervell, releasing Groves for special assignment. "Colonel Groves' duty will be to take complete charge of the DSM [Development of Substitute Materials] project as outlined to Colonel Groves this morning by General Styer."[50]

The story of Groves's visit to Bush on the afternoon of his appointment is described in almost every account of the Manhattan Project and is the usual point at which we are introduced to him. The scene has good dramatic elements and characters: the aristocratic Yankee and the "burly colonel," with the former ignorant of the latter's new appointment.

At three o'clock Groves and Nichols entered Bush's office at the Carnegie Institution of Washington, at 1530 P Street, Northwest.[51] Bush and Groves had never met. Quickly it became clear to both visitors that Bush was unaware that Groves had been appointed to head the bomb project. Bush, normally voluble, was uncharacteristically quiet. He refused to answer many of the direct questions about the project that Groves put to him. Bush was not about to divulge highly secret information to a person he had never met before, whose status on the project was a mystery to him. Furthermore, he simply could not believe that such an appointment had been made without his knowledge or approval. And he was horrified by Groves's bluster and apparent lack of tact. For his part, Groves did not hide his frustration at Bush's unwillingness to tell him what he believed he needed to know about the project.

As soon as Nichols and Groves left, a flustered and worried Bush immediately went to Styer's office. The general apologized for not informing him of the Groves appointment. To Bush, here was additional evidence that Somervell and the army were taking over. He had been outmaneuvered and there was little he could do about it now.[52] Nonetheless, he expressed his misgivings to Styer in no uncertain terms. Groves's arrogance, his lack of tact, and his brusqueness would offend many of the scientists and make it difficult to recruit technical personnel.[53] Styer listened to Bush's concerns for several minutes. When he finally got a chance, Styer acknowledged that Groves was indeed blunt but that he was a person who took charge and knew how to get things done.

But Styer's reassurance did not allay Bush's fears. Upon returning to his office, Bush penned a short note to Harvey Bundy, Secretary of War Stimson's special assistant for atomic energy development:

> I visited General Styer. . . . I told him (1) that I still felt, as I had told him and General Somervell previously, that the best move was to get the military commission first, and then the man to carry out their policies second; (2) that having seen General Groves briefly, I doubted whether he had sufficient tact for such a job.
>
> Styer disagreed on (1) and I simply said I wanted to be sure he understood my recommendation. On (2) he agreed the man is blunt, etc. but thought his other qualities would overbalance.
>
> Apparently Somervell saw General Marshall today regarding Groves. I fear we are in the soup.[54]

While waiting to receive his promotion to brigadier general, before taking over the project formally, Groves spent an extremely busy week taking initiatives and making decisions on a grand scale.[55]

The next day — September 18, his first in charge — he sent Nichols to New York City to see Edgar Sengier, the Belgian head of Union Minière, to arrange to buy almost twelve hundred tons of high-grade uranium ore that were already in the United States and to buy additional amounts in the Belgian Congo.

On Saturday the nineteenth Groves issued a directive to buy the Tennessee site and confronted the priorities problem head-on. Groves knew from two years of grappling with army construction projects that priorities can make all the difference to completing a project successfully and on time. If this wild idea had any chance of succeeding, it could only do so with the highest ratings, and Groves was determined to get them. Less than forty-eight hours after he had been given the job Groves marched into Donald Nelson's office at the War Production Board and told the chairman that an AA-3 rating was unacceptable. The work was too vital to the war effort; he required authorization to assign whatever rating was needed to get the job done.

Obtaining a high priority rating had been a goal of Marshall and Nichols.[56] With presidential support Nichols and Marshall met with Brig. Gen. Lucius D. Clay, ASF deputy chief of staff, for requirements and resources, on June 30 to request an AA rating for the project. On July 13 Clay offered only a blanket AA-3 rating.[57] His reasoning lay partly in the prohibition against using AA-1 and AA-2 ratings for construction projects. As a consolation, Clay allowed for the use of AAA ratings sparingly to overcome bottlenecks. They kept trying. On July 31 Marshall wrote,

> Nichols and I met Dr. Bush in his office. . . . also present were Dr. Conant [and] Dr. Stewart [of the Office of Scientific Research and Development] . . . Dr. Conant was willing to sponsor the memorandum handed him by Colonel Nichols which suggested that Dr. Bush go to see General Marshall. . . . it was finally agreed that Dr. Bush would see Mr. Nelson in person as soon as practicable and see if Nelson could and would direct the ANMB [Army Navy Munitions Board] to grant our request for higher blanket priorities for small items, subject to whatever restrictions Nelson or ANMB felt necessary.[58]

This timid approach was in sharp contrast with Groves's method, which was to arrive in Nelson's office with a letter in hand, to himself, lacking only Nelson's signature.[59] Nelson initially refused to sign, only to receive a service ace delivered by Groves.

> It was only after I brought the conversation to an abrupt close and started to leave the room that he asked me what action I was going to take. My reply was, "Recommend to the Secretary of War that the Project be abandoned on the grounds that Mr. Nelson refuses to carry out the wishes of the President." It was then that his resistance folded and he asked me to sit down and discuss the matter further and then signed my letter.[60]

The letter, which Groves produced on numerous occasions afterward as a badge of authority to pressure others, stated:

> I [Nelson] am in full accord with the prompt delegation of power by the Army and Navy Munitions Board through you [Groves], to the District Engineer, Manhattan District, to assign an AAA rating, or whatever lesser rating will be sufficient, to those items the delivery of which, in his opinion, cannot otherwise be secured in time for the successful prosecution of the work under his charge.[61]

On Monday, September 21, four days into the job, Groves had a second and more successful meeting with Bush, accompanied by Marshall and Nichols.[62] Bush apologized for the confusion and told them that the navy had been left out of the project at FDR's direction. Bush himself preferred dealing with the army and that settled the matter.[63]

The days were getting busier. On Wednesday the twenty-third Groves took the oath of office as a brigadier general and formally took charge of the Manhattan Project. At 3:05 P.M. there was an important meeting in the secretary of war's office. Attending were Stimson, Generals Marshall, Somervell, and Styer, Rear Adm. William Reynolds Purnell,[64] Bundy, Conant, Bush, and Groves. The purpose of the meeting was "to decide on the form and make-up of the policy supervision of America's atomic effort."[65]

Groves presented his views on how he expected the project to operate, saying that to avoid empire building, existing facilities of other government agencies should be utilized to the utmost.[66] This was followed by a discussion of how oversight would be facilitated. As Groves described it later, it was Stimson who proposed a committee of able men from OSRD, the army, and the navy. As we have seen, Bush's idea of a Military Policy Committee had been germinating for some days, and no doubt he discussed it with Stimson prior to September 23.[67] When Stimson suggested that it consist of nine or possibly seven members Groves vigorously objected, saying that would be too large and unwieldy. A committee of three was the right size.

As it worked out in practice during the months that followed, the three members were effective allies whom Groves drew upon to gain the support and cooperation of key individuals inside the government and out. Bush was selected as chairman, with Conant as his alternate, and Groves drew upon the two repeatedly for scientific resources. Rear Admiral Purnell was the navy representative. Purnell also served on Adm. Ernest J. King's staff and thus was able to get full support from the navy whenever it was needed. Groves was for a few hours the army representative, but later that afternoon Stimson replaced Groves with Styer. Styer, as deputy commanding general and chief of staff, Army Services Forces, supplied Groves with anything he needed from the ASF's vast resources.

Groves would be, in Stimson's words, "the executive head of the development of the enterprise from now on."[68] Bush later described the MPC

> as a sort of board of directors for General Groves. It met frequently, had no staff or secretary present, kept no formal records, but provided a point at which every important move could be discussed and closely examined.[69]

As it worked in practice normally, the MPC reacted to what Groves had already done rather than giving him orders that he then carried out.[70] Retaining the position as chairman allowed Bush to say that the scientists retained some influence in atomic policy making.[71] But Bush's authority was limited from the outset and decreased, in about the same proportion as Groves's influence and power grew, over the next two years.[72] Groves was little concerned with wiring diagrams of who was theoretically responsible to whom. As Nichols said, "Organizational channels meant nothing to Groves."[73]

The Rubicon had been crossed. A military-industrial effort was about to be forged by a relentless and driven taskmaster. Left behind was the small-scale scientific-technical model. Over the longer term into the postwar world the decision to have the army run the project and the marshalling of government money with science and technology was a far-reaching one with implications for how science would be funded for decades to come.

With the meeting nearly over Groves stood up, glanced at his watch, and told his superiors that if they had nothing more important to bring up he had to leave to see to the purchase of the Tennessee site. Bush must have felt some relief, for he knew that Colonel Marshall had been agonizing over this decision for the better part of four months. Here in less than a week Groves had learned all he needed to know to convince him to move ahead. Though he was undoubtedly pleased to see things move forward with such decisiveness, Bush must have sensed as well that control of the project was slipping out of his hands into those of the newly minted brigadier general who was on his way out the door.

Upon leaving the meeting, Groves headed directly for Washington's Union Station and boarded a five o'clock train for Knoxville. The next day, after inspecting the proposed Tennessee site, Groves took the initial steps to obtain the land and begin construction of the isotope separation plants. By the beginning of October condemnation proceedings were well under way.

Groves returned from Tennessee on Friday to new offices in the New War Building, specifically Rooms 5120 and 5121. He also received word from the Army Navy Munitions Board that the Manhattan District had been granted blanket AAA priority to be used at Groves's discretion. Groves was warned that indiscriminate use of the authority to the detriment of other war projects would lead to a withdrawal of the rating. It was to be used only in emergencies. Groves believed that he would now be able to proceed using the baseline AA-3 priority, supplemented with his emergency AAA authorization. Later when he became convinced that the combination would not work, he asked that the base rating be changed to AA-1, the highest base rating possible. By order of the Joint Chiefs of Staff, such a rating was forbidden for construction projects, but in March 1943, the Manhattan District was issued an AA-2X rating, a special rating for high-priority industrial operations. Groves's persistence in trying to get a still-higher base rating finally paid off in mid-1944.

In his efforts to establish the highest priority possible for the Manhattan District, Groves periodically appealed to Undersecretary of War Patterson and even to Secretary Stimson. Time and time again Groves used Patterson and Stimson to hammer home to the War Manpower Commission that the Manhattan Project had priority for labor over all other programs.

When the War Production Board established the Controlled Materials
Plan in late 1942, rank ordering AAA requests from various defense projects, Groves made sure that the Manhattan Engineer District was given
first place above all other programs, including the Landing Craft Program,
Aircraft Programs, Ordinary Munitions, and Aviation Gasoline. Much to
the consternation of those in other programs, they were repeatedly forced
to give way to the Manhattan Project, even though the only thing they
knew about it was that it was secret.

Groves's actions during these first few days on the job showed extraordinary decisiveness. As we have seen — but it is worth noting again — he
was not a novice to the project. For three months he had been observing
Col. James Marshall, and there were the several initiatives that Nichols had
under way that could quickly be concluded, such as buying the ore from
Sengier, finalizing the Tennessee site, and dealing with Stone & Webster.
This got Groves off to a running start, and he never looked back or
second-guessed himself. As with previous projects, he threw himself into
Manhattan with enormous energy and no letup.

At some point during these first few days, Groves must have said to
himself that if he was going to make his mark in this war, it would have
to be with this risky project, and not on the battlefields of North Africa,
Europe, or the South Pacific. Though the prospects were uncertain, if the
president, secretary of war, chief of staff, and a number of Nobel
Prize–winning scientists thought there was a chance, then it would have
to be tried. He would give it his all. Halfway measures would not do.

And so over the next few months as he chose the sites and the key figures, and began to commit huge sums of money and resources, there
would be no turning back, for him, for the Military Policy Committee,
for anyone. Whether they liked it or not, he had committed them to a program whose proportions they may not earlier have fully grasped. To make
the gamble pay off required larger and larger commitments so that the earlier amounts would not be wasted. This dynamic bound all of the key
people together. A juggernaut was building up speed and Groves was the
central and crucial figure, driving it forward, ever faster, racing toward the
finish.

PART FOUR

The Manhattan Project

· *1942–1945* ·

General Groves planned the project, ran his own construction, his own science, his own Army, his own State Department and his own Treasury Department.

WILLIAM A. CONSODINE[1]

His Own Construction: Atomic Factories and American Industry, Oak Ridge and Hanford

The Manhattan Project was, among other things, a gigantic industrial and engineering construction effort, run by the military under great secrecy, rapidly accomplished, using unorthodox means, and dealing in uncertain technologies. Overemphasis of the roles played by science and the scientists has distorted our understanding and appreciation of these features.

Groves ran the Manhattan Project in precise and resolute ways, recruiting some of the giants of American industry to build and run his atomic factories. A closer look at Groves the administrator and builder will supplement — and partially supplant — the scientific story for a fuller understanding of how the atomic bomb was built.[2] Without Groves's organizational and managerial skills, and construction know-how, the project would have taken longer to accomplish, or perhaps even failed. Individuals do make a difference, and in this instance Groves was indispensable to the project's success.

The formal order that Groves received from Lieutenant General Somervell on September 17, 1942 specified his responsibilities:

> 2. Colonel Groves' duty will be to take complete charge of the entire DSM project as outlined to Colonel Groves this morning by General Styer.
>
> e. Draw up the plans for the organization, construction, operation and security of the project, and after approval, take the necessary steps to put it into effect.[3]

The words *take complete charge* and *take the necessary steps* did not have to be repeated or explained. In carrying out his duties Groves took an expansive interpretation of the order. His reading of them would be closer to: "Do whatever it takes to build the bomb as quickly as possible and thereby end the war."

Groves created the organizational structure and the lines of command of the Manhattan Project with himself at the top as commanding general. Normally he would have reported to General Reybold, the chief of engineers, but the lines of authority instead ran directly to Secretary of War Henry L. Stimson and Chief of Staff Gen. George C. Marshall.

The relationship between Groves and Stimson was one of mutual trust and support. Stimson had already had a long and distinguished career, as secretary of war under President Taft, secretary of state under President Hoover, and numerous lesser positions. At the age of seventy-two Stimson was called upon by President Roosevelt to serve his country once again. He was, as one biographer has called him, "the founding father and patron saint" of the foreign policy establishment.[4] Stimson had gone to the right schools — Andover, Yale, and Harvard Law School — belonged to the right clubs, and was at the center of a circle of influential people for almost half a century. He was proper and upright to a fault and was able to inspire loyalty in others. The group of close aides he recruited at the War Department — George L. Harrison, Harvey H. Bundy, John J. McCloy, Arthur W. Page, and Robert A. Lovett — was thoroughly dedicated to him.

There was some social distance between the upper-class Stimson (and his loyal aides) and the middle-class Groves, twenty-nine years his junior. Groves was aware of this but was never intimidated by it. In fact, Stimson embodied many of the things that Groves valued most: intelligence, honor, and patriotism. There was a physical toughness to Stimson as well. He had been in the army and believed that "the code of the officer and gentleman was my own code."[5] With regard to building the bomb Stimson acted as the chairman of the board with Groves as chief executive.

Groves's other superior in the formal command channel was General Marshall. Over the three-year period their relationship was always excellent, based on mutual trust and respect.

> One reason why we were successful was non-interference from above. General Marshall never interfered with anything that was going on. He didn't ask for regular reports: he saw me whenever I wanted to see him and instructions were very clear. . . . The same thing was true of Secretary Stimson, with the exception of interfering on the target of Kyoto; never once was there any interference or intrusion; never once did I have to explain why; never once did I have to talk about approval for money appropriations.[6]

The Manhattan Project was officially established on August 13, 1942, to develop and construct an operable atomic bomb for military use. The Manhattan Project consisted of the Manhattan Engineer District (MED), the Santa Fe Area, and the Manhattan Project Headquarters and Staff. Often the terms *Manhattan Project* and MED are used interchangeably though they were not precisely the same. The initial arrangement had a special district engineer — Col. James C. Marshall — as head of the project, with the office located in New York City. With Groves's selection as commanding general, he became senior to the district engineer and had his headquarters in Washington.[7] Marshall remained district engineer, and the MED office remained in New York, but there was little question who was now in charge. Marshall continued to carry out his duties until August 1943, when Groves eased him out and he was reassigned to command an engineer training center.[8] At that time Groves appointed Col. Kenneth D. Nichols district engineer and moved the Manhattan District office to Oak Ridge, Tennessee.

Groves kept his Washington headquarters staff small. The two rooms on the fifth floor eventually grew to five or six, but by almost any standard the office was minuscule. Later he said that his model was Gen. William T. Sherman, who, during his march to the sea, limited his headquarters to what would fit into an escort wagon.[9]

A handful of people oversaw the activities and functions of tens of thousands in the field. Within project headquarters was the Washington Liaison Office. Liaison Officer Capt. (later Lt. Col.) Allan C. Johnson was in charge and, with his deputies, procured anything that was needed at Oak Ridge, Hanford, or anywhere else, using the assigned priority rating — up to and including AAA if need be. The purpose of the liaison office was to obtain the special consideration that was necessary to speed the project along through close contact inside and outside the MED. Johnson's staff focused on specific parts of the project and worked with the Services of Supply (later Army Service Forces, ASF), the War Production Board (WPB), and other agencies, departments, and offices to see that the MED maintained "the preferred position" and got whatever it needed. For example, K-25 Liaison Officer Maj. Albert C. Roeth Jr.'s job was to work with the WPB, ASF, and contractors to ensure that whatever was needed at K-25 was provided on a prompt and efficient basis. Other Johnson deputies were liaison to Y-12, to X-10, or to corporations like General Electric or Westinghouse. They broke through the red tape and expedited the flow of supplies, labor, or whatever, invoking the highest authority and the topmost priority ratings. In the office they were known as expediters.

As we have seen, one of Groves's first acts, within two days of getting the job, was to make sure that he had first priority on anything that he might need, with approval of the War Production Board, the "supreme industrial mobilization control agency," of the war effort.[10] But the AAA rating he forced Nelson to assign to the MED was not an all-purpose solution that could be used every day. It was something to be held in reserve, to be used sparingly, only when absolutely necessary. For normal daily activity Groves had to push for the highest lesser rating he could get. Nothing came easy or was permanently settled; pressure had to be kept on all the time.

Initially on July 23, 1942, the project was assigned an AA-3 priority, under the new rating system that had just gone into effect.[11] Rating inflation set in again with the new system, and Groves had to twice press for a higher priority for the MED. On March 22, 1943, the AA-3 was raised to AA-2X (midway between AA-2 and AA-3), and on July 1, 1944, this was raised to AA-1, with AAA used when needed.

Many government bureaucrats involved in the priority system, as well as construction and materials contractors, who were ordered to give their unqualified support to a highly secret program, knew nothing about the purpose of the MED, and sometimes there was resistance. Groves, and his liaison officers, had at their disposal letters and directives signed by some very important people that served to get people's attention, break logjams, and hasten the process. There was the Bush-Conant report of June 17, 1942, signed by President Roosevelt.[12] There was Nelson's letter to Groves of September 19, 1942, and two particularly strong letters from Undersecretary of War Robert P. Patterson, who was Groves's highest official contact in the War Department for industrial matters, to see that the Manhattan Project moved forward expeditiously.[13] These usually did the trick.

Problems also arose from other high-priority secret programs competing for resources and manpower. The managers of the high-octane gasoline, synthetic rubber, or landing craft programs thought their projects no less important than the mysterious Manhattan District and fought vigorously for higher ratings to achieve their goals. Groves's challenge was to navigate the MED through the vast wartime labyrinth of boards, offices, bureaus, divisions, branches, and commissions that controlled the industrial mobilization for war. With his knowledge of the structure and the personalities, the full support of his superiors, and a healthy dose of self-confidence, Groves was more than up to the task as he manipulated the system to the MED's advantage to get the job done.

At the next tier down was the Manhattan Engineer District, originally in New York City but after September 1943 in Oak Ridge, Tennessee. The MED was organized into half a dozen operating and staff units, permitting Colonel Nichols to administer them from Oak Ridge and allowing Groves direct access to his many outposts from Washington, if need be. The two largest operating units were the Hanford Engineer Works (HEW), headed by Lt. Col. Franklin T. Matthias, and the Clinton Engineer Works (CEW), headed by Lt. Col. Thomas T. Crenshaw. Various divisions and area offices were the administrative vehicles to carry out Groves's and Nichols's directives. There were area offices located in close proximity to each district's university laboratories. Lt. Col. Arthur V. Peterson was Chicago Area engineer (and also Nichols's brother-in-law). Maj. Harold A. Fidler was the Berkeley Area engineer, and Lt. Col. James C. Stowers, followed by Capt. Lawrence L. Grotjan, headed the Columbia office.

Area offices were also established to work with the corporations that were building or operating the MED's plants. Following the Corps of Engineers principle of decentralization, they also provided a direct link to General Groves, if need be. A phone call from the general to the area officer could convey an order or learn what was going on.

On the corporate side the firm selected a project manager — several of whom had worked with Groves in the past — to be in charge of construction or operations. They would work directly with the area and/or the district office, and they too were only a phone call away from Groves, who checked on them constantly and visited the sites often. Albert L. Baker of Kellex was project manager for K-25, August C. (Gus) Klein of Stone & Webster directed all production work on Y-12, while Frederick R. Conklin of Tennessee Eastman oversaw its operation. These three had enormous energy and got on well with Groves. He liked people like himself.

The expanding range of activities connected to the K-25 gaseous diffusion plant, overseen by the New York Area Office, eventually led to creating three smaller offices in Decatur, Illinois, Milwaukee, and Detroit. The Boston Area Office served as Groves's eyes and ears to Stone & Webster. The Rochester Area Office worked with Tennessee Eastman, in charge of the operation of the electromagnetic plant at Oak Ridge. The Wilmington office worked with Du Pont. There was one area office in Canada — at Trail, British Columbia — where heavy-water production was carried out.

The Madison Square Area Office was located at 261 Fifth Avenue (at Twenty-ninth Street) in Manhattan. After the district office moved to Oak Ridge in August 1943, the Madison Square Area Office served as a double-blind procurement center under the direction of Lt. Col. John Richard Ruhoff.[14] Six smaller area offices served under it: Beverly, Colorado, St. Louis, Tonawanda, Wilmington, and Murray Hill. The Tonawanda Area Office near Buffalo, New York, oversaw the Linde Air Corporation, which was responsible for the processing of uranium ore. The St. Louis office worked with Mallinckrodt Chemical Company, which processed uranium.

Most of the materials requisitions, regardless of the ultimate destination, were issued from this office. Thus, the supplier or manufacturer had no idea of the ultimate destination of the shipment. The shipment would then be rerouted to its destination as originating from the Madison Square Area Office, blinding the recipient from knowing the location and in some cases the identity of the supplier. In 1946 GAO auditors must have been initially mystified to understand how hundreds of millions of dollars of acquisitions of every kind were sent to a tiny office in a Manhattan office building. An area office in Los Angeles, Calexico, served a similar purpose supplying the Los Alamos and Berkeley labs.[15]

There were offices outside the United States as well. Groves opened a Montreal office to oversee Canada's involvement in Manhattan Project operations. Horace Benbow was the liaison officer among the MED, the Canadians, and the British. As we shall see, Capt. Horace K. Calvert was sent to London in January 1944 to head the foreign intelligence section.

One of the most secretive offices was the Murray Hill Area Office, located in the Murray Hill section of Manhattan, on the East Side between Thirty-fourth and Forty-second streets. As Groves gained greater knowledge of the project's technical problems, he rightly concluded that having a plentiful supply of uranium ore was going to be crucial to its success. A corollary of this was that if the United States could control the world's supply of uranium and/or thorium ore, it could maintain a monopoly on the bomb. The Murray Hill office was responsible for surveying the world's supply of uranium and thorium ores and pinpointing those regions the United States should try and control one way or another.

Groves approached this project as he had approached the hundreds he had run before. There were certain tried-and-true procedures; stick to them and you increase your chances of success. First and foremost, find and recruit the best people you can. Intelligent people with ability go a long way in getting the job done. Do the same with companies. Either find those that you have worked with before and become familiar with their performance, or use the biggest and the best American industry has to offer. Second, run it on a military basis with tight vertical lines of authority, delegating responsibility to those you know and trust.[16] Do not get swamped in detail; let only the largest and most important issues rise to the top to resolve. All lesser decisions should be taken care of down below. Pay close attention to warning signs. Part of Groves's success can be attributed to his ability to spot trouble early and head it off before problems became serious.

Finally, a bedrock procedure that Groves adhered to was to use already existing military and civilian offices and departments of the U.S. government to perform important functions. There was no need to duplicate something if an already functioning organization could do the job.[17] An excellent example of this was his use of the Real Estate Branch of the Construction Division of the OCE as the vehicle to acquire more than half a million acres for the major and minor Manhattan installations. Groves knew the branch well and had used it hundreds of times to acquire even more millions of acres during the mobilization period while he was deputy chief of construction. Beginning in 1940 the secretary of war had delegated responsibility to the Real Estate Branch to administer the acquisition of land for military purposes. After the transfer of the Construction Division from the OQMG to the OCE, the Real Estate Branch operated under the decentralized procedures of the Corps of Engineers. Each division engineer was responsible for acquiring real estate within the geographical boundaries of his division.

Groves had a long-standing and harmonious relationship with Col. John J. O'Brien, who had become chief of the Real Estate Branch in late 1941. O'Brien had extensive experience in the real estate field. He had served as negotiator and appraiser for the Federal Land Bank (to make long-term, low-interest amortized loans to farmers) in St. Paul, Minnesota, and then joined the Department of Justice in 1936, where in five years he rose to be general assistant to Norman M. Littell, assistant attorney general in charge of the Lands Division. O'Brien was commissioned a colonel in 1941 at the age of thirty-seven.

"[O]ur cooperation," said Groves, "was so complete that there was never any question as to whether he was [my subordinate] or not. Instructions given to him by me, as was my custom, were almost always given in the way of advice, not only on general problems but on specific problems. This was the general policy on all matters and only rarely did I issue a direct order."[18]

Needed land and facilities were either publicly owned or privately owned. Well over half of the total land and facilities acquired by the MED had been privately held — nearly all of it at Clinton, more than half at Hanford, and several thousand acres at Los Alamos. Acquiring publicly owned property was fairly straightforward, and it was Groves's preference if possible. More difficult were the acquisition of private farms, residences, and commercial properties. Under existing regulations there were several methods available to the Real Estate Branch to acquire the property for the MED. These included purchase, condemnation, donation or transfer from another government agency, lease, easement, war restriction agreement, or similar legal instruments.[19] All of these methods were used, though direct purchase and condemnation was the most common. Direct purchase was preferred, because it had lesser chance of stirring up local opposition than condemnation. At Hanford the unwillingness of some property owners to accept the price set by the government appraisers forced the War Department to resort to condemnation in many instances, but as we shall see below, it did not quell the opposition.

As Groves began his new job, he brought with him several trusted aides with whom he had been recently working. First and foremost among this group was his secretary, Jean Marley O'Leary.

Jean Marley was born at the stroke of midnight, between December 31, 1909, and January 1, 1910. To make herself seem a year younger she chose the latter date as her birthday.[20] Her father, William Marley, was born in England and fought in the Boer War, after which he came to America. There he met Ellen Dillon, a widow with two children. They married and had four daughters. Eugenie (Jean) was the oldest. After Mrs. Marley died in about 1925, William had to raise the four girls. Jean went to Sacred Heart Academy in New York City, graduated, and went to work for *Time* magazine. In 1929 she married William O'Leary, who sold advertising for the *Jersey Journal* newspaper in Jersey City. Their only child, Coni, was born on March 2, 1931. In the mid-1930s they moved out of New York to West Englewood, New Jersey.

William O'Leary died of leukemia in March 1940, leaving Jean to support herself and her nine-year-old daughter. She moved to Jackson Heights, Queens, and took a job as secretary to Asa Smith Bushnell, head

of the Intercollegiate Athletic Association of the United States (predecessor to the NCAA) in Manhattan. At the same time she applied to work
for the federal government. In early 1941 she was offered a position, came
to Washington, and lived in a rooming house off Connecticut Avenue, near
Chevy Chase Circle, in upper Northwest. On March 12 she began work
in the typing pool of the Office of the Quartermaster General.

Groves had a reputation for his brusque treatment of secretaries. The
practice of the day was that if you needed something typed or some other
task accomplished, you got someone from the pool. When the task was
finished the secretary went back to the pool. Only higher-ranking officers
had permanent secretaries. Groves's habit was to acquaint himself with the
new faces, to be on the lookout for talented people with whom he might
be able to work. But those assigned from the pool to work with Groves
on a permanent basis never lasted long.

In the spring of 1941 Jean O'Leary was one of the new faces. Their first
encounters were anything but pleasant. She responded to his gruff, teasing
sense of humor in kind and then refused to speak to him at all. She was
not fazed or cowed by his rank, his position, or his manner. This probably
got her the job. As a general rule Groves respected people who stood up
to him. There were exceptions, of course, but normally it showed a person
with some backbone and character.

In June 1941, more likely at his request than by chance, she was assigned
to him. O'Leary quickly gained Groves's trust, and her secretarial and
managerial skills soon made her invaluable and irreplaceable. He found
that she shared most of his worldview and was as keen a judge of talent as
he in fulfillment of their common purpose. While they teased each other
unmercifully, they worked together very well, and she quickly took over
the running of the office.

It was hardly noticed when at the end of September 1942 Brigadier
General Groves and Mrs. O'Leary disappeared from the offices of the chief
of engineers to set up quarters in two small offices on the fifth floor of the
New War Department Building.[21] Over the next three years it would be
from these few rooms that General Groves, Jean O'Leary, and their tiny
staff ran the Manhattan Project.

Those who worked under Groves began to refer to her as Major
O'Leary. To see her initials, *JOL,* on a directive meant it was authoritative.
She not only ran the office, she ran Groves as well. She would remind him
of the normal office schedule regarding appointments and phone calls, and
type out his detailed travel itineraries. She would remind him of personal

things as well, such as taking his medicine or remembering to give Mrs. Groves a valentine or to write to his son at West Point. Several times she composed such letters herself; all father Groves had to do was sign. If he was not around or was too busy, she sometimes even signed his name herself.

She was probably the only other person who knew all, or almost all, of the details of the Manhattan Project. She was the only secretary permitted to takes notes on the deliberations of the Military Policy Committee, the Combined Development Trust, and other groups that met in Groves's office. She kept his office appointment book (sometimes referred to as his "diary") and transcribed almost every phone call.

Groves had a practical reason for giving her enormous responsibility and sharing everything with her: If something should happen to the general — if his plane should crash or he die suddenly of a heart attack — at least one other person would know the workings and secrets of the Manhattan Project. That person could then work with Groves's successor to complete the project.

During the war Groves did not have a chief of staff or an executive officer.[22] He asked Jean if she wanted to be commissioned in the WACs as a major so that she would be accorded the rank officially, but she declined. Though she did not have the titles of *chief of staff* or *executive officer* she performed most of the functions, and when the general was not in the office she was effectively the director. Her official designation was *administrative assistant,* which hardly described the scale and scope of her job. On a number of occasions Groves sent her on trips to Oak Ridge, Hanford, and Los Alamos, and sometimes used her as a courier to carry top-secret information.[23]

Groves had amazing stamina. His workday was normally twelve to fourteen hours long. Six-day weeks were routine, with an occasional Sunday thrown in when necessary. Even July 4, Thanksgiving, and New Year's Day were opportunities to get things done. He never seemed to lose sleep over decisions. The minute his head hit the pillow, he was asleep.[24]

Groves was extraordinarily effective in using his time to full advantage, asleep or awake. There was barely a minute wasted throughout the day. Just reading the daily log of calls, visits, and visitors in his appointment book makes one weary. Part of his secret — one that he no doubt shares with other leaders who have huge responsibilities — was his ability to focus on an issue or problem and work through it until it was solved or a decision made. Groves was faced, of course, not with just one issue, problem, or decision but with dozens of them in a never-ending stream. His ability to compartmentalize them in his mind, figuring out a solution and then moving on, kept the long list from overwhelming him.

Over a period of seven years Groves did not take a vacation and was never sick. There was an occasional day off here or there, but certainly not anything serious or sustained in the way of rest and relaxation. The intense period of mobilization and early wartime construction work from 1940 to mid-1942 was followed by the even more intense Manhattan years from 1942 to 1945, building the bomb.[25]

Throughout the latter period Groves was out of Washington more than half of the time each month. Initially he had to familiarize himself with the status of the project, mainly at university campuses. The next step was to decide upon the location of the major sites. Groves selected the three main areas in Tennessee, New Mexico, and Washington State in September and November 1942 and January 1943, respectively. By the spring of 1943 construction was well under way everywhere, and the purpose of his visits was to speed the projects' completion through constant, unrelenting pressure.

Much of his travel was by railroad aboard some of the celebrated trains of the day: the Santa Fe Super Chief, the Baltimore & Ohio Capitol Limited, the New York Central Wolverine, the Southern Pacific Streamliner, and the Pennsylvania Railroad Liberty Limited, to name a few. Washington's grand Union Station was his point of embarkation to Chicago, Knoxville, San Francisco, Pasco, Washington, Santa Fe, Detroit, Boston, New York, Wilmington, Pittsburgh, Montreal, and elsewhere. A transcontinental trip could take three or four days. With no interstate highways and only two- or (sometimes) three-lane roads, travel by car was time consuming.

Often Groves would go to Wilmington, Delaware, to see Du Pont executives in the morning and return in the evening. Day trips to New York City and back were frequent as well. For his many trips to Oak Ridge he normally took the 4:30 P.M. train from Washington, which arrived in Knoxville at 6:55 the following morning, and was met by a car and driver. He would then spend eight hours or so conferring with Nichols, visiting the sites, anticipating potential problems, and urging everyone to work harder and faster. He then took an afternoon train back to Washington, arriving early the next morning and heading directly to the office for a full day of appointments.

Groves used the train as his mobile office. He normally traveled in civilian clothes, and his roomette became the temporary headquarters of the Manhattan Project. To cram more work into a day, upon leaving Washington Mrs. O'Leary or another aide would occasionally accompany him partway while he dictated letters, gave instructions, and planned his busy schedule. After an hour or two the assistant would get off, take another train back to the office, send off the letters, file the reports, and schedule his future appointments. Sometimes an aide might travel to meet him as he was returning. For example, Maj. Robert F. Furman, in charge of several intelligence matters, might go to Chicago, meet the general's train coming from Santa Fe, and return with him the rest of the way to Washington, briefing him on developments while he was gone, bringing him reports, mail, and news from the office. By the time the train pulled into Union Station, many hours of work had been accomplished. Another time saver was to schedule short meetings at the railroad station while en route to somewhere else. For example, the train trip to New Mexico went through Chicago; a brief meeting there with Area Engineer Major Peterson could be used to solve problems and make decisions.[26] When he could, Groves used Sundays to travel. To save valuable time on weekdays the general left in the late afternoon or evening and traveled overnight, arriving at his destination in the morning fresh and ready for action.

Groves occasionally traveled by commercial civilian aircraft, though he preferred the train. By the spring of 1945 Groves's schedule was becoming extraordinarily busy, and too much time was wasted traveling. In April 1945 Groves asked that a military plane and crew be assigned to him. Secretary Stimson quickly approved the request and provided Groves with a plane, one that was being prepared for the secretary of war.[27] Even then, of course, a transcontinental flight could take the better part of a day.

Groves's intimate knowledge of the strengths and weaknesses of construction firms across the country meant that he could choose the best contractor for a particular task. His years of overseeing the construction of army cantonments, munitions plants, and airfields, along with earlier experiences dating back a decade, had prepared him well. With army construction having peaked in mid-1942, he had his pick of the best engineers and the best firms. He also knew particular individuals whom he could call upon to work on the highly secret project.

In fact, Groves had already been involved in another large and difficult secret project: the Holston Ordnance Works (HOW), in northeastern Tennessee, at Kingsport.[28] Several of the techniques that he used to build it were borrowed and adapted to the Manhattan Project, most notably the process of designing, building, and operating plants virtually at the same time.

The HOW was a munitions plant for producing RDX. RDX (Research Department Explosive) was a much more powerful explosive than TNT.[29] The British method of making RDX, known as the Woolwich process, required huge amounts of nitric acid. Dr. Werner E. Bachman, an organic chemist of the University of Michigan, had found an alternative method in March 1941, the "combination process" requiring (among other chemicals) acetic anhydride.

In January 1942 the National Defense Research Committee asked the Tennessee Eastman Corporation to undertake experimental work on RDX, and to build a pilot plant. On February 26, 1942, the Wexler Bend Pilot Plant, in Kingsport, produced small quantities of RDX. On this basis a huge plant was authorized to produce Composition B (the mixture of RDX and TNT), in quantity. Construction of the HOW began on July 4, 1942; on April 20, 1943, the first explosives were produced. Seven more lines were opened during June and July, and — in a supplemental contract — ten additional lines were functioning by mid-March 1944. The HOW had two separate areas, one for the manufacture of the raw materials (acetic acid and acetic anhydride), the other for the manufacture of RDX and, when mixed with TNT, Composition B. In addition to its widespread use in conventional bombs Composition B was also used as the high explosive (HE) in the atomic bombs.

Much like the Manhattan Project that would soon follow, Holston was literally designed, built, and operated simultaneously.[30] Given the enormous scale and complexity of the project Groves created a special Holston district, with its own district engineer to supervise construction by the contractor. Groves would soon call upon Tennessee Eastman again for another difficult job. The HOW covered more than sixty-five hundred acres, with 242 permanent buildings erected, 141 magazines, thirty-one miles of railroad, fifty-nine miles of roads, and a total cost of $128.7 million. It was calculated that it took almost thirty-nine million man-hours to build. The large labor demand — at its peak eighteen thousand — necessitated that most workers be brought in from outside and housed.[31]

> Problems were many: a dearth of design information, an element of friction between the architect-engineer-manager and the principal subcontractor, the district engineer's seeming reluctance to crack the whip, a weak priority rating, a scarcity of equipment, and a persistent shortage of labor.[32]

Holston was producing at full design capacity of 170 tons of RDX daily by July 1943.

> The powdered RDX is mixed with melted TNT, and the resulting hot liquid Composition B explosive is passed through specially designed equipment, dropping in tiny globules onto a slowly traveling steel belt. These globules, cooling as they travel, become the solid pellets known informally to the operators as "Jap kisses," which are the end-product of the plant.[33]

By January 1944 Holston was producing and shipping about 570 tons a day of Composition B. Its total output of Composition B for the war effort was 434,000 tons, almost a billion pounds.[34]

Groves would follow the Holston model on a grander scale as he set about to simultaneously design, build, and operate Oak Ridge and Hanford. The Military Policy Committee had decided upon three methods to produce fissionable materials: the pile method for plutonium production, and the gaseous diffusion and electromagnetic methods for uranium enrichment.[35] Originally all of the work was to be done at Oak Ridge, but a decision was arrived at early on to produce plutonium elsewhere.

Groves approached Oak Ridge and Hanford much as he had approached many projects in the past: Put together an integrated group of architectural, engineering, and construction firms working in concert with the army engineer area office. The universities and the scientists that had had the earlier research contracts would provide advice and opinion about the special technologies involved. Groves's Washington office, with its ability to use whatever priority it needed for men and materials, could supply anybody or anything, anytime, anywhere.

Groves wasted no time in acquiring the Oak Ridge site and getting under way. On September 19, two days after he had been selected, he ordered the purchase of 56,200 acres along the Clinch River in the hills of eastern Tennessee. The location — first designated the Clinton Engineer Works (CEW) — had been under consideration for several months as the site to build the uranium and plutonium plants, but no final decision had been made. It was isolated with a dispersed population; close to a good-sized labor pool in Knoxville, eighteen miles to the east; and had an adequate source of water. Col. James Marshall had begun to assemble a team and to choose the contractor. He assigned Maj. Thomas T. Crenshaw as area engineer at Berkeley and Capt. Harold A. Fidler as his deputy to assist Ernest Lawrence in the electromagnetic venture. This was a start, but now things must move faster.

Groves quickly assembled a team to build the electromagnetic plant that would become known as Y-12.[36] By December 23, 1942, Groves, Lawrence, and John R. Lotz, the president of Stone & Webster, in a meeting in Berkeley, were able to agree on the general size and materials of construction so that orders could be placed. On January 4, 1943, Groves contracted with Westinghouse Electric and Manufacturing Company of Pittsburgh to provide the tanks, ion sources, liners, and collectors; with General Electric Company of Schenectady to provide the high-voltage equipment; with Allis-Chalmers Manufacturing Company of Milwaukee to fabricate the magnetic coils; and with the Chapman Valve Company for precision vacuum valves.[37]

On January 5, 1943, Groves succeeded in recruiting the Tennessee Eastman Corporation (TEC) of Kingsport, Tennessee, to operate the plant. Groves had decided that Stone & Webster should design and build the plant, but not operate it. For Groves TEC, a subsidiary of Eastman Kodak, was his first choice, largely based on its building of the Holston Ordnance Works earlier that year. Groves had first contacted the general manager, James C. White, in late December 1942 to arrange a meeting in Washington. On December 30 Groves told White that the United States needed the skills of TEC to operate a plant for the military, though the purpose or location of the plant was not discussed. A subsequent meeting was held on January 5 in Rochester, where Groves argued his case that TEC was the best qualified to get things done. When Groves put pressure on corporate executives to do something he wanted, it was almost impossible to resist him. Later that day Kodak and TEC approved. On January 23 they agreed to prepare the feed materials and the chemical processing as well.

On January 13, 1943, Groves, in a meeting in Berkeley with Lawrence and the Eastman Kodak people, set up the schedules. The first Alpha racetrack was to be operational on July 1, less than six months away; all five would be up and running by the end of the year. Ground was broken on the first Alpha building on February 18, 1943; a month later Groves froze the design for the first five (of an eventual nine) Alpha racetracks. Each had ninety-six tanks in either an oval or a rectangle. The two largest Alpha buildings were 540 feet long and 300 feet wide.

The decision to build a second stage to further enrich the material from the Alpha tracks was not an automatic one; not even Lawrence was sure it would be needed. As the idea was discussed in February and March 1943, Groves thought it might be a good idea; he felt it might even be used in conjunction with the gaseous diffusion plant. As the authors of one of the official histories of the Manhattan Project say,

> Resolving problems such as these required much more than technical knowledge; it demanded an intuitive sense or "feeling" for the process, a boldness and yet a reasonableness which Groves seemed to possess in the right proportions.[38]

On March 17 Groves made the decision to build four Beta calutrons, each about half the size of an Alpha track. At this point it was uncertain how productive the tracks would be or even how much enriched uranium would be needed for a bomb. In July Oppenheimer reported that the original estimate of the amount of U-235 needed had tripled. Lawrence proposed improvements and finally convinced Groves in September to build four Alpha II tracks, at a cost of about $150 million. By every measure, building the Y-12 complex was a colossal undertaking; it was eventually comprised of some two hundred buildings spread over 825 acres in Bear Creek Valley.

A famous example of the MED's insatiable appetite for materials, recounted in most books about the Manhattan Project, was the need for a conductor for the coils of the magnets. The solution shows to what lengths the project would go to get the job done. In the summer of 1942, when Colonel Marshall was still in charge, the preliminary plans for the electromagnetic plant had called for five thousand tons of copper. Copper was in short supply, due to its use for other war-related needs and to strikes in the industry.

Silver was suggested as a substitute, because it has the highest electric conductivity of any natural substance. On August 3, 1942, Colonel Nichols visited Undersecretary of the Treasury Daniel W. Bell with a request for a large amount of silver. When Bell asked how much he needed, Nichols replied, "Six thousand tons," to which the secretary replied rather indignantly, "Young man, you may think of silver in tons, but the Treasury will always think of silver in troy ounces."[39] Eventually 14,700 tons of "free" silver was loaned to the MED, worth some four hundred million dollars at the time. In characteristic fashion Secretary of the Treasury Henry Morgenthau was not told for what purpose the silver was desired.[40]

The bullion was shipped from the Treasury's West Point depository to the Defense Plant Corporation of Carteret, New Jersey, where the silver bars were melted and cast as cylindrical billets, eventually more than seventy-five thousand in number. The billets were then extruded and rolled into strips forty to fifty feet long by the Phelps Dodge Copper Products Company at Baywater, New Jersey. The strips were cooled, trimmed, coiled, and sent by rail to the Allis-Chalmers Company in Milwaukee, where they were wound around the steel plates of the magnet casings. The completed magnet coils, resembling twenty- by twenty- by two-foot-square doughnuts, were then shipped on flatcars to Oak Ridge and installed at Y-12. A total of 940 magnet coils was fabricated between February 1943 and August 1944.

The magnets were estimated to be one hundred times larger than any magnets previously built — so powerful that they pulled on the nails in workers' shoes, making walking difficult, and caused tools to fly out of their hands. Special nonferrous tools and equipment had to be produced.

When it came time to return the silver to the Treasury after the war, every ounce was scavenged. In the final accounting, of the 14,700 tons borrowed, only a minuscule fraction of 1 percent was missing.[41]

On visits to Berkeley, Y-12, or elsewhere Groves often chaired meetings to better learn about the status of the projects. The transcripts show his fixation with trying to accelerate things, emphasizing "the always overriding necessity for speed," to use his words.[42] He pressed each individual on his schedule and asked what problems were standing in the way. Why is it behind schedule, where is the bottleneck, what can I do to make it easier for you to do your job? He encouraged everyone to have their say and to be open and honest. He asked them for their best judgment about things so he could make the best decisions. As Groves noted, things did not always go well.

> From beginning to end, the problems encountered in building and running the Y-12 plant, in their variety and often plain unexpectedness, would have taxed the ingenuity and industry of Hercules.[43]

One Herculean problem had to do with the magnets. During test runs of the first Alpha racetrack in late October and November 1943 electrical shorts caused wide fluctuations in the magnetic field strength, a problem that threatened the entire Y-12 process. Nichols reported the problem to Groves in early December. Initially it was determined that sediment of rust and dirt in the oil coolant inside the coils was the cause. Upon opening a magnet, however, other causes were discovered, including moisture in the insulation and, more seriously, the silver bands that were too closely wound. A very unhappy Groves arrived on the scene on December 15 to discuss how to solve the problem with project engineer Klein and the Allis-Chalmers officials.

Many had failed to foresee the hazards, including Groves. There had been warning signs, but they had not been addressed at the time. Heads rolled as Groves reorganized the electromagnetic team. At Groves's urging Stone & Webster brought in Frank R. Creedon to run its operations as general manager. Creedon, a tough no-nonsense manager, had been one of Groves's closest associates during the mobilization period, when they built numerous ordnance plants together. As for the magnets, Groves ordered them shipped back to Milwaukee to the Allis-Chalmers plant for cleaning and rewinding, a process that took about three months. The magnets for the subsequent racetracks were quickly redesigned, and higher standards of cleanliness were imposed.

Groves always kept close watch on his plant managers to see if they were pushing hard enough. As he told Conant, he had his doubts about Frederick R. Conklin, the TEC manager, and Maj. Wilbur E. Kelley, chief of operations. He thought they were "too similar in disposition" and complained that neither of them was a

> hard driving, optimistic executive. Instead of setting an impossible goal and then breaking their hearts to almost achieve it, they set a nice, comfortable goal making plenty of allowances for difficulties and then feel very proud of themselves for having been proven right in their pessimistic outlook.[44]

While the electromagnetic process was elegant in theory, it was not very efficient in practice. All agreed that at the theoretical level the science was sound. Lawrence's laboratory cyclotrons and mass spectrographs confirmed this, but only specks of enriched uranium were produced, when kilograms were required from machines a hundred times larger. In scaling up to operate the industrial-sized calutrons a host of problems emerged. One was to decide on the best feed material to use for the effective and efficient separation of isotopes. After much study, the chemists decided that a gaseous form of uranium, uranium tetrachloride, was the answer.

A second problem was to devise a method to retrieve the tiny amounts of U-235 that adhered to the inside of the Alpha and Beta stages. When the feed material was introduced into the Alpha tracks, the desired isotope did not always end up in the receiver bin or tank. In fact, more than 90 percent of it scattered and splattered throughout the innards of the calutron. Elaborate and labor-intensive procedures were implemented to recover the precious material. A huge chemical area was built to wash and scrub with steam and acid the parts that were removed from the tracks.

Dozens of other afflictions buffeted the temperamental machines, causing them to break down. The first half of 1944 was a time of widespread pessimism about whether this process would ever deliver enough material in time. In those dark days, when nothing was going right at Y-12, Groves challenged Lawrence by saying, "Dr. Lawrence, you know your reputation's at stake." Lawrence shot back, "No, General, my reputation is made. Yours is at stake."[45]

The other isotope separation process was the gaseous diffusion method.[46] In those months at the end of 1942 and the beginning of 1943, as he assembled the Y-12 and Hanford teams, Groves also formed a group to design, build, and operate the gaseous diffusion complex. He started with the M. W. Kellogg Company of Jersey City, New Jersey, a firm that specialized in chemical engineering projects. Kellogg had done work on gaseous diffusion under an OSRD contract, designing a small ten-stage pilot plant. Columbia University, where research on gaseous diffusion had begun in 1940 under Harold Urey, was an obvious participant. Researchers had built a twelve-stage pilot plant in Pupin Hall that furnished useful data.

Kellogg accepted a letter contract on December 14, 1942, to design, supervise the construction, and procure the equipment for a gaseous diffusion plant, designated K-25. For security reasons a separate corporate entity was created to carry out the project, known as the Kellex Corporation. The company assigned one of its vice presidents, Percival C. "Dobie" Keith, as the executive in charge.[47] Keith, a Texas-born chemical engineer and MIT graduate, had been serving as a member of the OSRD S-1 planning board and thus already knew the purpose. Chosen as project manager was Albert L. Baker, Kellogg's chief mechanical engineer, a huge man of enormous energy who was organized and methodical, qualities that Groves respected.

At the height of its activities in 1944 Kellex employed some thirty-seven hundred people, a majority in the New York area. Kellex had its offices in the Woolworth Building in Lower Manhattan, at 233 Broadway, an address that Groves visited many times during trips to the city. Kellex employees also worked at the laboratory in Columbia University's Nash Building, in Jersey City, and at Oak Ridge.

To oversee the Kellex contract, the New York Area Office opened in January 1943. Groves selected Lt. Col. James C. Stowers as area engineer. His office was also in the Woolworth Building and had a staff that eventually grew to about seventy by early 1945. As the K-25 project grew, three smaller area offices were established to relieve Stowers of some of his workload. They were located in the cities of three important K-25 sub-contractors. The Decatur Area Office in Illinois oversaw the firm of Houdaille-Hershey, which manufactured the barrier material used at the K-25 plant. The Milwaukee Area Office monitored the Allis-Chalmers plants where the pumps for the gaseous diffusion process (and the magnets for the Y-12 plant) were manufactured; the Detroit Area Office worked with Chrysler in the design, and then the manufacture, of the diffusers for K-25.

To build K-25 Groves selected the J. A. Jones Company, a firm that he knew well. Edwin L. Jones had worked with Groves during the mobilization period, building more army camps that any other contractor in America.[48] Edwin Lee Jones was born in 1891, the eldest son of J. A. Jones of Charlotte, North Carolina, who had started a construction company there. Jones was the just the kind of man Groves liked and could do business with. He neither drank nor smoked; he went to church every Sunday, was punctual and orderly, and never missed a day of work.[49]

On May 18, 1943, Groves contracted with the J. A. Jones Company to build the largest steam-electric power plant ever constructed, to supply electricity for K-25. Within a fortnight Jones men were at Oak Ridge; ground was broken on June 2. Ten months later the plant was completed, providing a generating capacity of 238,000 kilowatts, or twice that of TVA's Norris Dam. Its three giant boilers each produced 750,000 pounds of steam per hour. Groves had decided that he did not want to have to rely on TVA, whose power might be subject to interruption from natural causes or sabotage.

But Groves had bigger plans for Jones: He recruited the company to build the K-25 gaseous diffusion plant, signing a contract in September 1943. Construction started on the main plant on October 20. The schedule called for completing it in sections, and on April 17, 1944, the first six-stage cell was ready for testing. The additional sections were completed either on time or ahead of schedule. The main building resembled a squared-off letter U. Each side was 2,450 feet long and 400 feet wide. The total area of the three floors and the basement was 5,568,000 square feet, almost as large as the Pentagon. The tiny windows, found only on the top floor, gave it a sinister appearance. The entire complex supporting K-25 included seventy additional buildings spread over six hundred acres. The cleanliness standards for the inside of the K-25 building rivaled those of an operating room — it was to be totally free of dirt and dust. Anyone entering, including General Groves, had to wear special clothes.

At Dobie Keith's suggestion Groves chose Union Carbide & Carbon Corporation to operate the plant. On January 18, 1943, James A. Rafferty, a vice president, accepted a letter contract to operate the plant and serve as a consultant. Rafferty would supervise the project along with his assistant, Lyman A. Bliss, a vice president of Linde Air Products, one of Carbide's companies. Dr. George T. Felbeck was project manager in charge of operations.

The gaseous diffusion process posed formidable technological challenges. Even with the best advice he could gather it often fell to Groves to decide which method or process to follow. Often the evidence and arguments seemed equally balanced; Groves had to use his intuition and instincts to make a choice. Sometimes decisions had to be reversed. Groves's role in designing and manufacturing the barrier is a good case in point.

The principle of the gaseous diffusion process is based on Graham's Law, which states that if two gases of unequal density are placed in a porous container surrounded by an evacuated space, the lighter gas will tend to escape at a more rapid rate than the heavier. By controlling the gas through a series of stages, an almost pure concentration of the lighter component can be achieved. All of which was fine in theory, but working out the technical and engineering problems in practice was another matter. And the gas, in this instance, was uranium hexafluoride (UF_6), an extremely corrosive gaseous compound of uranium.

The most crucial component of the process was the barrier, which had to separate, effectively and efficiently, the U-235 and U-238 isotopes in the hexafluoride gas.[50] Of all of the many technological-engineering problems that had to be solved during the Manhattan years, designing and fabricating the barrier was perhaps the most difficult. The barrier was a thin metal sheet or membrane with millions of submicroscopic pores of uniform size (each smaller than two one-millionths of an inch) per square inch. The sheets were formed into tubes, which went inside airtight vessels known as diffusers. The barrier material had to be resistant to the corrosive gas that passed through it. Researchers at Columbia had been searching for the right design and the tough metal or alloy for two years. At the end of 1942 Edward Adler of Columbia and Edward Norris came up with a design made out of a corrosion-resistant nickel, which seemed promising but still had drawbacks. In June 1943 a Kellex scientist named Clarence Johnson came up with an improved design using powdered nickel. Over the next six months each team worked feverishly to correct the defects. But Groves could not wait to see which one was the better before building a plant to manufacture the vast quantities that would be needed.

In April 1943 the Houdaille-Hershey Corporation, which worked with finished nickel, was tapped by the MED to build a plant to produce the barrier material. The company chose a site in Decatur, Illinois. Exactly what the plant would be producing had not yet been decided, but Groves hoped that improvements in the Norris-Adler design would be found along the way and that, by the time the plant was completed, they would be ready to mass-produce it.

By November 1943 the time was drawing near to choose a method — the still-imperfect Norris-Adler design or the newer Johnson-Kellex design, which looked encouraging in the laboratory but was still a question mark for mass production. Each had its defenders, Urey and his Columbia team versus Keith and his Kellex group. On November 5 Groves announced his decision: Both designs would go forward temporarily. He continued to hope that the Norris-Adler design would be improved and, if not, the Johnson model could be the backup. Two more months of deliberations by the scientists and engineers led to a decisive meeting on the morning of January 5, 1944, in the Woolworth Building. After listening to a heated discussion about what to do, Groves made up his mind.

On Saturday afternoon, January 15, Groves left Washington by train to travel to Decatur, Illinois, to inspect the almost completed plant and to meet with representatives of Houdaille-Hershey, Kellex, Columbia, and Carbide. He missed his connection in Cincinnati and telephoned the local district engineer to request a car and driver to take him to Decatur. When the car arrived at the railroad station at midnight Groves told the driver,

"Now I am going to talk to you all night long, just to be sure that you'll stay awake."[51] After driving all night across Indiana and half of Illinois, Groves arrived in time for the nine o'clock meeting. As he entered the conference room, he announced, "Well, let's get the meeting going."[52] After a tour of the new factory he announced his decision: The plant would be immediately stripped of the just installed equipment to produce the Norris-Adler design, and new equipment to manufacture the Johnson-Kellex barrier would replace it. With the order given, the meeting was adjourned, and Groves left for Washington.

Stone & Webster Engineering Corporation was originally responsible for planning and building the town of Oak Ridge, but given its responsibility for the Y-12 plant, other firms were found for the job.[53] The Boston architectural firm of Skidmore, Owings and Merrill took over the task of planning the site. The John B. Pierce Foundation of New York designed the initial houses. Originally the architects were told to plan for a community of twelve thousand (some three thousand families), though they were not told what the residents would be doing. The early plans were soon obsolete. New requirements called for a town of forty-two thousand, and further expansion pushed Oak Ridge's on-site population to seventy-five thousand by the end of the war. All of the essentials to support town life were built, from schools to churches, libraries, hospitals, recreation facilities, and a commercial shopping center. The *Oak Ridge Journal* newspaper reported local information, providing everything from what was playing at the movie theater to the next meeting of the Boy Scout troop.

After August 1943 Oak Ridge became the location of the Manhattan Engineer District and the district engineer. Groves had decided to move the growing administrative offices from New York City to Tennessee and appointed Col. Kenneth D. Nichols to replace Col. James C. Marshall as district engineer. From headquarters in Oak Ridge the district was responsible for all design, construction, and operation of the plants at Oak Ridge and Hanford, the feed materials plants located in the United States and Canada, the acquisition of domestic uranium ore, and the administration of research efforts at the universities, as well as supporting Los Alamos.

Nichols was eleven years younger than Groves and had graduated fifth in the West Point class of 1929. Commissioned in the Corps of Engineers as a second lieutenant, he was assigned that same year to Nicaragua, where he first met Groves. He served in Company B of the Twenty-ninth Provisional Engineer Battalion, at Greytown, whereas Groves commanded Company A. After leaving Nicaragua he got a master's degree from Cornell, studied in Berlin for a year, earned a Ph.D. at the University of Iowa, and taught engineering for four years at West Point. Prior to joining Colonel Marshall, Nichols was area engineer in charge of construction of the Rome Air Depot, Rome, New York, and area engineer for construction of a TNT plant at Williamsport, Pennsylvania.

The relationship between Groves and Nichols was always a bit strained. Nichols captured some of the tension after the war:

> First, General Groves is the biggest S.O.B. I have ever worked for. He is most demanding. He is most critical. He is always a driver, never a praiser. He is abrasive and sarcastic. He disregards all normal organizational channels. He is extremely intelligent. He has the guts to make timely, difficult decisions. He is the most egotistical man I know. He knows he is right and so sticks by his decision. He abounds with energy and expects everyone to work as hard or even harder than he does. . . . if I had to do my part of the atomic bomb project over again and had the privilege of picking my boss I would pick General Groves.[54]

Whatever their differences and views of each other, they worked well together. Most of Groves's directives to Nichols were verbal, reducing excessive paperwork. Nichols supplied Groves with short monthly progress reports that were hand-carried by courier to Washington. They were usually only three or four pages long and summarized the status of each plant in concise terms.[55]

While the Army Corps of Engineers was the ultimate authority, the job of actually running the town of Oak Ridge was contracted to the Turner Construction Company, which established a subsidiary named the Roane Anderson Company after the two counties over which the Clinton Engineer Works were spread.[56] The project manager was Clinton N. Hernandez. He, his staff, and (by war's end) his ten thousand employees provided the basic services necessary for the town to function. They oversaw the housing assignments, collected the rent, and maintained and repaired the houses, dormitories, trailers, barracks, and hutments. They provided custodial services for the public buildings and ran the cafeterias, bus system, motor pool, and laundry. They negotiated and managed the concessionaire contracts for all the business and commercial services: the grocery stores, drugstores, and movie theaters. They picked up the garbage, kept the water flowing, kept the electricity on, and delivered coal and wood for heating.

Roane Anderson performed a very important function from the perspective of the army: It absorbed and deflected much of the criticism when things did not go well. Groves did not mention Hernandez or Roane Anderson in his book; nor were they recipients of the Army-Navy "E" Award given to the other major Oak Ridge contractors on September 29, 1945. Groves never explained why.

By the time Groves returned to Washington on October 11, 1942, from his tour to Chicago and California, he had decided on a tentative strategy for turning a small-scale laboratory process to produce plutonium into a large-scale industrial one.[57] From the first day he saw that if this project were to be successful, it would have to be conducted on a grand scale.

Groves approached Du Pont's chief engineer, Everett G. Ackart, on September 28, at Compton's urging, to propose that the E. I. Du Pont de Nemours Company design, construct, and operate the separation plant.[58] Its professional and businesslike approach to solving tough problems was just what was needed. Du Pont was a special company that emphasized research and engineering to a greater degree than its competitors.[59] Groves had had good experiences with Du Pont in building several munitions plants during the mobilization.[60] Morale was good, standards were high, and there seemed to be good communication among a dedicated workforce.

At the time Du Pont was practicing what came to be known as the critical path method (CPM), or critical path scheduling. The CPM can be detected in procedures that Du Pont used to build large munitions and chemical plants during the mobilization period from late 1940 to 1942, specifically the Belle Plant, near Charleston, West Virginia, and another at Morgantown, West Virginia. Groves had signed many of these contracts, worked with Du Pont, and recognized and appreciated its methods as effective ways to get a job done quickly. As Matthias recounted it:

> I had learned how they'd do that because Groves at one time had me review some of the construction jobs they were doing, long before he was even with the Manhattan District. And he got me to review some of their ways of making reports, because they were so far ahead of any of the other builders in the country — he just wanted to know

how they did it. So I had gone through that kind of a drill on my own — that's one of the personal assignments Groves gave me.

They — for the first time — they didn't talk about it then, but they were using the program of critical path.[61]

Building and operating Hanford was accomplished by using CPM procedures on a grand scale. In the postwar period the CPM, or related practices, became the modus operandi of civil engineering, adopted by almost every architect-engineering-construction firm, and are routine today.[62] The essence of CPM is a detailed, efficient procedure for every step of the project, with every worker at every level knowing precisely what is to be done. The engineers would lay out every task to every foreman for the day ahead and supervise the implementation, keeping to a coordinated schedule. Du Pont was meticulous in preparing its designs, checking them thoroughly beforehand, performing quality assurance checks along the way, and monitoring its vendors' products as well as its own construction.

On September 28, 1942, Groves met Du Pont's board of directors to request their assistance for design and procurement of a semiworks (or prototype) separation plant, then planned for Chicago. On October 3 a letter contract agreement was signed. But Groves had much larger plans in mind for Du Pont. A week after the semiworks agreement he requested that it design and construct the nuclear piles itself.

Initially, Du Pont resisted. President Walter S. Carpenter told Groves that company personnel had absolutely no experience in nuclear technologies (but then, who did?), and that they were already over their heads with other commitments for the war effort. They were also reluctant to get involved with a project that had the potential for public controversy and outright failure. Du Pont was still smarting from the congressional investigation of 1934 by Senator Gerald P. Nye of North Dakota, in which it was revealed that during World War I Du Pont had made excessive profits in selling munitions to France and England. For this it had been dubbed "Merchant of Death" by the press.

Groves was unrelenting. When he wanted something it was almost impossible to resist his pressure, and he wanted Du Pont. On November 1 a key meeting was held in his office with two Du Pont vice presidents and James Conant. Groves and Conant explained the situation up until then and ended by telling them that they wanted Du Pont to develop the plutonium processes. The executives said they would have to check with President

Carpenter and the other members of the executive committee. On November 12 Carpenter and the executive committee agreed to undertake the engineering and construction of the nuclear piles, at a location yet to be determined.[63] The initial thinking was that the plant would be at Clinton, Tennessee, but on December 1 Du Pont recommended that it be in a remote location. This was approved by the Military Policy Committee on December 10.[64]

Though Groves was making decisions one after the other, there were still serious disagreements over which of the techniques being explored for isotope separation should be pursued and over the viability of using nuclear piles to make plutonium. There were even arguments about whether a bomb would work, even if enough fissionable material were produced.

In order to settle some of these questions — but, more important, to convince Du Pont to take the job — Groves, after consulting with Conant, appointed a committee of experts to visit the laboratories in New York, Chicago, and Berkeley at the end of November 1942 to review the situation and make recommendations. Known as the Lewis Committee after its chairman, chemist Warren Kendall Lewis,[65] it reported back on December 4. It endorsed the program for the manufacture of plutonium, was enthusiastic about the possibilities of gaseous diffusion over the other means for producing U-235, and recommended caution about Lawrence's electromagnetic separation process. This seal of approval by the distinguished committee members, coming two days after Fermi's chain reaction in Chicago on December 2, 1942, gave Du Pont some hope that its commitment was not in vain.[66]

Groves turned to the Military Policy Committee, which had sanctioned his initial moves and now supported the steps he took after the Lewis Committee's recommendations. Groves wanted to proceed differently. The first of these steps was to give Du Pont full responsibility for the industrial design, construction, and operation of the fission piles for the production of plutonium. Du Pont was also given responsibility for developing the plutonium separation facilities. But the MPC ignored the Lewis Committee's recommendations when it endorsed Groves's intention to construct the factories needed for the manufacture of plutonium without first building pilot plants and semiworks. Groves believed — and the MPC agreed — that there was no time to proceed in the normal fashion, moving from small-scale experimental plants to large production ones. Full-scale plants would have to be built, and built right, from the beginning.

Groves wasted no time in initiating a search. During the last two weeks of 1942 the site where the production reactors would be built was chosen. The selection was based on eight engineering and safety criteria established by Groves, Du Pont, and the Met Lab scientists.[67] The criteria were:

- An abundant water supply with at least twenty-five thousand gallons of water per minute.
- A large dependable electrical supply of at least one hundred thousand kilowatts of available power.
- A rectangle of land approximately twelve by sixteen miles for the hazardous manufacturing area.
- A laboratory area located at least eight miles from the nearest pile or separations plant.
- An employee village located no closer than ten miles upwind of the nearest pile or separation area.
- At least twenty miles between the piles and separation areas and the nearest existing community of one thousand or more inhabitants.
- No railroad or main highway closer than ten miles from the piles or separation areas.
- A climate that would not affect the process.[68]

Groves picked a close associate, Lt. Col. Franklin (Fritz) T. Matthias to lead a survey team to find the site. Matthias had grown up on a farm in Wisconsin and attended the state university at Madison, receiving a B.S. in 1930 and an M.S. in civil engineering in 1933. He had taught engineering for two years and worked for the Tennessee Valley Authority as a construction engineer for four years. He also was commissioned as a reserve second lieutenant in the Army Corps of Engineers in 1930 through the ROTC program; when ordered to active duty on April 15, 1941, he was assigned to the Construction Division of the Corps of Engineers as a first lieutenant. Matthias worked closely with Groves building the Pentagon, and on special matters during the first months of the Manhattan Project.[69]

On December 14 Groves sent Matthias to Wilmington to see Du Pont officials and Met Lab scientists, telling him to remember everything he heard. The following morning Matthias returned to Washington and was met at the train station by Groves, who drove him home. In the twenty-minute drive the general explained to the thirty-four-year-old lieutenant colonel the real purpose of the project, which Matthias had not known up until then. He ordered Matthias to assess areas that met the criteria for a plutonium plant and then to visit them. The Corps of Engineers prepared a preliminary list of twenty sites, but the number narrowed to a few in the Pacific Northwest and California because of the power sources to be found there.

Matthias and two Du Pont engineers, Allan E. S. Hall and Gilbert P. Church, arrived in Seattle, Washington, to begin their survey on December 18. They spent the next few days considering sites in Idaho and eastern Washington, especially areas near Grand Coulee and the tiny village of Hanford. On December 22 Matthias flew over Hanford and was convinced as soon as he saw it that this was the place. Visiting it on the ground with his Du Pont companions settled the issue for the team. That night Matthias called Groves to tell him that they had probably found the site.[70] The Columbia River could supply large amounts of pure water, the Bonneville Power Administration could supply power, rainfall was minimal, the winter climate was mild, and relatively few people would be displaced. The ever-cautious Groves asked if there were other candidates; Matthias said there were three in California.[71] Groves told him, "You have to go down to the other sites; otherwise someone in the future can say we went off half cocked. You might miss something down there, or we might have difficulty getting one of the sites, so you better go down and look at them."[72] They spent the next week in California but found that each of the sites failed to meet all of the criteria.

The team wrote its report on the nineteen-hour flight back to Washington on December 31, and Matthias delivered it to Groves on New Year's Day, 1943. Groves settled on Hanford, but before finalizing such a monumental decision he wanted to see it for himself. His engineer's mind and eye saw real estate in terms of topography, water and power sources, access to rail lines, and weather conditions.[73]

Over the next ten days Matthias reviewed his findings with Du Pont and with Nichols and left again for Hanford on January 11th to begin to arrange for real estate appraisals. Groves set in motion procedures for the government to purchase the land, visited the site on January 16, and confirmed his initial decision.

Groves no doubt viewed the vast and remote desert environment in south-central Washington as an isolated wasteland, perfect for the secret military mission. Like his engineer colleagues, and most other Americans of the day, land was seen as something from which resources could be extracted, nature as something to be tamed and shaped to serve man's purpose.[74] In the early 1940s there was little appreciation for the environmental consequences and ecological disturbances that might occur as a result of vast building projects. Occasionally Groves showed some concern about the dangers surrounding his venture.

> We knew, too, that in the separation of plutonium we might release into the atmosphere other highly toxic fumes which would constitute a distinct hazard . . . I was more than a little uneasy myself about the possible dangers to the surrounding population.[75]

But building a weapon for use in the war was the first priority. Groves, more than any other single person, created a certain culture that infused the Manhattan Project. His relentless drive to accomplish the military mission was all-consuming. It took precedence over everything else and the practices, mind-set, and worldview that were established left a lasting imprint on the agencies and departments that followed over the next four decades throughout the Cold War. Though civilian run, the Atomic Energy Commission and the Department of Energy operated in similar fashion to their military predecessor. The Russian threat replaced the German threat as a rationale and raison d'être. The invocation of "national security" prevented democratic accountability and oversight, and environmental and safety issues ran a poor second to building bombs.[76]

On February 8 a War Department directive was issued authorizing acquisition of more than four hundred thousand acres at the Hanford site.[77] In early March Groves picked Matthias to be area engineer and run the project at Hanford.[78] His deputy area engineer was Lt. Col. Harry R. Kadlec, who had been responsible for building the massive Detroit Tank Arsenal with the Chrysler Corporation in 1940 and 1941, and later served as a troubleshooter for Groves during the army construction period.[79] Du Pont chose Matthias's traveling partner Gil Church as field project manager, and he arrived by the end of March.

Eventually Hanford would grow to more than 428,000 acres — five hundred square miles in area, half the size of the state of Rhode Island. When the acquisition program began in early 1943, almost all of the land was utilized for some type of agriculture. More than 88 percent consisted of sagebrush rangeland, chiefly used for grazing some eighteen to twenty thousand sheep during the winter and spring months. Another 11 percent was farmland, but not all of it was under cultivation due to a lack of water. Less than 1 percent was the small communities of Richland, White Bluffs, and Hanford.[80] More than 225,000 acres belonged to private individuals or corporate organizations, with the balance federally, state, or county owned.

The plan called for dividing the site into three sections. The first section was a tract where the main production facilities would be located, with a safety belt surrounding it. This would be taken by the government and cleared of all persons. The second section was also taken outright by the government. The owners could lease certain portions, but no one could live there. In the third section parcels were either purchased and then leased back or easements were obtained that gave the government the right to evict anyone, at any time, for any reason. As Groves later said, "This third section of the tract caused us much trouble before we were through with it."[81]

The first tract of land was acquired on March 10. Organized opposition to the land acquisition grew throughout the spring. Groves, Matthias, and the Real Estate Branch officials met with the Justice Department to try to work out procedures that would avoid litigation. Their hopes went unrealized, because farmers were dissatisfied with the low prices the government appraisers had set for their land. An equal area of disagreement was the question of how much money should be paid for the crops growing on the land at the time of acquisition. The major crops of the area were cherries, apples, pears, peaches, asparagus, and alfalfa — and 1943 was shaping up to be a bountiful year. The farmers argued that if they were not permitted to remain on their land until their crops could be harvested, then they should be fairly compensated for them as well. Ways were found to permit farmers on some of the acquired land to harvest their crops, but unavoidably security and construction demands required that other farms be cleared before the harvest.

The land acquisition problems caused Groves and Matthias many headaches, took up an enormous amount of time, and were not even completely settled by the end of the war. It did not take long for the repercussions from the local controversies to reach Washington. Rumors circulated that this land grab was all for the benefit of Du Pont. The National Grange made sure that the farmers' complaints reached President Roosevelt, who did look into the matter. A letter was prepared, with Groves's assistance, explaining that the government was doing everything possible to protect the agricultural interests at Hanford. Congress became involved as well. The Truman Committee, alerted by letters from local residents, began to look into the matter.

In early June the committee sent written inquiries to Du Pont president Walter Carpenter and to Julius Amberg, special assistant to the secretary of war, asking for precise information about the Hanford project. The committee requested information on the factors governing the choice of the location, the estimated cost and status of the project, and an explanation of why so much land was needed. Immediately upon being notified of this unfortunate development, Groves went to see Bundy on June 11. They agreed that Secretary Stimson should talk with Senator Truman and have him cease his investigation of Hanford. On June 17 Stimson called Truman and informed him of the importance of the project and the need for secrecy. Stimson recorded in his diary that Truman was "very nice" and told him "that was all he needed to know" and that he trusted him "implicitly." Truman agreed to stop further investigations.[82]

On the same day at a cabinet meeting in the White House the chairman of the War Manpower Commission, Paul V. McNutt, brought up the Hanford land problems, causing President Roosevelt to ask whether another location might be found. Stimson, having just put out one fire with Truman, talked with Groves that afternoon and found that great care had been taken in selecting the site, and that there was no other place "where the work could be done so well. It required isolation for safety's sake, enormous quantities of pure water and power, and a railroad independent of mainline traffic." With those assurances Stimson "called up the President and satisfied his anxiety." He also called McNutt to caution him about bringing up the subject at cabinet meetings.[83]

But the litigation and all the problems that went with it were just beginning for Groves and Matthias. A series of condemnation trials got under way in the fall of 1943. In early October the first trial began at Yakima, Washington, and cases followed at regular intervals until early March 1944. In a full-scale affront to Groves's security procedures the juries were taken to Hanford to inspect the tracts of land under adjudication. This was most troubling. In Groves's opinion none of these people had any business knowing anything about what went on at Hanford. But the legal procedures required the visits, and until Groves could change them, all he could do was fret that someone was going to see and say too much. An added problem with all of these trials was the number of stories they generated in the local newspapers, further compromising security.

Finally, to add insult to injury, the juries consistently awarded the landowners payments greatly in excess of what the Corps of Engineers appraisers had established.

Of course Groves and Matthias opposed the payments, which they believed to be excessive. But for the moment, like the visits, they could do little to stop them. Of great concern was the pace of land acquisition, slowed by the trials, and potentially even interfering with the project itself. The court was averaging seven cases a month, and there were twelve hundred tracts under dispute. A way had to found to speed this whole business up and complete it.

A partial solution was to request the Department of Justice to assign more judges and to end the jury inspection visits. On April 24, 1944, Groves, Matthias, and Real Estate Branch officials conferred with Assistant Attorney General Norman M. Littell, in charge of the Lands Division, in a meeting at Yakima arranged at Groves's behest. Littell, who was from the state of Washington, had earlier been critical of the condemnation procedures. Some part of Littell's actions in this entire affair no doubt was motivated by his aspiration for a political career in his native state.

A tentative agreement was worked out between Groves and Littell to expedite the acquisition process. So that more cases could be settled out of court, special Justice Department attorneys were empowered to make adjustments in the appraised value, with increases up to 20 percent being sanctioned. Littell also agreed to a second court and additional judges.

While the agreement led to some improvements, it did not end the problem. Settlements increased to an average of seventy-five per month, but Littell continued to cause trouble for Groves. Littell decided that the root of the problem was inadequate and faulty appraisal work by the War Department from the beginning. He sent one of his staff to reappraise a sample of the remaining tracts of land. The evaluation, perhaps not performed with absolute objectivity by the staff member, convinced Littell that he was correct in his theory.

In the trials that followed during September 1944 the jury awards increased, in some instances exceeding the amounts demanded by attorneys for the owners. Littell was still not happy; in a September 28 letter to Patterson he said that he was going to recheck the valuations again.

In October Littell was in his home state to participate in the 1944 presidential election. On the thirteenth, in a move that caught Groves off guard, Littell suddenly appeared before the district court in Yakima. He requested of Judge Lewis Schwellenbach that no more condemnation cases be brought to trial until the Justice Department had reappraised the tracts still pending. Once again he ascribed blame to the Corps of Engineers, which had used faulty and incomplete appraisal methods from the beginning. His remarks before the court were given wide publicity in the local newspapers, along with much editorial comment and many letters to the editor.

Although Groves knew Littell was making a trip to the Hanford area, he did not know of his intention to appear before Judge Schwellenbach. When he learned of it, three days after it had occurred, he was not pleased. As he wrote in a letter, prepared for Patterson's signature, Littell's actions were

> obviously incompatible with essential military security, the need for which had been carefully explained to him. His statement to the court has resulted in a considerable amount of undesirable publicity concerning a project which the President has personally directed should be blanketed with the utmost secrecy.[84]

A further unfortunate aspect of this affair was that it gave the false impression that the Justice and War Departments were at odds with one another over land acquisition. For Groves it was time to end this matter once and for all, to bring Littell to heel, and to have him stop interfering in a project of the highest national importance. Groves mobilized the resources at his disposal, which were considerable. The first prong of his attack was to have a new statement of the War Department's position, which Patterson could then send to Attorney General Francis Biddle.

Groves prepared the statement, noting Littell's utter disregard for security and the one-sided character of his conclusions about the appraisal policies. While the War and Justice Departments had consistently tried to cooperate to resolve these issues, Littell's "public airing of alleged differences between [them were] in contravention of expressed executive policy."[85]

Simultaneously Groves initiated a security investigation of the Littell incident. He sent his top security officer, Lt. Col. John Lansdale, to Hanford at the beginning of November to survey the situation. Lansdale submitted a report and took measures to curb the newspaper publicity.

The third prong of Groves's attack was to undermine Littell's supposition about the original War Department appraisals. On November 21 Groves sent his trusted aide and troubleshooter Gavin Hadden, with two other investigators, to Hanford to interview people, examine records, and visit typical properties, covering land acquisition from February 1943 to date. But before the team could finish its report the Littell affair took an unexpected turn.

Attorney General Biddle had had a long-standing feud, predating Hanford, with Littell over his administration of the Lands Division. This incident was the final straw. On November 18 Biddle requested Littell's resignation, but was rebuffed. Littell even counterattacked by charging Biddle with maladministration, and submitted grievances to the Mead (formerly Truman) Committee. On November 22 Biddle requested Littell's resignation a second time, and was ignored. The attorney general had no recourse but to have President Roosevelt remove Littell from office. On the twenty-sixth Littell was dismissed for "insubordination."

Groves was pleased at the outcome, because normal relations resumed between the War and Justice Departments. The problem of jury visits was resolved in negotiations between Judge Schwellenbach and Colonel Matthias. By the spring of 1945 the settlement rate had increased to one hundred per month. Nevertheless, by the time the MED transferred control to AEC on December 31, 1946, two hundred tracts remained under litigation.

The numbers at Hanford, as with most things connected to the Manhattan Project, were staggering: 540 buildings, more than 600 miles of roads, 158 miles of railroad track, vast quantities of water, concrete, lumber, steel, and pipe. Approximately 132,000 people were hired over the period (working 126 million man-hours) — eight times the number that had built the Grand Coulee Dam, and almost as many as had worked on the Panama Canal. Peak employment occurred in June 1944 at fifty-one thousand. The living quarters at the Hanford construction camp, forty miles from the Pasco railhead, were barracks for twenty-four thousand men and five thousand women, nine hundred huts for ten thousand more, and a trailer camp with forty-three hundred units. This temporary city housed the construction workers and their families and was segregated by sex and race. A bus service transported them to work locations. Peak bus use was in September 1944, with 904 buses in service. The Hanford camp was abandoned in early 1945 after construction was completed; it was dismantled the following year.

Groves and Du Pont had to build a second town to house the six thousand employees, and their families, who would operate the finished reactors and plants. Richland, initially a farm community of two hundred located twenty-five miles south of Hanford, near Pasco, was by 1945 transformed into a town of seventeen thousand with forty-three hundred houses, a hotel, schools, stores, theaters, a bank, a hospital and medical center, a fire station, a bus depot, and churches.[86] The total cost to build and run Hanford during the war was $358 million, or $4.65 billion in 2001 dollars.[87]

Du Pont put some of its best scientific and technical talent to work on the project, veterans who knew one another's capabilities from years of working together. Some knew Groves from earlier projects. A dozen or so Du Pont people worked closely with him over the next three years and were instrumental to the project's success. A key Du Pont person in Wilmington was Roger Williams, who was appointed director of a new manufacturing division called Explosives-TNX that Du Pont formed to manage the plutonium work. Groves's appointment book is filled with calls to Williams; he was also one of the people he saw on his frequent trips to Wilmington.

A second key person was Crawford H. Greenewalt. Greenewalt had managed the technical development of nylon for Du Pont and had had experience in translating pure research into full-scale production. Greenewalt was the liaison between Du Pont headquarters in Wilmington and the Met Lab in Chicago. In concert with Compton, Greenewalt played a major role in transforming the Met Lab's scientific ideas and advice into engineering realities, as well as smoothing the feathers of the temperamental scientists.

The scientists were initially suspicious of the army's intrusion into their domain, and now industry was taking over. Many thought these military-industrial efforts were misguided and overly ambitious.[88] Some saw Du Pont's rigorous procedures as wasting too much time and money. A few had the unrealistic notion that an adequate amount of plutonium could be produced in the laboratory. These naive thoughts soon passed, but in passing left some people with hard and bitter feelings, which gave rise to certain myths about and caricatures of General Groves that have lasted until this day.

Working in concert with TNX was Du Pont's Engineering Department, where additional groups were formed under the Design and Construction Divisions to design and build Hanford. Granville M. "Slim" Read headed the Construction Division in Wilmington and visited Hanford periodically. The "gruff . . . hard-boiled, cigar-chomping engineer" and Groves knew each other well, having worked on munitions plants together. Alike in many ways, they had great respect for one another; it was partially because of Read that Groves had wanted Du Pont for the job.[89]

One of the factors in Hanford's success was favorable labor conditions.[90] Lt. Col. Clarence D. Barker was chief of the Labor Relations Branch of the Corps of Engineers' Construction Division and had worked for Groves during the mobilization period. In September 1942 Groves informed Barker that vast numbers of laborers would be needed for a secret project. Barker and his staff set about recruiting them through hiring halls and advertising. Later, when schedules were not being met, Barker went to the War Manpower Commission to get top priority. Officials of labor unions — bricklayers, carpenters, painters, electricians, plumbers, and lathers, among others — were contacted and told that they were needed on a secret project that had the highest priority in the country.

At Hanford few hours were lost due to work stoppages.[91] Matthias formed good relations with local labor officials, got a no-strike pledge, and worked out an agreement with the craft workers. Because the Building Trades Council of Pasco could not supply enough craft workers, it agreed to set aside jurisdictional boundaries. Thus, plumbers permitted boilermakers and machinists to work at plumbing and pipefitting, and machinists let carpenters work as millwrights. While most of the workers did not know precisely what they were working on — one cover story was that it was an RDX plant — being told it was an emergency effort that could shorten the war was a convincing reason for the secrecy. Everyone had a son, brother, father, relative, or friend in uniform. To work hard on something that might bring the loved one home sooner was incentive enough, for most.

The Hanford workweek was 108 hours, two shifts of six 9-hour days, with Sundays and overtime added as needed. The managers' workday was anywhere from ten to twenty hours long, with rarely a Sunday off. Pay scales were about $1.00 an hour for laborers, up to $1.85 for skilled workers such as plumbers, steamfitters, electricians, and bricklayers.[92] Efforts were made to keep morale high and turnover low. Attention was paid to serving good food — despite the rationing — and in quantity. There was a big beer parlor, slot machines, a bowling alley, movie theaters, shows, and recreation.[93] There were also occasional homicides, fights, suicides, automobile accidents, drunkenness, gambling, and some prostitution.[94]

Sometimes workers were killed on the job. From the beginning to the end of the war there were nineteen fatalities at Hanford, all during the construction of the vast plants. At Oak Ridge there were forty-two fatalities, most during construction, but some during operation. Workers fell to their deaths, others were crushed, and several at Oak Ridge were electrocuted.[95] Groves's diary indicates that he was notified about some of the accidents, especially when there were multiple deaths.[96]

According to a recent calculation, had the Hanford project proceeded according to the traditional rational/sequential rather than the parallel method, the first plutonium bomb would not have been ready to test and use until May 1948, almost three years after its actual completion.[97]

With the main site selected, the design of the full-scale production reactor had to be chosen. It was also decided to build a pilot plant, as well as a separation facility that would supply experimental quantities of plutonium to Los Alamos and provide training. The small reactor, to operate at the forty-megawatt level, was never conceived as a model for the larger Hanford ones. The designs under consideration for the full-scale reactor and the pilot reactor were air cooled, helium cooled, or water cooled. Greenewalt, with his Du Pont colleagues and Met Lab advisers, weighed the pros and cons of each. For the pilot plant, an air-cooled design was chosen in mid-February 1943, after Groves had issued a contract on January 4. Originally the plant was to be built in the Argonne forest twenty miles west of downtown Chicago, but safety issues necessitated that it be located at Oak Ridge. The reactor, designated X-10, was built by Du Pont but managed by the Met Lab and began operation in December 1943, less than ten months after the signing of the contract.[98] By the spring of 1944 small amounts of plutonium, about a gram a day, were being shipped to Los Alamos for research purposes.[99]

On February 16, 1943, Greenewalt chose Eugene Wigner's water-cooled, graphite-moderated design for the full-scale reactors, with a power level of 250 megawatts. Three were built on the south bank of the Columbia River — within 100 Area — and designated B, D, and F.[100] Each reactor block was a cube about forty-six feet high by forty-six feet wide by forty feet deep, encasing an inner core, weighing a total of nine thousand tons. The eighteen-hundred-ton inner core — the actual block of graphite (made up of some one hundred thousand stacked blocks) — was thirty-six feet high by thirty-six feet wide and twenty-eight feet long. The graphite block was surrounded by six thousand tons of cast-iron and steel shielding. The front face was where the fuel was inserted ("charged"), and the rear face was where it was discharged.[101] The uranium fuel was sealed in aluminum "cans" or "slugs," each eight inches long and one and a half inches in diameter. Thirty-two slugs were placed in each of the 2,004 tubes, which penetrated the pile from front to back. A "once-through" cooling system was devised in which water from the Columbia River circulated through piping in the graphite. Before the water was returned to the river, it would remain for about eight hours in cooling ponds to reduce the amount of radioactivity. After the slugs were irradiated, they were pushed out the back end of the tubes into a basin of water twenty feet deep. Using remote-control tongs, the fuel elements were transported to a storage pool and then loaded into casks and transported on special railcars to the chemical reprocessing separation plant in the 200 Area.

Building the reactors forced Groves, the Du Pont engineers, and the Chicago scientists to confront a perpetual tension that existed at Hanford and elsewhere. On the one hand was Groves's goal to build the bomb as quickly as possible. Speed was of the essence if the weapon was to be a factor in the war. On the other hand was the degree to which the engineers should adhere to precise specifications and design standards. Certain corners could be cut to speed up the work, but if too many were disregarded the reactor, or whatever, might not function at all. How to weigh the continuing challenges of this dilemma was ultimately Groves's responsibility. His decisions were arrived at with superior scientific and engineering counsel to advise him, but in the end it was he and he alone who had to decide to do something one way or another.

A key example of this had to do with how many fuel tubes there should be in the pile. The Met Lab scientists had proposed 1,500 tubes (using two hundred tons of uranium), while Du Pont's George D. Graves, assistant manager of the Technical Division of Explosives TNX, insisted on an additional 504.[102] The Met Lab physicists argued against the additional tubes, claiming they were a waste of time and money. Groves settled the matter by backing Du Pont's conservative position. "It was simply normal sound engineering practice to provide a margin of safety."[103] This proved to be a wise choice, because the unforeseen happened when the reactor first went critical on September 27, 1944.[104] The initial start-up of the B reactor used the central 1,500 tubes, leaving the outer 504 tubes empty. Shortly after criticality, enough neutron-absorbing xenon-135 had built up to "poison" the reactor, halting the chain reaction. Greenewalt, Fermi, and Wheeler quickly discovered the cause. Over the next six weeks fuel slugs were placed in the other 504 tubes (requiring sixty tons of uranium), and the

reactor was restarted. The extra amount of fuel proved enough to overcome the poisoning effect.[105]

If a new larger reactor had had to be built, with enough tubes to overcome the effects of xenon poisoning, it would have taken eight to ten
months, pushing a start-up date to the spring or summer of 1945. There
would not have been enough plutonium for the Trinity test in July, the
Nagasaki bomb in August, or any subsequent bombs that were being
readied.

On March 17, 1943, it was decided to build two separation areas, East and
West, located about four miles south of B reactor in the 200 Area. Because
chemical separation is the most hazardous process in the production of plutonium, the 200 Area was located on a plateau five miles from the Columbia
River and much higher above the water table.[106] Initially there were to have
been six separation plants, one for each reactor. In May this was reduced to
four and later to three, one for each reactor. At 200 West there were two separation plants, and at 200 East one separation plant. Each separation plant
consisted of several buildings that performed the process that separated the
plutonium from the irradiated fuel slugs.[107] The largest building was the
canyon, 810 feet long, 102 feet wide, and 85 feet high, also known as the T-
Plant or building 221-T. Ground was broken on June 22, 1943, and the first
batch of irradiated fuel rods was processed on December 26–27, 1944. A
second canyon, 221-U, was built in 200 West (but not used) and a third, 221-
B, was built in 200 East, which became operational on April 13, 1945. Inside
the windowless concrete canyon were forty cells, arranged in pairs, where
the uranium was dissolved and the plutonium extracted using a bismuth
phosphate chemical batch process, which had been decided upon in June.
Chemist Stanley Thompson, a colleague of Glenn Seaborg, invented the
process. Given the high radioactivity, everything was done with shielded
cranes, periscopes, television, and special remote-control equipment.[108]
Further processing of the plutonium, to concentrate and purify it, took
place at other buildings within 200 West or 200 East. The final product
shipped to Los Alamos was plutonium nitrate, originally a wet paste (like
molasses) but later a solid that was further worked upon there and shaped
into its final form. The first shipment left on February 3, 1945, taken by
Matthias to Los Angeles, where it was turned over to a Los Alamos officer.

As massive as they were, the installations at Oak Ridge and Hanford were
but a part of the full operation that Groves created and ran. By 1945 there

were factories, laboratories, and mines in thirty-nine states, Canada, and Africa supporting the operations at Oak Ridge and Hanford.

In all, between 1942 and 1945 the Manhattan Project entered into agreements with more than two hundred prime contractors, who in turn engaged thousands of subcontractors. Before the end of the war, approximately 600,000 people had worked on the project, which at its peak employed more than 160,000.

It is worth asking whether the Manhattan Project had to be as big as it was. Could the few bombs that were made been produced from a smaller complex? Some have suggested that the sheer vastness of the Manhattan Project was an indication that the United States had intentions, even during the war, to build bombs on the scale eventually achieved at the height of the Cold War. There is no convincing evidence that suggests such foresight or intent. No one at the time knew how much or how long it would take to produce the two kinds of fissile material for what turned out to be two types of bombs.

As with most things about the project, it comes down to Groves. With his conservative and redundant approach, building on a large scale was seen as more likely to achieve the goal than trying to calculate just the right size. As big as the project was, it might have been even bigger. The original plans for Hanford, for example, called for six reactors, each with a separation plant. It would be several years after the war before the United States decided to build bombs in an assembly-line fashion. An additional complex of reactors, plants, and facilities was built in the 1950s to accomplish this.[109]

The Manhattan Project is often cited as a prime example of how the U.S. government mobilized the talent and resources of American society to achieve a goal effectively and quickly. One recent study examined seven collaborative efforts — of which the Manhattan Project was the paradigmatic case — and the factors responsible for their success.[110] The authors emphasized the dynamics of the group effort rather than the efforts of a single individual, though certain individuals do make a difference.

The study provides no real surprises in its list of lessons derived from studying the seven "great groups." To achieve success, start with superb and gifted people. It is best that they produce something tangible as opposed to working on an abstraction or an idea. Young people are normally more energetic, confident, and curious and thus are more likely to work harder and longer. It is all the better if the undertaking is driven by moral purpose.

Put this special population in an isolated spot without any distractions. Living in Spartan conditions makes work the focus, with no distractions. This tendency to escape into the work may result in ignoring or not having the time to reflect on what is being produced. The cooperation of the many parts toward realizing the overall goal is essential. Ensure that those below have faith in their leaders, and make sure that the leaders have faith in those below. Though Americans like to believe in the triumphant individual, meeting challenges and overcoming adversity, it is really a blend of individual and collective effort that gets things accomplished. The leader finds greatness in the group and also helps them find it in themselves. All of this sounds familiar in Groves's running of the Manhattan Project.

The Manhattan Project's success in having a usable bomb ready by mid-summer 1945 can in large part be attributed to the administrative ability, leadership, and style of General Groves.[111] Groves was the major contributor to the atmosphere, rules, and culture of the project. Many of the methods he used he brought with him from the Corps of Engineers. Many of the key personnel choices he made were of people whom he had known personally from working with them on other projects. Groves had been schooled by the army to command, and when he achieved a top-level position with virtually unlimited authority, he did whatever it was that had to be done. Certain elements were essential to lead: a strong backbone, a certain toughness, and even meanness when necessary, as he once said.[112]

Total program authority was vested in Groves. He had the complete support of the president and the other high officials of the administration. The project was initially understood to involve possible national survival against an evil enemy bent on world domination that may have been trying to build an atomic bomb of its own. The objective was clear, unmistakable, finite, and well defined. Compartmentalization, in addition to maintaining security, kept people focused on their assignment to achieve it. Each element had its own task, and all were carefully allocated, assigned, and supervised so that the sum of the parts resulted in the accomplishment of the mission. Command channels were clear-cut, well understood, and direct. Authority was invariably delegated with responsibility. Large staffs were avoided, especially in Groves's Washington office. People at the higher levels knew one another from past experiences and could quickly communicate to solve problems and make decisions. Written communication was kept to a minimum. Most business was done verbally by phone or face to face. Groves's decisions were not based on staff studies, committee reports, written opinions of consultants, or the like.[113]

Groves always projected an optimistic attitude, which inspired others. Morale could only be sustained if everyone thought that the thing could be done. If Groves showed any doubt, hesitation, or fear, it might infect the others and undermine the project. Groves normally set completion dates that he was sure his subordinates could not meet. Keep the bar high and people will work harder to jump over it.[114] Success is not a matter of luck "but the result of mental and physical capacity, of endeavor, of determination, and in large measure, of competent management."[115]

On several occasions Groves set forth the tenets he thought were fundamental to the speed, efficiency, and success of momentous undertakings by the government. In 1958, as the United States was in the midst of a crash program to develop and deploy ballistic missiles, General Groves provided his blueprint for how to do it, pretty much the one he followed.[116]

CHAPTER TWELVE

His Own Science: Oppenheimer, Los Alamos, and the Scientists

Groves had contradictory ideas about scientists, and the scientists who worked for him had contradictory ideas about him. While he greatly admired their intelligence, he found some of them to be terribly impractical, overly optimistic, and basically unrealistic about what it took to complete a task. Groves ran the secret Manhattan Project along military lines, and that included the laboratories. From the outset the military culture of discipline, censorship, and security clashed with the free exchange of ideas that is the academic and scientific ideal. Trouble was bound to come, and it did, but his relationships with the scientists had many different facets and layers.

The first contact between Groves and the scientists came less than three weeks after he was chosen to head the Manhattan Project. At the beginning of October Groves set out to tour the university laboratories in Chicago, Berkeley, and New York. On this trip he made several major decisions regarding the organization of the project, he met Robert Oppenheimer, and during his meetings with a few of the scientists, impressions took root (on both sides) that would be long lasting.

On Monday, October 5, 1942, Groves went to the Metallurgical Laboratory at the University of Chicago.[1] Since the beginning of the year Arthur H. Compton had been coordinating the theoretical and experimental work on fissionable materials for a bomb. This included work on uranium isotope separation, as well as a design of a graphite pile that would produce plutonium. In the morning the general inspected the laboratory facilities and met with the president of the university, Robert M. Hutchins. Groves underscored the importance of the work for the war, and Hutchins pledged his wholehearted cooperation.

In the afternoon Groves acquainted himself with the leading members of the team. He and Compton entered a room in Eckhart Hall and sat down during a meeting of the Technical Council, which was already under way. The Technical Council was composed of approximately fifteen senior scientists at the Met Lab;[2] that day they were discussing which of four methods to pursue. Sensing discord, Groves spoke up and informed them of how important the War Department considered the project. In what would be a hallmark of the way he operated, he told them that if there was a choice between two methods, then build them both. A wrong decision that brought some results was better than no decision at all. Time was more important than money. Groves told them to have a decision about which methods to pursue ready for Compton by Saturday.[3]

Much would happen that day that would set the tone for the relationship between the military and the scientists, a clash of cultures that never was resolved.[4] For Groves's part he was especially struck by the scientists' concern with small-scale laboratory techniques, apparently unable to grasp the huge engineering-industrial effort that would obviously be required to build the bomb. He was particularly appalled by the arrogance of some of the Chicago scientists, who resisted turning construction of the nuclear reactors and separation plants over to Du Pont on the grounds that the engineers and industrialists would never understand the principles necessary to make proper decisions.

The scientists, on the other hand, were horrified at Groves's obvious lack of intuition concerning scientific practice.[5] Physicists often rely on order-of-magnitude calculations to make sure their techniques have given them reasonable results. But to an engineer, an uncertainty of one order of magnitude is more than the difference between a viable structure that will stand and one that will collapse. In this sense Groves and the scientists were talking past each other. Groves was struck that no one could answer the question of how much material would be needed for a bomb. The best they could offer was an estimate that varied by a factor of ten — maybe it would take ten pounds, maybe one pound, or maybe a hundred pounds. As he later said of them, "They were simply not accustomed to moving with courage and rapidity. . . . None of them were go-getters: they preferred to move at a pipe-smoking academic pace."[6]

One person he met in Chicago that fateful day was Hungarian refugee Leo Szilard. Szilard was born in Budapest in 1898, had studied engineering and physics, and received a doctorate in physics from the University of Berlin in 1922. He taught there until 1933, when he fled Germany to escape Nazi persecution. He first settled in London and in 1938 moved to

New York. Upon learning of the discovery of fission, he saw that uranium might sustain a chain reaction and was instrumental later that year in bringing this news to the attention of President Roosevelt — by drafting a letter for Albert Einstein's signature — and warning him that Germany may be involved in such research. Perhaps more than any other single individual in this early period, Szilard was responsible for encouraging the American effort to explore the military use of atomic energy.

Szilard, the European elitist intellectual, considered Groves an uneducated boor, totally unqualified to make substantive decisions about this project. Likewise, Groves had very strong feelings about Szilard. He despised him from the minute he met him. He believed that Szilard would not be willing to give his life for the good of the whole, the basic duty of a good soldier. He typified scientists, and the academic world in general, who were "willing to have other people killed and wounded to protect their own interests but they were unwilling to participate in the dangerous occupation of a soldier."[7]

Groves mentioned Szilard only once in his book: "the brilliant Hungarian physicists Eugene Wigner and Leo Szilard."[8] In his private papers he had many unflattering things to say about the man. In annotations made while reading Margaret Gowing's book on the history of the British involvement in the bomb program, he says of Szilard, "He was a parasite living on the brains of the others." Groves doubted that Szilard and Wigner could have drafted the letter to Roosevelt that they got Einstein to sign; "it was too cleverly written and indicated a knowledge not only of international affairs but also what would appeal to President Roosevelt." "Szilard was not a most distinguished scientist." He "did not have a shred of honor in his make-up."[9]

Groves thought Szilard a voluble busybody who stirred up differences and distracted other scientists from their work, threatening a delay in the entire program. In the spring of 1945, or possibly earlier, Groves discussed "the Szilard problem" with President Conant of Harvard and "asked him if he would make Szilard a member of the Harvard faculty, and thus get him out of my hair." Groves even offered to pay Harvard to take him. Conant's reaction was raucous laughter, replying "that [Groves] couldn't pay him enough to take on the headache of having Szilard at Harvard," and added that he did not know of any other university that would take him either.[10]

Groves considered Szilard "an enemy alien" and, soon after meeting him, attempted to have him imprisoned. He went so far as to draft a letter, dated October 28, 1942, to the attorney general for the secretary of war's signature, that proposed Szilard be apprehended and "interned for the duration of the war."[11] Stimson refused on the grounds that, as he said, as far as he was aware the nation still operated under the general guidelines of the Constitution. Groves later said, "This was the answer I expected but I thought that there was no harm in trying."[12]

Throughout the war Szilard and Groves harassed one another at every opportunity. Groves never let Szilard visit Los Alamos, had him followed wherever he went, and kept a well-documented record of his security infractions. While Szilard was untrustworthy and uncooperative, to fire him would only increase the security risk, so he was retained and watched.

Szilard was not shy and was quick to take credit for everything he could. With the war just over and the bomb no longer a secret, Szilard, Groves wrote,

> now claims for himself practically sole responsibility for the initiation of the project in the United States, as well as major credit for the successful completion of the Hanford process. There is some credit due him as a promoter in the initiation of the project but none for the success of the Hanford project. In this he was actually a detriment. He has been of no real value to our work for the last three years.[13]

In that one meeting at Chicago in early October 1942, in the minds of some scientists the impression of General Groves as an uneducated military martinet was sealed. They were horrorstruck to think that the project on which they had been working so hard was now to be taken over by the army, and what was worse to be run by this particular general. To make matters worse, apparently in order to impress them, Groves announced that his scientific training included knowledge of calculus.

> There is one last thing I want to emphasize. You may know that I don't have a Ph.D. Colonel Nichols has one, but I don't. But let me tell you that I had ten years of formal education after I entered college. *Ten* years in which I just studied. I didn't have to make a living or give time for teaching. I just studied. That would be about equivalent to two Ph.D.'s wouldn't it?[14]

The quote first appears in an excellent work by Stephane Groueff published in 1967. The context of the quote is that one of the scientists was at the blackboard writing figures and formulas about how much material would be needed for the bomb. Groves caught a mistake in the copying of a figure from one line to the next and pointed it out. Groves suspected that he was being set up to look the fool and that it had been done on purpose. It was at this point that he apparently uttered the above, more so as a boast than defending an insecurity. Some time later he told an interviewer:

> I made quite a point of letting these scientific people know and I told [James] Marshall and Nichols that they must do it, how much education we all had in the course of our career . . . I pointed out that after all if he [Marshall] had been taking Ph.D's he would have had two Ph.D's. The whole idea was to build up in their minds that we were not only engineers but also we were educated people and while our education might have been in a different field, we were just as smart academically as they were.[15]

After Groves had left the room, Szilard exploded with indignation, "You see what I told you? How can you work with people like that?"[16]

For many of the scientists, this caricature of Groves would prevail throughout the war years and long afterward, shaping the way his role was depicted in many of the histories that were written of the Manhattan Project. Many never understood that behind the bluff and bluster was a man capable of quickly grasping the essentials of most technical problems and whose experience and performance in the administration of grand engineering projects had few equals.[17]

Groves's chief aide for security and intelligence matters, John Lansdale, observed him at close range and saw these two sides.

> It is true that General Groves, like many of us, had a very adequate appreciation of his own abilities. The problem was he had no hesitation in letting others know of his own high opinion of himself and his abilities. This is the origin of the feeling that he was arrogant and the reason why many people disliked him.[18]
>
> Unfortunately, it took more contact with him than most people had to overcome a first bad impression. He was in fact the only person I have known who was every bit as good as he thought he was. He had intelligence, he had good judgment of people, he had extraordinary perceptiveness and an intuitive instinct for the right answer. In addition to this he had a sort of catalytic effect on people.

Most of us working with him performed better than our intrinsic abilities indicated.[19]

———

Some writers have accused Groves of being anti-Semitic, using his views and his treatment of Szilard as evidence.[20] The fact is that many found Szilard insufferable or worse, and Groves had plenty of other reasons to despise Szilard beyond his religious or ethnic background. His main complaint about Szilard had to do with his unwillingness to abide by security rules. Groves was used to having people obey his orders. With the scientists he was faced with a group totally at odds with the military regimen. Groves was too practical a person to let prejudices interfere with his job. He had no difficulties with the other émigré scientists, many Jews among them, including Edward Teller, Eugene Wigner, and Hans Bethe, and he even tolerated the antics of Richard Feynman.[21] If he thought a person could help him get the job done, then that was what counted, not his ethnic identity. The best example of this is his choice of Robert Oppenheimer as scientific director. Groves never wavered in this choice, and they remained friends and carried on a cordial correspondence until Oppenheimer's death in 1967.

The security issues about the refugee scientists on the project did cause Groves concern. Those concerns came in several varieties; given a choice Groves would have excluded many of them, if he could have. He was even troubled by American citizens with strong European accents, but his natural instincts were tempered by his pragmatic recognition that certain individuals such as Enrico Fermi, Emilio Segrè, Eugene Wigner, Edward Teller, and Hans Bethe were not only loyal but also indispensable.[22] Segrè has recounted that once he, Fermi, and two other of his countrymen were having lunch at Fuller Lodge, at Los Alamos, when they slipped

> into Dante's tongue; as usual, talking loudly. General Groves was nearby, and he let us know that he did not like us speaking Hungarian (!) in public; he delicately hinted that if we wanted to speak foreign languages, we had better go into the woods.[23]

While issues of racism and prejudice were acknowledged, they did not prevent attempts to recruit the best people, if it was thought that they could help. One example was a discussion between Groves and Oppenheimer about whether to recruit Subrahmanyan Chandrasekhar to work at Los

Alamos. In 1944 Chandrasekhar was a thirty-four-year-old assistant professor of physics at the University of Chicago. Apparently earlier efforts to persuade him to work on the project were not successful. Groves and Oppenheimer discussed how much he might be needed now, raising the matter of his race and the fact that he was not an American citizen. The brief synopsis of their conversation is not definitive, but it suggests that they believed that if he were really needed he should be urged to join.[24]

Whatever degree of racism or anti-Semitism may have been present in Groves was probably shared by most white Anglo-Saxon Protestant males of his day.[25] This is not to excuse it, but only to underscore that these attitudes and prejudices were unexceptional.

Groves was very judgmental about people, and sometimes he let them know it. This attitude started early and lasted a lifetime. His standards were difficult for most people to achieve. Of the traits he valued most, self-discipline probably headed the list. Through discipline one could focus one's intelligence and energy toward a goal and accomplish it. His own discipline had steered him to West Point and then on to a career that aimed toward the higher reaches of the army. Groves readily accepted people if he thought them intelligent and competent, and pretty much dismissed those who were not. He liked people like himself: organizers, people who were decisive and not afraid to make difficult choices and, at times if necessary, even take risks. He sized people up quickly and decided that they either were acceptable or not. There were few second chances.

During the Manhattan Project Groves's relationship with Szilard was the exception. With most of the others with whom he worked, Groves got on well. His closest contacts were with Vannevar Bush, James Conant, and Richard Tolman.

Vannevar Bush was also the son of a minister with a great sense of public service. As we saw in a previous chapter, he was a prime mover in the decision to build the bomb and, through his establishment of the National Defense Research Committee and later the OSRD, at the center of war-related research. He had a management style that fit well with Groves's. They agreed that allowing subordinates independent initiative was the best way to get the most out them. Try to find the most able man you can, one with

> a thorough knowledge of his field, a standing in his profession, a
> vision of possibilities, a courage to attack the unknown, a patience

that is inexhaustible, and a kindly humanity that will cause his coworkers to rally about him with enthusiasm.[26]

With the project under way Bush retreated from the day-to-day affairs, leaving these matters to Conant and, of course, to Groves. Bush kept abreast of developments through the MPC and was available to assist whenever any help was requested of him.

Bush got on well with Groves. They were both engineers, practical men who were focused on the job at hand. Bush fully supported Groves's imposition of secrecy and security and was a man of action when it came to trying to halt the German bomb. He was an early advocate of aerial bombing and later fully supported the Alsos mission to gather information about the German bomb program, supplying resources and personnel. He saw his role as one of giving the president his best advice, looking out for his interests and keeping him briefed on the status of the bomb. There was often frustration, because Roosevelt had a penchant for making secret promises and agreements and not informing his advisers. One danger point, Bush thought, was Roosevelt's too-chummy relationship with the wily Churchill. Bush was Groves's equal in his distrust of the British, and he did his best to dilute Anglo-American cooperation.

Bush never had any hesitancy about using the bomb, or any qualms about it afterward. He was concerned about what effect the bomb might have on the postwar world. With his alter ego, Conant, he was among the first to try to get the secretary of war and the president to think about some of the issues just over the horizon, after the bomb was used. Two things were certain: America's monopoly would be limited, and there were no real secrets about the bomb. Any nation with enough resources would eventually duplicate the American effort. What the United States did in the interim period might either promote some form of international control or ignite a dangerous arms race. A corollary concern of Bush's was the future role and relationship of science, the government, and the military.[27]

The relationship was amiable with a single exception. At one point Bush discovered that Groves's security agents were tailing him. He confronted the general about it.

> When he ducked my question I asked him out flatly, "Have you, or any of your people, attempted to find out in any way whatever, anything about my affairs except those that I have told you, or that you might properly consider to be part of *your* affairs as well?" When he

admitted that something of the sort might have occurred, I told him, "You take steps to see that it doesn't occur again."[28]

At the outset Groves knew very little about nuclear physics and the complex technologies involved. But whatever the topic, Groves always did his homework. It was not his style to leave anything to chance if it could be avoided. He knew that in many cases he was going to have to rely on the judgment and advice of others. Therefore, he would have to learn at least the basics and find scientists whose opinions he could trust.

Groves took up the task of educating himself about the subject at hand. An early step was to direct a series of thirteen questions on technical and scientific matters to Conant and Oppenheimer.[29] Over the next several months Groves discussed myriad questions with Conant and with Richard C. Tolman, a theoretical physicist and former dean of the graduate school at California Institute of Technology who served as vice chairman to Conant on the NDRC.

Richard Tolman played a significant role in the Manhattan Project, one that has not been fully appreciated. Groves relied upon him extensively. Tolman was a respected scientist in his own right and knew all of the major personalities in the field. He assisted in the constant recruitment that sought the best scientists. He traveled to all of the sites, smoothed the relationship between the scientists and the military, and drafted countless memos and policy papers for Groves.

On June 11, 1943, Groves formalized the arrangement with Conant and Tolman, asking them to serve as his scientific advisers. In response to his probing questions, Conant and Tolman provided Groves with detailed and comprehensive primers on all aspects of nuclear technology. Armed with this kind of advice, Groves was able to make several fundamental decisions on technical matters that affected the basic course of the project.

Conant and Tolman were not the only sources of counsel and opinion for Groves. The three of them were able to draw upon some of the best scientists in the world. Eight on the project had already won a Nobel Prize, and more than a dozen others would do so after the war.[30] But after all was said and done, when a big decision had to be made it was Groves's call.[31] Often it fell to him to make a choice between what appeared to be equal alternatives, each backed by sound scientific advice. Some of these decisions involved the ways that hundreds of millions of dollars would be spent, or wasted.

Groves's most important relationship with a scientist was the one with Robert Oppenheimer. They met on Groves's first tour of the laboratories. Oppenheimer remembered that it took place after a luncheon in President Sproul's home in Berkeley, probably on October 8.[32] From the first they hit it off very well.[33] Their discussion centered on the kind of establishment that would be required to design and assemble the bomb once there were sufficient quantities of uranium 235 and plutonium 239.

By October 1942 Oppenheimer had been directly involved in research on bomb design for about a year, since June as head of Compton's theoretical group. In October 1941 Lawrence felt Oppenheimer had "important new ideas" about fast neutron reactions and urged Compton to invite Oppenheimer to a meeting of the Advisory Committee of the National Academy in Schenectady, New York, on the topic.[34] When Compton organized the Metallurgical Laboratory in early 1942 he put Gregory Breit in charge of coordinating fast-neutron research at several universities and institutes, with Oppenheimer directing the effort at Berkeley. Breit, a theoretical physicist from the University of Wisconsin, was a very difficult character, quick to take offense and worried about security to the point of paranoia.

Breit's personality so grated on everyone that on May 18, 1942, by mutual consent, he resigned, citing the lax attitudes of his colleagues toward matters of secrecy and security. Compton immediately chose Oppenheimer to take Breit's place. His task was to coordinate the theoretical calculations on basic nuclear reactions with the experimental data in order to estimate how much material would be needed for a weapon and how efficient it would be.[35]

It was decided that Oppenheimer would stay at Berkeley, and the new assistant he had requested, John H. Manley, would remain at the Met Lab in Chicago to coordinate the other scientists who were working on fast neutrons.[36] Oppenheimer lost no time in arranging a conference to take stock of where the research stood. Several prominent physicists, almost all of whom would soon go to Los Alamos, attended the Berkeley summer conference held in July.[37]

Groves was impressed with Oppenheimer's clear sense of the kind of laboratory it would take to accomplish the task. He was intrigued with his suggestion that rather than locating the laboratory at, say, Chicago or Oak Ridge, they should find a remote and isolated site where free communication among the scientists could go on, but secrecy and security could be assured.

Their second meeting took place a week later as Groves wound up his initial tour and headed home. Groves asked Oppenheimer to join him in Chicago to continue the discussion about the laboratory. With many issues still unresolved and the 20th Century Limited about to depart, Groves invited Oppenheimer to accompany him on the New York–bound train. Nichols and Marshall were also on board, and after dinner all four of them squeezed into a tiny roomette to discuss in some detail how and where to create a laboratory.[38]

Clearly by the end of October, if not before, Groves had decided to establish a laboratory along the lines that he had discussed with Oppenheimer. The next tasks were to pick a director and a site. Groves ruled out appointing Lawrence or Compton. They already had their hands full running their own labs. Harold Urey was out of the question. Groves's scientists came in all shapes and sizes, with their peculiarities and their range of abilities. Whereas someone like Szilard was, in some ways, too strong a personality, Harold Urey, in Groves's mind, was too weak.[39] When he met Urey at Columbia University, Groves decided — Nobel Laureate or not — that he was almost totally lacking in the kinds of administrative skills needed if the project were to succeed. A second black mark against him was Urey's advocacy of liberal causes, all the more a problem after 1944, when Urey began exploring ways of promoting international control of the atom. Throughout the Manhattan years, his judgment in this matter was repeatedly confirmed. In Groves's dealings with Columbia University and the work he ordered them to do he disregarded any objections posed by Urey. But as with Szilard, to have dismissed Urey from the project would only have fed the fires of resentment against military control and endangered security, so he was kept on.[40]

Lawrence strongly recommended Edwin M. McMillan, one of his protégés. Groves, however, had become convinced that the right man for the job, indeed the only man for the job, was the thirty-eight-year-old Oppenheimer. Few he asked agreed with the choice, though Bush and Conant concurred after much discussion.[41] Oppenheimer's fellow scientists objected on the grounds that he was a theoretical physicist with no experience running an experimental laboratory. They also objected that he had had no administrative experience. There was nothing in his career to suggest that his talent lay in this direction. Finally, they pointed out that unlike Urey, Compton, and Lawrence, he had not won a Nobel Prize and therefore would not command the kind of respect required for the job that Groves had in mind.

None of this changed Groves's mind in the least. He apparently saw then, in the fall of 1942, what others did not. Some of the things that Oppenheimer's critics saw as drawbacks Groves saw as strengths. In retrospect we know what those qualities were, and others would see them displayed over and over throughout the next three years. His colleagues and writers who have discussed Oppenheimer's leadership of Los Alamos have described them in some detail.[42] He was inspiring, he could grasp the essentials of a problem with lightning speed, and he could be charming and supportive. He had a remarkable ability to listen to conflicting opinions and then synthesize them in such a way that the different points of view melded together and everyone felt they had shared in the solution. Years later Oppenheimer was asked how he accounted for the general choosing him, and he replied immodestly that Leslie Groves "had a fatal weakness for good men."[43]

Groves no doubt made other calculations in making his choice. He knew Oppenheimer had "a very extreme liberal background," but as he told the Personnel Security Board in 1954, he usually thought it was better to keep a doubtful person on the job and under surveillance than to expel him, leaving him free to disclose what he knew. Oppenheimer, after all, "was already in the project," and "he knew all that there was to know about" the bomb computations.

Oppenheimer's complex personality and influential life have already been the subject of several biographies, and several more are on the way.[44] The chief concern here is his relationship with Groves. The traditional approach has been to see the differences in the two men and wonder how this "odd couple" ever got along. The way they looked — the rotund military man and the thin, almost gaunt professor in the porkpie hat — supports the odd-couple analogy. Yet while the two were dramatically different in some ways, they were alike in others.

That Oppenheimer and Groves should have worked so well together is really no mystery. Groves saw in Oppenheimer an "overweening ambition" that drove him. He understood that Oppenheimer was frustrated and disappointed; that his contributions to theoretical physics had not brought him the recognition that he believed he deserved. This project could be his route to immortality.[45] Part of Groves's genius was to entwine other people's ambitions with his own. Groves and Oppenheimer got on so well because each saw in the other the skills and intelligence necessary to fulfill their common goal, the successful use of the bomb in World War II. The bomb in fact would be the route to immortality for both of them.

They treated each other in special ways. Oppenheimer could at times be sarcastic with students or colleagues who could not keep up with his quick mind. Not so with Groves. He patiently answered whatever query the general asked. On Groves's part he treated Oppenheimer delicately, like a fine instrument that needed to be played just right. Groves's normal approach with most of his subordinates was to push them as hard as he could. The pressure was a test to see what they were made of. The more they took, the tougher they were. The good ones would make it through; those who broke would be transferred, demoted, or replaced. The general saw that this approach would not work with Oppenheimer. Some men if pushed too hard will break.

In late October 1942 Groves took the first steps toward establishing the laboratory/factory that would design and build the bomb itself.[46] Groves concurred with Oppenheimer's suggestion that the laboratory be located in an isolated region of the country.

Isolation was not the only criterion. The site also required a climate that would permit work to go on year-round. Access by air, road, and rail was necessary. Adequate power, water, and fuel supplies must be available. The questions of who owned the land and how fast it could be acquired were also important.

Late in October Lt. Col. John H. Dudley, an engineer with the Syracuse district, was ordered "to make a survey for an installation of unnamed purpose." The district engineer's office in Albuquerque surveyed sites in New Mexico, California, Utah, Arizona, and Nevada.[47] Most of the sites were rejected as inadequate, leaving five possibilities in the vicinity of Albuquerque. Upon closer analysis, three of these were rejected, leaving Jemez Springs and Otowi. On Sunday evening, November 15, Groves flew to Albuquerque, arriving early on the morning of November 16. Later, with Oppenheimer, McMillan, and Dudley, he visited the Jemez Springs site, arriving midmorning.[48] After inspecting it, Groves disapproved of it at once. It lacked space and would have required the taking of a number of small farms owned by Indians, something that would have been vetoed by Secretary of the Interior Harold Ickes if he had known about it. At Oppenheimer's suggestion they then traveled by car to the Otowi site where the Los Alamos Ranch School was located, arriving at about three o'clock, just as classes were ending.[49] McMillan recounted the scene:

> [It] was late in the afternoon. There was a slight snow falling. . . . It was
> cold and there were the boys and their masters out in the playing field

in shorts . . . I remarked that they really believed in hardening up the youth. As soon as Groves saw it, he said, in effect, "This is the place."[50]

Wasting no time, by November 23 the right to begin surveying the Ranch School was granted, and two days later the War Department directed the acquisition of the land, almost fifty thousand acres. On December 7 the formal eviction notice was received.[51] District Engineer Col. Lyle Rosenberg and the Albuquerque Engineer District supervised the design, engineering, and construction contracts.[52] The choice for prime construction contractor was the M. M. Sundt Company of Tucson, Arizona, and for architect-engineer the firm of Willard C. Kruger and Associates of Santa Fe.

Groves, ever security conscious, was concerned about having an outside contractor, and even the Albuquerque Engineer District, involved in the program; he thus took steps to tighten security. After completing the first major phase of construction in the fall of 1943, the Sundt Company withdrew its personnel and equipment and the post commander prepared to assume responsibility for any future minor construction and for maintenance. This did not last very long, because Los Alamos's population continued to grow and new construction firms were recruited.[53]

No precise date is recorded in the official histories, or in Groves's book, as to when Oppenheimer was appointed scientific director of Los Alamos. By mid-December 1942 Oppenheimer was actively recruiting personnel from other laboratories and universities to join him in New Mexico. In a letter to Hans Bethe and his wife, Rose, at the end of the month, he went into great detail about the plans he had for "setting up this odd community."[54]

From the outset Groves established dual lines of authority, to the military commander at Los Alamos, and to the scientific director. Both reported directly to Groves and functionally did not go through the district engineer's office. With regard to Los Alamos Groves assumed many of the functions of the district engineer and the area angineer. The direct lines permitted Groves — by phone, Teletype, or frequent visits — to exercise broad policy control over the bomb program and to intervene in the day-to-day operations of the laboratory.

> With respect to Los Alamos, it was directly my responsibility in every way, everything that happened. The orders were issued direct. We tried to keep Nichols informed to such extent as was necessary. So

from a practical standpoint, though not on paper, the chain of command was direct from me to Dr. Oppenheimer.[55]

And again,

> Due to the magnitude of the District I retained personal direction of
> the Los Alamos bomb laboratory and took personal charge of the
> development of the weapon from the point where fissionable mate-
> rials were supplied through and including the military operations.[56]

Groves called Oppenheimer, Ashbridge (and later Tyler), Parsons, and
Maj. Stanley L. Stewart in the Los Angeles procurement office frequently
to ensure that things were running smoothly. These were supplemented by
Teletype messages, letters, memos, and — about once every two or three
months — a visit that lasted two or three days. In between Parsons would
visit Groves in Washington on a monthly basis. Conant and Tolman would
make periodic visits to Los Alamos and report on progress to Groves.
Occasionally Oppenheimer would come to Washington to see Groves or
meet him in Chicago or in Berkeley.

The military commander of the post oversaw all military personnel
and was responsible for housing, food, and supplies for the entire population. He was like a mayor or city manager who also had to supply
libraries, hospitals, theaters, and recreation facilities for the populace.
Security was the top priority; the commander oversaw an extensive
police force to prevent trespassers from getting in and ensured that the
inhabitants followed authorized procedures when they left the site. The
guard force patrolled on horseback and in jeeps twenty-four hours a day,
sometimes with dogs.

Three commanding officers served during the war, and two after the
war. All five were selected by Groves. The first, Lt. Col. John M. Harmon,
lasted only about four months, from mid-January to mid-April 1943.
Among Harmon's deficiencies were a fondness for alcohol and an inability
to deal with nonmilitary personnel. Lt. Col. Whitney Ashbridge, a reserve
officer engineer, replaced Harmon.

Ashbridge had served under Groves in the Construction Division,
working for him on special projects and as a troubleshooter.[57] In late 1942
and early 1943 he was one of Groves's assistants, handling various odds
and ends. Coincidentally he had gone to the Los Alamos Ranch School
and knew the area.[58] Groves ordered Ashbridge to Los Alamos to see if

the friction between the army and the scientists could be overcome or at least minimized. Ashbridge's three-page memo listing his recommendations convinced Groves that Ashbridge should be the commanding officer and replace Harmon, which he did in May 1943.

The prime goal of the post commander and his military officers was to keep the community running as smoothly as possible, "and trying to help General Groves keep the scientist's noses to the grind stone," as Ashbridge put it.[59] Living conditions were difficult for the scientists and their immediate families. Most were accustomed to an academic environment with physical comforts, cultural entertainment, and ready access to relatives and friends. Transplanted to a barren mesa with a shortage of everything and thrown together with strangers under the watchful eye of the military was bound to cause problems.

And special problems arose. For example, the state of New Mexico insisted that all the scientists and military personnel at Los Alamos have New Mexico driver's licenses and New Mexico license plates. Ashbridge went to see Governor Dempsey to explain that it was impossible for security reasons to have a list of names in the New Mexico motor vehicle office. An arrangement was worked out whereby licenses were issued, each with a number, and Ashbridge kept a list of the matching names in his safe. If he were notified that number 234, for instance, were involved in a traffic violation or an accident, he would deal with the person and resolve the matter.

Ashbridge, perhaps because of the constant pressure, had a mild heart attack at the Amarillo Airport, where he fainted. This eventually led to his being replaced in October 1944 by Col. Gerald R. Tyler, who remained until November 1945.[60]

The estimates of how large and fast Los Alamos would grow were always low, and thus everything was always in short supply. Complaints about inadequate housing, poor food, shortages of essentials, and few opportunities for diversion or recreation were endemic. The population grew from a few hundred in the spring of 1943 to thirty-five hundred by January 1944. By the end of 1945 there were eighty-two hundred, and a year later ten thousand.[61] Contributing to this growth was the birth of children. Groves was none too happy with this trend but was unable to control the cause.

On the technical side Los Alamos was organized into divisions for theoretical physics, experimental physics, chemistry and metallurgy, and ordnance. Within each division there were a number of groups that were further split into branches. Branch leaders reported to group leaders, who reported to division leaders, who reported to Oppenheimer. And Oppenheimer reported to Groves, about matters large and small. For the first year or so Oppenheimer relied upon a governing board of some seven to ten members to help coordinate the work of the laboratory. As work progressed throughout 1944 and 1945, and as the program shifted from research and experimentation to engineering, fabrication, and testing some groups completed their projects and disbanded, while others merged to form new divisions.

The recruitment of scientists never stopped. In the course of developing the bombs problems arose that required specialists. Groves had several talent scouts always on the lookout for the particular person who had the required knowledge and expertise. Samuel T. Arnold, a dean at Brown University, was repeatedly in touch with Groves to help recruit mainly younger personnel at educational institutions.[62] A second was Merriam Hartwick Trytten, for a time at the NDRC, and then with the War Manpower Commission, recruiting physicists for the war effort. At times Groves had to intervene directly to convince university presidents to release one of their faculty.[63]

An important supplement to the prestigious scientists who were recruited were the draftees of the MED's Special Engineer Detachment (SED). The SED was composed of generally young scientists, engineers, and technicians who had been drafted into the army. Rather than being sent off to combat, they were transferred to Los Alamos, or elsewhere,[64] to fill a personnel shortage and perform a variety of scientific and technical tasks. Almost 30 percent of them had college degrees. Many had been in graduate school when they were called, and some had completed their Ph.D.'s. By the end of 1943 nearly 475 SEDs had arrived; by 1945 the unit included 1,823 men.

Los Alamos, for all of the outward military routine, retained an inner civilian core in part because of the influence of I. I. Rabi. A leading physicist who was a close friend of many scientists at Los Alamos (although he never worked there himself), Rabi forcefully protested a plan to require the scientists to join the army and opposed restrictions on the scientific staff. Rabi went so far as to threaten that if these conditions were not met, he would actively advise prospective scientists not to serve at Los Alamos.[65] Groves, after consulting with the Military Policy Committee, had little choice but to agree.

This was further underscored in a February 25, 1943, letter to Oppenheimer cosigned by Conant and Groves stating that the scientists did not have to join the army until January 1, 1944, at the earliest.[66] The main purpose of the letter was to spell out matters of responsibility and organization. The military intention of the effort is addressed in the first section. "The laboratory will be concerned with the development and final manufacture of an instrument of war, which we may designate as Projectile S-1-T."[67] Oppenheimer was informed that he could show the letter to prospective recruits, who would clearly be impressed with the sentence, "The laboratory is part of a larger project which has been placed in a special category and assigned the highest priority by the President of the United States."

The precise details of how the University of California was chosen to administer Los Alamos remain something of a mystery. By Christmas 1941 the university had a contract with the Berkeley Radiation Laboratory. Probably through the efforts of Lawrence and Oppenheimer, the university was requested to enter into a contract with the new laboratory.[68]

As secretary and treasurer of the board of regents, Robert M. Underhill's duties included making all banking arrangements for the university, managing its investments, handling all acquisitions of real estate, and overseeing patent matters. He also signed all contracts to which the university was a party.

Underhill has recounted that he met with Oppenheimer in 1942 about administering Los Alamos; in early 1943, "after it had been agreed that this new project would be undertaken," he inquired as to where it would be.[69] At first Underhill did not know what state the project was in; after pressuring for a reply, he was told it was New Mexico.

On Saturday, February 13, 1943, General Groves came to Underhill's office to discuss the various arrangements. They met in New York a week later in a room at the Biltmore Hotel, along with an army lawyer, Oppenheimer, and five or six nuclear physicists. Underhill then went to Washington and on February 22 agreed to take the contract.

Underhill first visited the site in mid-March 1943 and met with Oppenheimer. Final negotiations were discussed during a trip east to New York, and on April 20, 1943, Contract No. W-7405-ENG-36 between the U.S. government and the Regents of the University of California was signed at the MED's New York office on Fifth Avenue. Col. James C. Marshall as district engineer signed as the contracting officer, and Underhill, the secretary of the university's board of regents, signed on behalf of the university. The contract was "for the conduct of certain studies and experimental investigations at a laboratory located at a site which has or will be informally made known to the Contractor." At this point Underhill still did not know the purpose of the project and did not learn of it until November, when Lawrence informed him. The term of the original contract was from January 1, 1943, through June 30, 1944, with the option of extending it for "a further period or periods not exceeding the duration of hostilities with the Axis powers, plus six (6) months." Almost sixty years later the University of California remains the contractor.

On that first get-acquainted trip in October 1942, Groves also met Ernest Lawrence of the University of California, Berkeley. Ernest Orlando Lawrence was, among other things, an energetic scientific entrepreneur.[70] He recognized in Groves a man with the kind of drive that could supply him with the personnel, materials, and facilities he needed to solve the riddles of electromagnetic separation. For his part Groves was impressed with Lawrence's enthusiasm and willingness to do whatever it took to get the job done, though he was skeptical of his confidence that electromagnetic separation was sufficiently assured that the project could ignore other techniques. Groves thought,

> Ernest Lawrence was by far the ablest man of the four directors of the Manhattan Project. I doubt if he had the intelligence possessed by Oppenheimer but he had a driving capacity and an ability to organize that was outstanding. . . . He demanded and received the utmost cooperation. He was almost a domineering manager but at the same time was beloved by his subordinates. On the other hand Oppenheimer was a very smooth leader who did not force people into his way of thinking but rather led them into it by his sheer force of intellectual capacity.[71]

On many things the two men agreed.[72] Lawrence was a hard worker; he inspired others with his optimism and enthusiasm and was willing to put everything into winning the war. The question of a scientist's true purpose was a crucial determining factor in Groves's judgment of him. Did a scientist want most to win the war or to pursue scientific research? Groves believed there was no time to contemplate the mysteries of the cosmos. What counted was doing everything possible to build this weapon, to end this war. If that was not your goal, then Groves had little use for you.

Lawrence had great respect for Groves's abilities, as he said when congratulating him on getting a second star.

> As my association with you has progressed I have steadily developed an ever greater admiration for you, as a man and in respect to your handling of manifold and complex problems requiring vast common sense and an equally vast store of driving energy.
>
> It goes without saying that you are largely responsible for pushing through this extraordinary undertaking. I look forward to the day in the not distant future, when I trust the President and Congress will place a third star on your capable shoulders in commemoration of the laying of the first "egg."[73]

Arthur Holly Compton was the head of the Metallurgical Laboratory at the University of Chicago. During the three years that he and Groves worked together they got on very well. Born in 1892 in Ohio, Compton was the son of a Presbyterian minister; he wondered later if this was one of the reasons why they got along.

> I wonder now whether this common aspect of our cultural heritage may not have been a substantial help in enabling us to understand each other's language and point of view. . . . The idea that the good life is one devoted to service has always been for our family a living force, as evidently it had been also for General Groves. As a corollary it was part of our tradition that a person owes it to his God that he shall keep himself physically and mentally fit to serve at his highest efficiency. In Groves I found a man with the same spirit. He was keeping himself in strict physical training, like an athlete preparing for the great contest. He had not chosen to build atomic weapons. The task had been assigned to him, and as a good soldier, loyal to his country, he would put into it everything he had.[74]

After earning a Ph.D. in physics from Princeton, Compton worked as a research engineer at the Westinghouse Lamp Company in Pittsburgh for two years and during World War I helped develop aircraft instruments for the signal corps, experience in practical matters that would later serve him well. But Compton was best known for his theoretical contributions, winning a Nobel Prize in 1927.

Compton was in on the ground floor of the U.S. explorations of atomic energy. In 1941 he was appointed chairman of the physics department and dean of the physical sciences division at the University of Chicago; later that year, as we have seen, he headed a National Academy of Sciences committee on the military potential of atomic energy. Early in February 1942 the Met Lab was organized to pursue the many questions surrounding isotope separation, plutonium, and the bomb's design.

Compton saw Groves's strengths and weaknesses quite clearly and described them well in his book, *Atomic Quest.*

> The nation was fortunate indeed in the selection of General Groves for this task. A devoted servant of his country, a man of brilliant mind, unlimited courage, and tireless energy, he was a skilled and able, though not always smooth, administrator. He was competent to operate on whatever scale the conditions might demand. Moreover, from the very beginning he was convinced that the atomic project would succeed in time to be of important effect in the war. In his temperament and method of operation, he compares with some of the most effective industrial leaders with whom I am acquainted. A handicap that Groves only partially recognized was his unfamiliarity with scientists, their motivations, and their way of thinking. This occasionally led to misunderstanding with the men with whom mutual confidence was most essential. But such misunderstandings were in large measure offset by his respect for what the scientists were accomplishing, and was much more than counterbalanced by his own understanding of what is required to make great enterprise bring results. Especially important was his ability to overlook irrelevant personal traits and to select men on the basis only of their competence and loyalty. Thus he soon had working with him the best men anywhere available for carrying through their various assignments.[75]

In many ways Niels Bohr was for Groves the typical scientist. He had no concept of security and was much too voluble. Nevertheless, Groves had great respect for him, realizing the awe that many of his colleagues had for him, and the influence he had on morale at Los Alamos.

For security reasons and "[b]ecause he was so absentminded and so childlike in his approach in his wanderings [around Washington] I [Groves] felt it necessary to have him followed by security agents."[76] This was as much for his own safety as it was to "look for anything that would indicate that he was not trustworthy." One agent recounted to Groves the difficulty of tailing him. After a visit to the Norwegian Embassy on Massachusetts Avenue, Bohr started toward town. Without looking right or left he darted across the busy street during rush hour. The agent marveled at how he remained alive. The joke in Groves's office was that they knew when Bohr was coming because they heard the screeching of brakes outside.

On his first visit to Los Alamos at the end of 1943 Groves and Tolman accompanied Bohr and his son, Aage, on the long train ride from Chicago to Lamy. To keep the journey a secret, Groves tried to confine the two in connecting staterooms. Meals were served in the compartment; Tolman and Groves took turns staying with him to keep him from wandering. Groves spent many hours in the compartment trying to understand Bohr's "characteristic whispering mumble." The tactic did not work, as they later found out. Bohr and his son had gone to the dining car for breakfast both mornings.

CHAPTER THIRTEEN

His Own Intelligence: Domestic Concerns

Leslie Groves knew how to keep a secret. Secretary of War Stimson said of him that he had never known a man who was so security conscious. His aide in charge of security, John Lansdale, called him obsessive. As a result, the Manhattan Project, under Groves's direction, contributed to the "intelligence revolution" that occurred during World War II in important ways that have not hitherto been recognized or appreciated. Security practices and procedures that Groves helped develop were later adopted in the formative years of the Cold War, and persist to this day.

The Manhattan Project's relationship to Congress, with its secret budgets and lack of legislative oversight, make it in effect, the first large-scale "black" program, to use a more recent term. It was also one of the recruiting fields for a group of people who went on to careers in the Central Intelligence Agency,[1] and other sectors of the intelligence community after the war. The Manhattan Project established new levels of security consciousness and awareness. It was unprecedented in exacting information control not only among military and civilian government employees but those at universities and private corporations as well. The Manhattan Project was a turning point, a watershed in national security policy that served as a model for the postwar system, and Leslie Groves was its key architect.

With regard to secrecy in atomic matters Groves listed eight major objectives.[2]

- To keep knowledge from the Germans and, to a lesser degree, from the Japanese.
- To keep knowledge from the Russians.
- To keep as much knowledge as possible from all other nations, so that the U.S. position after the war would be as strong as possible.

- To keep knowledge from those who would interfere directly or indirectly with the progress of the work, such as Congress and various executive branch offices.
- To limit discussion of the use of the bomb to a small group of officials.
- To achieve military surprise when the bomb was used and thus gain the psychological effect.
- To operate the program on a need-to-know basis by the use of compartmentalization.

Groves is not often thought of as a contributor to modern intelligence practices, but his widespread use of compartmentalization as an organizing principle was novel and significant.[3] While Groves did not invent compartmentalization, he implemented it on a scale not previously seen. In his hands this organizational scheme was at once the prime method to limit information — and thus enhance security — and a major source of his power and influence. In government bureaucracies, especially ones heavily involved with secrecy, knowledge is power, and by knowing more one is able to shape the substance and pace of a policy or project. Groves had no agenda of his own. He was merely carrying out the decisions of the senior-level civilians to whom he reported, and he was in perfect agreement with them. There is not a hint that he ever abused his power, but by the same token, it has not been fully recognized how much power he had, how he acquired it, and what he did with it.

From the moment Groves took control a top priority for him was to establish and maintain a security system that would brook no violation of secrecy. The most important single secret about any weapon program is the fact that it exists, just as the most important single danger to that program is the possibility an enemy will build it first. This meant that the Manhattan Project had to be cloaked in the utmost secrecy.[4] If the Germans had even a hint that the Americans were involved, and making progress, they might begin a program to build a bomb of their own, if they were not doing so already. Second, if the enemy knew of the atomic project, it might find a way to sabotage the effort and delay progress. Finally, secrecy was essential to ensure maximum surprise when the bomb was ready, achieving shock that could cause the enemy to surrender.[5] Tight security was thus required to match the scale of the effort, and Groves put as much energy into this as he did into the industrial and scientific aspects of the project.

Keeping the existence of the program secret meant that everything about it was secret — nothing could be talked about openly. Compartmentalization had another purpose: to limit knowledge held by any individual so he could not betray it to an enemy and — just as important — to limit discussion of the program to a few top officials with authority to decide how and when to use the bomb. As he said,

> Compartmentalization of knowledge, to me, was the very heart of security. My rule was simple and not capable of misinterpretation — each man should know everything he needed to know to do his job and nothing else. Adherence to this rule not only provided an adequate measure of security, but it greatly improved over-all efficiency by making our people stick to their knitting.[6]

Wallace A. Akers, the initial scientific director of the British bomb program, was a critic of compartmentalization. After a conversation with Groves, Akers wrote to his superior:

> We have always thought that [the American] degree of subdivision was too great for efficient progress, but Groves is determined to go much further in that direction.
>
> In fact, he states that his intention is to divide the work into as many separate compartments as can be devised, and to fill each of these compartments with as many people as can possibly be used efficiently therein.
>
> He explains that only one person in each cell will be able to see over the top, as he expressed it, and that person will not be able to see into more than a minimum number of other cells.
>
> The only people who will be allowed to have any general knowledge of the work will be the group leaders, who are members of the S.1 Committee.[7]

After the war, when the Alan Nunn May, Klaus Fuchs, and David Greenglass/Rosenberg spy cases were revealed, Groves pointed out that the damage would have been much less had his views on compartmentalization prevailed. In a memo for the file written in 1965, Groves reviewed certain facts in the Greenglass case.[8] At the time trained civilian machinists were in great demand, he noted. To fill some positions at Los Alamos, army records were searched and a few machinists selected. The names were sent to the FBI for clearance to see if there was any record of Communist or German sympathies. In retrospect the FBI clearly erred in letting Greenglass slip through, given his sister Ethel's and brother-in-law Julius Rosenberg's affiliations. That was bad enough, but once at Los Alamos, Groves asked, how did the low-level Greenglass manage to learn so much? He attributed it to

> numerous violations of security rules by scientific personnel at Los Alamos. Probably no man did more to set up the background under which such a violation could take place than did Dr. Edward Condon during the short period he was Associate Director. It was he who persuaded Oppenheimer to establish colloquia . . . [which] tended to make individuals at Los Alamos less security-minded and less apt to guard their conversation within the laboratory.[9]

In Groves's estimation the worst violators of security in the Greenglass case were those men who told Greenglass things he should not have been told. Even though the FBI had failed, strict compartmentalization could have prevented Greenglass from knowing some of the things he told to the Russians.

In Groves's hands compartmentalization was more than just a technique for keeping secrets. It was as well the prime method by which he consolidated his own substantial power and control over the entire project. In general, when secrecy is involved and compartmentalization the technique, consolidation of power at the top is an inevitable consequence, whether it is the Manhattan Project or its descendants, the National Security Council, the Central Intelligence Agency, or the Department of Defense. In Groves's case all the boxes of the wiring diagram connected to him, and to him alone.[10] The cliché that "knowledge is power" was never truer. Knowing more allowed Groves greater control of the events as they unfolded.

To clarify this point, it is worth asking who knew what about the bomb and when. The categories of knowledge of the Manhattan Project might be listed as follows, from the general and broad to the specific and detailed:

> • Those who knew that there was a highly secret project under way to exploit atomic energy for military purposes.

- Those who knew that it was located at three major and many minor sites, and what general activities were taking place at each.
- Those who knew who the key personnel were.
- Those who knew about the German program, or lack thereof.
- Those who knew that a special air force unit was being trained to deliver the bomb when ready.
- Those who knew the technical details of bomb design, how uranium was being enriched and plutonium produced, and how much of each material was required for the two types of bombs.
- Those who knew the schedule and timetable for the bomb test, for its use, and against which targets.

Each of the hundreds of thousands of people associated with the Manhattan Project fell somewhere on the above list in the amount and kinds of knowledge they possessed. Most only had a very fragmentary appreciation of the project, its scope and scale, its evolution and timeline. Tens of thousands of workers at Hanford, Oak Ridge, and elsewhere did not even know what they were working on, as was the intention. Often even a fairly senior person at one of the sites knew very little, or perhaps nothing, of what went on at other sites. Groves did not let Hanford Area Engineer Franklin Matthias visit Los Alamos, for example. No one was to have access "solely by virtue of his commission or official position," and this meant, in Groves's mind, members of the executive branch, members of Congress, and military personnel.[11] Everyone, and that meant everyone, needed to "stick to their knitting."

Several thousand people probably knew that the government was involved in some kind of atomic energy program. Several hundred may have known what went on at the individual sites.[12] Many fewer than that knew who the key personnel were, and fewer still knew of the preparations and timing of the military mission. A small group of senior policy makers, civilian and military, in Washington had an integrated knowledge of what was transpiring, but even they were ignorant of much that was happening in the field. At the uppermost level of the pyramid were the senior-level officials, Stimson and Marshall, members of the Military Policy Committee (Purnell, Styer, Bush, and Conant), some close advisers (Tolman, Farrell), plus a few others who knew almost everything, though perhaps not every last detail.[13] Alone at the very top, and omniscient, was General Groves.[14]

Senior people do not need to know everything. To be mired in detail is time consuming and takes away from other responsibilities. This was especially true for Stimson and Marshall, Groves's immediate superiors. They had the entire war to wage, not just the atomic bomb program, and were involved in an extensive range of diplomatic and political activities as well as military ones. They left the bomb program to their tireless subordinate, Groves.

The view of Groves as merely a functionary has contributed to his marginalization in the Manhattan story. But as close students of bureaucracies are aware, in the implementation of any policy crucial decisions are made by subordinates that give it shape, focus, and content. The basic policy guidance given to Groves was never that specific or detailed. In fact, it was expansive and open ended. "Colonel Groves' duty will be to take complete charge of the entire DSM project. . . . Draw up the plans for the organization, construction, operation and security of the project, and after approval, take the necessary steps to put it into effect." It was left to him to amass the power he needed and to make what he thought were the necessary decisions. With the element of speed overriding everything, major decisions on how to implement the general policy fell to him by necessity or default.

The roots of compartmentalization, at least as practiced during the Manhattan Project, lie with the National Defense Research Committee (NDRC), established formally on June 27, 1940, to mobilize science to serve the military's needs.[15] To overcome the skepticism of some in the military and to gain their trust and confidence, the committee set up procedures to demonstrate that they could keep secrets, chief among them compartmentalization.

Each NDRC member took an oath of allegiance to the United States. Clerical personnel and each person accepting an appointment in any division or section did the same. All received letters from the committee stressing the need for secrecy. Procedures to handle classified information were adopted, generally following army and navy regulations. If there was a discrepancy between the two services, the more restrictive was chosen. As adopted and defined by the Manhattan Project, there were initially three security classifications. A fourth category, "Top Secret," was added in 1944.[16]

Security clearances from the army and navy were required for NDRC employees and for contractor personnel.[17] Photographic badges were adopted, as were guard services and burglar alarms at certain buildings, where needed.

When the NDRC and its Committee on Uranium were incorporated into the OSRD on June 28, 1941, the classification and security arrangements were continued.

Even earlier the scientists had introduced secrecy into their work on the atomic bomb. From the moment that fission was discovered many realized that weapons of enormous destruction were possible.[18] Some scientists were concerned that discussions of the physics breakthroughs and technical advancements, in journals and at conferences, might lead others to try their hand at making a bomb. As Enrico Fermi said,

> . . . contrary to perhaps what is the most common belief about secrecy, secrecy was not started by generals, was not started by security officers, but was started by physicists. And the man who is mostly responsible for this extremely novel idea for physicists was Szilard.[19]

Over the next few months Leo Szilard and a few of his colleagues pleaded with nuclear physicists in America and in Europe to refrain from publishing new studies on aspects of fission.

Soon more formal security procedures were put into place. At an April 1940 meeting of the Division of Physical Sciences of the Academy's National Research Council, Gregory Breit proposed formation of a Reference Committee to review papers in all fields of military interest, with a special focus on fission.[20] The procedure had editors forward articles to the committee for review. After a determination by the members they recommended as to the advisability of publishing it or not; a substantial number were withheld.[21] In practice, over the next few years voluntary restraint by individual scientists lessened the need for the more formal arrangement.

Breit, a member of the committee and chairman of the subcommittee on uranium fission, was a vociferous advocate of restrictions. When Breit moved to Washington in the summer of 1940, he introduced the cubicle system, a system of compartmentalization in which the information provided to a working group was restricted to what they needed to know for the problem at hand. They were not supposed to discuss their work with colleagues in other groups unless they had permission from Lyman Briggs or Breit; even then, discussion was to be limited to very narrowly defined topics. Alternatively, Breit sometimes decided that the discussion could take place only with him as mediator. Many scientists working on the project did not take this issue very seriously and were somewhat bemused at Breit's intensity over the matter of compartmentalization and secrecy. When Breit resigned from the work of the Uranium Committee in the spring of 1942, he complained bitterly about the irresponsibility of his colleagues with regard to matters of security. Breit was not alone in his concern about the security of the project.

The first involvement that the army had with the atomic bomb was in the area of intelligence. In February 1942, four months before the army took full control of the bomb program, the assistant chief of staff, G-2, Brig. Gen. Raymond E. Lee, as the result of a discussion with James B. Conant, ordered Capt. John Lansdale to report to Conant for a briefing.[22]

John Lansdale was born in Oakland, California, in 1912. He graduated from Virginia Military Institute in 1933 and received a commission in the U.S. Army Reserve as a second lieutenant, assigned to field artillery. He graduated from Harvard Law School in 1936 and joined the Cleveland law firm of Squire, Sanders & Dempsey. On June 10, 1941, Lansdale reported for active duty as a first lieutenant in the Investigation Branch, Counter-Intelligence Group, Military Intelligence Division, Assistant Chief of Staff, G-2, War Department General Staff. For the next year Lansdale worked on investigating potentially subversive elements in the army, mainly Communist sympathizers or Nazi adherents. He also served as the G-2 representative on a board responsible for releasing individual Japanese Americans who had been confined to the internment camps. One of the officers who worked for him at the time was William A. Consodine, a lawyer from New Jersey, whom we shall meet again.

Conant informed Lansdale about the nature and status of the fledgling project, as well as about a possible German program; he also shared his concerns about security matters, particularly among Ernest Lawrence's group at Berkeley. Conant said he was turning to the War Department for help, and he sought Lansdale's advice about what to do. Lansdale suggested he (Lansdale) go to Berkeley under cover to assess the situation. Conant arranged for him to be given temporary membership at the faculty club, and in the middle of February 1942 Lansdale left for California to spend about two weeks on campus. He easily obtained a complete description of the work being done in Lawrence's laboratory, to the point of being able to steal blueprints for the cyclotron, if he had wanted to. He met physicists and chemists who, with little encouragement, told him of their work trying to separate uranium isotopes, among other sensitive topics.

Lansdale returned to Washington in early March and briefed Conant on the lax security situation he had witnessed. Lansdale returned to Berkeley, this time in uniform, and gave the assembled scientists a stern lecture on the need for security. In May Lansdale met with Arthur Compton at the Met Lab in Chicago to discuss the need for better security there as well.

Lansdale's next contact with the project was late September, a few days after Groves had assumed command. Groves came to Lansdale's Pentagon office with the news that, with the consent of Gen. George Veazey Strong, assistant chief of staff, G-2, he was to be responsible for the security of the Manhattan Project.[23] To implement this Lansdale created within G-2 a separate organization that operated outside regular military channels, kept separate records, and had its own chain of command.

One result of General Marshall's March 9, 1942, reorganization of the War Department was the delegation of administrative duties previously performed at the staff level to subordinate commands. In particular, certain counterintelligence duties once performed by G-2 were delegated to the newly established nine service commands.[24] Lansdale chose an officer to head his organization within the security and intelligence division of each service command. They reported to him in Washington and he reported to Groves and, if need be, to General Strong. Lansdale's organization at the service command level did the investigative work in clearing people for access to secret information and investigated potential espionage.[25] Soon there were several hundred officers and agents in Lansdale's "nameless adjunct," as he called it.

This initial arrangement suited Groves's goals of keeping his own headquarters small, limiting knowledge, and utilizing the existing army organization as much as possible. But as the Manhattan Project grew, the need for a more closely supervised and integrated intelligence and security organization grew as well, and new procedures were initiated.

In February 1943 Lansdale transferred Capts. Horace K. Calvert and Robert J. McLeod to the Manhattan Project, to organize an Intelligence Section to make and enforce security policy, with Calvert designated district intelligence officer. The commanding generals of the service commands were notified of this new Intelligence Section and were requested to designate an officer from their intelligence divisions to be the liaison with the new MED organization. By late spring MED CIC officers were stationed at Oak Ridge, Chicago, St. Louis, Los Alamos, and Berkeley. Additional branch intelligence offices, as they came to be known, were established over the next year, and after consolidation and reorganization, eventually numbered eleven, to more or less parallel the geographic sub-divisions of Army Service Forces. Each branch intelligence officer was responsible for all intelligence and security within his geographical area.[26] Functions once conducted by army G-2 and the service commands — investigation and clearance of personnel, plant protection, safeguarding military information, and shipment and courier responsibilities — were eventually transferred to the Intelligence Section (later the Intelligence and Security Section and, after February 1944, the Intelligence and Security Division), initially at MED headquarters in New York City and after August 1943 at Oak Ridge.

Toward the end of 1943 Lansdale's "nameless adjunct," still within G-2, "was becoming so large it was almost impossible for it to operate outside of regular channels".[27] It should be noted that at the same time the CIC itself was coming in for a great deal of criticism. Its activities were not uni-versally popular either within or outside of the army. Among other things, tradition-minded army officers disliked enlisted personnel investigating them. This led to an investigation by the inspector general. Its November 1943 report was a devastating critique of the CIC's operations and organ-ization.[28] So as not to get caught in the maelstrom Lansdale recommended that his organization be moved to the Manhattan district, and Groves and Strong agreed. A Special Detachment of the CIC was created in December 1943, and a unit of twenty-five officers and 137 enlisted men was formed.[29] Captains Calvert and McLeod selected the special agents, a group of lawyers, technicians, linguists, auditors, and other specialists.

On January 7, 1944, Lansdale and several of his staff transferred from the Pentagon to Groves's office and moved in to the fifth floor of the New War Building. Lansdale served as special assistant to Groves and had full responsibility for all intelligence and security matters affecting the project. From this point until he left the project in December 1945 Lansdale saw Groves virtually every day they were both in Washington; they frequently traveled together.

The two objectives of MED counterintelligence were to prevent infiltra-tion of enemy personnel who might supply vital information about the bomb program to America's adversaries, and to prevent sabotage of district plants or facilities.[30] To attain these objectives MED personnel conducted many types of investigations and activities. These ranged from background checks and surveillance to the safeguarding of information and materials.

The personnel who were recruited or hired to work at MED facilities were investigated more or less thoroughly in rough proportion to their importance, though there were exceptions. Low-level workers, who did not even know what they were working on, would be given more cursory background checks than those with had access to classified material. Basic clearance requirements included positive identification, personal history statements, fingerprints and background checks, and proof of citizenship or alien registration. Companies receiving MED contracts were also checked, a new security practice. In a few cases loyalty considerations were waived if the person was considered extremely important or, in still fewer cases, irreplaceable. Nevertheless, even if the check were waived, MED agents kept constant vigilance. From the beginning of the project until August 1945, approximately four hundred thousand employees and six hundred companies were subjected to basic investigations.[31]

When there was suspected espionage, certain measures were taken. The CIC special agents assigned to espionage cases became proficient in their jobs, taking pictures with use of a telephoto lens or with tiny cameras concealed in their hand. They also used eavesdropping equipment to listen in on conversations, and tapped telephones. They impersonated men of all occupations: hotel clerks, bell captains, electricians, painters, exterminators, gamblers. Certain subjects were shadowed when traveling on planes or trains.

Every mechanical failure, equipment breakdown, fire, or accident was investigated to see if the cause might be sabotage. General Groves ordered that any suspicion of sabotage be reported to him immediately. The record appears to be clean. Where problems did occur they were attributable to accidents, incompetence, or, in one instance, disgruntled employees.

Groves was even concerned that German prisoners of war might escape from prisoner-of-war camps near MED facilities and sabotage them. In a memo to General Somervell Groves wrote that it was his understanding that a detachment of German POWs was doing maintenance work at Bruns General Hospital in Santa Fe, thirty-six miles from the district's Los Alamos installation, with the intervening country open and sparsely inhabited. Likewise, he was aware that some German prisoners were to be placed in a hospital in Walla Walla, only sixty-seven miles from Hanford. "It would be more than unfortunate if an escaped prisoner of war committed an act of sabotage at any one of these highly important sites."[32]

———

At the MED's industrial facilities security procedures were instituted, most based upon strict compartmentalization. A second procedure was to place counterintelligence agents in each plant disguised as workers to observe and listen to what was said. There were also agents posing as bus drivers, or as bartenders, waiters, and waitresses in the local restaurants and bars frequented by workers. Those who were overheard being too inquisitive about the nature of their work were almost never given a second chance. They found themselves without a job by the end of the day. This technique proved extremely effective in keeping mouths closed.

The need for security was impressed upon the workforce through constant reinforcement and repetition. Lectures about security were repeatedly given, and films shown. Posters, handbills, circulars, notices, and leaflets were displayed and distributed: "No Loose Talk," "Zip Your Lip," "Is Your Safe Locked?" Frequent editorials in project newspapers stressed the importance of security.

Another goal of Groves's compartmentalization scheme was to limit the public's knowledge of MED operations as much as possible, and to confine what was known or suspected to a region of the country. Groves and Lansdale knew they could not keep secret from local residents the fact that something very big was happening nearby, at Oak Ridge, Hanford, or Los Alamos. They did make it difficult for anyone to realize that Oak Ridge, Hanford, and Los Alamos were part of the same project. For example, an agency that recruited labor for Hanford was instructed that it could go anywhere in the country for workers, except Tennessee. Conversely, the skilled and unskilled labor for Oak Ridge was recruited from everywhere except the state of Washington. Workers who applied for work at one MED plant and had worked at another were never hired. Groves and Lansdale allowed cover stories and rumors to spread, as long as they had nothing to do with atomic energy.

The security strategy was somewhat different at the university laboratories, all of which, with the exception of Los Alamos, were located in urban centers. Lansdale's experience at Berkeley, confirmed by information he received from other intelligence agencies, convinced him that Communist organizations there were actively attempting to find out about the scientific and technological operations. Agents were assigned to follow particular individuals thought to be security risks. Lt. Col. Boris Pash, head of counterintelligence of G-2, Western Defense Command, organized an elaborate system for tapping the telephones at the Radiation Laboratory.[33] Throughout the war nearly all telephone conversations were monitored. Lansdale had to warn the voluble Ernest Lawrence periodically that he talked too much. According to Pash, Lawrence was not aware that his phone was tapped.[34] In order to eavesdrop on suspected Communists Pash and his agents altered the telephones in such a way that conversations nearby could be heard, as well as the calls.[35] There were several cases in which an individual should have gone to jail but only at the cost of divulging the clandestine operation. The second best solution was to draft him into the army and send him to an isolated location.

Pash was a colorful character. Son of the head of the Russian Orthodox Church in the United States at the time of the Russian Revolution, his anti-Communist sentiments were highly developed. After coming to the United States he was with the YMCA and then head of the athletic department of Hollywood High School, also taking a reserve commission in the army. When called to active duty in 1938 he went to the Presidio and began counterintelligence work.

It is sometimes difficult to know in Pash's case whether his words and actions were just bravado and bluster or whether they were true. An example is a memorandum for the FBI detailing CINRAD (an FBI acronym for "Communist Infiltration of the Radiation Laboratory") activities, dated July 23, 1943. The agent says,

> Pash has been negotiating for authority from Washington to obtain a boat for the purpose of Shanghaiing various Communists employed in the Laboratory and taking them out to sea where they would be thoroughly questioned after the Russian manner. [Deleted] stated that he realized that any statements so obtained could not be used in prosecution but apparently Pash did not intend to have anyone available for prosecuting after questioning."[36]

At Los Alamos security matters were handled somewhat differently. A prime reason for locating the laboratory in such isolation was to make it difficult for the scientists to have any contact with outsiders.[37] Oppenheimer's conviction, shared by his colleagues, was that strict compartmentalization at this laboratory would be counterproductive and unacceptable. To solve problems required knowledge and/or information from colleagues working on related problems.[38] Many heads were obviously better than one. The solution was to gather them together, put a fence around them, and let them go about their work.

Measures were taken at Los Alamos (and at all MED facilities) to physically secure the site through the use of fencing, lighting, alarms, guards, and troops patrolling the perimeter, and checks of identification cards and badges at access gates. The road from Santa Fe eventually reached the main east gate; it was another mile or so to the center of town.[39] Army security guards carefully checked the papers and IDs of those entering. Within an outer barbed-wire fence was a second fence encircling the Technical Area, or T Area, where the most sensitive matters were discussed and carried out. Those wearing the proper badge were allowed in, which meant they were permitted fuller, but not total, access to information.[40]

Interaction, or interchange as it was called, between sites was another matter and one strictly regulated by General Groves. Travel by individuals between sites had to be authorized by Groves. His appointment log is filled with calls from the scientists, corporate figures, or others asking for permission to visit a site or to travel between them. Only if Groves gave the okay was it allowed. There are numerous memos in the MED files justifying and explaining why certain officers must visit this site or that. Some of them concern why travel by air rather than rail is essential. The MED at times had to call upon the Air Transportation Command to use its planes.

An extensive set of principles and regulations was established between Los Alamos and the Met Lab. In June 1943 Richard Tolman drafted a three-page memo on the subject for Groves's consideration, which he slightly revised and then adopted.[41] For Groves the dilemma in dealing with compartmentalized information was the tension between secrecy and speed of accomplishment. Greater interchange might accelerate the work, but it might also increase the danger that information would leak out and jeopardize security. Groves normally erred on the side of security, but made exceptions if he thought it would speed up the work.

An additional reason for compartmentalizing the sites was to stimulate competition so as to make them work harder. As Norman Ramsey acutely observed:

> Generally General Groves would deliberately give Los Alamos excessively optimistic reports as to what was being done at Oak Ridge. In fact, that was one of the reasons he didn't let people from Los Alamos go to Oak Ridge; we might find out how slow they were on their

schedules. Likewise, he'd give the Oak Ridge people excessively optimistic reports as to how things were going at Los Alamos, with dominantly, I think, the laudable reason that in this fashion he could make both groups really work hard, since each group would think it was a bottleneck and therefore things would get done faster.[42]

Soon after the end of the war Szilard made some extravagant claims that compartmentalization had delayed completion of the bomb by eighteen months. With the wisdom of hindsight Szilard asserted that the Americans should have realized that uranium 235 could be made in quantities sufficient to make a bomb much earlier than they did. Fortunately, says Szilard, the British, who were not compartmentalized, put two and two together and communicated this to the United States in the middle of 1941. Had the Americans gotten going in the fall of 1940, most likely, bombs would have been ready before the invasion of Europe, Szilard surmised.[43]

The counterintelligence program was essentially effective against German and Japanese penetration. Neither Berlin nor Tokyo had an inkling of the extensive atomic bomb program that was under way. There are no recorded cases of successful espionage directed toward the Manhattan Project carried out by Axis agents. There were occasional reports about American research that did filter back. In one instance Groves framed a message of disinformation to deceive the Germans. It came about when some German agents, who had come through Portugal, were apprehended as soon as they arrived in the United States. Groves was told of this German espionage attempt at the Military Policy Committee meeting of June 21, 1944. He deliberated with his security people over what to do. They wanted to use the agents to send the message back to Berlin that nothing was being done on atomic energy in the United States. Groves overruled them, and a message was crafted saying that certain people at certain universities were doing certain work. Obviously, those who were mentioned had nothing to do with the project. Groves's reasoning was that by acknowledging a minimal effort, rather than none at all, it would more likely convince the Germans that nothing could come from such a small academic program.[44]

If the true scale of the American program had been accurately grasped, it would have aroused concern, and larger German and Japanese efforts would likely have followed. But as we now know, their programs remained minuscule and undeveloped.

As later came to light, the counterintelligence program was not nearly as effective against America's wartime "ally," the Soviet Union. The full extent of the penetration by Soviet intelligence agents or by Americans passing information to them is still not fully known. The most famous cases are those of Klaus Fuchs and the Rosenbergs. New revelations add details about them through disclosures from the Soviet archives and a more open Russian press. Of equal, or perhaps of greater, import has been the case of Theodore Alvin Hall, which only became publicly known in 1996.[45]

Peer de Silva was another of General Groves's many interesting personnel choices. De Silva was born in California in 1917 and graduated from West Point, class of 1941.[46] After six months with the quartermaster corps at Fort Ord, California, he was assigned to the G-2, Fourth Army, Western Defense Command, in San Francisco, under Boris Pash, performing counterintelligence work. Groves, during a trip to the West Coast in the fall of 1942, met de Silva and selected him to take over from the FBI the espionage cases at the Radiation Laboratory for the Manhattan Project. Among de Silva's responsibilities was to investigate whether or not Robert Oppenheimer should be given a security clearance. De Silva headed the investigation and wrote a memorandum to Pash, dated September 2, 1943, that reviewed Oppenheimer's indiscretions and flatly accused him of "playing a key part in the attempts of the Soviet Union to secure, by espionage, highly secret information which is vital to the security of the United States."[47] Pash had already come to the same conclusion in late June, recommending Oppenheimer be "removed completely from the project and dismissed from employment by the United States government."[48]

These negative findings about Oppenheimer put Groves in a very difficult and delicate position. Earlier Lansdale had investigated Oppenheimer's past and interviewed him. In examining Oppenheimer's relatives and friends Lansdale found that his wife, brother, and sister-in-law were probably Communists and that Oppenheimer had engaged in a host of leftist activities in the late 1930s and early 1940s. Normally such a person would not receive a clearance; another person would be found.[49] But Lansdale concluded that "he was completely loyal" and that his ambition and desire for a place in scientific history were overriding incentives to remain loyal to the United States.

There is some question as to how seriously a replacement was considered, when Groves and the others were faced with this information. Lansdale felt that Groves's account makes it seem as though the effort to find someone else was not very vigorous. Lansdale's recollection was of "a rather aggressive search for someone else."[50] An interesting FBI memo supports Lansdale, raises the question of Oppenheimer's prospects in mid-1943, and suggests that Groves's commitment to Oppenheimer might have wavered a bit, but soon firmed up. Writing to headquarters about general CINRAD matters in June 1943, the agent then turned to Oppenheimer:

> Recognizing the fact that J. Robert Oppenheimer is more or less in a class apart and at the present time is regarded as indispensable, it is reported that plans are under way to find a physicist of equal caliber and attempt to work him alongside Oppenheimer for a period of time until the indispensable knowledge held by Oppenheimer is also held by the planned substitute. When this point is reached, according to local officials of G-2, Oppenheimer himself will be relieved of his duties in connection with the project."[51]

Groves finally put the matter to rest. In a directive to the district engineer dated July 20, 1943, he wrote:

> In accordance with my verbal directions of July 15, it is desired that clearance be issued for the employment of Julius Robert Oppenheimer without delay, irrespective of the information which you have concerning Mr. Oppenheimer. He is absolutely essential to the project.[52]

Groves solved his differences with de Silva in a characteristic and clever way. De Silva could hardly have been more suspicious and distrustful of Oppenheimer, as evidenced by his strong memo to Pash.[53] Groves's solution was to place de Silva as head of security at Los Alamos. This would give him ample opportunity to see if Oppenheimer was really disloyal. De Silva's position became an important one at Los Alamos. He was the only army official who regularly attended the colloquia and was thus apprised of all technical and scientific developments. For a short period of time — about two months — after Ashbridge left and before Tyler began, de Silva was the commanding officer of Los Alamos.[54]

Apparently, despite his watchful security de Silva found nothing objectionable about Oppenheimer's leadership at Los Alamos. Upon leaving he wrote to Oppenheimer on April 11, 1945:

> Upon my transfer from duty at the project, I want you to know my sincere appreciation of the support and encouragement which you

have personally given me during my services here. In spite of your many more urgent problems and duties, your consideration and help on matters I have brought to you have been gratifying and have, in fact, contributed much to whatever success my office has had in performing its [security] mission."[55]

The relationship between the MED and the FBI, between Groves and J. Edgar Hoover, was born out of convenience but was never a very close one. Groves and Hoover never met, though Hoover's assistants, detailed as liaison to the MED, agents Whitson and Reynolds, came to see the general on one or two occasions. As a normal practice Groves was satisfied to have the resources of other government agencies, such as army G-2 and the FBI, do his work for him while keeping his headquarters small. But as we have seen, at a certain point this became impossible; new arrangements had to be worked out with regard to security. In a meeting in the War Department on January 20, 1944, Lansdale, representing the MED, and Whitson and Reynolds, representing the FBI, agreed to basic policies on how the two organizations should function on security matters. It was agreed that the MED would have complete jurisdiction over all investigative matters at Hanford, Oak Ridge, and Los Alamos, and that any matters that fell within the criminal jurisdiction of the FBI would be turned over to the FBI field office nearest the installation.

At this point, in February 1944, Lt. Col. William B. Parsons replaced Capt. Horace K. Calvert as chief of the Intelligence and Security Division at Oak Ridge.[56] Parsons, soon after taking over, planned to subject FBI agents to a thorough search of their cars and luggage by MED guards, and to require them to obtain special passes. The FBI apparently protested these regulations as disrespectful of the bureau, and they were never fully enforced. But other requirements were imposed, such as obtaining special stickers and the painting of serial numbers on the inside of glove compartments. FBI agents often criticized the MED for withholding information, complaining that it sometimes took a week or two for them to transmit it. They also protested that they were not informed about the movement of suspects under FBI surveillance. What the FBI resented, with reason, were MED efforts to restrict FBI movement on MED territory. The bureau felt that its jurisdiction was being compromised, and it was not able to perform its job, which often had nothing to do with the project.[57]

In an embarrassing incident for the FBI one of its agents was caught trying to infiltrate Oak Ridge as an employee when he failed a lie-detector test. Groves never raised the matter with the FBI, but merely discharged the man and let the FBI presumably punish him for his stupidity.[58]

Lansdale had his own criticism of the FBI. He did not think much of its reports. They consisted of a cover sheet and a poorly done summary, accompanied by unedited reports by field agents. They were a mixture of speculation, hearsay, and solid information with no analysis or comment. Fact and fancy were presented to the recipient with no attempt to say which was which. Lansdale has said that there were some very good people in the bureau, but "I personally formed a very low opinion of its ability as an organization and a feeling bordering on contempt for the way it discharged (or failed to discharge) its responsibility."[59]

Groves worried that he might be a target of espionage. To test the hypothesis, he had CIC agents follow him for two days during a trip to the West Coast, "to determine whether or not he was the subject of surveillance by person or persons unknown."[60] On October 1, 1943, three CIC agents from the Los Angeles Branch Office established contact with Groves, or "Mr. Starr" as he was code-named, as he arrived at Union Station at 12:15 P.M. PWT (Pacific War Time).[61] They followed him throughout the day. First he went to the Biltmore Hotel. The agents took measures to see "that no one in the lobby or on floor seven was armed. . . . This action was accomplished through observation of bulging pockets, etc." Mr. Starr then had lunch, met his sister Gwen, and proceeded with her to Bullock's Department Store, where they met his brother-in-law, Spencer. "The trio proceeded to Melody Lane Cafe at 744 S. Hill Street, to engage in light refreshments," finishing at 5:00 P.M. They then returned to the Biltmore, sat in the lobby and talked until 6:00 P.M., "whereupon party entered the Grill Room of the Hotel to partake of the evening meal." After dinner the trio took a brief walk outside "observing displays in shop windows." Mr. Starr was then driven to Union Station by his brother-in-law, whereupon he boarded the Southern Pacific Lark at 8:55 P.M. PWT. "Subject was observed reading a newspaper in comfort in roomette 5, car 84 as train departed at 2105 PWT for San Francisco. No unusual incident or circumstance was involved in this surveillance. No one was seen to have been interested in Mr. STARR's movements other than soldiers who were unarmed and conspicuous of the presence of the important-looking Mr. STARR in their immediate vicinity," the agents concluded.

Their counterparts in San Francisco picked up the subject's trail when he arrived the next morning at 9:05. Colonel Lansdale and Colonel Pash met him at the train station and drove to Berkeley, where the subject met Captain Fidler (head of the Berkeley Area Office) and his driver, Miss Luten. With Fidler, Groves proceeded to the Berkeley campus, where he spent the day with Lawrence and others. The general left Berkeley on the City of San Francisco train late Saturday afternoon for the long ride to Chicago and then on to Washington. The agents noted in their memo, "During the surveillance particular attention was paid to the activities of persons in the vicinity of the Subject, but nothing of a suspicious nature was observed."[62]

The Manhattan Project generated huge quantities of classified material in the form of papers, documents, directives, and reports, as well as various kinds of materials, some of which were radioactive. By March 1945 shipments of radioactive materials had become so numerous that there was concern over potential health hazards to the couriers. Film badges were issued to the couriers for each trip and then forwarded to the Medical Section for analysis. General War Department policies to safeguard military information were supplemented by special procedures developed by the Manhattan District to handle its unique items.

Agents of the CIC, later supplemented by officer personnel, established a special courier system to transport "Top Secret" documents and other classified material. The couriers frequently traveled in civilian clothes and carried the material in ordinary luggage, in order to avoid attracting attention. They were instructed to keep the material in their physical possession at all times.

Groves himself often carried highly secret papers when he traveled. He placed them in a plain brown government envelope with his name and a War Department address on the outside. When traveling by train, he kept the papers with him at all times. He sat on them in the dining car and placed them under his mattress in his compartment. Before going to sleep he tied the door handle so that anyone trying to enter would awaken him. Groves carried a small automatic pistol in his trouser pocket.[63]

Trains were the preferred method of transportation, though planes, trucks, and cars were also used. Factors such as delivery time, size, weight, shipping costs, government and commercial restrictions, and the value of what was being transported determined the degree of security.

Sizable quantities of uranium metal, ores, oxides, special chemicals, and other metals, for example, were carried by rail freight under armed guard. During July 1945 air transport was used on three occasions to expedite the delivery of U-235 from Oak Ridge to Los Alamos. Military planes flew from the Knoxville airport to the Santa Fe airport carrying the unique cargo that ended up inside the Little Boy bomb.

An extensive program was established to safeguard information relating to any and all aspects of the Manhattan Project. Each of the four categories of classified information ("Top Secret," "Secret," "Confidential," and "Restricted") had regulations prescribing how to prepare, transmit, reproduce, account for, store, and destroy documents and materials.

Teletype messages to and from the office were handled with equipment provided by the Army Signal Corps, with the codes constantly changed. For phone conversations concerning matters of high secrecy, Groves employed individual codes with specific people. Each was told to carry his own code in his billfold and report it immediately if it were lost. Only Groves and Mrs. O'Leary had all the codes. When needed they might just spell out a few words, which were supplemented by double-talk and references that no one else could know. The basics of the code were simple: a ten-by-ten checkerboard with numbers across the top and down the side and letters filling the squares, with an occasional empty space. Combining a side number with a top number located a square and provided a letter. A number designating a blank space indicated a break between words or was intended to confuse a potential interceptor.[64] For officers going overseas special codes were devised, containing one hundred words or so, which were represented in cables by other words, often cities or states. A one-time code was used for Alamogordo and the dropping of the bombs on Japan.

Code names and code words were used to describe principal project sites, people, and basic materials. Site "Y" was Los Alamos, Site "W" was Hanford, and Site "X," and sometimes "Dogpatch," was Oak Ridge. Nicholas Baker was Niels Bohr, Henry Farmer was Enrico Fermi, and A. H. Comas was Arthur H. Compton.

CIC agents were detailed as bodyguards for some of the more important scientists. Oppenheimer, Lawrence, Compton, and Fermi were accompanied almost constantly. Given the amount of time they spent with their wards, these agents were selected on the basis of being good companions as well as being able bodyguards. Compton spoke highly of his traveling companion, Julian Bernacchi, a law graduate and member of the Chicago police force. He assisted Compton in many ways; they became firm friends and remained in touch after the war.[65] Fermi, whose English was uncertain, had an Italian-speaking American as a bodyguard, John Baudino. Fermi would introduce Baudino to his fellow scientists as "my colleague," and say, "Soon Johnny will know so much about the project he will need a bodyguard, too."[66] Groves ordered Lawrence to refrain from flying in airplanes and driving

> an automobile for any appreciable distance (above a few miles) and from being without suitable protection on any lonely road. On such trips a competent, able bodied, armed guard should accompany you. There is no objection to the guard serving as chauffeur.[67]

While the image is that the Manhattan Project was the biggest and best-kept secret of the war, there were actually dozens of news accounts about various aspects of it. On December 19, 1941, President Roosevelt established, by executive order, the Office of Censorship.[68] Byron Price, a widely respected newspaperman from the Associated Press, was appointed director and served throughout the war. The office's objectives were to prevent the transmission of messages into or out of the country that might be useful to the enemy, including those in the press, and to obtain from intercepted communications information of value in prosecuting the war. For the most part the press was willing to cooperate in the mild censorship that the office's Press Division imposed through a voluntary code of guidelines. Journalists, no less than other Americans, supported the war effort and knew that publishing certain information could hurt it. Nathaniel R. Howard, former editor of the *Cleveland News,* headed the Press Division. Jack Lockhart, a former managing editor of the *Memphis Commercial-Appeal,* replaced Howard in late June 1943, and Theodore F. Koop, a former AP and *National Geographic* employee, replaced Lockhart in May 1945.[69]

The Press Division was not informed of the bomb project until March 30, 1943. Initially the army called for a complete press blackout. Fearing that such an extreme step would arouse curiosity Howard recommended a cautionary approach. On June 28, 1943, to prevent stories about atomic energy from being written, the office issued a confidential letter to the nation's editors and broadcasters, that stated:

> Confidential and not for publication: The codes of wartime practice for the American broadcasters request that nothing be published or

broadcast about new or secret military weapons . . . experiments. In extension of this highly vital precaution, you are asked not to publish or broadcast any information whatever regarding war experiments involving:

Production or utilization of atom smashing, atomic energy, atomic fission, atom splitting, or any of their equivalents.

The use for military purpose of radium or radioactive materials, heavy water, high voltage discharge equipment, cyclotrons.

The following elements or any of their compounds: Polonium, uranium, ytterbium, hafnium, protactinium, radium, rhenium, thorium, deuterium.[70]

CIC personnel in the branch offices reviewed newspapers, magazines, and other publications to see if they violated the Censorship Code, or contained any information, however small, that might relate to the project. If a violation was found, it was forwarded to Groves's office, which in turn referred it to the Office of Censorship. Lansdale, Consodine, or other members of the staff would bring it to the attention of Howard, Lockhart, or Koop. If the item were deemed a violation, letters were written to the editor or broadcaster explaining the violation and requesting future cooperation.

While the directive helped, it did not end the problem. Articles about "atom smashing" appeared occasionally, and the Tennessee and Washington sites were so large that curious journalists were bound to inquire. Even Los Alamos was not immune. Under the headline FORBIDDEN CITY, a March 15, 1944, article in the *Cleveland Press,* by John Raper, called the project near Santa Fe "Uncle Sam's mystery town directed by '2d Einstein.'"[71] Raper identified it as Los Alamos, once "a private school for boys," and said it now had a population of between five thousand and six thousand people. Raper made it to the gate only to be stopped by armed guards. "The Mr. Big of the city is a college professor, Dr. J. Robert Oppenheimer, called 'the Second Einstein' by the newspapers of the west coast." Raper speculated on what went on there. "Thousands believe the professor is directing the development of chemical warfare so that if Hitler tries poison gas Uncle Sam will be ready with a more terrifying one." Another belief was that he was developing ordnance and explosives. The Scripps-Howard Newspaper Alliance was asked not to move the story to other newspapers, the reporter was interviewed about his sources, and the editor of the *Press* visited the Office of Censorship to discuss the problem.[72] Groves considered drafting the reporter and sending him to the

South Pacific, an option that was dropped when it was found that he was in his sixties.[73]

On August 15, 1944, Arthur Hale, a commentator who had a nightly radio program titled *Confidentially Yours* implied on the air that research on atomic energy had succeeded in creating a weapon and that actual use was imminent. Hale mentioned two MED contractors by name — Columbia University and Du Pont.[74] The program was broadcast from Harrisburg, Pennsylvania, over 156 stations of the Mutual Broadcasting System to about two million listeners. Hale was immediately apprehended, ordered to Washington, and subjected to an all-day interrogation on August 17. The script had been prepared by someone else, who claimed he could not remember who he received the information from, and offered no excuse as to why he did not submit it to the Office of Censorship prior to broadcast.[75] Groves was furious and wanted Hale and his staff prosecuted. Lansdale and Consodine wanted him to drop it but Groves persisted, backing off only when it was pointed out to him that he might be liable to prosecution for kidnapping.[76]

In September 1944 Groves's staff compiled 104 press references to the Manhattan Project or related subjects over a fifty-eight-month period beginning in November 1939 and sent it to the Office of Censorship to underscore the problem. Seventy-seven references had come after the June 28, 1943, directive.[77]

Groves intervened from time to time to impress upon the owners and editors of the nation's newspapers the need for strict secrecy. An August 24, 1944, *Minneapolis Tribune* editorial-page column discussed the promise of uranium, "yielding to science's quest for a key to release sub-atomic energy. A race was on between Axis and United Nations scientists before war broke, and success had rewarded British-American efforts on a highly experimental basis." After efforts by the Office of Censorship and Price failed to convince the *Tribune* and Cowles Newspapers, the owners, General Groves visited John Cowles to convince him of the seriousness of the matter. No more articles appeared in any of the Cowles papers.[78]

In the end Groves was largely pleased with the office's ability to suppress information about the bomb project and wrote Price an appreciative letter on August 14, 1945. As seen in the files of the office, the numerous articles about the bomb during the war undercut the myth that there was total secrecy. But the articles revealed no technical details and did not connect the sites in Washington, Tennessee, and New Mexico; moreover, the violations were inadvertent rather than deliberate. Unlike the censors, the average American may have only noticed a few, if any, of the articles. Even reading all of them together would not have told very much.

Twice a day an MP and a mail clerk, both armed, drove from Los Alamos to the Santa Fe post office to pick up the mail addressed to P.O. Box 1663, Santa Fe, New Mexico, and to bring mail that was being sent to the outside world.[79] P.O. Box 1663 was the address used by the scientists at Los Alamos to receive their mail.[80]

The issue of postal censorship at Los Alamos came up early. Initially there was an honor system, but as the population grew other measures had to be taken. On October 28, 1943, Lansdale presented a list of regulations to the lab's governing board. Groves used the provisions of paragraph three of War Department Training Circular No. 15, dated February 16, 1943, as his authority to impose censorship. The provisions stated that censorship of communication was permitted at sites under military jurisdiction. By the end of 1943 censorship procedures were in place. For outgoing mail the regulations were aimed at preventing information that might arouse curiosity about Los Alamos. Obviously, the location and size of Los Alamos were forbidden topics, as were the names of personnel and what they were doing. Any photographs could show only people and not identifiable buildings or backgrounds.

An army unit of censor examiners, located upstairs in the Santa Fe post office, opened incoming mail. They did not know the nature of the work at Los Alamos and used the regulations as their guide. Even the censors were compartmentalized. Letters were opened, read, resealed, and stamped PASSED BY EXAMINER with the examiner's four-digit number. Because of the volume not every letter was opened, but all were stamped. No hand stamps or seals were affixed to outgoing mail, since they did not want outsiders to know that censorship was in effect.

Spot checks were conducted on the routine mail, while all letters to and from the more prominent scientists, "and those upon whom we had derogatory information," were read.[81] In the FBI's files are long lists of "mail covers" forwarded from Lansdale to Hoover through Whitson.[82] Mail sent to Box 1663 was recorded by addressee, sender/return address, date, postmark, and class (airmail, first-class letter, postcard). Everything was recorded, no matter how benign. The majority of the mail was routine and connected to the project: Dr. J. Robert Oppenheimer received a letter from the War Department, Box 2610 (that is, Groves), postmarked August 27, 1943. Some of the mail was connected to the lives they left behind: Professor Hans A. Bethe received a letter from the Cornell University Department of Physics postmarked August 18, 1943. Some of the mail was from relatives, banks, insurance agencies, and storage companies. Some of it was bills: a letter to Robert Serber from Abercrombie and Fitch and one to Mrs. Bethe from Bloomingdale's. Occasionally a suspicious item caught the attention of the inspector.

> On September 16, 1943 Subject [George E. Moore] received through the mail in Santa Fe, N.M. a copy of "Bread & Butter," which is apparently a Communistic periodical; copies of this publication were also received by Emilio Segre, Mrs. Paul Olson and Cyril S. Smith, all of which were addressed to P.O. Box 1663, Santa Fe, N.M. "Bread & Butter" is published by the Consumers Union of the United States, Inc., 17 Union Square West, New York, New York.[83]

Essential to Groves's compartmentalization scheme was to keep Congress and executive-branch departments from interfering with the mission. The Manhattan Project was the first large-scale "black" program, to use a more recent term. While there had always been small, secret programs here and there, financed with discretionary funds, the Manhattan Project broke new ground. As its budget grew, it became harder to hide the large sums under general categories, and thus a few key congressional leaders had to be informed.

On February 18, 1944, Stimson, Bush, and Marshall saw Speaker of the House Sam Rayburn, Majority Leader John W. McCormick, and Minority Leader Joseph W. Martin Jr. In June 1944 Stimson, Bush, and Maj. Gen. George J. Richards, the War Department budget officer, provided the same briefing to Senate Majority Leader Alben W. Barkley, Minority Leader Wallace H. White, Elmer Thomas, chairman of the Military Subcommittee of the Appropriations Committee, and the ranking minority member, Styles Bridges. In May 1945, with approval by President Truman, a five-member delegation from the House Appropriations Committee went for a two-day inspection of Clinton.[84]

Unauthorized attempts were handled differently. One congressman who had tried to force his way into Hanford in 1944 found himself in a windowless room under the glare of bright lights being questioned by Manhattan Project CIC personnel. Though the congressman was released after four hours of interrogation, Groves never felt the need to apologize.

Reporters and newscasters who ignored censorship rules with regard to the Manhattan Project also ran the risk of such treatment.

Lansdale also arranged for the airspace over Oak Ridge, Hanford, and Los Alamos to be restricted. When the occasion required he kept other departments of the government from interfering in Manhattan affairs. During 1943, for example, the antitrust division of the Justice Department began an investigation of Du Pont. Lansdale spoke to the assistant attorney general in charge of antitrust, and the matter was dropped.

After the war, Groves was asked whether in his army career he had dealt with matters of security. He responded, "Never before this thing started. We didn't deal with matters of security in the Army, really, until this time."[85] The fact is that there was not very much of a counterintelligence effort in the United States during the 1920s and 1930s or an intelligence/security infrastructure throughout the government. Some writers have made a great deal of the Espionage Act of 1917 as the taproot of what would become the American obsession with secrecy during the Cold War. While the act no doubt had an impact at the time, its influence was not sustained through the interwar years and largely dissipated.[86]

It was World War II that changed things in a fundamental and lasting way, and the Manhattan Project, "the best kept secret of the war" is a better source to look to as a progenitor of postwar practices. Under Groves's direction, the security system surrounding the atomic bomb grew to enormous proportions. He expanded and refined it, and it became the prototype for many postwar governmental departments, agencies, and programs.

CHAPTER FOURTEEN

His Own Intelligence: Foreign Concerns

For Groves fear and suspicion filled the void of his ignorance about whether Germany was working on an atomic bomb. Groves had an early appreciation of how important the bomb was going to be. If they possessed it, they could dictate their terms to the rest of the world.[1] This obsession drove Groves to race faster to build his bomb, and to stop the Germans by any means necessary from building theirs.

In the process Groves introduced some novel features into foreign intelligence operations and practices, supplementing his contributions in the domestic area. The most distinctive was the Alsos mission, a successful venture in "scientific intelligence." During the Cold War that followed, the role of science and scientists in the intelligence field would be greatly enhanced. Highly sophisticated means would be used to gather and analyze information.[2]

Several months before Groves took charge of the American project, an attempt was made to sabotage or interrupt Germany's suspected atomic bomb program. The initiative was Britain's plan, begun in the spring of 1942, for an assault on the German-controlled heavy-water plant in southern Norway.[3]

In their invasion of western Europe in the spring of 1940 the Nazis also seized Norway. The Norsk-Hydro plant at Vemork, eighty miles west of Oslo, manufactured heavy water and sold small quantities of it abroad to researchers investigating its properties as a moderator to slow neutrons in a controlled chain reaction. An enormous amount of effort goes into producing heavy water, or deuterium oxide (D_2O). Normal water contains a minuscule amount of heavy water, only 0.02 percent. Using the electrolysis process, forty grams of D_2O are extracted from each ton of water. The Germans estimated that about five tons were needed for a reactor or pile.

German interest in the Norwegian heavy water was keen even before its invasion. The giant German company I. G. Farben owned 25 percent of the shares in the plant. In February 1940 a Farben representative visited Oslo proposing to buy the current stockpile of about two hundred kilograms, worth about $128,000. At the time Norsk-Hydro was producing about ten kilograms per month. Farben urged that production be increased to one hundred kilograms per month, saying it would purchase all of this as well. After Nazi occupation, the Germans increased production even further; by September 1942 it was estimated that Germany was shipping approximately 120 kilograms a month to Berlin.

The French were the first to act, suspecting that the German interest in large quantities of heavy water indicated it was working on an atomic bomb. In March 1940 French officials got the Norwegians to agree to remove virtually the world's entire supply of heavy water — about 185 kilograms — from Vemork to France. Truck, plane, train, and ferry accomplished this, with the canisters arriving in Paris on March 16. Less than nine weeks later the Germans invaded France, and the water was in peril again. This time Hans Halban and Lew Kowarski, two assistants of Frédéric Joliot-Curie, took the heavy water and sailed from Bordeaux on a British coaler, arriving in England on June 21, 1940. While the water was now safe, the plant at Vemork was not.

The information supplied to British intelligence in the spring and summer of 1942 from its agents in Norway spurred the War Cabinet to request that Combined Operations destroy the plant. In concert with Special Operations Executive (SOE), an assault mission was planned. Code-named Operation Grouse, the plan was to have an advance team of four Norwegians parachuted in, to be followed about a month later by thirty-four British troops, who would land in two gliders towed by Halifax bombers. The latter mission, code-named Operation Freshman, was planned by Combined Operations. The Norwegian commandos would lead the British team to the plant, which would be attacked; all would then flee to Sweden, four hundred miles to the east.

On October 18, 1942, the Grouse commandos were dropped into Norway and spent several difficult weeks preparing for the attack team. On the evening of November 19 the Freshman troops took off from Wick, Scotland, on what turned out to be an ill-fated and tragic journey. Just before midnight, as they approached the drop zone, one of the Halifax bombers crashed into a mountain, killing the crew of four. The glider it had been towing also crashed, killing three. The fourteen survivors were captured the next morning and quickly executed by the Nazis. The second glider also crashed into a mountain after coming loose from the second Halifax, killing seven of seventeen and injuring the rest. An eighth man would die of his wounds shortly afterward. The nine survivors were captured by the Germans, questioned for a time, and then executed.

Operation Freshman had failed totally. Thirty-eight lives had been lost, the Norsk-Hydro plant was still operating, and the Germans were now alerted to how interested the British were in heavy water.[4] The plant had to be destroyed; the choices were another commando raid or aerial bombardment. This time SOE took total charge and decided upon another raid, code-named Gunnerside, using six Norwegians. After three months of extensive training, the team was dropped on February 17, 1943. After suffering some extreme weather, they met up with the Grouse men — who had never left — a week later. In an incredibly daring and dangerous assault the nine-man team attacked the plant on the night of February 27–28, blowing up the eighteen electrolysis cells that contained the equivalent of about 350 kilograms of heavy water.[5]

American officials knew little of this, and General Groves knew nothing at all.[6] He learned of it from an account in a Swedish newspaper two weeks later.[7] The general was not accustomed to reading about things like this in the press, and he was furious. On a matter of this importance he should have been involved in the planning of the mission and known all of the details. Wallace Akers had informed him on January 26 that the British were planning a commando raid, but few details were provided, and nothing had been received from them about whether the raid had actually taken place.[8] This was typical of the British, in Groves's view: They wanted us to share information with them and then they kept important secrets to themselves.

The Swedish news article of March 14 told of the successful sabotage of the Norsk-Hydro plant.

> All the apparatus, machines and foundation for the production of heavy water were blown up by the saboteurs.... Heavy water has for some years come into extensive use in scientific investigations, especially in attempts to break down the atom. . . . Many scientists have pinned their hopes of producing the "secret weapon" upon heavy water, namely an explosive of hitherto unheard-of violence.[9]

It was bad enough being kept in the dark; worse were the revelations of extremely sensitive information. Why in the world were these matters being discussed in the press? The newspaper's speculation about heavy water "caused [Groves] some headaches,"[10] and more were on the way as new details of the raid were revealed in the April 4 *New York Times*. The main headline read, NAZI "HEAVY WATER" LOOMS AS WEAPON, with a sub-headline, PLANT RAZED BY "SABOTEURS" IN NORWAY VIEWED AS SOURCE OF NEW ATOMIC POWER. Even uttering these words was a security violation, and here there was speculation about it in the newspapers. Harold Urey was mentioned in the article as the discoverer of this "queer chemical," and this led to him being asked his opinion about it. He did his best to deflect attention:

> So far as I know, heavy water's uses are confined solely to experi-
> mental biology. I have never heard of an industrial application for
> heavy water, and know of no way it can be used in explosives.[11]

Security concerns aside, while Groves felt these first reports were encouraging, he was concerned at how quickly the Germans might resume production. A British estimate by SOE in April was much too optimistic, claiming that the plant would be "ineffective for at least two years." A few days later this was altered to read "will not be fully effective for more than 12 months."[12] Even Field Marshal Sir John Dill, head of the British Joint Staff Mission in Washington, who had not known about the raid, complained to London about the widely different estimates and the lack of detail about what they meant.

Groves believed that as little as three months would be sufficient to start deliveries again. In fact, initial operations had begun in April, and full production of heavy water and shipments to Germany resumed in August.[13]

At Groves's behest General Marshall wrote to Dill, urging that Vemork be bombed "at the earliest opportunity" and that this recommendation be transmitted to Chief of Air Staff Sir Charles Portal.[14] Groves kept up the pressure over the next few months but the British did little, resisting in part because the raid would probably kill civilians.[15] The only choice left was unilateral American action. On August 13, 1943, Gen. George V. Strong, assistant chief of staff, G-2, wrote a memo to Marshall after a discussion with Vannevar Bush and Groves. Both of them, Strong went on, "consider it of the highest importance that the heavy water plant with adjoining power plant and penstock at Rjukan near Vemork, Norway, which have been restored to operation be totally destroyed."[16] Finally the pressure from Washington resulted in British agreement, and President Roosevelt ordered Gen. Ira Eaker, head of the Eighth Air Force, to plan and execute an attack.

On November 16, 1943, 174 B-17 bombers dropped thousands of pounds of bombs on the Norsk-Hydro complex. Actually, only two of the aircraft hit the electrolysis plant; the heavy-water cells were left untouched. Unfortunately, as feared, civilians were killed, eight men and fourteen women and children. But the point had been made to the Germans: Vemork was indefensible. They knew they were now subject to air attack, and if they rebuilt they would be hit again. As a consequence, they decided to dismantle the heavy-water apparatus, transport it to Germany, and erect a plant there.

For the saboteurs one final mission was left, and that was to prevent thirty-nine drums of heavy water — that had escaped the bombing — containing the equivalent of some six hundred kilograms, from getting to Germany. The plan was executed by the Norwegian underground, which placed explosives aboard the rail-ferry *Hydro,* which was to cross Lake Tinnsjo. On the morning of February 20, 1944, a Sunday, when fewer Norwegians were likely to be aboard, the ferry and the heavy water sank in deep water in four minutes. Of fifty-three people on board, twenty-six were killed, fourteen Norwegians and twelve Germans. Thus "ended the most sustained, and arguably the most effective, clandestine campaign of World War II."[17]

With the invasion of Europe a certainty at some point, active measures to find out about the German program had to be ready when the time came. Early in 1943 John Lansdale conceived of forming an intelligence unit comprised of combat troops and scientists that would apprehend European scientists thought to be working on nuclear physics, and seize their records. In later recounting this Lansdale noted that others had similar ideas, but he was the one who was successful in securing General Staff approval.[18]

On September 25, 1943, General Strong presented General Marshall with a plan for an intelligence mission to Italy to investigate secret German scientific developments.[19] In order to mask the true concern of the unit, the net was not restricted solely to nuclear matters but included other scientific and technological activities as well. The name given to the special unit was Alsos, Greek (αλσοσ) for "a sacred grove" (in this case the reference is to a small wood or forested area, and not to "groves" or "Groves"). General Groves was not happy with the choice, because it might betray a secret, but changing it might draw even more attention, so Alsos it was.[20]

Oppenheimer and his scientists were asked to help pinpoint things that the agents should look for. Evidence of raw materials, uranium, pure graphite, heavy water, and beryllium would indicate suspicious activity, as would certain-sized plants. "If the Germans are operating a production pile they will be operating it where water is plentiful and where the flow from the plant passes either through open country or through country inhabited by an 'inferior race' whom they do not mind killing off," he told Maj. Robert F. Furman, Groves's chief aide on foreign intelligence matters. If a few cubic centimeters of water could be collected from a river downstream of a suspected plant, the sample could be tested for radioactivity.[21]

In late November 1943 Groves, Bush, and Dr. Alan T. Waterman, deputy chief of OSRD, picked the scientific and military leaders. Lt. Col. Boris T. Pash was appointed chief of the project on the twenty-sixth, to serve as military commander. Pash, as we have seen, had headed the counterintelligence branch of the Western Defense Command and the Fourth Army and had sharp differences of opinion with Groves over Oppenheimer's suitability to be scientific director of Los Alamos. Pash knew his man when it came to Groves.

> I had had experience with General Groves while working on the Soviet espionage case. We had always come to a speedy meeting of minds — and there had never been a question as to whose mind was met![22]

Though there is no surviving record to prove it, it is quite likely that Groves solved his problem much as he had with Peer de Silva. He had Pash reassigned. Pash also turned out to be the right man for the job, as Groves may have surmised, in another one of his instinctual personnel selections. Full of derring-do and bravado, Pash carried out his duties vigorously. His reputation, at the time and ever afterward, was one of fearlessness, to the point of recklessness, and sometimes beyond.[23]

Alsos operations began in Italy on December 17, 1943, with a headquarters established in Naples. In January and February 1944 Alsos personnel located and interrogated a few Italian scientists in Sicily and southern Italy. But little information was gained about German scientific developments, and with Allied advances toward Rome stalled, Pash and the Alsos team withdrew from the Mediterranean and returned to Washington to prepare for the coming invasion of Europe.[24]

The information that Alsos had secured in Italy was sent directly to Groves's office in sealed envelopes, which were not even opened in G-2. It showed that the Germans were not using Italian resources for their program, which was valuable negative information in itself, but Groves wanted more hard facts about what was really going on. In a memo to General Strong about the Italian mission Groves recommended that "A similar scientific mission with the same general objectives be made ready for use in other European territory as soon as progress of the war permits."[25]

General Groves's responsibilities in the foreign intelligence field were not part of his original orders, but, like many of the jobs that filled his growing portfolio, they emerged as circumstances dictated.

In the fall of 1943 General Marshall asked Groves to take responsibility for the aspects of foreign intelligence having to do with atomic energy.[26] Apparently Marshall felt that the existing intelligence agencies might not be as sensitive to, or aware of, certain kinds of information that they received as would a smaller, more focused organization, like the MED. In addition there was always the question of limiting the number of people who knew about this sensitive topic. The creation of new sections or branches within existing offices could only increase the chances that information might leak out. Better to let Groves handle it and to convey it to him verbally, as was customary for many Manhattan orders.

Groves tells us in his memoir that it was at this point that he discovered the jealousy and competition among the different intelligence offices of wartime Washington. If Groves was going to get what he wanted, he would have to make his way carefully through these turf-conscious bureaucracies. He already had a good relationship with army intelligence, having shifted John Lansdale, and some of his people, to work for the MED. The assistant chief of staff, G-2, Gen. George Vezey Strong — known around town as King George for his grand manner — was the most powerful intelligence figure in town, and he meant to keep it that way. He guarded his dominion tenaciously and did not like interlopers such as William J. Donovan and his newly formed Office of Strategic Services (OSS), a group of mostly civilian amateurs, Strong thought.[27] Groves, a bureaucratic fighter of the first rank himself, was sensitive to these dynamics, which constitute much of Washington politics in war and in peace, and he respected them.

In August 1943, in response to the continuing concerns of his scientists about a German bomb, Groves began to organize an intelligence effort of his own to find out what he could, and to take action if necessary. Groves chose Major Furman to head it.[28] Furman had briefly worked for Groves two years before helping to build the Pentagon. Groves apparently remembered him, thought him able, and called the twenty-eight-year-old to the office to give him a new job. Groves told Furman that he would be responsible for finding out what the Germans were doing and for working with the scientists.

While much could be potentially learned from Alsos, it would take time, time that the Germans might be using to build their bomb. While waiting for the invasion of Europe, other things had to be done to find out what was going on and maybe prevent it. While some information was gained from interrogating German prisoners of war who were being held in the United States, more had to be done.

It is an indication of just how concerned Groves was over the question of a German bomb that, several times in 1943 and 1944, he considered kidnapping, and perhaps even assassinating, Germany's leading physicist, Werner Heisenberg.

In a memo to the file written twenty years later Groves recalled trying out an idea on the chief of staff.

> At one time during the war, I think it was late 1943, it was suggested to me by someone in the Manhattan organization, I think a scientist, that if I was fearful of German progress in the atomic field I could upset it by arranging to have some of their leading scientists killed. I mentioned this to General Styer one day and said to him, "Next time you see General Marshall ask him what he thinks of such an idea." Some time later Styer told me that he had carried out my wishes and that General Marshall's reply had been, "Tell Groves to take care of his own dirty work."[29]

The "someone" who may have put the idea in Groves's head was probably two German refugee scientists, Victor Weisskopf and Hans Bethe. The date may have been earlier and the method, at least at the outset, less drastic, but it seems to have set Groves thinking.

Victor Weisskopf was born in Vienna in 1908 and studied with or worked with most of the major figures of twentieth-century physics.[30] He studied for a short while with Heisenberg in Leipzig. In 1937 he accepted an offer from the University of Rochester and from there went to Los Alamos. His colleague Hans Albrect Bethe received his Ph.D. in theoretical physics in 1928 and knew all of the major figures as well. Bethe, whose mother was Jewish, lost his position at the University of Tubingen in 1933 and left Germany. In 1935 he accepted a position at Cornell University and was one of Oppenheimer's first recruits for Los Alamos.

In late October 1942 news reached Weisskopf from his former teacher, Wolfgang Pauli, that Heisenberg was working in the Kaiser Wilhelm Institute in Berlin, and that he had been appointed its director on October 1. Pauli also wrote that Heisenberg was scheduled to give a lecture at the university in Zurich in December.[31] Weisskopf contacted Bethe to inform him of this alarming news. The two agreed that an opportunity presented itself. If the Allies wanted to impede the German bomb program, then kidnapping Heisenberg in neutral Switzerland would be a way.[32]

Weisskopf immediately wrote to Oppenheimer, informing him of all of this news and urged that something be promptly done, specifically organizing the kidnapping of Heisenberg in Switzerland.[33] Weisskopf thought that that is what the Germans would do if, say, Bethe or Oppenheimer appeared in Switzerland. Weisskopf suggested other possibilities, such as sending an appropriate person to speak to Heisenberg to find out his intentions. But very few people have the special qualities that would make that option a success, and thus "kidnapping is by far the most effective and the safest thing to do."

In his reply to Weisskopf Oppenheimer thanked him for his "interesting letter" and informed him that his information had already come to his attention and had been submitted to the proper authorities.[34] He went on to tell Weisskopf that he would forward his letter and suggestions, and that it would receive the attention it deserved. The same day Oppenheimer wrote to Bush, enclosing Weisskopf's letter. Oppenheimer told Bush, "I should not myself want to endorse any of [Weisskopf's suggestions], except to the extent of remarking that Heisenberg's proposed visit to Switzerland would seem to afford us an unusual opportunity."[35]

Groves apparently took Marshall's response as a green light to at least explore the possibility. For him multiple approaches to solving problems were better than relying on a single one. This was a trademark characteristic of his, whether it was pursuing methods of enriching uranium or discovering what the German physicists were up to. As one recent writer has said of the latter,

> Groves's intelligence policy was similar to that of a hunter who looses a pack of sturdy hounds on the same vague scent and waits to see which dog will scare up the fox.[36]

One of the hounds he set loose was Moe Berg, who joined the Office of Strategic Services on August 1, 1943, just in time to carry out several espionage missions for Groves.

Groves's diary shows that he and Major Furman went to William J. Donovan's office on October 13, 1943. It was probably at this meeting that Groves convinced the OSS head to help him find out about the German bomb program. Over the next two years OSS would supply General Groves with whatever intelligence it obtained in Europe or elsewhere about the German bomb program. The code name for this special, highly compartmentalized information was Azusa.[37] In addition to supplying information Groves also needed from the OSS an agent who would actively seek it out, and perhaps even do more. Donovan had just the man for the job.

Morris "Moe" Berg was a well-known figure in his day. After graduating from Princeton in 1923, he began a career as a baseball player and played with three professional teams until he ended his career on the day before war broke out in Europe, in August 1939. The events in Europe greatly troubled him, and he felt he should do something. He had a gift for languages: Besides English he was fluent in German, Japanese, Italian, French, Spanish, and Portuguese, and had limited knowledge of a dozen more tongues. The press adored him and wrote endlessly about the eccentricities of this unusual athlete. In three years during the off season he earned a law degree from Columbia University and, while still playing baseball, took a position as an associate with a Wall Street law firm. It was through one of the firm's law partners that he had come to Donovan's attention.

Donovan's chief of special projects, John M. Shaheen, assigned Moe Berg and another OSS agent, William Horrigan, to Project Larson in the second week of November 1943.[38] Soon after, Berg was called to the office of Col. Howard W. Dix, head of the Technical Section of OSS. Dix handled all of the Azusa information and directed it to Groves's office, normally through Major Furman. On that day Major Furman was in the office to look over the new recruit. Without actually uttering the forbidden words, Dix intimated to Berg the things that he was to find out. He was told what to look for without being given the reasons why. The ostensible purpose of Project Larson was to contact Italian rocket and missile experts, remove them by submarine, and bring them to the United States. Rockets and missiles, while interesting, were a smokescreen for the real purpose, which was to interview Italian physicists to see if they knew anything about a German bomb program.

While waiting to get permission from the theater commanders in Italy, and for the American army to liberate more of the peninsula, Berg bided his time in Washington, living at his favorite hotel, the Mayflower, reading up on physics, learning from scientists and engineers, and running up a substantial bill. Finally in early May the wait was over and Berg went off to Italy, via London, Portugal, and Algiers. He went alone, which was always his preference and because Horrigan had been given another assignment. Before leaving, Furman told him what he should find out: which German and Italian scientists were alive, where were they located, what were their travel plans, and if there were industrial complexes being built, among other things. Furman also wanted to know what sort of damage had occurred and if any deaths had resulted in the American bombing raids of the Kaiser Wilhelm Institute in Berlin in February and March.

Berg entered Rome a day or two after the liberation and contacted the physicist Edoardo Amaldi. Amaldi had little direct knowledge of what the Germans might be up to but suspected that if they were working on a bomb, it would take them at least a decade. Berg also contacted Gian Carlo Wick, who had studied under Heisenberg in the 1930s and was later Fermi's assistant for five years. He told Berg that he had done no atomic research for the Germans or anyone else. Berg learned that recently Heisenberg had sent Wick a postcard, and he read it with great interest. Heisenberg told of his Leipzig institute and his home being destroyed by bombings and that, though the Kaiser Wilhelm Institute in Berlin-Dahlem was still standing, he had moved his family to the Bavarian Alps. Wick also knew that Heisenberg had moved to the south. Berg sent this information to Washington and to Furman and Groves, who were pleased with these new facts. Throughout the summer Berg continued to visit other scientists, but little else was learned about the atomic program. Berg did accumulate a substantial amount of information about other types of German and Italian weapons and military technologies. Berg left Italy in September and went to London, where he kept in close contact with Calvert and with Furman, when he passed through. Soon there would be a new mission.

The idea of kidnapping or even killing Heisenberg seems not to have left Groves after Weisskopf and Bethe first brought it up in the fall of 1942.[39] By early 1944 the plan was under active consideration, and Furman requested that Donovan send him an able man to do the job. Donovan chose Col. Carl Eifler, "[b]eyond any doubt the toughest, deadliest hombre in the whole O.S.S. menagerie," as a senior OSS officer described him.[40]

Eifler was a huge bear of a man who had been the captain of an infantry company before being recruited by OSS. He was an extraordinary marksman and a pilot; he also considered himself "well qualified in the art of boxing." In May 1942 Gen. Joseph Stilwell approved the establishment of an OSS guerrilla unit to fight the Japanese in Burma, and Eifler was chosen to lead it. Eifler recruited other Americans and a number of Kachin tribesmen, forming Detachment 101. The exploits of the Kachin Raiders were legendary. By the end of the war they were credited with 5,447 Japanese troops killed and an estimated 10,000 wounded. The Kachin natives lost seventy lives and the OSS only fifteen.

Furman told Eifler that a German scientist was making a new weapon and must be stopped. He was told that he should be kidnapped "to deny Germany his brain." Over the next four months an elaborate and risky mission, worthy of James Bond, was concocted. Eifler would enter Germany, find and capture Heisenberg, and lead him out on foot to Switzerland. From there they would fly over the Mediterranean, ditch the plane, and parachute into the water, where they would be picked up by a waiting submarine. In late June Donovan called off the mission. Better to wait for more favorable opportunities.

And they were not long in coming, as it was learned that Heisenberg was to leave Germany to give a lecture in Zurich on or about December 15. Rather than sending Berg after Heisenberg, better to let Heisenberg come to Berg. Berg went to Paris to wait for instructions. There he was told that if there were indications that the Germans were working on the bomb, he should shoot Heisenberg, right there in the lecture hall if need be. "Nothing spelled out," Berg later noted, "but — Heisenberg must be rendered hors de combat."[41]

The lecture took place on the afternoon of December 18 before a small audience of professors and graduate students. Well placed to hear his quarry and to view the audience, Berg sat with a pistol in his pocket, for Heisenberg, and a cyanide tablet for himself, just in case. Berg's German was not up to understanding all that was said, and it was unlikely that Heisenberg, if he were working on a bomb, would be talking about it to this, or to any, audience. The lecture ended and Berg met with Heisenberg's Swiss host Paul Scherrer, the director of the Physics Institute and an invaluable OSS source, under the code name Flute. Scherrer passed on much useful information to Berg about Heisenberg's recent whereabouts, and invited him to a dinner he was giving for his German guest at his home later that week.

Berg listened carefully to the conversation that evening and after dinner walked with Heisenberg through the Zurich streets until they parted.

But Berg's work for Groves was not over. While it was evident that there was no German bomb program, Groves approved a plan to send Berg, and several other OSS agents, to Switzerland via London to prepare for a surreptitious entry to the villages of Hechingen and Bisingen in the Black Forest region of Germany, where it had been learned that Heisenberg and other German scientists were living. Ever cautious, Groves later canceled the operation, fearing that if they were caught they might reveal information about his bomb project.[42]

Detailed records of such intelligence missions are not readily available and, not surprisingly, the secretive Groves does not have a great deal to say about Moe Berg or the work that he did for him. The general made it clear that Berg worked for him, and the continuing instructions by Furman and Calvert are further evidence.

> Moe Berg was a multi-linguist under MED control always. Insofar as atomic intelligence went, MED had complete power and responsibility. Any OSS assistance was carefully coordinated and was invariably in accordance with our request.[43]

Another way to get Heisenberg and his colleagues, or at least to disrupt their activities, would be to bomb them from the air. Early on Groves had set about to find out where they were physically located. In a letter to Compton on May 30, 1943, Groves asked him to survey a limited number of members of his staff of foreign origin, or those who had lived in Germany, to pinpoint the sections of the Kaiser Wilhelm Institute (KWI) that would be likely working on atomic energy. The purpose of this request, Groves informed Compton, was to decide which sections should be selected for possible targets for either bombing or sabotage. He asked Compton to rank them "of vital interest," "might be of interest," or "of no significance," and supplied about three dozen street addresses of parts of the KWI. Compton cabled back two days later with his recommendations. The two targets of vital interest were the Research Institute of Physics and the Research Institute for Physical Chemistry and Electrochemistry, both in Dahlem, a suburb of Berlin. Based on Compton's information, Groves sent General Strong a memo the next day listing the sections of the KWI "which should be selected as targets for either bombing or sabotage."[44]

Destroying the laboratories inside the buildings was but one goal of the bombing mission; killing the occupants was the other, as General Strong bluntly put it in a memo to the chief of staff. After a discussion of these matters with Bush and Groves — who were the original initiators — Strong recommended that "the killing of scientific personnel employed therein would be particularly advantageous."[45] The most important "scientific personnel" at the institutes, of course, were Werner Heisenberg and Otto Hahn.

It took some time for the mission to be carried out by the Eighth Air Force.[46] On February 15, 1944, American bombers flying from RAF bases in England attacked the Dahlem section of Berlin where the Kaiser Wilhelm Institute was located. Some windows were broken; Max Planck's house was destroyed, as was the Institute for Physical Chemistry, where Otto Hahn worked. Hahn was away preparing his move to Tailfingen, but his scientific papers and personal possessions were lost. Heisenberg was in Berlin but escaped. A few weeks later Hahn's office was hit again.[47]

In a memo to the file written much later Groves referred to "the bombing of the Dahlem sector in Berlin which we undertook at my request to drive German scientists out of their comfortable quarters."[48] He also commented on his initiative in his annotations of David Irving's book. "Re the bombing of Berlin I had asked for certain bombing with a view to forcing the Germans to move from well-established laboratory facilities."[49]

It took some time to find out whether the bombers had found their targets, but reports were relayed to Furman in April and May to suggest they had.

> According to reliable Stockholm report dated April 20, Kaiser Wilhelm Institute for Chemistry (at Berlin) was practically destroyed in raids 15 February and 25 March. Most of personnel have gone to South Germany. Present activity unknown.[50]

Even after the scientists had been located in southern Germany, Groves apparently continued to urge that they be bombed. Goudsmit referred in his book to someone, probably Groves, who wanted to bomb Heisenberg in Hechingen.

> In fact, plans for the bombing were already under way, but we were able to prevent it by stressing the insignificance of the German project. Aerial photographs confirmed our statement that the project

could not be of major importance. The main laboratory was housed in a wing of a small textile factory.[51]

At Groves's direction the MED established and operated a foreign intelligence unit based in Britain. In December 1943 Groves sent Major Furman to London to make the necessary arrangements with the British government. In January 1944, at Lansdale's suggestion, Capt. Horace K. (Tony) Calvert was sent to London to open a liaison office to serve as the point of contact with British intelligence, with the Tube Alloys project, and with the G-2, European theater of operations, USA (ETOUSA). The office was at 31 Davies Street behind the U.S. Embassy.

A principal British contact in London on intelligence matters was Sir Charles Hambro, a member of a prominent British banking family. Hambro was the first head of the Scandinavian Section of the SOE. For a short time he was the head of SOE. In 1944 he was selected to be the deputy director of the Combined Development Trust, a Groves-inspired nonproliferation venture that will be discussed in chapter 16. Calvert also dealt with R. V. Jones, Lt. Cmdr. Eric Welsh of M.I.6, and Michael Perrin of the Tube Alloys project.

The British supplied the MED with intelligence and their assessments of it.[52] While Groves was generally pleased with the cooperative exchanges of intelligence with the British, he remained largely unconvinced about their conclusions. The British view of the German bomb was that there was no large-scale program under way.[53] Groves, on the other hand, adopted a worst-case analysis, assuming that until it was confirmed otherwise, Germany was working on a bomb at full capacity. He wrote,

> Unless and until we had positive knowledge to the contrary, we had to assume that the most competent German scientists and engineers were working on an atomic program with the full support of their government and with the full capacity of German industry at their disposal. Any other assumption would have been unsound and dangerous.[54]

Calvert's group concentrated on collecting information about German atomic activities, specifically on individual nuclear scientists, on the location of laboratories and industrial facilities, and on the mining of fissionable materials. Through canvassing German physics journals and questioning refugees Calvert's unit learned the whereabouts of the most important German scientists, and through periodical aerial surveillance of the mines at Joachimstal, Czechoslovakia, it monitored the mining of uranium ore. All of this was to be enormously helpful to the Alsos team, which was preparing to follow the invading Allied armies onto the continent in June 1944.

But before this happened, the new Alsos group needed to be established.[55] After the Italian mission, Bush had recommended that Alsos be continued in a letter to Groves of February 29, 1944. On March 10, 1944, Groves requested that the newly assigned army G-2, Maj. Gen. Clayton L. Bissell — who had replaced General Strong on February 7 — form the new Alsos group along the lines of the Italian mission. While Bissell agreed that a high-level scientific organization was needed, indecision on the part of the General Staff delayed action. At the end of March Groves, concerned by the delay, personally intervened with G-2; as a result, the deputy chief of staff approved an April 4 plan to reorganize the mission. Though the directive was worded in broad terms — Alsos was to secure "all available intelligence on enemy scientific research and development, particularly with reference to military application" — it was generally understood that the primary purpose was to uncover and analyze German atomic activities.

Groves and Bush took charge of selecting the military and civilian scientific personnel, with Bissell providing some of the intelligence and administrative staff. The new Alsos mission emerged as a much bigger and more ambitious organization than its Italian predecessor. It had its own advisory committee, a scientific director, and an enlarged staff of military and civilian personnel.

The five-member advisory committee was comprised of representatives from naval intelligence, OSRD, Army Service Forces, and two from G-2. Boris Pash was named military chief on April 4. Samuel A. Goudsmit, a nuclear physicist from the University of Michigan who had been on leave to work at the MIT Radiation Lab, was chosen by OSRD on May 15 to be scientific director. He arrived in Washington to report for duty on May 25 and left for London on June 6.

Goudsmit was born in the Netherlands, was educated in European universities, and knew several languages and many of the leading physicists, German and non-German. As he would soon learn, his parents had been taken to a concentration camp and put to death in a gas chamber. He knew a limited amount about the Manhattan Project, probably from his voluble

B1. General Brehon Burke Somervell (1892–1955). A formidable engineer officer who was Groves's boss and with Styer chose him to head the Manhattan Project in September 1942. Commanding General of Army Service Forces, March 1942–December 1945.

B2. Groves ran the Manhattan Project from his office on the fifth floor of the New War Building at 21st and Virginia, Northwest. Today it is part of the State Department.

B3. Mrs. Jean M. O'Leary, administrative assistant to General Groves.

B4. Groves reported to Army Chief of Staff General George C. Marshall and to Secretary of War Henry L. Stimson.

B5. Col. Wilhelm Delp Styer, C.E. (1893–1975), helped General Somervell choose Groves to head the Manhattan Project and later served as the Army member of the Military Policy Committee.

B6. Lt. Gen. Eugene Reybold (1884–1961), Chief of Engineers (October 1941–October 1945).

B7. Col. Kenneth David Nichols (1907–2000) served as Manhattan District Engineer from August 1943 to April 1946.

B8. Close aide Col. William A. Consodine worked on intelligence, legal, and public relations issues for Groves.

B9. Vannevar Bush, James B. Conant, General Groves, and Col. Franklin T. Matthias at Hanford (Hagley Museum and Library).

B10. Groves chose J. Robert Oppenheimer to be the scientific director at Los Alamos.

B11. Major Robert F. Furman (courtesy of Robert Furman).

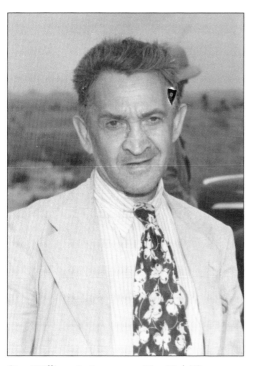

B12. William L. Laurence, *New York Times* science reporter.

B13. Col. John Lansdale, Groves's special assistant for intelligence and security matters (courtesy of John Lansdale).

B14. Major General Leslie R. Groves.

B15. New York Yankee Manager Casey Stengel, Moe Berg, and General Groves, 1959 (USMA Library Special Collections).

B16. Maj. Robert F. Furman and Dr. Samuel A. Goudsmit (courtesy of Robert Furman).

B17. Arthur H. Compton (1892–1962), director of the Metallurgical Laboratory at the University of Chicago, helped to design and build the Hanford reactors and the X-10 plant at Oak Ridge.

B18. Niels Bohr (1885–1962) served as an advisor to Groves and Oppenheimer in 1943.

B19. Enrico Fermi (1901–1954) directed construction of the world's first nuclear reactor.

B20. Ernest O. Lawrence (1901–1958) invented the cyclotron and designed the Y-12 plant at Oak Ridge.

scientist friends. Though he would have been a security risk if captured, he did not know the big picture or important details. Like everyone else under Groves's command, Goudsmit was briefed about what he needed to know to do his job and little else. Compartmentalization still ruled.

Pash arrived in London in mid-May, and headquarters were established in the quarters of the OSRD Mission. Additional personnel were recruited, some of whom were OSRD personnel attached to the London Mission. By the beginning of August Alsos was ready to move.

The impending invasion of Europe precipitated fears and concerns about possible German use of radioactive materials on Allied forces. Groves had alerted Marshall of this possibility as early as July 23, 1943, and urged that precautions be taken. These fears led to Operation Peppermint.[56]

Conant, Compton, and Urey wrote a report in the latter part of 1943 on the possible use by Germany of radioactive materials as an offensive military weapon and also of detection and defensive measures to deal with the threat. If the Germans had advanced as far as producing a pile, which some thought possible, then in addition to producing plutonium they also would have other radioactive materials at their disposal for use as a radiological weapon. Areas of a certain size could be contaminated; if exposure to penetrating gamma radiation were sustained long enough, or if concentrations were high enough, death would ensue. A second method of using the radioactive material would be in the form of a fine dust or smoke. Small quantities would be inhaled into the lungs, and the individual would eventually die of the radioactive poison.

The authors concluded that there would be many difficulties in devising munitions that would effectively deliver radioactive materials in this fashion. There was the problem of uniform distribution; also, if the second method were attempted, keeping the clouds of radioactive dust or smoke from dissipating to the ground before they were inhaled would be problematic. The bombs themselves would need to be shielded to protect the aircrews who dropped them.

While the chances that these methods might be used were remote, there were other uses radioactive materials could be put to, such as rendering unsafe and uninhabitable territory recently evacuated by German troops and thereby disrupting invading armies. This could achieved by distributing radioactive dust from low-flying airplanes, spraying buildings, streets, airfields, and rail yards from automobiles carrying lead-coated tanks filled with a solution, or burying lead-coated land mines and setting them off at a distance. The likelihood that the Germans might use some of these methods was greater, the authors felt, and they suggested that detection and defensive measures be considered.

Groves read all of this with great interest and, as the invasion date drew nearer, decided to warn Eisenhower directly of the potential dangers that might occur.[57] In late March, after getting Marshall's approval, he sent Maj. Arthur V. Peterson to England to brief the supreme commander. Peterson, a chemist by training, was Groves's Chicago Area engineer and as such knowledgeable about the pile program and its by-products. When Peterson arrived in April he briefed Eisenhower, his chief of staff, Lt. Gen. Walter Bedell Smith, and selected members of the Supreme Headquarters, Allied Expeditionary Force (SHAEF) and ETOUSA staffs. Eisenhower replied to Marshall in a letter of May 11 that while he was not going to take "those precautionary steps which would be necessary adequately to counter enemy action of this nature," he would take certain measures. These included briefing top U.S. officers of Theater Headquarters, Naval Forces in Europe, and Strategic Air Forces, and informing Gen. Sir Hastings L. Ismay, the British chief of staff.

The British were not initially interested in these preparations, so under ETOUSA a plan, code-named Peppermint, was prepared in late April. In May the British changed their minds and joint plans were implemented. Eisenhower directed that special equipment be earmarked for quick dispatch to the Continent if needed, and that medical units be informed about what symptoms would occur if there was radioactive poisoning.[58]

Certain invasion troops were provided with detection devices and were instructed to alert their superiors if they were activated.[59] Fortunately, the forces that landed on D-Day did not encounter any radioactive material, nor did they over the next months as they pursued the German army eastward.

With Allied forces now on the ground on the Continent and moving to the south and east, it was time to find out, once and for all, about the German bomb program. Boris Pash's first target was Frédéric Joliot-Curie, France's leading nuclear scientist. On August 9 Pash landed in France, but his first operation was largely unproductive. It was directed at L'Arcouest, the coastal village where Joliot had a summer home. When Pash arrived neither Joliot nor his wife, Irene, was there.

By the summer Tony Calvert, Groves's man in London, had become attached to the Alsos team. He supplied members with a considerable amount of intelligence that he had gathered, and with the fall of Paris imminent, he knew where Joliot would likely be. On the morning of August 25 Pash, Calvert, and two CIC agents, Gerald L. Beatson and Nathaniel W. Leonard, riding in two jeeps, accompanied by the Second French Armored Division under General Leclerc, led an Alsos advance party into Paris, the first Americans to enter the city.[60] Calvert captured much of this on film, while the small party dodged intermittent small-arms fire in the Parisian streets. Later that afternoon Pash and Calvert found Joliot in his laboratory at the Collège de France.

Goudsmit arrived two days later with the other Alsos team members; they set up their office at 2 Place de l'Opera.[61] In preliminary conversations they learned that German scientists had used Joliot's laboratory and cyclotron during the occupation, but beyond that there was no definitive evidence that they were working on a bomb.[62]

Throughout September and October Pash, Goudsmit, and the Alsos team operated in Belgium and southern France with better results. Many of Alsos' freewheeling movements throughout the different Allied military commands in Belgium, France, and Germany were made easier by the efforts of Col. George Bryan Conrad, assistant chief of staff, G-2, ETO. The problem of multiple British and American intelligence units, pursuing other technical objectives, often caused friction with Alsos that needed to be smoothed out. Pash referred to Conrad as "our godfather." Of course it did not hurt that he had been Groves's plebe roommate at West Point.[63]

In early September Pash learned of a Union Minière uranium refinery in Oolen, Belgium, twenty-eight miles east of Antwerp, and was ordered to find and seize any stocks of uranium ore there. After arriving in Antwerp, Pash further learned from officials of the Union Minière that more than a thousand tons of refined ore had been shipped to Germany; another seventy tons were still at Oolen in a warehouse. Pash also learned that the Belgians had shipped more than eighty tons to France on June 4, 1940, just days before the Nazi invasion. Where was all of this ore now? Pash alerted Washington of these discoveries. Upon learning of them, Groves dispatched Major Furman to meet up with Pash, help locate the uranium stocks, and get them under Allied control.

Furman and Pash went to Oolen on September 19 and, under German sniper fire, obtained samples and brought them back to Brussels to be analyzed by David Gattiker, a British scientist who worked for the Combined Development Trust. The following day they returned, were fired upon again, and began to arrange for the numerous small barrels of uranium ore to be sent to Britain, which occurred a few days later.

Soon a lead developed on the uranium that had been taken to France four years earlier. Pash and a small team rushed off to search for it. All that was known were the serial numbers of the seven railway cars that had carried it away. Groves sent one of his best technical aides, Maj. John E. Vance, to help look for the ore. The trail eventually led to the arsenal at Toulouse, in southern France. After searching around, the Alsos Geiger counter clicked, and the material was discovered in one of the warehouses. The barrels had markings that showed they had come from Belgium. The amount that was found totaled thirty-one tons, about three railcars' worth. Pash arranged for a special army truck convoy — borrowed from the famous Red Ball Express — with combat support, to haul the material to Marseilles. Helping to get the uranium safely to port was Major Furman, Groves's eyes and ears on the scene. On the evening of October 10 Pash briefed his unit that, in the morning, they might have to take the "certain vital material" they were after out of the arsenal by force; they should be ready for this possibility.

The next day the loading went forward without incident. The barrels were quite heavy, and occasionally some fine yellow dust seeped out of them as they were loaded onto the trucks. Perhaps, some thought, it was gold they were after. The convoy rumbled southward on its three-hundred-mile journey, arriving late that evening. Shortly thereafter a U.S. Navy transport ship took the uranium ore directly to Boston.[64]

From information gathered by Alsos in Rennes, Paris, and Holland, there were indications that some Nazi atomic research was being carried on in Strasbourg, the Alsatian city that had fallen under Nazi domination in 1936. Goudsmit, who studied the clues, was convinced that it was a target of enormous importance. Sixth Army Group was moving eastward and had Strasbourg in its sights for an assault planned for early November, which was eventually delayed by a few weeks. Pash moved forward from Paris to prepare. Intelligence supplied by Groves from Washington and from Calvert in London, combined with their own sleuthing, had made Alsos knowledgeable about their targets in Strasbourg.

On November 25 Pash, leading a small party, entered Strasbourg. Over the next few days the first contact with German scientists was made. Tops on their list was a German physicist, Dr. Rudolph Fleischmann, a professor at the university. When they got to his home a neighbor told them he had left the day before, but a search of his office yielded letters from interesting correspondents. The following day Fleischmann, along with six others on their list of German scientists, was rounded up, and a laboratory at the hospital was discovered. Von Weizsäcker had eluded them, but papers found in his office provided valuable clues.

Furman arrived on the scene, followed two days later by Goudsmit, who interrogated Fleischmann and the others. The newly seized documents were combed for clues. The pieces were coming together. The team learned that the research work originally conducted at the Kaiser Wilhelm Institute in Berlin had been evacuated to the small towns of Hechingen and Bisingen in Wurttemberg in the Black Forest region of southwest Germany. Precise addresses and phone numbers were given. Also mentioned was a cave in Haigerloch and Otto Hahn's laboratory in Tailfingen.

After two days of poring over the papers, grilling the scientists, and inspecting their laboratories, Goudsmit was quite certain that the Germans had no bomb program of any significance. As Goudsmit later said, "The conclusions were unmistakable. The evidence at hand proved definitely that Germany had no atom bomb and was not likely to have one in any reasonable time."[65]

With things still somewhat uncertain militarily in Strasbourg, Goudsmit and Pash went to Sixth Army Group headquarters at Vittel, 120 miles to the rear, to report this news to its commander, Gen. Jacob L. Devers, and to Vannevar Bush, who had come from Paris to learn what had been discovered. From there the message was sent directly to General Eisenhower, who forwarded it to Washington and to General Groves and his superiors.

Soon after, Furman returned with a full report for Groves, arriving in Washington at one-thirty in the afternoon of December 8. As Groves made his way back from Montreal via New York, he telephoned orders to his secretary to have Consodine and Maj. Francis J. Smith meet him at Union Station with Furman's report and go directly to the Cosmos Club to see Conant. Later Groves wrote Bissell about the Strasbourg material, "This is the most complete, dependable and factual information we have received bearing upon the nature and extent of the German atomic effort. . . ."[66] Alsos had clearly justified itself to produce this important intelligence.

The documents, letters, and notes captured by Pash and Goudsmit were initially translated and analyzed in the field and then shipped to Washington for further evaluation and interpretation. Groves had recruited Philip Morrison, a student of Oppenheimer's who had worked at the Met Lab and then moved to Los Alamos, to help evaluate the information, which had started to arrive in August. Furman, and his assistant Major Smith, kept track of processing the information. Richard Tolman assigned two of his technical assistants, William A. Shurcliff and Paul C. Fine, to work on translating and analyzing the German documents.[67] The deluge from Strasbourg kept them all extremely busy.

But while the Strasbourg material was important in showing that a German bomb was not an immediate threat, there was still much to be done and much to find out. Where were the other scientists, especially Heisenberg? Where were the secret pile, the heavy water, the laboratory, and the thousand tons of ore removed from Oolen? Was there a centrifuge project? There was even a suspicion that perhaps all of these papers were just a ploy, meant to mislead.

There was also the problem of targets too far east for Alsos to reach — specifically, suspect sites in the Russian occupation zone. Calvert cabled Groves in early February about instructions concerning "TA" (Tube Alloy) targets in the Russian zone. The main targets were laboratories in Berlin and the immediate vicinity, the uranium at Stassfurt, and German patents, especially secret ones.

This territorial inconvenience did not hamper Groves, who set about in early March 1945 to have a key installation bombed. The specific site was the Auer Gesellschaft Works, a chemical company located at Oranienburg, a town about fifteen miles north of Berlin. The Auer plant manufactured thorium metal and uranium metal, which Groves thought might possibly be used in an atomic bomb.[68] Groves dispatched Major Smith to explain the mission to Gen. Carl A. "Tooey" Spaatz, commanding general of U.S. Strategic Air Forces in Europe. In acquainting Spaatz with the highly secret matter the word *atomic* was never used, rather "as yet unused weapons of untold potentialities." Smith cabled Groves in code on March 13, "Reception by Salem [Spaatz] was excellent and he was delighted to undertake the contract which he guarantees will be carried through to full satisfaction." And full satisfaction was delivered. On March 15, the Eighth Air Force, using 612 B-17 Flying Fortresses, dropped 1,506 tons of high-explosive bombs and 178 tons of incendiary bombs on the Auer facility. Spaatz wrote to Marshall on the nineteenth describing the excellent results of the attack on the "special target," providing photographs and interpretive reports showing its virtual destruction. It was no longer a problem.

As the Alsos team hopscotched across Europe, information gained at one place would lead them to the next target, and then to the next. Discoveries made in late March 1945 in Heidelberg, and in Stadtilm two weeks later, proved a gold mine of information. Several key scientists were captured as well and interrogated, including Walther Bothe. The only functioning cyclotron was seized in Heidelberg. In Stadtilm/Thuringen, while Dr. Walther Gerlach, the administrator of fission research, had fled some time ago, he had once had his office and laboratory there, and documents of value were found and examined. The information confirmed what had been discovered at Strasbourg and expanded upon through yet other intelligence: that the Wurttemberg region, and particularly the towns of Hechingen, Bisingen, and Tailfingen, was of special interest. The problem was that the area had been assigned to the French armies, which were on the verge of moving into it and occupying it.

The State Department was told to try to work out a diplomatic solution, adjusting the zones with the French, but no progress was made. State had not been told the reason why this should be done, nor of course were the French informed. With the issue at an impasse, Groves decided to take matters into his own hands; he wanted to make sure that the Americans reached the Wurttemberg region first. On April 5 Groves and Marshall discussed the problem with Stimson and reached agreement that Alsos teams, accompanied by American troops, should move quickly into the area, question and capture any German scientists, remove records, and destroy any installations, all before the French arrived.

Marshall directed Groves to coordinate with OPD and SHAEF to plan for what became known as Operation Harborage. Groves sent Colonel Lansdale to Europe to assist in the planning and execution. After arriving in Paris on April 8, Lansdale immediately went to SHAEF headquarters at Rheims to see Lt. Gen. Smith, Eisenhower's chief of staff, who pointed out the delicate political and diplomatic issues — and the risks — involved with the French allies. The following day Lansdale and Furman conferred with Gen. T. J. Betts, Deputy G-2 of SHAEF, pinpointing the specific areas of interest in the area. On April 10 Lansdale, along with Pash, saw British Maj. Gen. Kenneth Strong, G-2 of SHAEF, at Rheims and discussed and confirmed the targets in the key area. Later that day they attended a conference, presided over by General Smith, at which the operational details of the mission were discussed. Several plans were proposed and one was chosen, but the fluid situation on the ground ultimately determined the timing.

The precise answer to the riddle of what had happened to the thousand tons of uranium ore that the Nazis had removed from Belgium also became clear. The ore was being stored in the Stassfurt area, and a newly arrived group of VIPs was anxious to find it. This group included Lansdale; Sir Charles Hambro, deputy director of the Combined Development Trust; Michael W. Perrin, administrative officer of the DSIR Division of Tube Alloys; and David Gattiker, British scientist and intelligence officer for the CDT.

On April 16 and 17 the ore was discovered at Stassfurt in several caves. U.S. army trucks removed it to Hanover, where RAF planes flew a portion out, with most of it hauled to Antwerp by rail for shipment by sea to Britain.[69]

Pash commanded his Alsos team, with a special unit attached to the Sixth Army Group that SHAEF had created for this mission. SHAEF's support, along with the presence of Hambro, Lansdale, and the others, testified to keen interest by Washington and London in what they were expected to find. On April 22 Pash, accompanied by Brig. Gen. Eugene L. Harrison, G-2 of the Sixth Army Group, led his forces across a bridgehead at Horb on the Neckar River and moved south and east to Haigerloch, which was seized the following day, just ahead of the French.[70] The following day, April 23, Groves's eyes and ears, Lansdale and Furman, proceeded to the town, along with the British contingent: Hambro, Perrin, Gattiker, Lt. Col. Percy Rothwell, Eric Welsh, British intelligence officer, and Wing Commander Rupert G. Cecil (on R.V. Jones's staff) of British Secret Intelligence.

The party located a cave in the side of a cliff. Inside was a large chamber with a concrete pit about ten feet in diameter. Within the pit was a metal contraption that was the Nazi version of an atomic pile. There was great excitement by the scientists as the "uranium machine" was examined. Under the metal cover in the center cylinder were graphite blocks. As Lansdale took measurements, others searched for the uranium fuel and heavy water that would go into the pile. Photographs were taken and dismantlement got under way, assisted by the accompanying combat engineers.[71]

Pash left Hambro in charge of the cave and with his task force moved on to seize and occupy Bisingen and Hechingen, without opposition. All of the extensive intelligence that had been gathered for the operation was now paying off, for the Alsos team quickly rounded up some twenty-five scientists and technicians, including Carl Friedrich von Weizsäcker, Karl

Wirtz, Horst Korsching, and Erich Bagge. Interrogation added two more targets. On April 24 Alsos moved into Tailfingen, a nearby village, where Otto Hahn and Max von Laue were captured and their research facilities seized.[72] The second target concerned Heisenberg. It was learned that he had left two weeks ago on his bicycle to visit his family in the mountain village of Urfeldt in the Bavarian Alps.

On the twenty-sixth Lansdale, Welsh, and Perrin interrogated the captured scientists, eventually learning from them the location of the heavy water and the uranium oxide for the Hechingen pile. The heavy water had been put in three steel barrels in the cellar of an old gristmill about five kilometers from Haigerloch. After a great deal of effort it was hoisted out of the cellar and onto a truck and sent to Paris, for transshipment to the United States. The uranium had been buried in a field on a hill overlooking the town and then plowed. The missing metallic uranium from the pile at Haigerloch, mostly in five-centimeter cubes, was dug up and also sent by truck to Paris for transshipment. The scientists were amused watching the surprised troops as they loaded the deceptively light-looking ingots into a two-cubic-foot stack weighing almost two tons.[73]

On April 27 the scientists were sent to Heidelberg for further interrogation. Just before leaving, von Weizsäcker disclosed that he had not really burned his papers, as he had earlier claimed, but that they could be found in a metal drum in a cesspool behind a house. The drum was retrieved, and Goudsmit and the others discovered a complete set of German reports about the progress, or lack thereof, of the German bomb program.

That day Harrison sent a cable to Devers, who relayed it to Eisenhower, who sent it on to Marshall and Stimson: "The Alsos mission . . . have hit the jackpot in the Hechingen area."[74]

What they found was final and definitive proof that the German atomic program was, in Goudsmit's words, "small-time stuff. Sometimes we wondered if our government had not spent more money on our intelligence mission than the Germans had spent on their whole project."[75] The program had never gone beyond a preliminary research stage. They had not attained a chain reaction; nor had they discovered an effective way to enrich uranium. They never seriously considered plutonium as an active material. Instead of being two years ahead, they were several years behind.

The final goal was to capture Heisenberg before the Soviets got him or he was killed by the SS or the "Wehrwulf," the fanatical Nazi underground. Pash and his team set off for Urfeldt, reaching it on May 2.

Learning from townsfolk where his prey was living, Pash and his group climbed up the last steep slope to Heisenberg's cabin to find him waiting for them on the veranda. Some ticklish moments ensued with some German troops, and this necessitated a temporary retreat from the village, but the next day they returned, seized Heisenberg and his documents, and left for Heidelberg. Two others on the "wanted" list, Walther Gerlach and Kurt Diebner, were rounded up by another team at the University in Munich; with Paul Harteck finally tracked down in Hamburg, then, all the key members of Hitler's Uranium Club whom Goudsmit had identified were now in custody.

It took a little while to figure out what exactly to do with this special group. In a discussion with Chadwick, Groves proposed one solution, though it is difficult to tell exactly how serious he was. Occasionally Groves would blurt out rough and raw words expressing his deeply felt opinions about certain matters. This time it was about the enemy quarry he had been after for so long. Groves told Chadwick, "the German scientists should all be shot as they were undoubtedly as great war criminals as anyone in Germany." He also recorded the British scientists' reaction: "This view was not relished by Dr. Chadwick although he understood the reason back of it."[76]

R. V. Jones has written that Eric Welsh, his colleague in British intelligence, told him "that an American general had said the easiest way of dealing with postwar developments in nuclear physics in Germany would be to shoot the physicists." Jones did not identify the American general, probably to shield him from embarrassment, but it is obvious from the above memo that it was Groves. Clearly the British felt that Groves might be serious. Jones went on, "Our resultant alarm was genuine enough for Welsh to suggest that I propose the physicists be moved to England for safe-keeping."[77]

Throughout May and June Major Furman, members of the Alsos team, and British intelligence officers were in charge of the German scientists as they moved them several times to places in France and Belgium. On May 2 Major Furman escorted six of the ten to a house at 75 Rue Gambetta in Rheims, France. A few days later, on May 7, the group was flown to Versailles and put in the detention center for Nazi scientists and industrialists known as the Dustbin, at the Château du Chesnay. The day after V-E Day Heisenberg and Diebner joined their colleagues, bringing the party to eight. Their stay at the Dustbin was brief; they were moved to a villa at Le Vesinet, a suburb of Paris, on the eleventh. That evening Major Furman brought Professor Harteck to join them. One more move took them to the Château de Facqueval at Huy, Belgium, on June 4. The number increased to ten on the fourteenth as Gerlach was brought from Paris. The next day permission was granted to take them to England, and on July 3 they arrived at Farm Hall.[78]

Farm Hall was a country house near Godmanchester, fifteen miles northwest of Cambridge, that was used by SOE and M.I.6. It served as a staging area and training school for agents going to Europe on missions. It also served as an interrogation center for agents and captives returning from the Continent. The ten captured German scientists were held incommunicado at Farm Hall until they were released and brought back to Germany on January 3, 1946, exactly six months later — the term that a British wartime law permitted someone to be held "at his majesty's pleasure."

Unbeknownst to these "guests" was the fact that the house was wired with microphones, and their more interesting discussions were listened to, recorded, and translated by a British team. Lt. Cmdr. Eric Welsh had earlier wired Farm Hall for the purpose of testing the loyalties of agents and captives who passed through. The Gunnerside commandos had spent three months there before their mission to Vemork. The microphones may have been removed, since Farm Hall was vacant prior to the Germans' arrival, and then reinstalled at the suggestion of R. V. Jones.

> Almost as an afterthought, I also suggested that microphones be installed, in case the German physicists might discuss among themselves items that they might not have revealed in direct interrogation."[79]

An eight-man team was responsible for listening, recording, transcribing, and translating what was said. Only relevant material having to do with technical or political matters was transcribed, perhaps 10 percent of the words that were recorded and then translated into English. The recordings were made on shellac-covered metal discs using six to eight recording machines. After use the discs were recoated and reused. They have not survived, nor apparently have the German transcripts.

Maj. T. H. Rittner and his assistant Capt. P. L. C. Brodie headed the team but did not listen or transcribe. The remaining six were native German speakers who had had several years' experience in this kind of work. The conversations were distilled into weekly or biweekly reports by Major Rittner and Captain Brodie and distributed to a limited readership. Copy number one of each report went to the Office of the Military Attaché at the American Embassy in London, where Lt. Col. Horace K. Calvert transmitted it to Groves's office in Washington. With Calvert's covering memos there are more than 250 pages of transcripts that make up the twenty-four reports. Groves read each report and flagged or underlined certain words, sentences, or sections that he thought noteworthy. There is a set of the Farm Hall transcripts in the Public Records Office at Kew, with some pages missing. A more complete set is filed in Record Group 77 in the National Archives, at College Park, Maryland. This set includes the copies that were sent to General Groves and some that were apparently retyped by Mrs. O'Leary.[80]

One of Groves's immediate concerns as he read the Farm Hall transcripts in the summer and fall of 1945 was what would happen to these scientists once they were released. They were safe while in Allied hands but defection was not out of the question; even returning them to their institutes was a problem, since some were in the French zone of occupation.[81]

Groves's book, published in 1962, was the first revelation that there were transcripts of conversations by the German scientists, and he quoted from them extensively.[82] This fact was interesting news to the German scientists, who up until this point had not known or suspected that their words had been recorded.[83] Goudsmit had excerpted several quotations from the conversations in his book fifteen years earlier, but did not mention that there had been recordings. In an introduction to a new edition of Goudsmit's book *Alsos,* published in 1983, R. V. Jones used Groves's verbatim extracts and argued for their release. In an earlier book Jones had drawn upon them as well,[84] but the transcripts remained unavailable to scholars and the public until February 1992, when they were finally released.

The Farm Hall transcripts demonstrate that the German scientists were truly shocked and astounded when they heard that the United States had dropped a bomb on Hiroshima. They had believed up until that point that tons of U-235 were needed. By August 14 Heisenberg had rethought the problem, and it became clear to him at last that kilograms were the answer.

This is not the place to review the heated debate that surrounds the Nazi atomic bomb program or to answer the questions about Heisenberg's role in it. Each issue and question is an enthralling research venture of its own. Were Heisenberg's misconceived physics calculations the result of ignorance or intent? How did his feelings of German nationalism influence his behavior before, during, and after the war? Did he prefer a Nazi victory or a defeat by the Allies? Scholars have wrestled with these questions and their implications at great length, and kept the topic interesting and timely. Over the past decade it has generated considerable interest, extending well beyond academia and producing a flood of books, newspaper, magazine, and journal articles, seminars and symposia, television documentaries, and even an award-winning play, titled *Copenhagen,* dealing with the famous September 1941 visit by Heisenberg to the Danish capital, at which time he saw Niels Bohr.[85] The fact that these questions arouse such emotion sixty years later is itself worthy of scholarly analysis.

Within hours of learning that the Americans had dropped an atomic bomb on Japan the German physicists at Farm Hall fashioned an argument about their past actions. Led by Heisenberg and von Weizsäcker, they quickly crafted a clever reinterpretation of their own role in the bomb project that had a scientific and a moral dimension. The scientific component necessarily had to exaggerate how much they knew about how to build a bomb. Through various twisted contortions they argued that actually they did know how. This was necessary so that their supposed moral stance against not helping Hitler get the bomb could follow. This argument has played especially well with their own countrymen, who were also in need of reinventing what happened under National Socialism: no shame in what they did, no apologies or responsibility for their actions.

In brief fashion we need only touch upon Groves's views about these matters. He thought there were fairly clear-cut reasons why the German program did not proceed very far. To help illuminate the Germans' failures he often contrasted them to the successes of his Manhattan Project. Their scientists were so full of the idea of German superiority that they falsely believed that since they could not build a bomb, no one else could either. Hence their shock and disbelief at hearing the news that Hiroshima had been bombed. Another reason was their attitudes about hierarchy and authority, the cultural context for the practice of science. The German senior scientists were viewed as, and thought themselves to be, infallible, whereas in the United States junior scientists — Groves used the example of Seth Neddermeyer and his ideas about implosion — were allowed to pursue their research, and to make major contributions.

But the most important reason for the German failure was that neither the government nor industry was ever in the picture in a central way. The nation's resources were never mobilized. The integrated and focused collaboration of the government, science, and industry — the hallmark of the

American effort — was absent in the German case. No one assumed overall leadership to give it firm direction. The scientists, Groves felt, lacked "drive" and "will," "nerve" and "courage," unlike their American counterparts. The question arises: Why was there so little drive? Was it because of a subtle moral stance the German physicists took, a kind of passive resistance against Hitler, or was it because they thought the problems insurmountable or too difficult to solve at the time? It is around these issues that much debate continues to swirl.

There is no doubt about where Groves stood on the matter. He thought it was just "plain bunk" that the German scientists held back in doing research because of Hitler. On Heisenberg, his

> statements in 1945 and since then have been pure lies. . . . I don't believe a word about the German scientific hesitancy as to moral propriety. This was an afterthought after the war was over when they all suddenly became anti-Nazis. . . . Heisenberg's present position is pure bunk. It is interesting to note that Heisenberg has been very quiet on this phase ever since my book was published for he knows that we have his entire views as expressed at that time down on paper.[86]

With the moral questions out of the way Groves examined a few of the technical reasons why they failed to go forward. They made miscalculations based upon not understanding some of the basic concepts that underlie the science of the bomb. This point is of course crucial, for it is central to the Heisenberg version that they understood how to make a bomb, but for moral reasons held back from pressing forward. Groves, having access to the Farm Hall transcripts, quoted Heisenberg's most glaring misconception — that tons rather than kilograms of U-235 would be needed for a bomb. In short, Heisenberg miscalculated the critical mass of uranium.

Groves said, "But the most surprising statement came from Heisenberg. He wondered how we were able to separate the two tons of U235 needed for a bomb."[87] At this point Groves went on to make a mistake of his own — though it does not change the basic fact that Heisenberg got the critical mass wrong. Groves, following Goudsmit, based his reasoning on the belief (later shown to be mistaken) that the Germans failed to understand that the reaction depended upon fast neutrons. In fact, Heisenberg *had* realized that fast neutrons were necessary. His confusion came in misunderstanding the efficiency of the reaction and concluding that tons of U-235 would be needed. Precisely how many tons were never quite clear, but making such a large amount would take too long to be a factor in the war. Heisenberg was a far cry from Fermi when it came to translating the theoretical into the practical.[88]

Groves's views were not highly original, and Sam Goudsmit no doubt influenced him. By 1947 there was already a spirited public and private debate under way. Heisenberg was engaging in a defense of his actions mainly with Goudsmit in a series of articles, letters, and interviews. Goudsmit was demanding some accountability about German science and morality, none of which he got from his onetime friend and colleague.

Groves was clearly aware of some of this, and it no doubt influenced his own thinking. Goudsmit's *Alsos* was published in October 1947. The manuscript was read in draft form by Groves and his assistants to look for security infractions.[89]

Groves also had a brush with journalist Robert Jungk, whose *Brighter Than a Thousand Suns* was published in 1956 and portrayed the German physicists in a morally favorable light. Groves felt that the statements he had provided to Jungk had been twisted and taken out of context, and that the book was untrustworthy.

As he did with all information, Groves carefully controlled the facts about the Germans' lack of progress in their bomb program. It is difficult to tell precisely when Groves became absolutely confident in his own mind that there was no threat, and to whom he may have confided this information. He feared perhaps that if it were known that the Germans had not advanced very far on their bomb, it might cause some scientists to stop working on the American project. After all, this had been a key inspiration for many of them to enlist in the project in the first place. He need not have worried. Clearly by the end of November 1944 the Alsos documents found in Strasbourg provided convincing evidence that the Germans had not progressed very far on a bomb. Even after the German surrender, in May 1945, not only did almost no one leave the project, but they redoubled their efforts to finish it in time to be used against Japan.

Goudsmit noted that in early 1945, as the full significance of the Strasbourg discovery sunk in, he said to Furman, "Isn't it wonderful that the Germans have no atom bomb? . . . Now we won't have to use ours." Furman replied, "Of course you understand, Sam, . . . if we have such a weapon, we are going to use it."[90]

CHAPTER FIFTEEN

His Own Air Force

Anticipation of what would be needed many months or even a year in advance was a Groves specialty. Be prepared when the time came by planning for it long before. An excellent example of this attribute was his initiative to form an Army Air Forces combat unit to deliver the bombs. After the bombs were developed, tested, and built, a capable aircraft would be needed to drop them on the enemy. Special crews would have to be trained and bases established at home and overseas. None of this could be left to the last minute. Groves's involvement in these actions placed him at the center of the operational use of the bomb.

Groves took a first step to prepare for the future combat mission at a meeting of the Military Policy Committee on May 5, 1943. He told them that he needed an ordnance officer and had been unable to find a suitable army candidate. Bush asked if he would have any objection to a naval officer. Groves responded that the only criterion was that he be good; Bush recommended Cmdr. William S. Parsons, who had recently served as his special assistant in OSRD for the proximity fuse project. Conant and Purnell, who also knew his work, concurred, and the admiral said he would have him in Groves's office the next day. Late in the afternoon of May 6 Parsons came to Groves's office. They soon discovered that they had first met in 1933 when Parsons was engaged in radar research and Groves was working on the anti-aircraft searchlight program.[1] This was another example of Groves's ability to almost instantly judge character and competence; as he later said, "[W]ithin a few minutes I was sure that he was the man for the job."[2]

Groves thought highly of Parsons, and was able to discern his strengths and his weaknesses. In an interview Groves contrasted Parsons to himself. Parsons would not have been effective at doing Groves's job. That required making big decisions on inadequate data and taking risks. Parsons tended to be cautious, to play it safe, and was too "tender-hearted." Groves thought Parsons did not have "command presence" — that certain tough-ness, even meanness, that was essential when the situation required.[3]

William Sterling "Deak" Parsons was born in 1901 in Chicago and moved to Fort Sumner, territory of New Mexico, when he was eight years old.[4] He entered the Naval Academy in June 1918, at sixteen years of age, with the class of 1922 and upon graduation ranked 48th of a class of 539.[5] A classmate, Hyman G. Rickover — who would also work for the Manhattan Project, at Oak Ridge, and afterward go on to create the Nuclear Navy — ranked 107th. A recent biographer has depicted Parsons as an example of a new breed of officer that began to emerge in the 1930s, one skilled in the technical and scientific matters that would transform the American military during and after World War II. Parsons's specialty was naval ordnance, the making and improving of guns, bombs, mines, and tor-pedoes. Prior to joining the Manhattan Project Parsons had worked for a short while on radar at the Naval Research Laboratory, and on developing and deploying the proximity fuse with Vannevar Bush, the OSRD, and the scientist Merle Tuve. Unlike most of his aloof navy counterparts, he was eager and willing to work with civilian scientists on pioneering projects. Parsons brought several other naval officers with similar dispositions into the Manhattan Project, including Frederick L. "Dick" Ashworth, Norris Bradbury, and John T. Hayward. After the war, Parsons, Ashworth, and Hayward would be instrumental in the first deployments of naval aircraft and atomic weapons aboard aircraft carriers, and other missions that soon followed.[6] Norris Bradbury would go on to serve as the director of Los Alamos Laboratory for twenty-five years.

After joining the Manhattan Project, Parsons's initial job was to work on the special gun-type bomb that was the sole design at the time. After the discovery of the predetonation problem with plutonium, Parsons worked on the implosion design as well. His personality fit in well at Los Alamos, where he became an associate director under Oppenheimer and helped lessen the civilian-military differences that were ever present there.

Parsons's most important job, as we shall see, was to be the weaponeer on the *Enola Gay.* The weaponeer for the second mission was also a navy man, Cmdr. Frederick L. Ashworth. As Ashworth later recounted,

> He [Parsons] and I had been personally selected by General Leslie
> Groves to act as tactical commanders. The General insisted that he

must maintain control of things by placing aboard the bombing planes his personal representative fully informed of the bomb's capability and technical operation.[7]

Parsons obtained Groves's agreement at the outset that he would be the person responsible for its combat use. As Parsons later said,

> The selling point was that by knowing this throughout the entire weapon development period I would be able to take a more personal interest in getting every detail exactly right. I had used a similar system with the radio proximity fuse.[8]

Groves took a second step to prepare for the combat mission by ensuring that the resources of the Army Air Forces were at his disposal. In June 1943 Groves met with Gen. Henry H. "Hap" Arnold, commanding general, AAF, to inform him of the progress to date on the atomic bomb and request that Arnold designate a member of his staff to serve as liaison to the Manhattan Project. Arnold delegated Maj. Gen. Oliver P. Echols to serve as his representative. Echols, in turn, designated Col. Roscoe C. "Bim" Wilson to serve as project officer and to be the liaison with Los Alamos. Many years later Wilson recalled that in the briefing he received from Arnold, he was told:

> Do what General Groves tells you to do; get things done and get them done promptly; offer no explanations which might reveal the nature of the project; if anyone argues with you, tell him to come to me immediately.[9]

There were many matters to be settled. First was to choose the correct plane to deliver the bomb, second was to form an air combat unit and establish bases for it, and third was to organize and train the crews. All of this would have to be done in close collaboration with Los Alamos, with Parsons and his division.

The person at Los Alamos with whom Wilson primarily worked was Norman F. Ramsey. Ramsey, son of an army officer, received a Ph.D. in physics from Columbia in 1939, and was drafted to work at the MIT Radiation Lab the following year.[10] After some time in Cambridge he came to Washington to work in Stimson's office on Army Air Forces projects for Edward Lindley Bowles, a consultant to the secretary. Ramsey's combination of expertise in physics and aerodynamics made him a prime candidate for Los Alamos. After an effective recruitment appeal by Oppenheimer in March 1943, Ramsey agreed to join the project. Oppenheimer left it to Groves to get Stimson to agree to release Ramsey. This took several months, because Bowles refused to let him go. Bowles and Groves were both used to getting their way, and who would get Ramsey turned into a matter of prestige and power. To men in powerful positions losing a bureaucratic battle, even a small one, could mean losing others in the future, and that must not be allowed to happen. Rather than push it to the hilt, Bowles found a way out by asking Ramsey what he wanted to do. Ramsey said he thought he should go to Los Alamos. To allow Bowles to save face Ramsey continued as a consultant from the secretary of war's office and was not an employee of the University of California, Los Alamos's contractor.

Ramsey was assigned to head the Delivery Group of the Ordnance Division and later served as deputy to Parsons.[11] His immediate tasks were to design the bomb casings that would carry the gun-assembly bomb and implosion bomb. By the end of 1943 it had already been established that the gun-type bomb — Thin Man — would weigh on the order of five tons. Ramsey assumed that the implosion bomb would weigh approximately the same. Given their size and weight, there were only two possible choices for an aircraft to deliver the weapons, the British Lancaster or the American B-29, which had begun production in September.

Ramsey favored the Lancaster and traveled to Canada in early October 1943 to meet Roy Chadwick, the plane's chief designer. Chadwick was in Canada to observe the initial Lancasters coming off the production line at the Victory Aircraft Works, Milton Airdrome, in Toronto. Ramsey showed Chadwick preliminary sketches of the large — thin-shaped and stubby-shaped — bombs and later wrote with more details.[12] Chadwick assured Ramsey that the Lancaster could accommodate them.

When Ramsey returned, he wrote to Parsons suggesting that the Lancaster be seriously considered and planned a memo to General Groves recommending that a modified Lancaster be used.[13] The bomb bay was thirty-three feet long and sixty-one inches wide. The depth was only thirty-eight inches, but this could be modified. The Lancaster's ceiling was 27,000 feet, its speed 285 miles per hour, and takeoff required only 3,750 feet of runway — a critical matter wherever it would be based.

Groves had not been informed about Ramsey's preference and was at a loss for words when he found out. It was beyond comprehension that Ramsey could consider using a British plane to deliver an American atomic bomb. Needless to say, Groves found an ally in General Arnold when he discussed this matter with him. The new Boeing B-29 Superfortress would carry the atomic bomb.

The first production B-29s were produced at the Boeing-Wichita plant beginning in September 1943. One was chosen to be modified; by December 1 it was at Wright Field for two months of modifications. The modifications were originally called Silver Plated, but this was soon shortened to Silverplate. On the same day General Arnold's office informed the commanding general, Material Command, that Silverplate modifications should be given "the greatest possible priority." The two bomb bays were made into one, and the four twelve-foot bomb bay doors were replaced by two twenty-seven-foot pneumatically operated doors. At this point the length of the Thin Man bomb was expected to be at least seventeen feet. Racks, bracing, and hoists were installed, along with the release mechanism and mounts for a motion picture camera to record training drops.

On February 20, 1944, the first hand-modified Silverplate B-29 flew from Wright Field to Muroc Army Air Base (now Edwards AFB) in the Mojave Desert of California. Drop tests of the Thin Man and Fat Man dummy bombs began on March 6.[14] Many problems were encountered and corrected. The ballistics for the Thin Man had been partially worked out at Dahlgren, where scale-model tests were conducted. The bulbous Fat Man shape was another matter; it proved more difficult over the next year to find the correct tail structure design to prevent it from wobbling. As the analysts sought their solution, they made another discovery. The standard tailfins used on air force aerial bombs seemed to have a flaw, which caused some of the fins to collapse as the bomb reached terminal velocity. This had an obvious effect on accuracy and would have been of great interest to the Ordnance Department. But when this information was passed on to Groves he suppressed it, not wanting to compromise security. It took another year for the Ordnance Department to find out.

As the tests continued, further modifications and refinements were made to the aircraft to arrive at a final standard. These mainly had to do with the bomb bay and the various frames, hoists, braces, and release assemblies that could handle the four- and five-ton bombs. One major change was to adopt a single lug to suspend the heavy bomb rather than twin release lugs, which had caused problems. Once premature release of a seventy-three-hundred-pound Thin Man caused severe damage to the bomb bay doors of the single B-29 in March. After repairs were made, testing resumed in mid-June.

By the summer of 1944 the design was fairly firm, and on August 23 the Glenn L. Martin–Nebraska Company received the contract to modify the first three B-29s — of a total of twenty-four — and selected its Fort Crook Modification Center in Omaha as the program site.[15] The delivery schedule was three planes by the end of September, the next eleven by the end of the year, and the final ten in January 1945. The initial fourteen were slated for test and training, with the other ten assigned as the combat unit. As with most parts of the Manhattan Project, schedules were accelerated and quantities increased. In February the number of Silverplate B-29s was increased to forty-eight, and on April 18 to fifty-three — fifty-four counting the hand-modified Muroc plane.[16] As Los Alamos continued to change and refine its design for the bombs, new instructions were repeatedly sent to modify the bombers. By the end of the war forty-six planes had been completed.

Aircraft 44-86292 would have a special place in history. It became known as the *Enola Gay,* named for the pilot's mother. It was delivered to the USAAF at the Martin Aircraft Factory, Omaha, on May 18, 1945, and after modifications was flown to Wendover on June 14.

During the summer and fall of 1944 representatives of the Army Air Forces and the Manhattan Project carefully screened candidates to command a soon-to-be-formed self-supporting unit — the 509th Composite Group. The person selected was Lt. Col. Paul W. Tibbets Jr., a seasoned, highly respected bomber pilot. The selection process was a collaborative one, with the air force making the initial selection and Groves's staff confirming it. On August 29, 1944, General Giles replaced an earlier nominee with Tibbets.[17] In an interrogation in the offices of Maj. Gen. Uzal G. Ent, commanding general of the Second Air Forces at Colorado Springs, Groves's intelligence chief Col. John Lansdale questioned the twenty-nine-year-old pilot and agreed with the choice. Tibbets was then briefed by Parsons and Ramsey, and was told by General Ent that he had been chosen to command and train a unit to drop an atomic bomb on Germany or Japan.

Tibbets had completed twenty-five combat missions in Europe and North Africa flying B-17s. He was a test pilot for the B-29 and was well acquainted with its characteristics and idiosyncrasies. Tibbets was chosen because he was a superb pilot. Groves, however, did not think much of Tibbets as an officer. While acknowledging his ability as a pilot, he felt that Tibbets was too young to be the head of the group. Much later Groves said of Tibbets: "[H]e showed that he didn't have the senior officer capabilities by the fact that with that back of him, he never got beyond BG [brigadier general]."[18]

General Arnold gave Tibbets carte blanche in his choice of personnel for the 509th, but Tibbets was forbidden to tell them the true nature of their assignment. Tibbets chose men he had flown with in England and North Africa, or in the B-29 test program, and when crews were selected he kept them together.[19]

General Ent chose the 393rd Bombardment Squadron (Very Heavy) as the operational component of the group and proposed three bases at which to train. Tibbets visited the squadron at Fairmont Army Air Field, in Geneva, Nebraska, where it had been training, and agreed with the choice. Tibbets also visited Wendover, whose facilities were adequate and whose remote location was ideal for the highly secret project. Soon after, the seven-hundred-man squadron was introduced to Colonel Tibbets but not told its true mission. Tibbets told the men that they would be engaged in a highly secret project that, if successful, would end the war twelve months sooner than anticipated. Absolute security was required of everyone. Later, when some type of explanation was needed, they were told that they were preparing for an aerial mining attack on the deep ravines of Formosa, a story probably not entirely convincing to ordnance personnel and aircraft crews handling the oddly shaped bombs.[20]

On December 17, 1944, the 509th Composite Group was activated and assigned to the 315th Bombardment Wing (VH) at Colorado Springs. At the same time other units were added to the group: the 390th Air Service Group, to train the unit in combat procedures, chemical warfare, first aid, the use of firearms, and camouflage; the 603rd Air Engineering Squadron, to repair the planes and rebuild the engines; the 1027th Air Material Squadron, to procure and issue all air corps, quartermaster, signal, chemical, and ordnance supplies and materials including food and clothing; the 320th Troop Carrier Squadron (soon to be known as the Green Hornets), for transport; and the 1395th Military Police Company (Aviation). On February 22, 1945, the First Ordnance Squadron, Special (Aviation), was added. The group had an authorized strength of 225 officers and 1,542 men.[21]

Muroc, already very busy providing final combat training for aircrews just prior to overseas deployment, could not handle the new combat unit being formed. In July 1944 Parsons, Ramsey, Wilson, and Tibbets went to Major General Ent to request a training field. Wendover Army Air Field, Utah, was chosen for its isolation and security.[22] Wendover was given the code name Kingman or Site K. It also went by the name of Project W-47.[23]

To gain experience in approaching and flying over coastlines, and ocean flying in general, aircrews of the 393rd Squadron, like many other AAF units, were sent to Batista Field in Cuba. The crews remained at Batista from January until the latter part of March, when they returned to Wendover.[24]

The unique shape of the Little Boy and Fat Man bombs demanded a great deal of testing to ensure accurate flight after being dropped from a B-29. The aerodynamics were difficult to understand; hundreds of tests were conducted to figure out the correct weight distribution and shape of the body and fins. What was needed were ranges with cameras and a soil that would not make it too difficult to recover the dummy bombs. The crews trained in airdrops at Wendover, Muroc, and Salton Sea, a test drop area in the Imperial Desert of California, six hundred miles from Wendover.

In November 1943 the navy established the Naval Ordnance Test Station (NOTS) at Inyokern, California, in the Mojave Desert.[25] The initial purposes of NOTS were to develop rockets and to improve aircraft ordnance. The rocket program had been conducted by Dr. Charles C. Lauritsen of the California Institute of Technology under a contract with the Office of Scientific Research and Development. Information about the secret work at Germany's rocket center at Peenemunde had forced an American response to explore the military possibilities.

Not unlike Los Alamos, Inyokern grew dramatically, and additional tasks were assigned to NOTS. By early 1945 some of the work was performed for the Manhattan Project, largely at the instigation of Oppenheimer and Parsons. Oppenheimer called upon his Cal Tech colleague Lauritsen to work with his Los Alamos scientists. Richard Tolman's role as vice chairman of the NDRC and the OSRD and as dean of Cal Tech clearly expedited the arrangement. The work included helping design reliable detonators for the implosion bomb, the production of high explosives, and the monitoring of airdrops of the special bomb shapes from B-29s of the 509th at Salton Sea. The code name for all of the technical work performed by Cal Tech at Pasadena and at Inyokern for the Manhattan Project was Camel.

The second task had to do with fabricating the specially designed high-

explosive lens that, when set off by the detonators, compressed the plutonium core of the Fat Man–type bomb to criticality. The work was done
at Los Alamos at S-Site, but, as General Groves recognized from his experience in building ordnance plants, it was a very dangerous place. "It was
like one of these firecracker plants that you read of blowing up every once
in a while."[26] While it would be unfortunate if people were killed as a
result of an accident, more important would be the loss of the sole production facility. In another instance of a trade-off — this time between
safety and time — in the end it was Groves's call to make. As he said,

> I ignored most of [the conventional safety standards] deliberately. It
> wasn't a case of not knowing any better or being reckless, or anything
> like that. It was just a cold-cut decision that it was worth taking a
> chance on . . . remember, time was all important to us.[27]

It was Parsons who suggested the need for a second high-explosive
plant, and also where it might be located. In early 1945 Groves and Parsons
flew to Pasadena to meet with Lauritsen, finalizing the arrangements for
Cal Tech and NOTS to build the plant at Inyokern. Like several of the
other deadlines Groves had set, this one was also very demanding: The Salt
Wells Pilot Plant was to be built and in operation one hundred days after
groundbreaking. Like other Manhattan ventures, construction went on
parallel with evolving design and manufacturing processes, solving unique
problems as they went along. For weeks into the effort Los Alamos
remained uncertain whether the pressing method or the casting method
was best to fabricate the high-explosive blocks.

Groves invoked his priority privileges to help meet the deadlines and
of course imposed strict compartmentalization to kept security tight. Of
the hundreds who worked on the design and construction of the Salt
Wells plant, only a few knew its purpose. Though Groves's deadline was
missed by fifteen days, it was still something of a miracle that the plant was
completed so quickly. On July 25, 1945, nine days after Trinity, the first
high explosives were melted, poured, and cast.

On Parsons's recommendation Lt. Cmdr. John T. "Chick" Hayward was
transferred to NOTS in August 1944 to be the experimental officer and
the following month was cleared to work with Los Alamos. Like Parsons,
Hayward got on well with the scientists and had a good technical background. He was also an experienced naval aviator with nineteen months
of combat experience in the South Pacific.

The reasons Brigadier General Groves hired me, he said, were that I could drive airplanes and wasn't in the air corps, the "Arrogant Corps," he called it. Marks against me were his hatred of my New York Yankees and his howling fits if Navy's football team beat Army. An able leader on the Project, from what I saw, he seemed to enjoy giving me the needle now and then, which, interestingly, created a friendly, frank rapport between us eventually.[28]

Hayward had several responsibilities. One of them was to fly Groves, Oppenheimer, Parsons, and many of the other Manhattan people from site to site and to and from Washington. In the process he got to know them all. He also flew F4U-4 planes to photograph Fat Man test bombs as they were dropped from B-29s over the test ranges. In early September he went to Hiroshima and Nagasaki to assess bomb damage for Cal Tech and turned in a 450-page report. In 1946 he participated in Operation Crossroads under Parsons, who was the technical director. In August 1947 he was appointed director of plans and operations for atomic warfare at the Armed Forces Special Weapons Project, the senior naval officer at Sandia Base there.

At the beginning of 1945 Groves and Arnold began preparations to set up a forward base in the Pacific from which the 509th would conduct its combat missions. In early February Groves dispatched Navy Cmdr. Frederick L. Ashworth to deliver a personal letter from Fleet Adm. Ernest J. King, chief of naval operations, to Fleet Adm. Chester W. Nimitz, commander in chief, Pacific Ocean Areas, informing him "that a new weapon will be ready in August of this year for use against Japan by the 20th Air Force" and asking for his assistance in establishing a base of operation.[29]

Ashworth discussed the problem of locating a forward base for the 509th with Nimitz's aides, and Tinian was recommended. Tinian had been captured from the Japanese in the summer of 1944. The tiny island was in the process of being turned into a gigantic B-29 air base. It also had navy construction battalions available that could be used. Ashworth visited the island, reserved an area near North Field for the unit, and made his report to Groves.[30]

Groves approved Ashworth's recommendation and in March sent Col. Elmer E. Kirkpatrick to construct the facilities needed by the 509th. Of course Kirkpatrick's true mission was a closely guarded secret, not even known to Tinian's commanding officer, Brig. Gen. Frederick von Harten

Kimble.[31] Kimble was not always receptive to Kirkpatrick's many demands to use navy construction battalions (initially the 67th CB, then the 110th CB) and requests that ships be promptly unloaded.

Kirkpatrick was Groves's eyes and ears on Tinian and one of the key individuals who helped Groves keep to his demanding timetable. Kirkpatrick oversaw the building of assembly buildings, warehouses, and Quonset huts. He was on the receiving end of the logistical network that was established to ensure that any and all bomb parts, tools, and specialized equipment were available on the island. Kirkpatrick was in direct contact with Maj. Jack Derry in Groves's office for anything that he needed.

From March to August Kirkpatrick received shipment after shipment of everything that would be needed. Secrecy sometimes undermined the smooth delivery. Since the shipping clerks did not know that the cargo was vitally important, it was often handled routinely. But more often than not the extra procedures that were put in place were followed. Ships would be loaded on the west coast so that the first things unloaded would be Kirkpatrick's items, going directly to Tinian without first stopping in Guam.

The first contingents of the 509th sailed from Seattle and arrived on May 19, 1945; others continued to arrive over the next two months.

Despite the size and scope of the air operation, Groves kept to his tenet of compartmentalization to prevent any breach of security. As late as the first week of July, the thirteen airplane commanders of the 393rd Squadron were largely unaware of their mission.[32] For the thousands of other Army Air Forces personnel not connected to the MED who were on Tinian, the true mission of the 509th did not become known until they heard about it on August 6.

In several respects Groves and Gen. Curtis E. LeMay were very much alike, including problem-solving skills and skills in leading large organizations. Both were no-nonsense, driven personalities, brusque and direct. LeMay would not truly come into his own until after the war, when he built the Strategic Air Command. As commander in chief for thirteen years, from 1948 to 1960, his personality heavily influenced the culture of SAC for decades to come.

LeMay had assumed command of XXI Bomber Command on January 7, 1945, replacing Haywood Hansell. The first bombing mission was January 23, attacking an aircraft engine plant in Nagoya.

LeMay first learned about the atomic bomb when Colonel Kirkpatrick arrived on Guam to inform him that he was going to set up some special facilities on Tinian. As LeMay later noted, when a field-grade army engineer officer presents highly classified information to the commanding general, "something big is in the wind."[33]

LeMay and most others in the Army Air Forces were convinced that it would be possible to defeat the Japanese without invading their home islands. "We needed to establish bases within reasonable range: then we could bomb and burn them until they quit. The ground-gripping Army and the Navy didn't agree. They discounted the whole idea." LeMay had no moral qualms about it then or later. He made no sharp distinction about how people were killed in wartime, whether with nuclear weapons or conventional weapons. In his direct and blunt style: "We scorched and boiled and baked to death more people in Tokyo on that night of March 9–10 than went up in vapor at Hiroshima and Nagasaki combined."[34]

Groves and LeMay had one conference, on June 20, 1945, in Washington, with Farrell also attending. Groves gave LeMay more of the details about the atomic bomb — when it would be ready, how large it would be, and the difficulties of using it. Groves first proposed that fighter escorts accompany the B-29s, but LeMay said he did not want to do it that way. He asked Groves how many planes were absolutely essential. Groves said three: the plane with the bomb, an instrument plane, and an observation plane, all stripped of nonessential armaments, except for a tail gun. LeMay said he would only use the three, flying them at the highest-possible altitude.

Without having to say it, Groves was sure that LeMay understood that it was not his prerogative to pick the targets or anything of that kind. But by the same token he would be the field commander when it came to the operation.

CHAPTER SIXTEEN

His Own State and Treasury Departments

For most of the war Groves served, in effect, as the assistant secretary of state for atomic affairs, another example of how his responsibilities expanded as issues had to be addressed and decisions made. The issues ranged from active measures to stop the suspected German bomb program, as we have seen in the previous chapter, to a vigorous effort to corner the world uranium market so as to ensure a U.S. supply and, as an added benefit, to prevent or delay an adversary's attempt to build the bomb. Primarily because of Groves the real State Department was late to learn that the government was working on a revolutionary new weapon that was destined to have profound consequences for diplomatic affairs. In January 1945 Secretary of State Edward Stettinius was informed of the atomic bomb project. This was done in preparation for the Yalta conference and because Groves needed the assistance of the department. Throughout all of these diplomatic activities Groves intended to serve American interests: Everything he did was legal, and every penny was accounted for. But by the same token, his strong views and arrogant manner caused controversy. After the war, they would help speed his retirement and diminish his reputation.

In the early part of the twentieth century uranium was used as a pigment for tinting glass and ceramics and was not the prime goal of the mining extraction. What was considered of value was radium. One European source was the mines at St. Joachimsthal in Bohemia, then part of Austria-Hungary.[1] But most of the world's supply of uranium at the time came from a mine at Shinkolobwe in Katanga Province of the Belgian Congo, run by the Belgian firm Union Minière and its director Edgar Sengier. Sengier played a key role in the early days of the bomb program and became a close professional associate and later a personal friend of Groves. They held each other in high esteem and worked closely together until Groves retired.[2] Sengier had access to the best-known supply of uranium in the world and was Groves's equal in his desire to keep all of these transactions secret.

After the Nazis struck westward in the spring of 1940, occupying Belgium, Sengier feared that the Congo might be invaded as well, and he ordered that an extremely rich stock of uranium from the Shinkolobwe mine be transported to New York via Portuguese Angola. Two shipments were made from Lobito in September and October 1940. The first, amounting to 470 metric tons, arrived on November 10, and the second, of 669 metric tons, arrived on December 19. For almost two years the material was stored in two thousand steel drums at warehouses of the Archer-Daniels Midland Company at Port Richmond, Staten Island, plainly marked URANIUM ORE and PRODUCT OF BELGIAN CONGO. In March 1942, and then again in April, Sengier informed Thomas K. Finletter, special assistant to the secretary of state, Division of Defense Metals, of the fact that he possessed the valuable ore and that it was stored in New York City. Somewhat surprised at the lack of interest by the American government, Sengier heard nothing more about it until Colonel Nichols called upon him on September 18.

As we saw, one of Groves's first decisions, within hours of being chosen to run the Manhattan Project, was to send Nichols to New York to see Sengier at his office about purchasing the ore. Ten days previously, on September 7, Nichols had first learned, in a telephone call from Finletter, that there was a supply of uranium already in the United States, though he did not know the quantity or quality of the ore. On September 18 Nichols went to New York to Sengier's office at 25 Broadway and worked out conditions to buy the ore on Staten Island, as well as another three thousand tons to be shipped from Africa to the United States.[3]

While this fortuitous windfall gave Groves a good head start, the issue of uranium supply was always present. There was uncertainty as to how much ore was going to be required. Adopting his normal conservative position, Groves concluded that the more of it there was, the better, and he set about trying to acquire as much of it as possible, as quickly as possible. The attempt to control the supply of uranium would soon come to have another purpose. In Groves's view monopolizing the supply could prevent, or at least delay, other countries from obtaining the bomb — an early, if unrealistic, attempt at nonproliferation. The effort ran aground on the fact that uranium was not scarce in nature, but it took some time for this realization to be understood and accepted.

The Manhattan Engineer District relied upon three sources of uranium during the war years. About two-thirds of the almost six thousand tons it accumulated came from the Belgian Congo (thirty-seven hundred tons). Slightly more than one-sixth, or eleven hundred tons, came from mines near Great Bear Lake in Canada.[4] The remaining eight hundred tons came from American ores, which, in reality, were the tailings from vanadium refinery operations.[5] The MED used three primary storage areas to store the ore: Senaca Ordnance Depot, Romulus, New York; Clinton Engineer Works, Tennessee; and Perry Warehouse, Middlesex, New Jersey, which in November 1943 also became a sampling, weighing, and assaying facility. The bulk of the Belgian Congo uranium ores and other ores were handled at Middlesex.

Throughout the first year of his tenure Groves unilaterally tried to control the world's uranium supply. This effort was to take a dramatic turn in the summer of 1943 through the collaboration with the British as a result of the Quebec Agreement.

In the months following his selection as head of the project, and up until January 1943, Groves, with Vannevar Bush and James Conant providing direction and support, effectively ended whatever collaboration there had been with the British in the atomic energy field. As Groves's compartmentalization policy took hold, the British were effectively cut off from most kinds of information.[6] As a recent scholar has described it, "Bush and Conant passed the sword to General Groves, who in turn resisted collaboration with a vehement passion that both men never came close to matching."[7]

Bush and Conant had long been wary of British motives and were fearful that the British were angling for a favorable postwar commercial/industrial position regardless of the fact that Americans in America had developed many of the innovations. There was also the question of the depth of Britain's wartime commitment to the bomb, as well as the issue of its "postwar strategic significance," to use Conant's words.[8] Groves was more than wary. He felt that any interchange with the British would divert resources from his primary mission of quickly completing the bomb, jeopardize security, and be one-sided.[9] Had he had his way, there would have been no agreement of any kind and no cooperation. Groves's deep-seated Anglophobia only intensified the other reasons he had to be suspicious about Britain's intentions.

Throughout the first half of 1943 the British sought to reverse direction on the issue of collaboration. They felt that the schism could only be overcome by Churchill's intervention with Roosevelt. The first opportunity was the Casablanca conference in January 1943, but more pressing matters crowded out discussion by the two leaders, and the issue went unresolved for months.

Eventually the British were successful: Roosevelt went against the advice of his advisers on atomic matters and chose to renew the full exchange of information with them, in the interest of strengthening the Anglo-American alliance to better fight and win the war.[10] Through much debate and discussion by officials and advisers on both sides, an agenda was worked out that led Roosevelt and Churchill to sign the three-page Quebec Agreement on August 19, 1943.[11] Only a few senior officials on each side of the Atlantic knew of the highly secret arrangement. For several years after the war the existence of the agreement was not known either to the American Congress or to the British Parliament. The terms of the agreement remained secret until April 1954 when they were disclosed by consent of the two governments.[12]

As with any agreement or decision of government policy, especially one containing language allowing for interpretation, the ways in which it was implemented largely determined the results and impact. In the case of the Quebec Agreement it was primarily General Groves who would be responsible for its implementation, which did not bode well for the British. Groves was most skillful at manipulating the British into thinking that he favored collaboration during the war and afterward, when in fact he did not.

If the British thought that they were going to have a free interchange of information on a full range of topics, they were about to learn otherwise. Key passages in the agreement itself provided just the kind of language Groves needed to withhold certain kinds of information from them. For example, the agreement distinguished between two kinds of information:

> In the field of scientific research and development there shall be full and effective interchange of information and ideas between those in the two countries engaged in the same sections of the field.

But when it came to the technical know-how, the nuts and bolts engineering that transformed scientific theory into the actual fissile material, there was another procedure, one that could be effectively controlled by Groves:

> In the field of design, construction and operation of large-scale
> plants, interchange of information and ideas shall be regulated by
> such *ad hoc* arrangements as may, in each section of the field, appear
> to be necessary or desirable if the project is to be brought to fruition
> at the earliest moment.

Needless to say, Groves had "such *ad hoc* arrangements" at the ready to
serve his purposes. One of his first steps was to share with the British a
progress report he had just submitted to President Roosevelt. The British
were staggered at the progress that had been made by the Americans and
the scale of the effort. The condescending British attitude toward the
American program between 1940 and 1942 suffered a rude shock. The
British had naively believed their project to be superior in those early
years. With the Manhattan locomotive gaining full speed it was evident to
everyone that there was really no need for British help at all. This was not
going to be a partnership of equals; still, to uphold good Allied relations
committees were formed and scientific teams sent.

The vehicle to implement the agreement was the Combined Policy
Committee (CPC). The American members were Stimson, Bush, and
Conant. Field Marshal Sir John Dill and Col. J. J. Llewellin were the orig-
inal British members, and Clarence D. Howe, minister of Munitions and
Supply, was the sole Canadian. Richard Tolman, James Chadwick, and C. J.
Mackenzie, president of the National Research Council, were the scientific
advisers. At the first meeting of the CPC, held on September 8, 1943,
Groves adopted what would be his attitude throughout; as an official his-
torian later wrote, he was "carefully and unenthusiastically correct."[13]

Though Roosevelt and Churchill had agreed that a contingent of
British scientists would come to America to work on the bomb, where
they worked, and thus what they were exposed to, was largely up to Groves
and the American side. By design, in his implementation of the agreement
Groves saw to it that no British scientists worked on the plutonium project
at Hanford. The contingent that came to the United States worked at Los
Alamos, Berkeley, Oak Ridge, New York, and Washington, mainly on isotope separation, a field in which they already had experience.[14]

The scientists in the British Mission, not all of them British, presented
Groves with a security problem that he took more seriously than the
British. When Groves was first presented with the list of the scientists who
would be coming, he insisted on doing a security check on all of them.
The British were deeply offended, and it was waived. In the end Groves

was forced to settle for a British assurance that the scientists' backgrounds had been carefully checked by British intelligence. One member of that delegation, of course, was Klaus Fuchs.

Much later, in a letter to Margaret Gowing, the official historian of the British program, Groves brought up some of the details of this matter. He asked about a statement in her book that a "qualified clearance was given to Klaus Fuchs." At the time Groves understood that the British Mission had been given an unqualified clearance. Groves had refused clearance until he got a definite statement as to their security status. Dr. W. L. Webster, of the British Central Scientific Office in Washington, initially supplied a list of names, which Groves rejected as inadequate. Webster then wrote that they were all reliable people; Groves rejected this as well. Finally Webster wrote to the effect that "Each man has been investigated by M-5 to the same extent that he would have been investigated if he were an American who was to have access to the same information." This statement was clearly false. The matter remains unresolved to this day. Did Webster lie or did someone else?[15]

One condition of the Quebec Agreement was that only British subjects would be allowed to come to the United States. This complicated matters for James Chadwick, who was responsible for assembling the team and was the head of the mission.[16] Several prospective members were not British subjects, among them Joseph Rotblat and Otto Frisch. While Frisch easily acceded to becoming a British subject, Rotblat, a Pole, refused to give up his citizenship. Chadwick appealed to Groves, giving him his word about Rotblat's integrity. Groves accepted it, and Rotblat came to Los Alamos without changing his nationality.

Much of the Anglo-American cooperation was filtered through the relationship of Groves and Chadwick. Chadwick was able to partially overcome Groves's ingrained Anglophobia and natural suspicions about the British. Groves saw in Chadwick a senior Nobel Prize–winning scientist who could argue calmly with his own best scientists. Chadwick also shared Groves's single-minded purpose of completing a bomb as quickly as possible. Chadwick was conciliatory and pragmatic, and, unlike his predecessor Akers, tried to assist in enforcing Groves's rigid security procedures, rather than evading them. Chadwick knew his limits with Groves and did not go beyond them. What Chadwick got out of it was limited participation in the American program, which resulted in invaluable experience that would be drawn upon to advance the British program after the war.

Chadwick and his wife, Aileen, moved to Los Alamos in early 1944 and lived in one of the better residences, a two-bedroom log cabin (now known as Baker House), chosen for them by General Groves, next to Fuller Lodge. When Joseph Rotblat arrived the following month, he lived in the second bedroom. On visits to Los Alamos General Groves would normally see Chadwick. On one occasion in March he came to dinner and, according to Rotblat, made the remark that the real purpose of building the bomb was to subdue the Soviets. In a later interview Rotblat recalled Groves saying, "'You realize of course that all this effort is really intended to subdue the Russkies.'"[17] The comment shocked Rotblat. Rotblat had few illusions about Stalin but felt that the sacrifices of the wartime ally deserved some respect.

As for Groves, he clearly had no love for the Russians and was as careful as he could be about preventing Communist espionage at the Radiation Lab and elsewhere. He knew the Russians were targeting the project, and he did his best to prevent it. As for Alsos, its purpose was as much to keep the documents and the scientists out of Russian hands as to capture them ourselves. Groves often said what he thought in a brusque and blunt manner. Subtlety was not his way. He was certainly aware that this bomb was going to have influence beyond its primary purpose of ending the war and saving American lives. Anyone who did not grasp this was quite naive.

The third member of the partnership was Canada. Prior to the Quebec Agreement Britain had established a laboratory in Montreal.[18] On November 10, 1942, a list of twenty-six scientists and technicians was submitted to Canadian immigration officials. Eighteen were British citizens; the remainder were refugees. In the former group was Alan Nunn May, and in the latter Bruno Pontecorvo. Both were later revealed to be Soviet spies. Hans von Halban, the Austrian-born scientist who had escaped from Paris to England with the heavy water, was selected as the director of research and was replaced in April 1944 by John D. Cockcroft. The main research focus of the Montreal laboratory was the use of heavy water to slow neutrons. A contingent of Canadian scientists joined them.

The Quebec Agreement did not basically change the relationship of the Anglo-Canadian laboratory at Montreal to the Manhattan Project. Before and after, Groves kept them at arm's length, although some modest exchanges took place.[19]

In the eyes of Bush, Conant, and Groves, the Montreal Laboratory was a symbol of all that was problematic with Allied atomic collaboration:

it did not advance the making of the bomb; it exploited American research and development for postwar military and commercial advantage; and the loyalty of some of its scientists, notably the "Free French" contingent, was suspect.[20]

Groves had serious doubts about certain French members, specifically Hans von Halban and Lew Kowarski. Halban and Kowarski had been colleagues of Joliot-Curie and had been on the cutting edge of fission research in the late 1930s. After the German invasion in the spring of 1940, they escaped from France to Britain with most of the world's supply of heavy water. Later they and several others were sent to Montreal and were a part of a British-Canadian-French group that designed the Chalk River heavy-water reactor, 150 miles northwest of Ottawa, built under the auspices of the Manhattan Project.

The Normandy invasion and the liberation of France in the summer and fall of 1944, as welcome as they were, presented Groves with a series of diplomatic and security problems that demanded his attention. The most immediate of them was what to do about five French nuclear scientists — von Halban, Kowarski, Pierre Auger, Jules Gueron, and Bertrand Goldschmidt — working on the Canadian program.[21] While Groves was content to allow them to get to Canada, he absolutely refused to let them enter the United States or have direct contact with American scientists. Keeping them in Canada was a way of maintaining control over them with a minimum risk to security.[22] No doubt some of them had probably discovered certain facts that would be of interest to America's enemies. With France now free they wanted to visit their country or to return home for good. For Groves this was too great a security risk. The war was not yet over, and their colleague Joliot, who had remained in Paris during the occupation, was a member of the Communist Party. At least one of the five would surely convey to Joliot crucial information about the American program, and he would waste no time in passing it on to Moscow.

But Groves's ability to control the French was limited. In May 1944 Pierre Auger had terminated his employment with Montreal, giving as his reason that he wanted to return to France and assist Joliot in rebuilding French science. Groves and the British agreed that neither Auger nor the others should be permitted to return to France. But the agreement came undone. When Auger did go to London in August 1944, the British reversed themselves and permitted him to go on to France.

In October 1944 Gueron requested permission to visit France on personal business. Groves objected strongly to the visit, but the British once again were supportive of Gueron. To Groves all of this appeared to be a clear violation of the Quebec Agreement, which forbade communicating "any information about Tube Alloys to third parties except by mutual consent" of the United States and United Kingdom. Groves represented the United States and he was not giving his consent.

Why were the British so seemingly lax when it came to the French? The American ambassador in London, John G. Winant, was asked to find out from the British, and in October he received a reply from Sir John Anderson. The basis of this relationship, Sir John explained to Winant, rested upon agreements that the British had negotiated with the French scientists before they went to Canada. Von Halban and Kowarski had negotiated an agreement assigning to the British their patent rights in exchange for Britain offering the French research center certain patent rights. Furthermore, when Auger, Gueron, and Goldschmidt had reached England they had worked out an arrangement with the British whereby they could return to France as soon as it was feasible. The British also had permitted them to retain their status as French civil servants and as adherents of General de Gaulle and the Free French.

These revelations by Sir John came as a shock to Groves and to other American officials.[23] The Americans had known nothing about such British-French agreements. Remarkably, Sir John had not bothered to bring them up while the Quebec Agreement was being negotiated. In Groves's view the former agreements were in direct contradiction of the latter ones. For Groves this was typical of the British — yet another example of their perfidy. But it was about to get worse.

Reluctantly, Groves consented to a British request in November 1944 to permit von Halban to visit London, with the understanding that he would not be allowed to go to France. No sooner had he arrived than Sir John went to Winant to ask that von Halban be permitted to see Joliot in France. Sir John said that the scientist was an honorable man who could be trusted.

Winant was in a difficult position. He asked Groves to come to London to talk to Anderson, but the general was much too busy with all of his many responsibilities. Under continuing pressure from Anderson, Winant finally consented and von Halban went to Paris. Before departing the British provided the French scientist with an agenda limiting the topics he could talk about with Joliot. Subsequently, Groves's intelligence agents learned, upon questioning von Halban after he returned, that he had ventured far beyond the agenda and provided Joliot with a great deal of vital information about the American program. They learned as well that sometime in November Joliot had discussed the topic of atomic energy with General de Gaulle.

From Winant Groves learned of von Halban's visit to Joliot on December 3, and was surprised to learn that he had been in Paris since November 24. Groves was furious. It was another example of the British attempting to take postwar commercial advantage at the expense of the Americans, to say nothing of the potential security dangers. Quebec Agreement or not, Groves felt he had to take assertive action. On December 14 he wrote to Stimson stating that "pending the receipt of instructions from you, I will take steps to safeguard the security of the DSM project by delaying insofar as practicable the passing of vital information concerning it to the representatives of any government other than our own."[24] Stimson called Groves to his office the same day, and, in a meeting with Harrison and Bundy, decided to take the matter up with President Roosevelt. Stimson had Groves prepare a complete account of the French situation and notified Winant to refer to Washington any future requests by the British for disclosures to the French.

Groves prepared a memo for the secretary, and the two went to see Roosevelt on December 30 — Groves's only meeting with the president during the war. Stimson and Groves explained the situation to the president, who asked what the French were after. They responded that they thought France wanted full partnership in a tripartite agreement. FDR dismissed this and the discussion turned to other matters, among them whether the bomb was to be used against Germany if the Battle of the Bulge went the wrong way.[25] Before leaving Groves showed the president his latest memo to General Marshall, stating that atomic bombs would be ready for use against Japan in six months.

Groves left the meeting convinced that he had the president's strong support to frustrate Sir John's efforts to bring France into the partnership and to prevent further disclosures of nuclear secrets to the French. From this point until the end of the war, the problem was contained. Groves said that his

> sole source of satisfaction in this affair came from a remark made by Joliot to an employee of the United States Embassy in Paris: while the British had always been most cordial to him and had given him much information, he said, he got virtually nothing from the Americans he encountered.[26]

Groves, Bush, and Conant all had felt that the agreement with Britain had been thrust upon them from above, which of course it had. With no choice but to collaborate in some fashion, Groves decided to take advantage of the situation and use the British to serve his ends.

At Groves's urging the CPC placed high on its agenda the identification and procurement of mineral lodes containing significant quantities of fissionable materials, specifically uranium and thorium. In order to carry out this task, the CPC created the Combined Development Trust (CDT) — a financial trust in the classic sense of the word.

Two legal problems had to be surmounted before the trust could be established. The first concerned the status of the organization under U.S. law. If the organization were established as a corporation, it would have to make its transactions public. This of course would be impossible. To maintain the necessary secrecy a common-law trust was suggested.

The second problem concerned whether the president had the authority to enter into this type of agreement. General Groves ordered three lawyers on his staff — Lt. Col. John Lansdale, Maj. William A. Consodine, and Pvt. Joseph Volpe — to research the legalities of the issue and provide an opinion. Secretary Stimson did the same with lawyers in the War Department. Also overseeing the diplomatic and legal arrangements was Brig. Gen. Edward C. Betts, the judge advocate general of the European theater of operations. All agreed that the agreement was within the powers of the president and that it should be in the form of a trust. A common-law trust was an ideal device to avoid the reporting requirements that would be required with an ordinary corporation.

Negotiations were carried out through the American ambassador to the Court of St. James, John G. Winant, and Sir John Anderson and carefully monitored in Washington by Secretary Stimson with help from Bundy and Groves. Various drafts were sent back and forth to London, where William Gorrell-Barnes, counsel to the Exchequer, handled them. Groves's aide Maj. Harry S. Traynor served as emissary and liaison.[27] Groves, with Stimson's approval, did not want the State Department involved or even knowledgeable about the matter.

By early June 1944 all of the problems were solved and negotiations complete. The agreement, signed by Churchill, and then by Roosevelt on June 13, established the Combined Development Trust to supervise the acquisition of raw materials in "certain areas" outside of American and British territory. Appointments to serve on the CDT were made, and approved at the September 19 meeting of the Combined Policy Committee.[28]

The chairman of the Combined Development Trust was General Groves. Representing the United States were George L. Harrison, Stimson's special assistant, and Dr. Charles K. Leith, a distinguished mining engineer. Representing the United Kingdom was Sir Charles Hambro, the deputy chairman, and Frank G. Lee from the British Treasury. George C. Bateman, a deputy minister, mining engineer, and friend of C. D. Howe, represented Canada. Groves's trusted aide Major Consodine served as one of the two joint secretaries.

Over the next two years the trust carried on extensive worldwide exploratory geological surveys and purchased the mineral rights to mine sources of ore — or, failing that, purchased the ore outright. Not surprisingly, Groves, as chairman, tended to dominate the trust's activities, as can be seen in the minutes of the meetings.[29]

The previous fall Groves had sent Capt. Phillip L. Merritt, a geologist, to the Congo to search for other sources of uranium. He found that even the tailings from Shinkolobwe were of value, varying in content from 3 to 20 percent. Groves pressured Sengier to reopen the mine to extract higher-grade ores. Now with the Combined Policy Committee as a lever to pressure Sengier, and the resources of the CDT at hand, negotiations with African Metals Corporation resumed. On June 30 Groves and Hambro met with Sengier to arrange for purchase of the rich Congo ore.

The contract between CDT and the African Metals Corporation was signed on September 25, 1944. It called for delivery of 3,440,000 pounds (1,720 tons) of uranium oxide. More would be needed, and further contracts were signed on October 27, 1945. The contracts involved only the recoverable uranium oxide. African Metals maintained ownership of the residue or tailings, which contained radium and other precious metals.

It was agreed by the Combined Policy Committee that the CDT would be initially capitalized with twenty-five million dollars, half from each country. The American share of $12.5 million was overseen by Groves. One problem was how to have ready access to the money, without revealing the purposes to which it would be put. Groves set the lawyers in his office to work on a solution and presented his plan to the CPC on September 19. The CPC endorsed it, and Groves set about putting it into effect.

The essential feature of Groves's plan was to establish a special fund, to be deposited with the secretary of the Treasury, from which he, or other designated American members of the CDT, could draw money as needed without further authorization.

On September 21 Undersecretary of War Patterson authorized a Treasury check to be made out to General Groves in the amount of $12,500,000. By the time Groves received the check, dated September 28, his legal staff had discovered that under existing law funds deposited with the secretary of Treasury were subject to handling and processing by many employees in the Treasury Department and the General Accounting Office. This would not do. Groves did not want even Secretary of the Treasury Henry Morganthau to know about any of this, much less his subordinates. Another solution was investigated whereby the money would be deposited directly in the Federal Reserve Bank in New York City or in a private banking institution. After further consultation with War Department lawyers, with Stimson, and with Harrison, his fellow trustee on the CDT, Groves concluded that these procedures would also require permission of the Treasury secretary.

On October 17 Groves and Harrison met in Stimson's office to try to resolve the CDT funds quandary. There appeared to be no legal way around the requirement that Secretary Morganthau give his approval to depositing the $12.5 million. Stimson was fearful that Morganthau would want full knowledge of the atomic bomb project before giving his consent, and Stimson was not about to provide it. He was bound to silence by orders of the president; without Roosevelt's permission he could not even tell a fellow cabinet member. Nevertheless, it was decided that Stimson should try to get Morganthau's permission without informing him of what it was about. Stimson called Morganthau, and an unpleasant exchange ensued. As Stimson later wrote in his diary,

> He at once told me that unless he was fit to be trusted with the secret in such a case he ought to be removed from the position of Secretary of the Treasury and he could not give his consent without knowing it. . . . I had no authorization from the President to communicate it further. He still withheld his assent and I closed the conversation. It made a rather painful impression on me and rather impaired the rest of my morning.[30]

A few days later Morganthau changed his mind.

> On Saturday morning Morganthau talked to me on the telephone from his farm at Beacon, New York, offering to let me make a deposit, about which I talked to him the other day, in the Treasury in

a secret account, the purpose of which will not be known to anyone in the Treasury or elsewhere. He evidently had been thinking over our telephone talk and this was an olive branch which I accepted with pleasure. It will solve that situation in a very satisfactory way.[31]

With the problem now solved Stimson made an appointment for Groves to go see the secretary in person. On the morning of October 26 Groves visited Morganthau to arrange for the $12.5 million to be put into one of his special secret accounts. Nearly a year later, just after the end of the war, an additional $25 million was deposited in the account, bringing the total to $37.5 million.[32]

In order to protect himself, Groves made sure that there was a long and deep paper trail documenting the creation of this account and the ongoing contact with Treasury Department officials.[33] Periodically Groves would withdraw several million dollars, in cash, from this account. One of his secretaries, Gerry Elliott, has reported being called into the office and seeing a table covered with piles of thousand-dollar bills. She was instructed to take this money, accompanied by a guard, for deposit in accounts in Groves's name at one of two banks.[34] A second secretary, Anne Wilson, was once called into Groves's office and told that the table full of cash represented a million dollars.[35] Almost all of the detailed records that would show how the money actually was used have been destroyed, but it is clear that Groves was using it to buy ore and mining rights in foreign countries for the CDT. Buying ore and mining rights in the United States presented another problem. Groves had the Union Carbide Company set up a wholly owned subsidiary called Union Mines Development Corporation to be the vehicle to buy domestic ores. In one case Groves purchased stock in a Canadian mining company using Andrew B. Crummy, who, before and after the war, was a law partner of one of Groves's most trusted aides, Lt. Col. William A. Consodine.

One of the commercial banks that Groves used for his transactions was the Fidelity Union Trust Company of Newark, New Jersey. This contact came through Consodine, who was from Newark, and his law partner Andrew Crummy. Crummy was also a director on the bank's board. Consodine wrote a memo to Chairman of the Board Horace K. Corbin introducing Groves, probably in 1945 or 1946. Though the details of all of these transactions are unclear, Consodine noted that Groves was a stockholder in the bank, and that "during the project, through the good graces of Mr. Crummy and Mr. [Roy F.] Duke [the bank's president], he used your bank on highly secret matters. Mr. Crummy has also acted in other highly secret matters pertaining to the project."[36]

From mid-1944 until he retired from the army in early 1948, Groves spent a significant amount of time on CDT efforts to gain control of the world's major sources of uranium. Groves was successful at keeping the Treasury Department mostly in the dark. He also wanted to ensure that the State Department did not know what was going on. His tricky diplomatic problem was how to negotiate agreements with foreign governments on behalf of the United States that did not require informing Secretary of State Cordell Hull, or his successor Edward Stettinius, who assumed office on November 27, 1944. Initially President Roosevelt assisted in this with the negotiations with the British over the CDT. Roosevelt directed his ambassador to the Court of St. James, John G. Winant, to act as his personal representative in any matters having to do with the trust, General Groves, or his representatives. Under no circumstances was Winant to act in his capacity as ambassador and report any of these activities to the secretary of state or to his representatives. Groves had Lansdale draft a legal memorandum to assure him that the executive branch had sufficient authority to negotiate agreements with other nations without resorting to the treaty process.[37]

In this way secret negotiations for ore rights were undertaken with the governments of Sweden, Spain, Portugal, Bolivia, and Brazil without the knowledge of the State Department. Secretary of State Stettinius was brought into the picture only in January 1945, in part because Groves needed his assistance.[38]

There was a second reason for informing the new secretary, which had to do with the upcoming Yalta conference and concerns about President Roosevelt's health. Groves had briefed the president at the end of December and had some doubts about Roosevelt's physical and mental condition as the Big Three conference in the Crimea drew near. To guard against Churchill swaying the president into some proposal that would not be in the best interests of the United States, as Groves saw it, he sent Lieutenant Colonel Consodine to Malta, which was the rear headquarters for Yalta, to be available in case any atomic matters arose and clarification was needed.[39] Consodine and Stettinius met on a British warship in Malta. In the course of their conversation the secretary was provided with further details about the project, and of the MED's suspicions of the Russians. Groves did not want to send Consodine to Yalta because the Russians would obviously wonder who he was — and the trail might lead back to Groves.

Groves displayed his characteristic urgency and determination in this area, as in all others. He felt that no time could be wasted; everything must be done to gain control of the uranium and thorium stocks as soon as possible. A memo to Bundy is typical.

> "We must not delay," he wrote as negotiations began with Brazil about thorium stocks. "There is no time to lose. . . . I can see nothing to be gained by delay and on the other hand I can foresee grave danger of complete disaster by temporizing and delay."[40]

As Groves prepared to retire, he made sure that he would be absolved of all responsibility should it not be possible to account for all of this money. His diary entry for December 3, 1947, read:

> I saw Mr. Sims this morning of the Bankers Trust Company and delivered a letter and check in the amount of $2,635,823.11 drawn in favor of the Bankers Trust and he in turn gave me a certified check in the same amount drawn in favor of the Sec of the Army.

On December 11 Groves saw Secretary Royall and turned over the check, and Patterson gave it to General Richards, the army's budget officer. Every penny, down to the last eleven cents, was accounted for.

Several years later, when the espionage activities of Harry Dexter White came to light, Groves was concerned that he may have known about the secret funds for the CDT. White had joined the Treasury Department in 1934 and was a source for Whittaker Chambers beginning in 1937.[41] No one had greater influence on Secretary Morganthau's thinking than White, who rose to be his assistant in 1941 and became assistant secretary of the Treasury in 1945. White was sympathetic to the Soviet Union, though not a party member. But he was one of the KGB's most valuable assets and passed valuable information to it for years. Groves feared that he may have known about the special fund and informed Moscow.

Still using his inside contacts, Groves had his old aide and troubleshooter Gavin Hadden, then working at the Armed Forces Special Weapons Project, check into whether White could have known about the funds. Hadden concluded that he did not.[42]

In November 1953, when further information about White was revealed, Groves wrote to Edward F. Bartelt in the Treasury Department to ask whether White or any of his associates had had any official or other knowledge of the special transactions he had had with him from 1944 to 1947.[43] Groves had the letter hand-delivered by his son Richard and asked Bartelt "to keep my present interest in all this confidential as I do not believe it would be in the national interests to have it talked about."

When the war was over Groves's actions necessarily had to become known to other parts of the U.S. government. The unenviable task of informing Undersecretary of State Dean Acheson fell to Joseph Volpe. Acheson was "absolutely flabbergasted," Volpe remembered almost forty years later, as he explained the situation. "Why, you sons of bitches set up your own state department," Acheson charged.[44]

David Lilienthal entered in his journal that Acheson complained to him that

> The War Department, and really one man in the War Department, General Groves, has, by the power of veto on the ground of "military security," really been determining and almost running foreign policy. He has entered into contracts involving other countries (Belgium and their Congo deposits of uranium, for example) without even the knowledge of the Department of State.[45]

Clearly Groves's supreme self-confidence, useful in some areas during the building of the bomb, blinded him in this instance. His views were based on arrogance and chauvinism about superior Western science and technology. His already strongly held opinions were sometimes impervious to contrary viewpoints. Several scientists had accurately predicted that the Russians would be able to build their own bomb in a matter of years, and others saw that the attempt to monopolize the raw materials was doomed to failure.

For all of Groves's concerns about security, his safeguards were not airtight. The Soviet Union did know about the Combined Development Trust, not through Harry White, but through its well-placed spy in Washington, Donald Maclean.

Donald Maclean was one of the notorious Cambridge spies, which included Guy Burgess and Kim Philby. In terms of the information he provided to the Soviets, he was perhaps their most valuable agent. This is especially true of his time in America.[46] Donald Maclean served at the British Embassy in Washington from May 1944 to September 1948. In April 1945 he was appointed first secretary and from February 1947 to September 1948 was the British secretary of the Combined Policy Committee. As Britain's man in Washington who kept abreast of atomic energy matters, Maclean had extraordinary access to and knowledge about highly secret matters. He had visited AEC headquarters at least twenty times, attended conferences, and knew important details about a range of American policies.

For example, on January 8, 1947, Groves met in his office with Mr. Makins, Mr. Donald Maclean, Dr. Leith, and Colonel John R. Jannerone, who replaced Consodine as secretary of the CDT, to hear of Mr. Anton Gray's recent inspection trip to the Shinkolobwe mine in the Belgian Congo. Gray also stopped in Belgium to see Mr. Sengier and to learn his plans for who would succeed him at Union Minière. No doubt all of this information was passed on to the Soviet Union by Maclean.

Recently it has come to light that the Soviets had another source of information about atomic intelligence matters. Melita Stedman Norwood, a member of the British Communist Party, worked as a secretary in the research department of the British Non-Ferrous Metals Association beginning in 1932. She was recruited by Soviet intelligence in 1937. The exact kinds of information that Norwood passed on remain unclear, but a recent assessment of her value to Moscow concludes that Norwood was "in all probability, both the most important British female agent in KGB history and the longest-serving of all Soviet spies in Britain."[47] In March 1945, after British Non-Ferrous Metals won a contract from the Tube Alloys project, Norwood gained access to documents of atomic intelligence that KGB records described as "of great interest and a valuable contribution to the development of work in this field."[48] With Maclean's information about Groves's efforts to control uranium and Norwood's material to supplement it, Moscow must have had a pretty good idea of what was going on.

Groves's attitude toward the British contribution was perhaps most candidly reflected in a document he had prepared just prior to his retirement titled *The Diplomatic History of the Manhattan Project.*[49] The document was prepared at the request of Undersecretary of State Robert Lovett to inform the department of what had transpired in the area of foreign relations and atomic energy up to January 1, 1947. Jannarone wrote the report itself, but it clearly reflected Groves's views; he oversaw its preparation and editing. More than 150 pages of letters and memos that constitute the central elements of the story supplemented the forty-five-page report.

In the concluding section Jannerone/Groves described a British contribution of modest proportions. Their most important contribution was encouragement and support at the highest official levels. They furnished a small number of scientific workers, a few of their top men, who were helpful in many cases, "but it became quite evident that their primary purpose was to follow our work and to obtain information from us." Anything that they did could have been done by Americans. "The scientific and technical information obtained in the United Kingdom laboratories was of so little value to us as to be practically negligible."

> In general, the scientific contribution of the British was far less than is generally believed. It was in no sense vital and actually not even important. To evaluate it quantitatively at one per cent of the total would be to over estimate it. The technical and engineering contribution was practically nil. Certainly it is true that without any contribution at all from the British, the date of our final success need not have been delayed by a single day.

CHAPTER SEVENTEEN

The Groves Family During the War

When Groves took control of the Manhattan Project in September 1942, the only visible sign of a change in his professional life was that he, Mrs. O'Leary, and their small staff now occupied a different set of offices in the New War Department Building. Groves continued to maintain the same hectic schedule. He traveled extensively. A steady stream of contractors and army personnel visited the office. Groves was still deputy chief of construction, the title he maintained during the war, and he continued to oversee some of the construction contracts that had not been completed at the time he began his new job, and to participate in functions as if there were nothing unusual. For example, at the end of January 1945 Groves, his wife, and his daughter traveled to Brunswick, Georgia, where the eightieth Liberty ship, the SS *Harold Dossett,* built by the J. A. Jones Construction Company, was launched at the shipbuilding plant, with Mrs. Groves as sponsor. Groves was able to have a couple of days' rest with his family at nearby Sea Island, Georgia.[1]

Other pre-Manhattan duties needed attention as well, especially some connected with the Pentagon. With Gavin Hadden he oversaw completion of an an athletic center with a gymnasium, squash courts, bowling alley, and exercise room. The center's original purpose, to provide athletic facilities for Pentagon officers, was postponed until after the end of the war while it served as a women's dormitory. Hadden was one of Groves's closest assistants and troubleshooters. He had graduated from Harvard in the class of 1910 and did advanced studies at Columbia.[2] During World War I he joined the army and was sent to France as an engineer. After the war he established a civil engineering firm in New York with a specialty in athletic facilities and playing fields. His knowledge of how to use steel and reinforced concrete was valuable when he joined the Construction Division of the quartermaster corps as a civilian in 1941, during the mobilization period, and came to know Groves. The two worked closely together for the next seven years until the general retired. For several years after his retirement Groves would ask Hadden to investigate a rumor or research a fact. Hadden was an excellent writer, drafting many of Groves's letters; under Groves's supervision he also wrote a comprehensive history of the Manhattan Project. Hadden's son John was a West Point classmate of the general's son.

For all intents and purposes, the fact that Leslie R. Groves headed the Manhattan Project escaped the notice of his professional colleagues, but it took considerable effort to maintain the fiction. Within the family, however, little effort was required. No matter how preoccupied Groves was with his work, he never carried it home and never talked about what he did with family, friends, or colleagues. During the Manhattan years his wife or children would occasionally come to the office. They would chat with Mrs. O'Leary and others, but they never asked, and no one ever hinted at, what was going on. It was a secret. When it was announced that an atomic bomb had been dropped on Hiroshima the most surprised person in Washington was no doubt Mrs. Leslie Groves.

The fact that Groves's responsibilities during the war required him to be away from home for days or even weeks at a time was not unusual. That had been the case ever since July 1940 when he had been detailed to be General Gregory's special assistant. Even when not traveling, Groves often stayed at the office and missed the evening meal. None of this meant that he was any more or any less attentive to family business than he had ever been. Even with members of his extended family — aunts, uncles, and cousins — he managed a short letter now and then. He wrote frequently to his sister Gwen, and kept in close touch with his brother Owen, now teaching in the English department at Adelphi University on Long Island. Groves's relationship with Owen's wife, Marion, was generally difficult. Groves was often quite outspoken at home about his politics and his views on society and its ailments. Those with opposing views, if present, were best advised to hold their peace. Marion, the family "Socialist," was feisty and outspoken in expressing her views. Groves sometimes referred to her as a "pinko." He did what he could to avoid her, but when they were together it made for a tense situation.

Recently declassified FBI documents reveal that Groves told the bureau about his concerns over Owen and Marion. In an April 23, 1947, memo the special agent wrote to Hoover that Groves "had observed that his brother and to a greater extent [deleted] have appeared to follow the Communist Party line pretty closely in their thinking."[3] The agent recommended a discreet investigation. A later memo stated,

In January, 1948, it further was noted that Groves had stated that college professors in general necessarily associated with individuals of
leftist tendencies. Groves' brother was a college professor and Groves
felt [deleted] wife appeared to follow the Party line in 1946.
Investigation by the New York Office did not develop any indication
of subversive tendencies by Groves' brother [deleted].[4]

For all of Groves's self-discipline there was one area where he did not win
the battle, and that was his persistent weight problem. This problem went
back to his childhood. Though never a serious factor in his professional
career, he was once warned that he was overweight and urged to take
action. As a child he often ate great quantities of food with a special com-
pulsion for sweets, especially chocolate confections. This had been a subject of concern to his mother and later of his stepmother. He was always
preparing to go on a diet, on one — of one sort or another — or just finished with one. The diets proved totally ineffective, either because he
cheated or because, after he did lose weight, he stopped adhering to them
and regained the pounds.

He made fun of his problem in occasional letters to family members.
Thanking his sister for the brownies she sent, "I am getting thinner all the
time, although I believe you would still recognize me."[5] And to his son,
"I am continuing to lose weight and am now merely a shadow of my
former self. I have almost reached the point where I can get on the scales
in front of my friends and in a short time will be able to do so in front
of my enemies."[6]

His corpulence could not mask the fact that he was a handsome man.
He looked out on the world with bright, penetrating, seemingly
unblinking, clear blue eyes. By the time he took charge of the Manhattan
Project his wavy dark hair had begun turning a color he later described as
iron gray. He wore a carefully cropped small mustache, and as a stratagem
to mask his ample waistline often wore a one-size-larger uniform, heavily
starched and pressed. Not many were fooled.

Among his papers at the National Archives are detailed diets that were,
at least in theory, intended to be followed. How closely he did follow
them is anyone's guess. There is some indication that he did stick to one
in August and September 1944. Occasionally transcribed with the daily
menus are notations of his weight. On August 7 we find that he weighed
227 pounds. By September 24 he is down to 212. There was a good deal

of tennis interspersed, often preceded with an ounce or two of semisweet chocolate to fortify him. As with most things he did, the menus were meticulously drawn up, down to the last calorie. For this particular diet the average number of calories per day was only 1,020. This regime, which would hardly have kept a bird alive, was routinely punctuated with "2 brownies — 200 calories," and occasionally — as on September 15 — "3 pieces choc cake — 300 calories." These calculations might be a little low, but then to sustain the workload Groves did on a thousand calories a day seems almost impossible.

Groves exerted firm control over the education and comportment of his children. It had been so in his own family — a competitive household where education and learning were valued above all else. The general's son Richard attended half a dozen schools through the ninth grade, in Delaware, Seattle, Nicaragua, Washington, D.C., and Fort Leavenworth. In September 1936 Richard entered Deerfield Academy in Deerfield, Massachusetts, as a sophomore and would spend four years there. His father had systematically gone about deciding what would be the best school for his son. He asked his brother Owen's advice and studied *Fortune* magazine's list of the ten best private schools. The first choice was Deerfield Academy, and after an interview with its legendary headmaster Frank L. Boyden, Richard was accepted.[7]

Richard's first-year grades were not very good due to inadequate preparation for the demanding curriculum, poor study habits, and playing tennis, soccer, hockey, and lacrosse, which took up a good portion of his time. Richard chose these sports rather than football, basketball, and baseball upon the advice of a classmate of his father's at Leavenworth, Capt. George Smythe. Smythe's guidance was a good expression of army doctrine, as it was taught at C&GSS at the time: Change the conditions of the problem so as to assure success. With little chance to succeed in the more popular sports, due to lack of size and speed, Smythe advised Richard to adopt one of the others, then practice and gain experience. When the time came to try out for those teams, he would stand out amid fewer competitors.[8]

Over the next three years Richard improved his grades, all the while being subjected to constant pressure by his father to do better. Even though his academic performance was eventually excellent, it could never be up to his father's standards. There was no satisfying the general. While Richard might receive mild praise (by letter) for one accomplishment, the moment his work slipped at all, his father's wrath, usually wrapped in sharp verbal jibes, arrived without much warmth or mercy, but without fail.

Richard recalls the moment that his grades came after the first semester of his second year. After a less-than-sterling first year, things began to improve.

> With considerable anxiety, I watched DNO open it; I waited for what seemed an interminable time while he read it. Finally, without a word, he handed it to me. It was a mixture of B's and C's — more of the former than the latter — and, for the first time, the head-master's words were both complimentary and encouraging. Although I made steady progress after that, ultimately getting all A's, I was never so pleased by a report — or so relieved — as I was by that one. DNO, apparently shared my feelings, for he smiled and told me that I had done well — one of the very few times I can recall his saying so.[9]

At Deerfield Richard's athletic prowess began to emerge, as Captain Smythe's advice started to pay off. It was clear that he was becoming an excellent lacrosse player. The good coaching at Deerfield, and later at Princeton, prepared him to play on West Point's A Squad (the varsity team) as a plebe. Eventually at the academy he was first-team All-American for the 1944 season — setting an all-time scoring record — and the 1945 season. Army won the national championship both years and some sportswriters touted Richard as the best player in the country. Groves congratulated his son, but with some good-hearted sarcasm added.

> All of us were very much pleased to see it; we were interested more-over in finding out the official title of the position you played. None of us can imagine that "inside home" is quite appropriate to one with your night-owlish tendencies.[10]

For years it had been a foregone conclusion that Richard would follow in his father's footsteps and attend West Point. The process of getting in began early. His father had him take the preliminary West Point physical several times.[11] At the beginning of his last year at Deerfield Richard took an eight-hour exam to compete for an appointment. While he did extremely well, he did not receive an appointment.[12] Upon turning seventeen, the month following his graduation from Deerfield, he was eligible to enter West Point, but the plan was to spend a year at Princeton.

At that time many boys entering West Point had a year or two of college; some even had graduated before embarking upon their four years at the academy.

With West Point his eventual goal Richard went to Sullivan's School for the month of August to increase his chances of passing the academy's entrance exam. Richard entered Princeton in September 1940. Though his freshman studies were interspersed with a busy social calendar, he managed to make the dean's list.

In November 1940 Richard was nominated as a first alternate candidate by Rep. John C. W. Hinshaw from the Eleventh District of California, using the Altadena house of his late grandfather as his address. Over Christmas break he took the West Point mental and physical examinations that were given in Lower Manhattan. The physical revealed that Richard had a condition called persistent albuminuria, and he thus was disqualified. For the next month Groves was in regular contact with Richard and with the surgeon general's office, finding out all he could about the condition and instructing Richard on the steps to take to relieve it, because in the judgment of the doctors it was only temporary.[13] Richard failed the physical examination again in April. Groves, even while urging Richard to do well in lacrosse and his studies to obtain a Princeton scholarship, was obviously disappointed. Undeterred, he was not about to abandon his efforts to get his son into West Point. To his close friend, West Point Professor T. Dodson Stamps, he wrote:

> My thought was that if the examination at Princeton showed that the condition no longer existed, I would ask for a reexamination if it would be of any use and that I would also continue to try desperately to get him an appointment from some unoccupied district. What I would appreciate knowing at the earliest time that such information could be discussed is the appointment status or possibly the examination results from the Senators from California as well as from Senator Reed and Congressman Lambertson, both of Kansas.
>
> Dick tells me that there is a Plebe-Princeton freshman game of Lacrosse at West Point sometime during May. You might take some time off to look at it and see if you don't believe that West Point is missing a bet from the standpoint of improving their Lacrosse squad.[14]

A second year at Princeton became necessary while plans continued to get Richard into West Point.[15] The goal eventually was achieved through the sponsorship of the Hon. Patrick J. Boland, representing the Eleventh District of Pennsylvania.

The impetus came from Maj. J. D. McIntyre, a member of the congressional liaison office of the War Department.[16] One Sunday afternoon in the summer of 1941 McIntyre invited General Gregory to a Washington Senators baseball game at Griffith Stadium. The general in turn invited his assistant, Colonel Groves, and Richard. Immediately after they sat down McIntyre inquired of Richard as to why he was not at West Point, to which Richard replied, "Because I don't have an appointment." The major said, "Do you want one?" and Richard of course said, "Yes." About a month later, presumably because of McIntyre's initiative, Richard was asked to come to Representative Boland's office in the Capitol, where he met Miss Guzey, his secretary.

Richard was nominated by the Pennsylvania congressman, as principal candidate, using the address of the Hotel Casey in Scranton. Unfortunately, the nomination was withdrawn seven months later. Boland had been embarrassed by an opponent in the primary who charged he was bestowing patronage upon people outside his district. After Boland won the primary, things were looking up again, but a few days later poor Boland had a heart attack and died, leaving Richard nowhere. A letter to Miss Guzey inquiring about his status brought the reply that, in going through Mr. Boland's papers, she had found his nomination among them. A few weeks later he received official notice to report to West Point on July 1, 1942, as a member of the class of 1946.[17] Upon later inspection and comparison of the correspondence, it appeared that Congressman's Boland's signature had turned decidedly feminine, indicating that Miss Guzey probably had come to the rescue.

On the eve of Richard's departure for West Point his father offered him some words to live by.

> In the Army you will have to depend on others and they will have to depend on you — sometimes for their very lives. The only lasting basis for that kind of interdependence is trust — complete trust that comes from always telling the truth, the whole truth and nothing but the truth. So always tell the people you are dealing with the truth, and insist that they tell you the truth. The worst thing that you can ever do — absolutely the worst — is to make a false official statement. If you ever utter a false official statement, you can and should be dismissed from the service.[18]

Throughout Richard's West Point years from July 1942 to June 1945 his father pressured him relentlessly to improve his grades, using a system of rewards and penalties, some very elaborate.[19] If Richard performed to a certain standard, he received a monetary bonus from his father. If he fell below the standard, he had to pay his father. There was never any question that payment was expected and exacted. Mrs. O'Leary — who occasionally interjected an editorial comment of her own, and even wrote a few missives herself — typed most of the letters. A constant stream issued forth throughout 1944 and 1945. One dated February 14, 1944, is typical:

> I am enclosing a slip showing your standing at the end of January. If you weren't so utterly putrid in math, you would be alright. What in the world is the matter with you? I wonder — so does your mother — so does Gwen — so does Mrs. O'Leary. I would not only advise, but urge, beg and entreat you to really get down and study it. Even if you have no regard for your own honor, for the honor of your old man, please buck up.

It is signed, "Disgustedly yours."[20]

Groves's approach to his son's performance was similar to the way he operated administratively on the job. He was direct, sure of what he wanted and of what was right, overbearing, demanding, quick to criticize, and reluctant to praise.

From time to time a sparing amount of praise was given. As Richard was about to start his final year, his father wrote, "I personally am very well satisfied with your last years standing although I hope that you can do enough better this year to bring your final standing up into the top fifty."[21]

Interwoven in the letters that survive, between father and son for these years, were other themes as well. There was much sarcasm, a wry sense of humor, teasing, fatherly advice, and — though sometimes it was hard to express it — love.

In a letter of January 27, 1944, after a paragraph berating Richard for poor math grades, the general informed him that he has just purchased some Reynold's Tobacco stock and registered it jointly; after graduation Richard could have his share or a cash settlement, "for a honeymoon or for other purposes."

> I hope not the former. Also, don't forget that I want to approve the girl in advance, not afterwards. For your own happiness, I would advise you

to pick one that you can at least have some measure of control over and not have to be a slave through life, as has been the case with me.[22]

On his grades:

I don't know when I will manage to get up to West Point again, although truth to tell I would feel a little diffident about walking along the sidewalk with a goat.[23]

He needles him about lacrosse:

I understand from your mother that you actually ran during the game she saw although the main purpose seems to have been to clout some passing player and thereby gain a rest in the penalty box.[24]

There is mocking self-pity:

I was completely forgotten on Father's Day by Gwen intentionally and by you either intentionally or otherwise, I am not certain which. Why mothers should be remembered and fathers forgotten in view of the fact that the fathers are the ones who have to slave their lives away, I cannot see.

With best wishes from your venerable father who was recently reminded that he was becoming extremely old.[25]

He reports on his teenage daughter:

Gwen is going to the Laurenceville [sic] prom this weekend and we are crossing our fingers in the hope that she will uphold the family honor. I understand she is wearing a new flame red dress guaranteed to catch any wolf's eye from the stag line.[26]

Even with his demanding schedule Groves tried to find time to see his son, arranging to meet him for an Army football game in New York City and going out to dinner and the theater.[27] He would also come to West Point occasionally, to see his son and to see his old boss, Francis B. Wilby, now the superintendent. In his official capacity as deputy chief of construction, Groves arranged for a great deal of new construction to be accomplished at West Point during the war years.

Richard met his future wife during the summer of 1941. His job that summer was digging graves in Arlington Cemetery, arranged by Quartermaster General Gregory, whose responsibility it was. After work he went to the Army-Navy Country Club to lounge around the pool and play tennis. It was there that Richard met Patricia Bourke Hook, daughter of navy medical officer Capt. Frederick Raymond Hook. Four years would pass before they were wed.[28]

One of the things Groves shared with his daughter Gwen was a love of tennis. No matter how busy he was, he seemed always to have time for his favorite form of exercise. His skill at the game always surprised people. Groves was a large man. Though not particularly tall, at just under six feet, his weight swung wildly between 210 and 250 pounds. Nonetheless, on the tennis court he was surprisingly fast and graceful, with good reaction time. In his daughter he had a more-than-worthy opponent. He had given her one of his old racquets when she was eleven years old, and she had steadily improved. They often played together in her teenage years during the war. She was very good — good enough to give serious consideration to the idea that she might train to play at Wimbledon. But for her father tennis was secondary to ensuring that she get a good education. They had moved to their present home to be near the National Cathedral School for Girls as Gwen began the sixth grade.

Unlike her brother, Gwen spent her high school years at home and went off to Bryn Mawr in the fall of 1946. But her father never missed an opportunity to tease her as well.

> I hope that you will make rapid progress with your Greek as you will probably need it some day in the future when you are a counter girl in a Greek restaurant.[29]
>
> It was a great pleasure to have you home for the weekend even if your visit was not to see your father but to visit your brat friends all over town.[30]

He tried to keep track of her spending by having the canceled checks returned to him and was constantly questioning what she had done with the money. After a visit home on Christmas vacation:

> I was surprised the day after you left to note that there were two dents in the right front fender of *my* car. Did you put them there. . . . Please

report immediately so that either your carelessness can be confirmed or your record cleared up.[31]

Groves was as careful in keeping track of the family's money as he was with the government's. Each member was required to keep a little bankbook to record what he or she had spent.[32] Gwen and Richard received an allowance for spending money and allocations for clothing and special things. The master of the castle kept track of all of it in a larger ledger.

With the general at the office, Richard at West Point, Mrs. Groves either working or volunteering at the USO, and Gwen out with her friends, the house on Thirty-sixth Street was often empty. Instructions and interrogatories dealing with household matters sometimes had to take place, especially between mother and daughter, by written note. During this period Grace's nickname for her daughter was Squirrel Girl — Squir Gir for short — and she drew little sketches of a squirrel to accompany the messages to her. "Get a dozen eggs at Taylors please"; "Drink your milk"; "There is some chicken for you in frying pan in ice box. Heat it slowly — (but eat it fast??) There are some acorns out in the lawn too. Nuts for the Nutty"; "DNO will be home tonight. Turn up heat." Sometimes they were in French or Italian. Occasionally there was a plaintive scribble, "Where is My Mate???"

The death of Grace's mother in Seattle, on February 2, 1943, brought on a family crisis. When Grace arrived unaccompanied in Seattle after a long train journey for the funeral, and to help settle the estate, she discovered that her sister, Mary, who had been caring for their mother, was planning to marry Nemesio Pascual, one of the Filipino servants who had worked for the family — the day after the funeral service, no less. Years of resentment exploded as Mary lashed out at her sister, complaining that while she was stuck taking care of their aging parents Grace was enjoying an interesting and comfortable life. Mary berated her that the least she could do was to relinquish her half of the modest estate. Grace, in a state of shock, agreed.

Mary, six years younger than Grace, was terribly nearsighted and wore glasses. From adolescence on she was overweight and not very attractive to boys. Unlike her sister, she was lazy and not particularly interested in languages, or the literature and music that her father tried to instill in his daughters.

On the return trip east Grace stopped in Cheyenne to scatter her mother's ashes on the Chaffin graves. Her husband met her at Union

Station and, upon being told of what Mary had done, was thunderstruck, appalled, and angry. Grace later recounted that she "never saw him angry except at that time."[33] "I was in a daze — I hardly knew what I was doing." He considered Mary's actions an affront to the memory of her parents, to Grace, and to her niece and nephew. In that much different time a relative's miscegenational marriage could cause family members embarrassment and shame. While Groves could not undo the marriage, he insisted that Grace keep her share of her parents' estate.

Later she reflected on the incident.

> I broke up completely — a nervous breakdown — to add to DNO's terrible burden.... I was not really responsible for anything I said or did.
>
> I had a lot of psychiatry during the following year or so. I got a job at Garfinkels, to help pay the doctors' bills and Gwen . . . and I lived on as best we could.
>
> With the great secret burden [DNO] was already carrying, he now had the burden of a sick, sick wife.[34]

Just what constituted Grace's "nervous breakdown" is not clear. It could perhaps be better described as depression. There is no reference to this incident in any of Groves's personal papers. According to notes that Richard appended to his mother's reminiscences, all his father would tell him is that his mother was distressed by events in Seattle and that he should show her a good time when she came to visit him at West Point, which occurred almost immediately after she returned to Washington. Richard went on to say that he had no idea if she enjoyed that visit, although photographs of her at the time show her to be smiling.

Grace saw a psychiatrist on a regular basis for more than a year and to help pay for it took a part-time job as a salesperson in a Washington department store — the Garfinkel's branch in Spring Valley, on Massachusetts Avenue in upper Northwest. Under normal circumstances in the traditionally based Groves household, Grace had not worked outside the home. In this instance taking the job probably had a therapeutic effect, because it was connected to paying for her visits to the doctor. Her husband supported the choices, and was probably pleased that his wife had something to occupy her while he was often away.

At the time seeing a psychiatrist was not a very common occurrence. As an indication that the crisis was on a minor scale, Grace was often the butt of family jokes about her visits. As for dealing with Mary and the family estate, Grace turned matters over to her husband, who saw that the papers were properly filed.[35] The issue never went away and was mainly dealt with by suppression and silence. Only after her husband's death did Grace undertake reconciliation with her sister.

All marriages and relationships have troubles, but those we know of in the Groves household seemed to be of the common variety. The fact was that Dick and Grace were devoted to one another and had a strong marriage of almost fifty years. On their forty-third wedding anniversary in 1965 Grace penned a tender note to her husband.

> I wish I knew how to write so I could tell you in some words that might express it how wonderful it has been all through these many years. How exciting it was, how much fun it was, how dear it was. How you never, never — even once — let me down, or disappointed me, or failed me. How you were always sound and wise and strong when I was weak and wavering; my haven and my home. Your side of it is different, I know. I have disappointed you and hurt you and failed you more than once and wish I might start over and do better. I can only say now that even at my worst I have always, always loved you. And always, always will.

The degree to which the general was able to shield his family from knowing the nature of his work is starkly revealed by examining the events that transpired at the Groves home just after the announcement of the bombing of Hiroshima. The evening before, the family had gone to dinner at the Army-Navy Club in downtown Washington. Several of the other diners that evening were in the inner circle and were aware of the fact that the *Enola Gay* had left Tinian for the Japanese mainland. Several times during the meal Groves responded to raised eyebrows from across the room with a slight, negative shake of the head. There had, as yet, been no word. After dinner Groves went to the office, finally did receive word of the successful attack, and spent the rest of the night preparing for the public disclosure.

The next morning, just prior to the first radio newscast carrying the announcement — it would occur at 11:00 A.M. — Groves called Grace to suggest that she listen to the radio. He also told her that he was sending a lieutenant from his public relations staff out to the house. He sent word to the secretarial school where Gwen was taking a typing course for her to leave class and get home as soon as possible. Colonel Consodine called Richard, attending Engineer School at Fort Belvoir, to warn him that journalists would soon want to talk to him but that he should not tell them anything.[36]

Mother and daughter were mystified about what was going on until they listened to the eleven o'clock radio broadcast. Within minutes they were besieged by members of the press, who found it virtually impossible to believe that the family had no idea about what the head of the household had been doing for the past three years.

But it was true. They had known nothing. As Mrs. Groves informed her sister-in-law a few days later,

> Yes, we were just as surprised as the Japs when Dick's bomb was dropped. Of course Gwen and I knew he was busy and fearfully burdened with responsibility but that was all. Now that the secret is out he seems so relieved and relaxed, even though he works as hard as ever. I am truly hoping that when the surrender is finally announced he can get away for a real rest.[37]

PART FIVE

Racing to the Finish

· 1945 ·

CHAPTER EIGHTEEN

Supplying Atomic Fuels: Enough and in Time

After a year of unrelenting pressure on the contractors to get the atomic factories built at Oak Ridge and Hanford, Groves turned his attention to ensuring that there would be an adequate and timely supply of uranium and plutonium for the bombs. Now that the factories were up and running, they could begin producing their unique concoctions. What Groves contributed was constant pressure on everyone to accelerate production schedules. With anyone less driven the atomic fuel would not have been ready when it was, and the bombs would not have been used when they were.

It is not exactly clear when Groves first learned that one of his bombs would not work — probably in early May 1944. In what seemed like a near catastrophe at the time the scientists at Los Alamos concluded, in despair, that a plutonium gun-assembly design was not feasible. In the gun-assembly design two subcritical pieces of fissile material were fired at each other inside a specially modified gun barrel to make a critical mass. To rescue the investment at Hanford — some $350 million — a new, but more difficult design was adopted. Under Groves's orders Oppenheimer quickly reorganized the laboratory for a crash program to develop a method of powerfully squeezing a subcritical mass of fissile material — plutonium — by the simultaneous detonation of conventional high-explosive charges focused inward — in effect, an implosion.

On the positive side, less plutonium would be needed due to greater efficiency. By the end of the year it was calculated that five kilograms of plutonium would be needed per bomb. As the Hanford reactors began to produce plutonium and the separation plants processed it, Groves repeatedly set ever–more-demanding schedules, urging Du Pont, the scientists, and his subordinates to find ways to meet them. Every gram was important and every day counted.

Originally, it had been assumed that the gun-assembly technique would be used for both the uranium 235 and the plutonium bombs.[1] For the first year Los Alamos was organized on this basis, and though many problems remained to be solved there was general optimism that they could be overcome. In March 1944 the first half-gram lot of plutonium produced in the Clinton reactor arrived at Los Alamos.[2] Experiments suggested that it contained an isotope of plutonium (plutonium 240) that fissioned spontaneously. Up until then, the properties of plutonium had been determined using microscopic samples created in the Berkeley cyclotron, and they had not exhibited spontaneous fission.

This new and unexpected finding meant that it would be impossible to assemble a critical mass of plutonium using the gun technique. Predetonation would occur, and the result would be a fizzle. The scientists working on this problem, mainly Enrico Fermi and Emilio Segrè, repeated the experiments throughout May and June to be absolutely sure of their conclusion, not always wanting to believe what they had found.

At some point this distressing fact was conveyed to Groves. At Los Alamos they expected the general to inform Compton and his scientists at the Met Lab, but apparently he did not. When Robert Bacher visited Chicago, from May 31 to June 8, 1944, he recalled that Compton, upon hearing the news about spontaneous fission, "went just as white as that sheet of paper."[3] Bacher called Groves to ask permission to inform the Met Lab scientists. Groves, though concerned with secrecy, now realized that by not telling them it might impede the project. He gave permission to Bacher to tell half a dozen people. Bacher was sure that within half an hour the whole lab knew.

This hard fact of science confronted Groves with the prospect that the money he had spent on Hanford might now be in vain. After enormous effort and expense the plutonium, soon to come forth in quantity, might be unusable.

On July 4 Oppenheimer presented the evidence for spontaneous fission in a colloquium at Los Alamos. Several further tests over the next few days confirmed beyond a shadow of a doubt that plutonium could not be used in a gun design. Various solutions were proposed. Faster gun assembly would not work because predetonation would still occur. Another idea was to separate out the bad isotopes from the good ones. While this was feasible, it would be enormously expensive and take too long.

On the afternoon of July 17 Groves flew to Chicago.[4] That evening he met with Oppenheimer, who had come from Los Alamos, and with

Conant, Fermi, Compton, and Charles A. Thomas[5] to discuss what to do next. As often happened during the Manhattan Project, a novel solution emerged just when it was needed.

Another technique was being explored by a few scientists based on the concept of using the shock wave from high explosives to squeeze a sub-critical mass of plutonium to a supercritical state. This technique, called implosion, had the added benefits of being more efficient and using less material.

While experiments aimed at perfecting this technique had thus far yielded disappointing and unsatisfactory results, it was, given the crisis now at hand, the only possible route. On July 20 Oppenheimer announced to the laboratory's administrative board that the implosion program should be given the highest priority.

Groves supported Oppenheimer's rapid reorganization of the laboratory, which began in late July. Groves made two quick trips to Los Alamos during this period to ensure that he understood the drastic changes being implemented and that he agreed with them.[6] Two new divisions were formed on August 14. Personnel from the Physics and Engineering Divisions, and new recruits, went to G (Gadget) Division under Robert Bacher to work on implosion, and to X (Explosives) Division, under George Kistiakowsky, to work on the novel kinds of explosives that would be needed for this difficult design.

Groves and Oppenheimer were now faced with perfecting two totally different methods of assembly, one for the uranium bomb and another for the plutonium bomb. Within a month, the number of personnel at the laboratory had increased dramatically, with much of the growth in the implosion group.

With the decision about the new design now resolved, there remained the issue of how much material would be needed. After a visit to Los Alamos in August 1944, Conant reported to Groves about the amounts needed for the various designs.[7] For the Mark I gun-type, uranium bomb — to achieve a yield of between ten thousand and twenty thousand tons TNT equivalent — thirty-nine to sixty kilograms of U-235 were required. (Approximately sixty-four kilograms of U-235 eventually were used in the Little Boy bomb.) For the Mark IV implosion bomb the amounts of plutonium needed varied, depending upon the potential yield. Seven and one-half kilos of U-235 should produce three thousand tons TNT equivalent, Conant recounted, while two and one-quarter kilos of plutonium should produce one thousand tons.[8] With many problems to solve, Conant felt that "it is not more than a 50-50 bet that Mark IV can be developed at all before the summer of 1945."

Robert Frederick Christy was a member of the Theoretical Division's Implosion Dynamics Group.[9] Christy proposed the use of a solid core of plutonium with an initiator and a "pusher" that would compensate for possible asymmetries during implosion — rather than a hollow shell, which had been the initial choice. These conservative, yet efficient, features were adopted, and the design became known as the Christy gadget or Christy pit. "On February 28, 1945, Oppenheimer and Groves decided provisionally on Christy's core design, using electric detonators and explosive lenses made of the explosives 'Comp B' and Baratrol."[10] The figure of 6.1 kilograms of plutonium was agreed upon as the amount to use in the Trinity/Fat Man bomb.

In February 1944, a year after construction of the Y-12 electromagnetic separation plant had begun, the first product — a mere two hundred grams of uranium 235 — was sent to Los Alamos. As late as Christmastime 1944 the exact critical mass of highly enriched uranium was still uncertain. There was also the question of determining the necessary isotopic purity of the U-235. Would a bomb work with uranium enriched to only 20 percent U-235, or would it need to be as much as 80 percent or higher? The situation was grave. The production rates at Y-12's racetracks were very low. By Christmas only about eight kilograms of enriched uranium had been produced. The K-25 gaseous diffusion plant was experiencing trouble with the barrier material and would not begin operations for several more months. If rates were not increased, there would not be enough fuel for even one bomb before the end of 1945. Again, as so often happened during the Manhattan Project, when a particularly difficult problem needed a timely solution, one emerged.

In April 1944 Deak Parsons, after returning from a trip back east, told Oppenheimer about some experiments in uranium isotope enrichment using the liquid thermal diffusion process that were being pursued by the Naval Research Laboratory at the Philadelphia Navy Yard. Oppenheimer wrote to Groves on April 28 informing him of this, and described the advantages of using the thermal diffusion process to supplement the other enrichment techniques.[11] The navy was building a one-hundred-column pilot plant and was scheduled to start up on July 1. The amounts of uranium produced, and the levels of enrichment attained, were expected to be modest — far less than what was needed for a bomb. But,

Oppenheimer realized, if enough of the slightly enriched product were used as feed for Y-12, it would result in a significant increase in the amount and the purity of the electromagnetic plant's final product.

For the first year or so Groves and his scientists had focused on single independent methods to enrich natural uranium, either by the electromagnetic process or by gaseous diffusion. Oppenheimer's insight was an important one: Add a third method and combine them, using the slightly enriched product of one as feed for another. It would soon be taken one step farther: The gaseous diffusion plant's enriched uranium would be used as feed for Y-12.[12]

At the end of May Groves appointed Lewis, Murphree, and Tolman to assess Oppenheimer's information. They quickly confirmed his conclusions. Groves wrote to Oppenheimer on June 3 to inform him that the committee had examined the feasibility of his idea, and that he would soon let him know what was planned.

Over the next three weeks Groves had Lewis, Murphree, and Tolman investigate further the costs and requirements of a plant. On June 24, 1944, Groves decided to proceed — in a typical way. He would not merely expand and operate the navy plant at Philadelphia. Rather, he ordered construction of twenty-one copies of the hundred-column plant to be built at Oak Ridge. To provide the massive amounts of steam that would be needed, it would be sited next to the powerhouse for the K-25 plant, still some months from completion and thus not being used. On the same day just after noon Groves met with Mrs. Ferguson, president of the H. K. Ferguson Company of Cleveland, Ohio, since her husband's death six months earlier, and the company's chief engineer, Wells N. Thompson.[13] Groves had worked with Ferguson during his days as deputy chief of construction and considered it an "outstanding engineering firm."[14] It got the construction job, and at Groves's insistence a second contract was signed for Ferguson to operate the plant once it was completed.

Groves assigned a young corps engineer, Lt. Col. Mark C. Fox, to oversee the job. Work got under way on July 9, and ground was broken a few days later. Originally the construction schedule was six months, an optimistic goal. Within three weeks Groves set 120 days as the goal to begin operation. Still not satisfied, Groves then told Fox that he expected the first product in ninety days. When Fox objected, Groves cut it to seventy-five days. From the first day, Groves made sure that Ferguson's materials and personnel needs were given the highest priority; every order it submitted was promptly filled.

The thermal diffusion plant was code-named S-50. In Groves's office it was referred to as Fox's Farm. The main building was 522 feet long, 82 feet wide, and 75 feet high. It consisted of three parts, each containing seven "racks" of 102 columns each, for a total of 2,142 columns, twenty-one Philadelphia pilot plants strung together. Each column was forty-eight feet high and had to be manufactured to precise tolerances. Inside each column were three concentric tubes: a 1¼-inch nickel pipe, a slightly larger copper pipe, and a four-inch galvanized iron jacket. The nickel pipe was steam-heated, under great pressure, to 280 degrees Celsius, and the copper pipe water-cooled to 70 degrees Celsius. When uranium hexafluoride was made to pass between the pipes, the lighter isotope concentrated on the hot nickel and moved upward while the heavier isotopes concentrated on the cooler copper pipe and moved downward.

In the very first days of assuming command, Groves had visited the Naval Research Laboratory and discussed the thermal diffusion method with Ross Gunn. Later Groves had an interesting comment about Gunn:

> Gunn's animosity towards the project dated from before the time that I was placed in charge. He apparently made himself persona non grata to other scientists and particularly to Bush and Conant who arranged to have him released . . . because he was non-cooperative and ultra secretive. If Gunn had been cooperative I think that we would have been able to have built a satisfactory thermal diffusion plant to bring about purification at the lower stages and we would have done so by the summer of 1943. How much this would have shortened the war is hard to say, but we undoubtedly would have gotten the uranium bomb sooner. At a wild guess by as much as three weeks.[15]

The Philadelphia pilot plant was used for training S-50 personnel. In September a serious accident killed two young chemical engineers and injured several others, causing temporary concern about the validity of the design. Under Groves's unrelenting pressure Fox finished the job under schedule. On September 17, 1944, sixty-six days after ground had been broken and eighty-three days after the Ferguson Company had taken the job, one-third of the plant was complete — enough to begin preliminary operations. The following month the first product was drawn. Application of high steam pressure caused many leaks, but over the next three months, as construction neared completion, the problems were corrected.

As Oppenheimer and the scientists had predicted, the slightly enriched uranium that emerged from S-50 caused a significant increase in efficiency when used as feed for the Y-12 plant. In April 1945, when the gaseous diffusion plant began partial operation, the S-50 output was used as feed for it. The K-25 product, enriched to about 23 percent U-235, was then used as feed for the Y-12. Y-12's final product, which was used in the Little Boy bomb, was enriched, on average, to about 84 percent U-235.[16]

It was only in this way that enough highly enriched uranium, sixty-four kilograms,[17] was produced in time to make the one gun-assembly bomb that was dropped on Hiroshima. It would take until October for there to be enough highly enriched uranium for a second bomb, and until December for a third. As it was, Oak Ridge was scoured for every bit of U-235 that could be found.

Even the design of Little Boy reflected this speed. The projectile and the target insertion were both composed of rings of HEU instead of one solid piece. As HEU was received from Oak Ridge, it was fashioned into one or the other type of ring. The rings themselves were not of a standard size, nor were target and projectile equal in weight. Sixty percent of the total HEU was in the projectile, made up of nine large washers bound together with a hole in the middle.[18] They were completed first and shipped aboard the USS *Indianapolis* to Tinian. The six rings that made up the target insert were smaller in diameter to fit into the hole of the projectile. These target rings were cast later and flown, two each, as the sole cargo aboard three C-54 Green Hornet flights that departed from Kirtland Field, Albuquerque, on July 26, and then to Fairfield Sunsun, California, and on to Tinian in the Pacific, arriving on July 28.[19]

The S-50 plant was scrapped on September 9, 1945, exactly one week after Japan signed the articles of surrender. It had not been fully completed.

With the reactors and the separation plants nearly completed by the end of 1944, Groves turned his attention to pressuring Du Pont to speed up the production schedule.[20] Though in the summer of 1944 there was still uncertainty as to precisely how much "Hanford product" was needed, in general the more of it there was, and the sooner it was in hand, the better. In late July, just after the decision that the gun-assembly method would have to be abandoned, estimates ranged from two to twelve kilograms for the "more difficult and problematical" implosion design.[21]

On October 18, 1944 (while the B reactor was being loaded with the additional tubes, after the "poisoning" experience), Du Pont informed the Wilmington Area engineer that monthly plutonium production would begin with two hundred grams in February 1945 and increase gradually to six kilograms a month by December 1945.[22] Under this schedule there would not have been enough plutonium for the first "unit" (using six kilograms) until the end of August 1945. It would have taken additional time to transport the material to Los Alamos, to further process and shape the core, and then be sent to Tinian. Under the original Du Pont schedule a completed first test device would not have been ready until early to mid-October, with a combat-ready bomb available a month or so later. This schedule was totally unacceptable to Groves. From November 1944 through March 1945 Groves repeatedly pressured Du Pont officials to find ways to accelerate the schedule. Groves informed Roger Williams that he needed five kilos of plutonium as quickly as possible, and then soon after, five more. Various plans were proposed to achieve these goals based upon different operating assumptions.

Basically there are three ways to speed up the process of plutonium extraction. The first is to increase the power levels of the reactors to beyond their design rating. The second is to push the uranium billets out of the reactor more quickly than normal. The third is to shorten the time interval between when the billets are pushed from the reactor to cool in the ponds, and when they enter the separation facility. If the billets are pushed early, the amount of plutonium extracted from any one billet is reduced. To make up for this shortfall, more billets are pushed than would otherwise be the case. This results in a large amount of waste. As for the cooling, under normal circumstances the billets should remain under water for a period of about 120 days, depending upon several factors, to allow the most volatile and dangerous radioisotope — iodine 131 — to decay to what are considered safe levels. Shortening the number of days and beginning to extract the plutonium sooner saves valuable time, but one cost in this frenzy is safety, with the potential of an increased exposure to radioactivity. Any lingering iodine 131, of which there will be more after 60 days than 120 days, will be expelled into the atmosphere when the billet's aluminum jacket is dissolved in acid, the initial step in the extraction process.

After reviewing the emissions from the period 1944 through 1947, one scholar has concluded that health and safety concerns were of lesser import than the production of plutonium. Questions about which isotopes were the most problematic — xenon, neptunium, iodine — and how hazardous they might be to workers at the plant or people downwind were not adequately addressed in these early days. As for Groves's role, he

. . . was concerned both about the security implications of informing employees of the hazards and about the effect of this information on their willingness to continue working at the plant. "Leaks are bound to occur; people are bound to talk," he later wrote, "but the less the better and the less disclosure there will be to unauthorized persons."[23]

Groves employed the three above-mentioned ways from March to August 1945 to save time and to increase the amount of plutonium available, in what became known informally as the "speed up program." Matthias sometimes referred to it as the "fast schedule" or "the super-acceleration program."

After months of incessant pressure by Groves on Du Pont officials, on February 14, 1945, Williams wrote the general with a schedule to produce five kilos by the end of June, and a second five kilos by the end of July, noting that the accelerated pushing would result in wastage.[24] Groves responded that he was willing to tolerate high amounts of waste if need be. He told Williams to do better, and said that the dates "must of course be improved if it is possible to do so."[25]

By the beginning of March, Matthias and Du Pont had agreed to a speedup schedule calling for five kilos by June 14 and another five by July 12. Groves was still not satisfied and continued to press.[26] After a March 22 meeting in Groves's office with Du Pont officials, he wrote to Roger Williams:

> The first priority of production is for a cumulative total of five kilograms of material. The above production is about to be realized insofar as the processing through the pile is concerned. The second priority is for a total cumulative production of 10 kilograms of material. After the production of the 10 kilograms . . . it is desired to obtain the maximum amount of material which can be produced without undue risk. The importance of securing the first ten kilograms is great enough so that no action should be taken which would jeopardize unduly the realization of that amount on the dates now planned. While tonnage of material consumed is of importance it is not to be considered at this time as a determining factor. It is understood that your present power rate of 250,000 kilowatts will produce approximately 500 grams of material per day.[27]

He also requested that Williams "make a study of the production processes to determine what maximum increased power rate can be employed without undue risk and that you advise me of the results of your study with your recommendations."[28] On the same day Groves wrote to Oppenheimer informing him that he now had firm commitments from Du Pont for five kilograms by June 1, and a second five kilograms by July 5.[29]

Two days later Groves arrived at Hanford for a daylong inspection of the operation. He outlined the new schedule to Matthias and his assistants, Major Sally and Lieutenant Colonel Rogers, and the Du Pont officials, plant manager Walter O. Simon, Fred A. Otto, and Gil Church.[30]

On May 3 Matthias recorded in his diary that the speedup was under way and achieving results.

> Reviewed the can delivery and found that 20 cans have been turned over and are on the way. This is approximately six days ahead of the latest official prediction as of April 21st. It appears that the gain is being made in the reduction of process time cycle and in the time of storage before solution.[31]

Six days later he recorded:

> Due to the reduction of initial cooling periods between the 100 and 200 areas production results have been extremely good and the time of delivery of units has been faster than originally planned.[32]

Hanford records showed that during the speedup most of the billets were allowed to cool for approximately sixty days, but correspondence between Groves and Williams indicated that in June, the time was shortened considerably.[33] By July Hanford was producing plutonium at a rate approaching three bombs per month.

In the very compressed time schedules of these crucial months of 1945 it is important to recognize the other steps and extra time necessary for completing and delivering a combat-ready bomb. The above schedule and dates were when the extracted plutonium was ready to be shipped from Hanford. Next came an approximately fourteen-hundred-mile road journey to Los Alamos, where the plutonium underwent further processing before final pressing into a finished core. The last step was to transport the plutonium core — and the non-nuclear assembly — to Tinian to be mated, and then delivered by the 509th.

The first batch of Hanford plutonium was produced beginning in December 1944. On February 5, 1945, it went by car from Hanford to Portland, Oregon, then by train to Los Angeles, and finally to Los Alamos.[34] A second batch went the same way, with a Los Alamos courier in Los Angeles to meet the shipment. Hanford records show that by April 21, 1945, 731.2 grams had been shipped to Los Alamos.[35] Larger quantities began to be produced beginning in May. About every two weeks the B and D reactors each produced fifteen hundred to nineteen hundred grams of plutonium, after cooling and separation.

Most deliveries to Los Alamos were made in unmarked olive-drab, one-and-a-half ton panel trucks that were converted army field ambulances.[36] The convoys were the responsibility of Capt. Lyall E. Johnson and closely adhered to Groves's rules about compartmentalization and security.[37] A normal convoy was three trucks, with a lead and a rear escort car. Each vehicle carried two men, heavily armed with shotguns and revolvers, and a submachine gun for each vehicle. Up to two dozen containers were carried in each truck on racks. The containers were wooden, about sixteen inches square, and not very heavy. Inside was a metal container, and inside that was a stainless-steel flask. Each flask held eighty grams of the bluish green slurry of plutonium.

One convoy would pick up the containers at the Gable Mountain vaults at Hanford and drive to Fort Douglas, in Salt Lake City, a distance of some seven hundred miles, a journey of two or three days. A similar convoy from Los Alamos drove an almost equal distance through Colorado and Utah to Fort Douglas. At a specified site within Fort Douglas the Hanford security officers delivered the containers to the Los Alamos security officers. The Los Alamos couriers would drive their precious cargo back to Los Alamos and the Hanford couriers would return, taking with them any empty containers brought from Los Alamos. Many of the Hanford couriers did not know what they were carrying or where it was bound. Likewise, many of the Los Alamos couriers were equally ignorant, not knowing what they received or where it had come from.

On July 4 Matthias told Groves that the "total quantity shipped to date is a little over 13½" kilograms with about another 1.1 kilogram ready to be sent over the next week.[38]

As time grew tight, there was one shipment of six units by air, apparently on August 5,[39] too late for the Nagasaki bomb but available for the third bomb.[40] Captain Johnson went to Stockton Field and arranged for two C-47s to transport the containers from Richland to New Mexico. He commandeered enough parachutes for each of the containers as well.

Once at Los Alamos the plutonium was purified and processed, reduced to metal, and fabricated into a "core" by the Chemistry and Metallurgy Division, numbering some four hundred scientists and technicians in 1945.[41] The plutonium arrived in the form of a relatively impure nitrate solution at D Building, where most of the division's work was conducted.[42] Plutonium was a very unusual element, as the scientists working with it found out. At room temperature it was brittle and difficult to work with. To make it malleable enough to shape it into the two hemispheres that made up the tennis–ball–sized core, it was alloyed with gallium. Then it was shaped into a hemisphere by hot-pressing. To prevent oxidation and contain radioactivity, it was coated with nickel.

In late July one of the hemispheres had to be remelted and recast because it was underweight. Provision had been made to produce three hemispheres, so two usable ones remained. The combat spheres were completed on July 23 and shipped out July 26.[43]

CHAPTER NINETEEN

Groves and the Use of the Bomb I:
The Target and Interim Committees

By the spring of 1945 Groves had dealt himself a very strong hand, and he played it deftly to achieve the goals he had set for himself. First and foremost among these goals was to complete one or more usable bombs and to keep control of how, when, and where they were to be used. His trump card, as always, was compartmentalization and the secrecy that was its raison d'être and result. Groves, sitting atop his security pyramid, was the only person who knew everything about the bomb project — more than the chief of staff, more than the secretary of war, more than the president. Truman, Stimson, Marshall, and the other officials were busy with a full range of complicated issues of their own as the end of the war drew nearer. Groves, on the other hand, was singularly concerned with the bomb, with getting it finished, tested, and used, and his superiors deferred to him time and again to make the choices that would make this happen. Not only were his cards good ones, but the deck was stacked in his favor as well, stacked against any alternatives to using the bomb.

More than thirty-five years ago two authors set out to write about the decision to use the bomb. They came to this conclusion:

> . . . there seems no doubt that by noon on Sunday, July 22, the decision had been made. What is not clear, what remains a puzzle clouded by lack of documented evidence, the passing of years, the dimming of memories, is *how* that decision was made. Was it even made at Potsdam? Or was it perhaps made by not being made at all, by allowing the machinery already in motion to continue in the direction and on the schedule that had been set long before?[1]

We are no closer today to determining precisely how that decision was made or when it was made, and for good reason. Even to use the word *decision* distorts the process by which the use of the bomb occurred. *Decision* implies thought, analysis, and discussion about a range of possible courses of action, a weighing of alternatives, and then, after the deliberation of the pros and cons of each, a choice being made. Clearly this is not what happened with regard to use of the atomic bomb.[2] As Groves said,

> One detail, and it seems to me just a detail, has been discussed by writers of recent years who were familiar with the workings of the project, and that was the directive for the actual use of the weapon. They have seemed to think that there would be a formal paper on which the President of the United States wrote: "The bomb will be dropped on such and such a place after such and such a date." *That is not the way it was done.*[3]

New insights into what happened are possible by examining the "machinery already in motion," that is, those military preparations that "had been set long before" and proceeded concurrently with the policy and diplomatic deliberations and maneuvers about how to force Japan to surrender unconditionally, how to deal with Russia's entry into the war, and the numerous other issues waiting just over the horizon in the postwar world.

Demonstrating Groves's role throughout this process can enrich our understanding of the events and lead us to a new awareness of certain issues in the ongoing debate about the bomb. Focusing on the counterfactual arguments of what might have been has diverted us from looking in more obvious places, at the "action channels," the actual military plans and preparations that did happen.[4] Those plans and the procedures for military use of the weapon were set in motion at an early stage and began to be forces of their own irrespective of any diplomatic considerations and maneuvers in the spring and summer of 1945. By this time a vast infrastructure of facilities, military units, and tens of thousands of people, largely controlled by Groves, was proceeding apace, a deus ex machina ready to appear in the final act, inexorably driven to carry out the mission.

The military action channels did not move in tandem with the diplomatic policy channels. The policy issues, much debated in the later literature, were formulated quite apart from the preparations to use the bomb. Whatever peregrinations occurred over policy matters, they had little effect on the basic assumption that the bomb would be used when ready; nor did they interfere with the juggernaut that Groves had set in motion. Only a decision not to use the bomb would have had an impact on the military plans, and apart from a few scientists, soon silenced, no one considered it.[5]

It may have been possible to stop the use of the bomb, but it would have taken some compelling reasons and some very forceful personalities in positions of power, up to and probably including the president, to do so. But from the vantage point of the spring and summer of 1945, it is difficult to see what would have prompted Truman and the others to make any other choice.

On the morning of April 13, 1945, his first full day as president, Truman met with his military advisers to hear their opinions on the course of the war. They told him that the war would go on for at least six more months in Europe and eighteen more months in the Pacific. Of course the war in Europe ended less than four weeks later, and the bomb was still two months away from being tested. Surprisingly, no one mentioned the bomb at this meeting; Truman had to wait until later in the month to receive a full briefing about it.

Truman was not totally ignorant of the bomb even in those first few days. After the first cabinet meeting, on the evening he became president, Secretary Stimson asked to speak to Truman "about an immense project that was underway — a project looking to the development of a new explosive of almost unbelievable destructive power." The following day Jimmy Byrnes provided a few details "about perfecting an explosive great enough to destroy the whole world."[6]

On the morning of April 25 Secretary Stimson and General Groves went to the White House to brief the president about the status of the bomb. Groves entered the White House through a private entrance so as not to arouse press interest, joining the two in the Oval Office about five or ten minutes after their conference had begun. As the secretary started to introduce the general, Truman interrupted him and said, "Oh, I know General Groves very well. I have known him for many years."

Groves presented a twenty-five-page memorandum, dated April 23, to the president. Truman said he did not have time to read reports, but after some gentle urging he did read it in its entirety. The information was staggering in its scale; indeed, the bomb's far-reaching implications for the future concerned no less a matter than the very survival of modern civilization. After reading the memo, Truman said he approved of what they had done; at no time did he criticize the weapon itself or plans for its use. He asked some questions, and as Groves noted later, "A great deal of emphasis was placed on foreign relations and particularly on the Russian situation."[7] Groves was embarrassed upon leaving because he did not know which door to use to get out of the Oval Office, saying, "Mr. President, how on earth do I get out of here?"[8]

Conceivably the war in the Pacific against Japan might also have ended before the bomb was ready. As it turned out, one bomb was ready on the last day of July and another in early August. They were used as soon as they were ready and the weather allowed. As we have seen, it was largely through the administrative and managerial skills of General Groves that the bombs were finished when they were. While it is easy to imagine how it could have taken longer, it is hard to see how the feat could have been accomplished any more quickly.

Groves had at least half a dozen reasons for using the atomic bomb on Japan by the summer of 1945. The dominant one was that it might contribute to ending the war and saving American lives. By the spring of 1945 Japan was clearly defeated, though fighting continued with costly battles at Iwo Jima and Okinawa.[9] In the summer of 1945 several methods were being used to force Japanese leaders to recognize their defeat and to surrender. The Army Air Forces (AAF) area bombing campaign and the naval blockade were both taking their toll. The impending Russian entry into the war and an American land invasion of Japan scheduled for November would bring additional pressure. It was only natural that Groves and his superiors would think atomic bombs might halt the slaughter, avoid an invasion, and end the war.

A second reason for Groves wanting to use the bomb was the sheer momentum of the project.[10] By mid-1945 nearly two billion dollars had been committed to the bomb, and hundreds of thousands of people had been forged by Groves into an effective instrument to complete it. Groves saw President Truman's role as not interfering with plans that were already in place and quickly moving forward. "As far as I was concerned, his [Truman's] decision was one of noninterference — basically, a decision not to upset the existing plans."[11]

Later he said,

> The responsibility taken by Mr. Truman was essentially, I think, the responsibility taken by a surgeon who comes in after the patient has been all opened up and the appendix is exposed and half cut off and he says, "yes I think he ought to have out the appendix — that's my decision."[12]

For Truman the most logical thing to do was to continue the policies that had been initiated by Roosevelt and to continue to rely on a circle of advisers who were knowledgeable.[13] As Groves told one author, "Truman did not so much say 'yes' as not say 'no.' It would indeed have taken a lot of nerve to say 'no' at that time."[14] It would be more accurate to say that Truman's "decision" was a decision not to overrule previous plans to use the bomb.

A third reason for Groves's willingness to use the bomb centered on domestic politics and the consequences that would have almost certainly ensued if the bomb had been available but not been used. Hiroshima avoided this danger. Two months after the surrender Groves confided to Styer, "One thing is certain — we will never have the greatest Congressional investigation of all times."[15]

Groves was also aware that the successful use of the bomb would enhance certain bureaucratic and organizational interests and was closely connected with the furtherance of individual careers. Many large bureaucracies, departments, and agencies had a stake in the successful outcome of the project. Considerations of protecting and advancing those interests often took precedence over "terminating the war with a minimum of gratuitous death and destruction."[16] At the personal level people in positions of responsibility might face ruined careers and tarnished reputations if the project failed. Conversely, if the bomb worked, prominence, praise, and influence would follow. A mix of self-interest and organizational interest clearly drove Groves; the Manhattan Project was his creation and with it he was going to make his contribution to ending the war — and make his mark on history.

For Groves a fifth reason for using the bomb was retribution and revenge, bound up at times with racism, feelings he shared with his countrymen. The Japanese during World War II were detested, and accounts of their brutality, wanton killing of civilians, and torture of prisoners were widely reported and abhorred. Their surprise attack at Pearl Harbor shocked and infuriated the American people. This initial sign of Japan's treachery and perfidy was reinforced by lurid accounts of Japanese atrocities, from starving prisoners to performing vivisections, emasculating them, decapitating them, even crucifying them and burning them alive.[17]

On August 7, when Groves reported to the chief of staff about Hiroshima, Marshall said, "we should guard against too much gratification over our success, because it undoubtedly involved a large number of Japanese casualties." Groves replied that he was not thinking so much about the Japanese casualties as about the Americans who had made the Bataan death march. Upon leaving the office Gen. Henry H. Arnold, commanding general of Army Air Forces, slapped Groves on the back and said, "I am glad you said that — it's just the way I feel." Groves went on, "I have always thought that this was the real feeling of every experienced officer, particularly those who occupied positions of great responsibility, including General Marshall himself."[18]

Finally, it is clear that several policy makers believed that the bomb could play a role in diplomatic relations with the Soviet Union after the war. While there is no record of Groves ever making such arguments, his actions implied that he was aware of it. Groves had no love for the Soviets and was suspicious of them. He was vigorous in trying to keep the bomb secret from them and preventing espionage. Others in high places clearly found compelling and seductive the notion that sole U.S. possession of the bomb would somehow make the Russians more tractable and manageable. This belief in a "winning weapon" turned out to be a delusion of the first magnitude.[19]

Among the circle of people who deliberated on the use of the bomb, each person was no doubt motivated by one or more of the reasons outlined here. They were not unique to Groves. The reasons were intertwined, they overlapped and reinforced one another, and each person may have ranked them in a different order. But whatever their motives at any given moment, they always wanted to go ahead, and Groves led the way.

In implementing any policy decision, regardless of what is "decided" at the topmost level, there are numerous instrumental and operational details that come into play. How these details are implemented very much influences what happens. This is true with the "decision" to use the atomic bomb.

If there is a precise point at which an authorizing decision document was drawn up for use of the bomb, it involved the report Groves prepared for General Marshall and Secretary Stimson in May 1945 about the targets, and the one-page draft of the directive. The directive, as its author later recounted, was then *"held awaiting the time when we could actually write in the final date as to when the bomb would be dropped."*[20]

General Groves's influence on events is clearly visible in his role on the Target Committee and on the Interim Committee, a high-level group whose purpose was to advise the president about key domestic and international issues surrounding the bomb once the war was over. Groves was the only person to participate in both committees' deliberations. He established the Target Committee, and though he attended only one of the four meetings he kept fully abreast of its work through his deputy general Thomas Farrell, along with his other aides. Though Groves was not one of the eight members of the Interim Committee he was an invitee, participated at times, and attended all of the meetings.

Groves devoted a thirteen-page chapter of his book to a discussion of "choosing the target" and his role in it. The assumption had always been that the bomb would be employed.

> From the day that I was assigned to the Project there was never any doubt in my mind but what my mission ... was to get this thing done and used as fast as possible, and every effort was bent toward that assignment. . . .[21]

Given the course of the war and the pace of building the bomb, target planning to drop the bomb on Germany never occurred. As the end of the war in Europe drew near the focus turned to Japan, and how it should be employed there. Over the next month or so at least six options were considered by various sectors of the government, at one time or another:[22]

- Use as a tactical weapon to assist in the invasion of Japan
- Use as a demonstration before observers
- Use as a demonstration against a military target
- Use against a military target
- Use against a city with warning
- Use against a city without warning

Groves was an advocate of the last option, use against a city without warning, the one that was carried out.[23] From his commanding position he was able to set the procedures that assured his preference would dominate, and as a result he was able to successfully outmaneuver or overrun any opposition. Groves did this, of course, with the support of several powerful patrons and allies. The most important of them, Stimson and Marshall, wavered at times over what option to pursue, but in the end supported or acquiesced in Groves's choice.

In a conversation with General Marshall in the spring of 1945 Groves suggested that an officer of the Operations Planning Division (OPD) of the

General Staff be designated to begin more precise operational plans. Marshall surprised, and no doubt pleased, Groves by saying, "Is there any reason why you can't take this over and do it yourself?" While it was "another job . . . dropped into our laps," said Groves, it also eased planning by keeping the circle in the know very small. Marshall's position, and Groves's, was to limit knowledge to as few people as possible. An additional benefit was that it avoided any potential opposition by OPD to the use of the bomb. The more widely known the plans, the more that opposition might emerge and build. Groves informed General Arnold and "[o]ur most pressing job was to select the bomb targets. This would be my responsibility."[24]

He said that for many months target criteria had been discussed with the Military Policy Committee and then with Oppenheimer, von Neumann, Farrell, and Brig. Gen. Lauris Norstad "to be sure that nothing had been overlooked."[25] Groves set five criteria that would largely determine the parameters on how the bomb eventually was used.

Groves said he had set as "the governing factor" that the targets chosen should be places that would most adversely affect the will of the Japanese people to continue the war. It was for this reason that Groves wanted to include the ancient city of Kyoto. As a secondary criterion, the target "should be military in nature, consisting either of important headquarters or troop concentrations, or centers of production of military equipment and supplies." A report that Groves sponsored after the war to assess the effects of the bombings described Hiroshima as

> a city of considerable military importance. It contained 2nd Army Headquarters, which commanded the defense of all of southern Japan. The city was a communications center, a storage point, and an assembly area for troops. To quote a Japanese report, "Probably more than a thousand times since the beginning of the war did the Hiroshima citizens see off with cries of 'Banzai' the troops leaving from the harbor."[26]

In the course of the war Americans had gradually ceased to distinguish clearly between what was military and what was civilian. Military targets were few and hard to hit, civilian targets were many and easy, and cities were the easiest of all.[27]

Of his third criterion, Groves said: "To enable us to assess accurately the effects of the bomb, the targets should not have been previously damaged by air raids." Connected with this was a fourth — that the target should be "of such size that the damage would be confined within it, so that we could more definitely determine the power of the bomb."[28]

Lastly, Groves felt that at least two bombs were going to be necessary, with the second dropped as quickly as possible after the first, to force an end to the war. Groves attributed the idea of two bombs to Admiral Purnell, the navy representative on the Military Policy Committee, though others felt more would be needed.[29] The logic was that the first bomb was necessary to show what it could do, and the second would convince them that the United States could produce bombs in quantity. This would give them an excuse to surrender.

To choose which cities to attack Groves convened a Target Committee to make recommendations. As a consummate bureaucrat, Groves knew that by controlling the choice of who is on a committee one could control the agenda and determine the outcome. He tells us in this regard he always followed the opinion of a wise and successful chief of engineers, Gen. Edwin Jadwin, who once said, "I have no objection to committees as long as I appoint them."[30]

The most important committee member that Groves selected was Brig. Gen. Thomas Francis Farrell. Throughout the last eight months of the war Groves gave Farrell broad responsibilities. Secretary Stimson had become concerned about what would happen to the Manhattan Project if Groves were killed in an accident or died suddenly. According to Groves the precipitating incident was a request around Christmastime of 1944 by the British for Groves to go to England to brief Churchill.[31] Stimson refused to let Groves fly because it was too dangerous, and a boat trip there and back would take at least two weeks. As it turned out, Groves did not go, but it forced the question of a replacement. It was decided to recruit a deputy who would know everything and carry on if Groves were not there. Groves drew up a list of about six Corps of Engineers officers and, after consulting with Nichols, decided upon his old colleague Farrell.[32]

Groves and Farrell had first met at Fort DuPont in Delaware in the summer of 1928, where Farrell, then a reserve officer, came for summer training.[33] Groves recruited Farrell for the first time in early 1941 to be his executive officer in the Operations Branch of the Construction Division during the mobilization. In November 1943 Farrell became chief engineer, Services of Supply, China-Burma-India theater. With Gen. Lewis A. Pick they supervised the building of the famous Ledo Road, one of the major combat engineering feats of World War II.[34] Stimson said that Groves could have anyone in the army irrespective of duty. Farrell was on leave in the United States in January 1945 and it was doubtful that he would return to the CBI theater. Another reason that Groves chose him was that General Marshall knew him, from their time during World War I, and was favorably impressed by him. If Groves were no longer around, Farrell would already have the confidence of the chief of staff, an important asset. He was also well known to Somervell and Styer. Upon taking on his new job his principle assignments were to familiarize himself with every aspect of the project in the event that he might assume command and to act as Groves's substitute on all matters concerned with bombing operations.

Soon after Farrell accepted the job, Groves took him to Los Alamos to introduce him to Oppenheimer and the scientists. After a two-hour briefing Oppenheimer spoke approvingly of Farrell to Groves, noting that he had not asked a single foolish question, a sign he would have the respect of the scientists.

Other members of the target committee included Maj. Jack Derry from Groves's staff, J. C. Stearns, Col. William P. Fisher and Dr. David M. Dennison from the AAF, and technical experts from Los Alamos. In total about a dozen and a half participants attended one or more of the three Target Committee meetings.

At the first meeting, held on April 27, Groves briefed the group emphasizing the need for secrecy and recommended that four targets should be decided upon.[35] Groves relayed General Marshall's opinion that ports on the west coast of Japan should be considered, because they were vital to communicating with the mainland. General Norstad's remarks concerned the resources of the Twentieth Air Force; he offered anything that might be needed. After these introductory remarks Groves and Norstad departed, leaving Farrell in charge.

Much of the meeting, which lasted until 4:00 P.M., was taken up with establishing criteria upon which to choose the targets. There was a lengthy discussion of Japanese weather conditions and cloud cover. Poor weather such as rain or fog would inhibit the blast effect as well as making it more difficult to find the aim point. Good weather and visual bombing only during the day were essential to enhance the bomb's effects. Groves was determined that the bomb be shown to its best advantage. Being able to predict the number of good days over the potential targets would help to determine the date of the mission. An eleven-year span of weather records suggested that conditions would be worst from June through September, with perhaps as few as five good flying days in August. Of course, it would be impossible to predict ahead of time which days they would be.

A related problem was that there were few undamaged cities left to bomb. Beginning in March 1945 the AAF had begun in earnest strategic incendiary raids intended to inflict maximum damage on the population centers of Japan.[36] The results were devastating.

> In all, 178 square miles were razed, amounting to 40 percent of the urban area of the 66 cities attacked. Twenty-two million people, 30 percent of Japan's entire population, were rendered homeless. 2,200,000 civilian casualties were inflicted, including 900,000 fatalities. These more than exceeded Japan's combat casualties in the Pacific of approximately 780,000.[37]

For the atomic attack it was decided that there should be one primary and two alternate targets for each attack. A consensus quickly emerged favoring urban-industrial areas.[38] Of the available targets, the group commented that Hiroshima was the largest untouched target not on the Twenty-first Bomber Command's priority list.[39] "Consideration should be given to this city." The minutes went on to say that:

> . . . the 20th Air Force is operating primarily to laying waste all the main Japanese cities, and that they do not propose to save some important primary target for us if it interferes with the operation of the war from their point of view. . . . The 20th Air Force is systematically bombing out the following cities with the prime purpose in mind of not leaving one stone lying on another: Tokyo, Yokohama, Nagoya, Osaka, Kyoto, Kobe, Yawata & Nagasaki.[40]

After a particular city was chosen, more precise aim points would be determined.[41] At the end of the meeting General Farrell assigned certain tasks to various participants to report on for the second meeting. Dr. William Penney and Dr. John von Neumann were to correlate the data "on the size of the bomb burst, the amount of damage expected, and the ultimate distance at which people will be killed."

The second round of Target Committee meetings was held in Los Alamos in Oppenheimer's office on May 10 and 11. Major Derry and Dr. Ramsey prepared an informative seven-page memorandum for General Groves detailing the deliberations and results.[42] The agenda focused on eleven issues including proposed altitudes for detonating Little Boy and Fat Man so as to maximize destruction on the ground. As Groves later explained, the optimum height of burst was entirely governed by the explosive force. If the altitude of burst were too high or too low, the area of effective damage would be reduced. If it were "much too high, . . . all we would produce would be a spectacular pyrotechnical display which would do virtually no damage at all."[43]

In May there was still some uncertainty over the possible yields of the two types of bombs. For Little Boy they calculated a yield between five thousand and fifteen thousand tons of TNT equivalent. The latter figure turned out to be approximately right. For Fat Man, at this point, the yield was calculated between seven hundred and five thousand tons. These rather low estimates would be revised upward after the Trinity test, and the eventual yield was calculated at twenty-one thousand tons. Because of the yield uncertainty fuses were designed for four different altitudes, from one thousand to twenty-four hundred feet, with fourteen hundred feet the most likely for both bombs.[44]

Dr. Stearns, the air force representative, then presented three criteria for target selection: "(1) they be important targets in a large urban area of more than three miles diameter, (2) they be capable of being damaged effectively by a blast, and (3) they are likely to be unattacked by next August." Stearns listed "five targets the Air Forces would be willing to reserve for our use unless unforeseen circumstances arise."

The preferred city was Kyoto, described by the Target Committee as "an urban industrial area with a population of 1,000,000. It is the former capital of Japan and . . . [f]rom the psychological point of view there is the advantage that Kyoto is an intellectual center for Japan and the people there are more apt to appreciate the significance of such a weapon as the gadget."[45]

Second on the list was Hiroshima, "an important army depot and port of embarkation in the middle of an urban industrial area . . . such a size that a large part of the city could be extensively damaged. There are adjacent hills which are likely to produce a focusing effect which would considerably increase the blast damage. Due to rivers it is not a good incendiary target." The third meeting provided additional information in which it was noted that Hiroshima was a major port of embarkation for the Japanese army and a convoy assembly point for the navy. Its local army headquarters had some twenty-five thousand troops, and there were railway yards, army storage depots, a port on the eastern side, and heavy industrial facilities. The population was approximately 255,000, about the size of Providence, Rhode Island, or Dallas, Texas.

Third was Yokohama, "an important urban industrial area which has so far been untouched" whose "activities include aircraft manufacture, machine tools, docks, electrical equipment and oil refineries."

Fourth was Kokura Arsenal, at more than eight million square feet one of the largest in Japan and surrounded by urban industrial structures. "The arsenal is important for light ordnance, anti-aircraft and beach head defense materials . . . if the bomb were properly placed full advantage could be taken of the higher pressures immediately underneath the bomb for destroying the more solid structures and at the same time considerable blast damage could be done to more feeble structures further away."[46]

Last on the list was Niigata, "a port of embarkation on the N.W. coast of Honshu. Its importance is increasing as other ports are damaged. Machine tool industries are located there and it is a potential center for industrial dispersion. It has oil refineries and storage." The May 28 meeting also cited an aluminum reduction plant, a large ironworks, an important oil refinery, and a tanker terminal.

The possibility of bombing the emperor's palace was discussed, but the committee agreed not to recommend it at this time. More information was needed to determine how effective the bomb would be against the palace.

Psychological factors in target selection were of great importance. The purpose should be to obtain "the greatest psychological effect against Japan" and to make "the initial use sufficiently spectacular for the importance of the weapon to be internationally recognized when publicity on it is released."

Stimson, Marshall, and Groves agreed that a strategy of "shock" was the most effective manner to employ the bomb. As Stimson later recounted in his memoir, "I felt that to extract a genuine surrender from the Emperor and his military advisers, they must be administered a tremendous shock which would carry convincing proof of our power to destroy the Empire." Among his senior military advisers, "General Marshall particularly was emphatic in his insistence on the shock value of the new weapon."[47] Absolute surprise and lack of warning were essential so as to maximize the "shock." Shock was necessary because the Japanese were not on the verge of surrender, as the intercepted diplomatic messages known as Magic showed.[48] Far from it. The Japanese were making extensive plans to repel an American invasion, in their firm conviction that they would either defeat or bloody the Americans enough to break their morale.[49] Only something dramatic could force Tokyo to surrender. By the end of the second meeting the committee recommended four targets: Kyoto, Hiroshima, Yokohama, and Kokura Arsenal.

The third and final meeting was held in Room 4E 200 of the Pentagon on May 28. General Farrell opened the meeting by pointing out that General Arnold and General Groves had concluded that "control of the use of this weapon should reside in Washington rather than be released fully to the theater."[50] Tibbets and Parsons covered various training and scheduling matters. Colonel Fisher announced the three cities that the XXI Bomber Command had reserved, Kyoto, Hiroshima, and Niigata.

Dr. Stearns presented more precise targeting conclusions about the three cities. Specific aim points should be left to a later time when weather conditions would be known; the industrial areas of these cities should not be the pinpoint targets since they were small, on the edge of the cities, and quite dispersed. Instead, planners and pilots should "endeavor to place first gadget in center of selected city."

With the Target Committee's recommendations, Groves prepared the plan of operations for General Marshall's approval.[51] But before he could deliver it to the chief of staff, Secretary Stimson intervened. While Groves was in the secretary's office on the morning of May 30 on another matter, Stimson asked him about the choice of targets and asked to see the report. Groves felt that it should first go to Marshall and was planning to see him the next day, but Stimson persisted. "This is a question I am settling myself. Marshall is not making that decision." Stimson then asked that the report be brought over. Groves demurred, claiming it would take some time. "He [Stimson] said that he had all morning and that I should use his phone to get it over right away."[52]

While they waited for the report to be brought over Stimson asked about the targets and learned that Kyoto was the committee's first choice. He immediately said he would not approve it. Groves tried to change his mind, arguing that a city of over a million must be involved in war work, but Stimson remained firm. Stimson called Marshall in from the office next door and explained why he was removing Kyoto from the list.

Groves has said that Stimson's reasons were that Kyoto was the ancient capital and a historic city with religious and cultural significance to the Japanese. Stimson had twice visited Kyoto — in 1926 and again in 1929 — and had been impressed by its culture.[53] Groves noted another reason.

By preserving at least one city from American air power Stimson could uphold a claim to American humanitarianism, something that would be needed in the coming postwar struggle with Russia.[54]

Groves was nevertheless persistent in trying to get Kyoto back on the target list. Throughout June and well into July Groves kept urging Stimson to include Kyoto. His reasoning was that "it was large enough in area for us to gain complete knowledge of the effects of an atomic bomb. Hiroshima was not nearly so satisfactory in this respect."[55] Less euphemistically, this meant that the bomb would have been shown to full advantage by killing more people and causing more destruction. Had the bomb been dropped on Kyoto, a city of approximately a million, including refugees from burned-out cities, instead of Hiroshima, a city one-quarter the size, or Nagasaki, a city one-fifth the size, it is quite possible that up to a half a million casualties might have occurred.[56] Groves also wanted to hit Kyoto for the shock value. Because Kyoto was an intellectual and cultural center, destroying it, he believed, would have a deeper impact on those (surviving) Japanese who might be able to bring about surrender.

Groves, on a number of occasions, possibly up to a dozen times, continued to urge Stimson to restore Kyoto, but the secretary was adamantly against it.[57] Even after Stimson was in Potsdam Groves kept pressing him to target Kyoto. In a cable on July 21, no doubt inspired by Groves, though sent by Harrison, he tried again.

> All your local military advisors engaged in preparation definitely favor your pet city and would like to feel free to use it as first choice if those on the ride select it out of four possible spots in the light of local conditions at the time.[58]

Stimson replied within hours, "Aware of no factors to change my decision. On the contrary new factors here tend to confirm it."[59]

The next morning Stimson called on Truman and discussed with him the recent interchange of cables with Harrison. "As to the matter of the special target which I had refused to permit, he strongly confirmed my view and said he felt the same way."[60] Just after noon Stimson reviewed the issue with General Arnold, who agreed about not striking Kyoto but got Stimson to permit General Spaatz, the theater commander, to "make the actual selection of targets" for each atomic mission and "coordinate his decision" with Groves.[61] Arnold sent an aide, Col. Jack Stone, back to Washington with instructions to add a new city to the target list — Nagasaki.

On Kyoto Stimson sent another cable on July 23, which ended the matter once and for all. "Also give name of place or alternate places, always excluding the particular place against which I have decided. My decision has been confirmed by highest authority." At this point Groves finally gave up.[62]

Groves said in retrospect that he was glad that he was overruled, "and that through Mr. Stimson's wisdom, the number of Japanese casualties had been greatly reduced."[63] Of course, Mr. Stimson's wisdom had little to do with it. If skies had been clear over Kokura on the second mission, it is quite possible that casualties would have been much higher. No one, neither Stimson, Groves, nor anyone else, had an accurate estimate of how many casualties there might be as a result of dropping Little Boy or Fat Man on any city.

Excluding Kyoto, the target list continued to undergo revisions throughout May and June, with the air force, Groves, and Arnold rearranging the cities. When air force field commanders tampered with the list Groves reasserted control, with General Arnold's support. Air force field commanders had little appreciation of this bomb, Groves believed, and they could not be trusted to show it to its best advantage.[64] On June 14 Groves submitted a revised target list to Marshall that included Kokura, Hiroshima, and Niigata.[65] The Twentieth Air Force was ordered to set aside its own plans and reserve these cities for attack only by the fission weapons of the 509th. Finally, by the end of July, the list was fixed: Hiroshima, Kokura, Niigata, and Nagasaki.

During his meeting with the new president on April 25, Stimson noted that steps were under way to establish a select committee to focus on some of the issues that the bomb presented.[66] A memo from George Harrison on May 1 suggested that "a committee of six or seven be set up at once to study and report on the whole problem of temporary war controls and publicity, and to survey and make recommendations on post war research, development and controls, and the legislation necessary to effectuate them."[67]

Groves had a keen interest in the issues listed by Harrison, and in one more as well. Compton alerted Groves that certain scientists were beginning to question how the bomb might be used. A group at Chicago, organized around James Franck, had begun to raise questions about the bomb and its role in the postwar world. Was the United States going to try to monopolize the weapon, was it going to keep as much as possible secret, or was it going to engage in a cooperative venture to control it? To head off the potential problems that this group might cause, Groves saw that

Stimson's committee could be used to diffuse the issue and run interference while final preparations went forward that could address the issues.[68]

Stimson finally acted and appointed an Interim Committee to assist him in making recommendations to the president.[69] As Harrison's memo makes clear, the purpose of the Interim Committee was not to recommend whether or not to use the bomb but rather to fashion a public relations strategy about the bomb when it was no longer secret, as well as to plan for its role after the war was over.

There were eight original members, all civilians. Secretary Stimson was chairman; his special assistant, George L. Harrison, was alternate chairman. Representing Secretary of the Navy James Forrestal was Undersecretary of the Navy Ralph A. Bard. Representing Acting Secretary of State Joseph C. Grew was Assistant Secretary of State for Economic Affairs William L. Clayton. The other members were Vannevar Bush, James Conant, president of MIT Karl T. Compton, and, representing the president, James F. Byrnes.[70]

The letter of invitation made no mention of wartime use, focusing rather on making "recommendations on postwar research, development and controls, as well as legislation necessary to effectuate them." The name *Interim Committee* was chosen to underscore the intent that it would serve only until Congress established a permanent body.

The first meeting was held on May 9 — the day after Germany surrendered — at 9:30 A.M. in Stimson's office. Seven more meetings were held in the critical period before Hiroshima, the last on July 19. Though General Groves was not a member of the committee, he attended all of the meetings as an invitee and participated upon occasion.[71] From these meetings Groves took his guidance in carrying out his job. According to Groves:

> The purpose of the establishment of the Interim Committee was not so much to obtain its advice but rather to make certain that the American people as well as the leaders of other nations would realize that the very important decisions as to the use of the bomb were not made by the War Department alone but rather that [they] were decisions reached by a group of individuals well removed from the immediate influence of men in uniform. It was for this reason that I was not a member of the Interim Committee, although it is true that I sat with the committed [sic] as an advisor and as the one who knew the most about the entire problem, at every meeting they had. But there was never any question in my mind or the minds of any members of the committee that they were the jury and I was a witness.[72]

At the second meeting, on May 14, the Interim Committee invited four scientists to serve on a scientific panel. Sigal argues persuasively that the panel was formed in response to the rising discontent among some MED scientists, "to act as a surrogate for the scientific community" and to "bless an option for using the bomb already chosen in military channels."[73] The four — Robert Oppenheimer, Enrico Fermi, Ernest Lawrence, and Arthur Compton — were encouraged to discuss technical issues as well as political aspects of the problem. Groves says that the panel members were suggested by him and included the heads of all the major labs except Urey.[74]

The committee members also decided to invite the views of several prominent industrialists involved in the project who could offer advice on other nations' potential to mobilize to build a bomb and catch up to the United States. Four Manhattan contractors, all of whom had been recruited by Groves, presented their views on June 1: Walter S. Carpenter, president of Du Pont; James White, president of Tennessee Eastman; James A. Rafferty, vice president of Union Carbide; and George H. Bucher, president of Westinghouse.

The most important of the eight Interim Committee meetings were the two held on May 31 and June 1.[75] Secretary Stimson chaired the first of these, and the agenda was a full one. It began with a discussion of the current and future stages of nuclear weaponry, including what would soon be called hydrogen bombs. Oppenheimer calculated they could have an explosive force of up to one hundred million tons of TNT. Next, views were presented on the scale of the future domestic program, with the consensus being that the production complex should be expanded to "[b]uild up sizable stock piles of material for military use and for industrial and technical use." Though this was not a specific directive as such, for General Groves it was guidance enough for him to continue production even after the bomb was used.

The problems of control and inspection were discussed with "the question of paramount concern ... the attitude of Russia." There were clearly two points of view on this among the members and invitees. Bush, Conant, at times Stimson, and even General Marshall thought that international control should be pursued and that the sharing of atomic research was a valid approach. Differing with these views were Byrnes, Lawrence, Karl Compton, Arthur Compton, and — from other evidence beyond the committee's notes and minutes — Groves. Though we do not have a verbatim transcript of the meeting, it is clear from Arneson's notes that Byrnes argued most vociferously in defense of his position and won the day.[76] Byrnes was acting as the personal representative of his close friend, colleague, and now new president, Harry Truman. His goal was to protect him from bad advice, whether offered by eminent scientists or high officials. With the shared perspective of their congressional backgrounds Byrnes was conscious of how all this would look to the Senate and the House, once they found out about it.[77] Groves found a like-minded ally in Byrnes in making sure that prompt use of the bomb would avoid any postwar investigations and recriminations.

Dr. Arthur Compton articulated what he thought were compatible positions for a program for the next decade but in fact he unknowingly anticipated one of the major stimuli for the fifty-year arms race that was about to commence.

> Dr. A. H. Compton stressed very strongly the need for maintaining ourselves in a position of superiority while at the same time working toward adequate political agreements [regarding cooperative understanding with Russia].

The Russians, as it turned out, would never quiescently settle for second place, and the United States would never settle for being anything but first.[78] The result was a recipe for the Cold War arms race.

Though the topic of the use of the bomb was not formally part of the Interim Committee's agenda, it did come up. As Compton said, ". . . throughout the morning's discussion it seemed to be a foregone conclusion that the bomb would be used."[79] A brief lunchtime discussion ensued about how the bomb would be used. Compton brought up the possibility of a nonmilitary demonstration, but the drawbacks were quickly identified. It might fail, it may not impress enough, and if there was prior warning it might be interfered with.[80]

> After much discussion concerning various types of targets and the effects to be produced, the Secretary expressed the conclusion, on which there was general agreement, that we could not give the Japanese any warning; that we could not concentrate on a civilian area; but that we should seek to make a profound psychological impression on as many of the inhabitants as possible. At the suggestion of Dr. Conant the Secretary agreed that the most desirable target would be a vital war plant employing a large number of workers and closely surrounded by workers' houses.

General Groves entered the discussion at this point regarding Oppenheimer's proposal of attempting several strikes at the same time. Groves doubted the feasibility, pointing out three objections:

> (1) We would lose the advantage of gaining additional knowledge concerning the weapon at each successive bombing; (2) such a program would require a rush job on the part of those assembling the bombs and might, therefore, be ineffective; (3) the effect would not be sufficiently distinct from our regular Air Force bombing program.

The matter was settled then, and the conversation turned to the minor current of opposition that was forming among some of the scientists. Groves stated that "the program has been plagued since its inception by the presence of certain scientists of doubtful discretion and uncertain loyalty." All knew whom he was referring to — especially Byrnes, who three days earlier had received Leo Szilard, Harold Urey, and Walter Bartky at his home in Spartanburg, South Carolina, to listen to their arguments for not dropping the bomb and for international control.[81] Their pleas fell on deaf ears, in Spartanburg and in Washington.

The notes of the meeting went on to say,

> It was agreed that nothing could be done about dismissing these men until after the bomb has actually been used or, at best, until after the test has been made. After some publicity concerning the weapon was out, steps should be taken to sever these scientists from the program and to proceed with a general weeding out of personnel no longer needed.

For the dissident group, how the bomb was to be used would have important consequences for the postwar period. Their predominant concern was that long-range government funding for science in general, and for atomic energy in particular, might be jeopardized after the war if the bomb were used in such a way as to set off an arms race. That would doom any chance of international control, keep the bomb largely in the hands of the military, and perpetuate the secrecy that they believed was so antithetical to science.

All agreed the bomb had to explode in some fashion, either as a demonstration or as a weapon. A successful explosion would validate the government's funding of the project throughout the war, and future federal support for atomic energy would be more likely. International control was the lynchpin. Wartime control under the military carried out in the greatest secrecy could not be the model for the postwar world. If that came to be, scientists would have to remain under the government's restrictions and do its bidding under layers of classification and secrecy. The more atomic energy could be put under international control, the scientists thought, the better chances were that other nations' desires to acquire a bomb could be curbed.

Concern about these basic issues would inspire the founding of a scientists' movement after Hiroshima. As we shall see, it was Groves who bore the brunt of their criticism, representing all that they wanted to change.

The Interim Committee meeting concluded with a request to members of the Scientific Panel to present as speedily as possible their views on the subject. Their views would help counter some of the criticism of the dissident scientists and help legitimize the decision to use the bomb on Japanese cities without warning. The panel members submitted their recommendations on future technical prospects and the immediate use of nuclear weapons. First among their technical recommendations was a call to develop thermonuclear weapons. If accomplished, "the energy release of explosive units can be increased by a factor of a thousand or more over that of presently contemplated fission bombs." Of equal importance, they urged development of breeder reactors, which could increase the amount of fissionable materials manyfold. The panel also called for exploring the availability and use of intermediate-grade uranium ores, and for support of basic research in nuclear physics in the university and in industry.

On initial use the panel noted that opinions among scientific colleagues ranged from those who wanted a technical demonstration to those who felt that direct military application could induce surrender. The panelists sided with the latter view and saw "no acceptable alternative to direct military use."

On the advice of the scientific panel the committee endorsed Groves's plan to release, as soon as the first bomb was used, sufficient information describing the Manhattan Project and its activities, while protecting what were considered vital secrets about the bomb. The committee reviewed early drafts of press releases prepared under Groves's direction. After making some suggestions, the committee delegated to Groves the responsibility for preparing the final drafts and for releasing the information.

The committee also considered several drafts, provided to it by General Groves, of legislation for postwar supervision and regulation of atomic energy. Many of these drafts were later incorporated into the May-Johnson Bill — legislation that would be defeated, as we shall see, largely as a result of an intense lobbying effort on the part of Manhattan Project scientists.

Thus we have seen that through the "workings of the project" Groves played a central role in shaping the decisions to use the bomb. Groves was too professional an army officer ever to contemplate going against the wishes or policies of his civilian superiors. He never needed to. Groves was carrying out their wishes, in a more determined fashion and at faster pace than they may have realized. He always kept his superiors informed; there is no suggestion that he ever exceeded his authority or ignored his orders.

The bomb was not necessary to end the war, but it was critical in ending it when it did. Had the bombs taken longer to prepare, history might have turned out quite differently. The war might have continued another few months, the Soviet Union might have attacked Japanese forces throughout China, the United States might have invaded the home islands, and peace, when it came, might have been followed by a joint occupation of Japan by the Soviet Union and the United States. But that is not the way things turned out.

What we do know is that Groves succeeded in building atomic bombs by July 1945; that the two dropped on Japan concentrated certain tendencies and forces at work within the ruling circles of Japan; and that the war ended on August 14.[82] All the rest is speculation.

CHAPTER TWENTY

Groves and the Use of the Bomb II:
Trinity, Hiroshima, and Nagasaki

From mid-July to mid-August 1945 was the busiest period of Groves's busy life as he scheduled the test, saw to completing the first two bombs, had them shipped to Tinian, and ensured they were delivered by planes of the 509th Composite Group on targets in Japan. Following his actions in this sequence of events reveals his dominant role. Groves had direct access to the special delivery unit. He wrote the orders for the mission and interjected himself into the command channels. Through the subordinates he placed at key junctures, at Tinian and elsewhere, he had a decisive role in how and when the bomb was used. He kept Stimson, Marshall, and the president informed while they were in Potsdam, and was the first person in Washington to know of what happened in the Pacific.

As if this were not enough he also devoted time to ensuring that the Smyth Report was completed, that press releases and statements were ready, that a preliminary atomic energy bill was drafted for Congress's consideration, and that the Combined Development Trust kept at its work.

Well before the assembly crisis made it imperative, the idea of testing the implosion bomb was proposed by the Los Alamos scientists. In a memorandum in February 1944, Oppenheimer argued that the "implosion gadget must be tested in a range where the energy release is comparable with that contemplated for final use." A full-scale test was essential "because of the incompleteness of our knowledge."[1]

Groves was originally opposed to a test, fearing that the precious, yet-to-be-produced plutonium would be lost if the test were a failure. He reluctantly agreed to a test only if the plutonium could be recovered. The Hanford reactors and chemical plants were months away from completion; no one knew how productive they would be or even how much plutonium would be needed for each bomb. Nevertheless, however much there might be, there had to be a way to recover it if the test failed.

Oppenheimer named Kenneth Tompkins Bainbridge to organize the test effort. Among Bainbridge's first tasks was to select a site to conduct a test. As with the other Manhattan land acquisitions, Groves was heavily involved. The criteria were that the area be flat, so as to minimize the blast effects and facilitate measurement; it should also have good weather and be far enough from population settlements to avoid potential radioactivity problems, but near enough to Los Alamos to minimize travel time. Other considerations had to do with security, ease of construction, and the difficulty of transporting Jumbo (see below). One other stipulation came from Secretary of the Interior Harold Ickes. The secretary stated that under no circumstances should a single Indian be displaced.[2]

Eight different sites were considered from map data. Auto trips were made to five, and aerial surveys were made by one or another of the group, which included Bainbridge, Robert Henderson, Maj. Wilbur A. "Lex" Stevens, and Maj. Peer de Silva. The choices narrowed to a desert training area near Rice, California, or the Alamogordo Bombing Range in the Jornada del Muerto Valley, in central New Mexico. After consultation with Maj. Gen. Uzal G. Ent, commanding general of the Second Air Force, on September 7, 1944, the commanding officer of the bombing range was approached and an area roughly eighteen miles by twenty-four miles in the northwest corner of the range was selected. The nearest towns were twenty-seven to thirty miles away, and the prevailing winds were westerly. The fact that it was already federal land appealed to Groves, who gave his agreement. Later in the month Bainbridge visited the site and staked out ground zero and three observation sites.

Plans for a base camp were drawn up in October, as was a survey of proposed scientific measurements. The documents were transmitted to General Groves on October 14, 1944, requesting his approval. Groves approved the plans; early in November contracts were signed through Major Stevens. The initial plan was to eventually house and feed 160 military and civilian personnel at the base camp. The camp was completed in December and a small detachment of about a dozen military police under 1st Lt. Howard C. Bush, commanding officer of Trinity base camp, took up residence. By July 1945 the number approached three hundred, but it fluctuated with constant coming and going. The precise number of personnel at the test site at the time of the detonation has not been documented. Film badge records show that approximately 355 people were at the test site sometime during July 16.[3]

At some point the name *Trinity* began to be used to describe the site and the test. It is generally agreed that Oppenheimer picked the name, but he could never quite explain why. In a 1962 letter to Oppenheimer Groves asked how and why it was selected.

"There is no rational answer to your question about the code name Trinity," Oppenheimer wrote back. "I did suggest it. . . . Why I chose the name is not clear, but I know what thoughts were in my mind. There is a poem of John Donne, written just before his death, which I know and love.

> As West and East
> In all flatt Maps — and I am one — are one,
> So death doth touch the Resurrection

That still does not make Trinity; but in another, better known devotional poem Donne opens, 'Batter my heart, three person'd God; — .' Beyond this, I have no clues whatever."[4]

Groves's decision to build Jumbo is another example of his cautious and conservative approach to confronting problems that faced him. Of the several ways that the impending test might fail, one of the more serious would be if just the high explosive detonated, causing the plutonium to scatter and disperse. This could not be allowed to happen. Groves thus set the Los Alamos scientists and engineers to work investigating several methods of recovery. The only option that showed promise was to place the test device inside a massive steel vessel, which would contain the force of the fifty-three hundred pounds of Composition B high explosive (TNT and RDX) used to implode the thirteen pounds of plutonium. Initial plans called for a thirteen- to fifteen-foot sphere weighing eighty tons. The container was immediately dubbed Jumbo.[5] A search began for a company to make it and a way to transport it to the test site.

As improved designs evolved, Jumbo eventually grew to a length of twenty-eight feet, a diameter of twelve feet, eight inches, and a weight of 214 tons. In early August 1944 Babcock and Wilcox of Barberton, Ohio — a company with experience in the construction of high-pressure containment systems — agreed to take the job. To ensure that it was supplied with everything it needed, Groves used the AAA preference rating to obtain certain items from subcontractors.

After the Alamogordo test site was approved in September, Groves confronted the problem of how to transport Jumbo from Ohio to New Mexico. As he had done many times before, he drew upon already existing organizations; for this job the War Department's Office of Transportation was contacted. For "high, wide and heavy" loads Clearance Chief Capt. Alexander Whilldin Jr. was the army's man in charge. He knew every potential transportation bottleneck in the country, from low bridges to narrow tunnels, soft railbeds, shallow waterways, and obstructing power lines. Whilldin's job was so novel that toward the end of 1944 *Liberty* magazine scheduled an article describing some of his more unusual assignments. When Groves's security officers learned that one of the examples would be Whilldin's search to find a route for Jumbo they pressured the magazine to remove the sensitive paragraph and were successful.[6]

Meanwhile, typically Manhattan-sized preparations were being pursued to transport Jumbo on its fifteen-hundred-mile journey. Jumbo was to be shipped on a special railcar — the only one of its kind, and the property of the Carnegie-Illinois Steel Corporation. The twelve-axle car was used to transport huge ladles of molten steel in the mill and had a depressed center section between its oversized wheels. Jumbo's weight of 214 tons, plus the 157-ton weight of the special car, limited what railroad tracks were available.

Whilldin found a route, though the security-conscious Groves did not divulge the final destination, implying that it was Los Angeles, the terminus of the Atchison, Topeka & Santa Fe Railroad. The train started from Barberton, Ohio, in early April, traveling to Griffith, Indiana, to Joliot, Illinois, and then south along the Mississippi to New Orleans, never exceeding thirty miles an hour. Only after it left New Orleans in mid-April 1945 did Groves inform the secret convoy of its true destination. Though anyone seeing it ride by would have been hard pressed to guess its purpose, Jumbo was covered with a tarpaulin so as not to arouse curiosity. Jumbo's train journey ended at a railroad siding at Pope, New Mexico, thirty miles from ground zero. It rested there for a few days on a special spur that had been built for that purpose.[7] For the final leg Jumbo was loaded onto a sixty-four-wheel trailer, specially built for the job.[8] Groves's subordinates had gone to B. F. Goodrich for wide-bottomed tires that would not sink in sand to carry the terrific load across the desert.[9] Four Caterpillar tractors pulled and pushed the seventy-three-ton trailer, with Jumbo aboard, across the desert at a speed of three miles per hour.

Soon after Jumbo reached the Trinity site in early May it was decided that it would not be needed. Adequate amounts of plutonium were being produced on schedule, there was greater confidence that the implosion design would work, and setting the gadget off inside Jumbo would prevent some crucial diagnostic measurements from being recorded. So on May 15

Groves, in consultation with Oppenheimer and his advisers, decided not to use Jumbo. Nevertheless Jumbo was upended and placed vertically in a steel scaffold twenty-four hundred feet from ground zero. While the July 16 blast demolished the scaffolding, it left Jumbo unscathed.

Though Jumbo turned out to be a twelve-million-dollar white elephant, it is a good example of Groves's approach, ever cautious and conservative. It was better to have a hedge to fall back on if needed than to be caught short when the time came. The challenge of building something unique for a unique purpose was commonplace for the Manhattan Project. Under Groves's relentless pressure specialized companies were found to build the behemoth and to transport it, the largest thing ever shipped by rail.

A confluence of factors including international politics, the weather, and the sheer ability to complete the bomb in time determined the date of the test.

It is clear that the opening of the Potsdam conference was deliberately postponed by Truman until after it was known whether the test was successful or not. In early May Truman had agreed to meet with Stalin and Churchill on July 1. Groves was being pressured by his superiors to conduct the test as soon as possible, and he in turn pressured Oppenheimer and the scientists to speed up their preparations.

In early March, July 4 was set as the test date, but it was clearly an unrealistic target. As certain problems were solved, others emerged. In a mid-June memo to his group leaders Oppenheimer sought to clarify the schedule. By agreement of the Cowpuncher Committee,[10] which had been formed to "ride herd" on the implosion program, he noted that July 13 was the earliest possible date, but he warned that it could be as late as July 23.

> In reaching this conclusion we are influenced by the fact that we are under great pressure both internal and external, to carry out this test, and that it undoubtedly will be carried out before all the experiments, tests, and improvements that should reasonably be made, can be made.[11]

In a phone call from Groves, the same day, Oppenheimer told him that there was a very small chance of making July 15 but no reason "why we could not make it by the 25th of July." Oppenheimer went on to lay out schedules:

> If the necessary things went by air, the Little Boy could be at destination August 5th; if by sea, August 20th. That this was pretty tight for

the Fat Man, but JRO thought August 10th if the things went by air and August 25th if by sea.[12]

By the end of June the date slipped to July 16. On July 2 Groves called Oppenheimer to discuss the test schedule. Oppenheimer proposed the seventeenth to be safe but told Groves

that the 14th was possible but was not sure. Dr O thought the wisest thing was to schedule it for the 17th in which case they would be fairly sure of getting the thing done within a few days of that day. . . . GG [General Groves] said that he did not like the idea of a later date because of the various things that were involved.

The "various things that were involved" more than likely referred to the upcoming meeting of Truman, Churchill, and Stalin at Potsdam. Later in the conversation, "GG then told Dr O the reason why the earlier date had to be. Dr O said that they would meet the earlier date but it went against his own feeling but if the Gen wanted it that way they would do it."[13]

At 3:30 P.M. Groves called Conant, who agreed that the general was right to push for the earlier date and suggested that Tolman be contacted at Los Alamos to exert his efforts on Oppenheimer and the others to meet it. An hour later Groves spoke with George Harrison, telling him "that he had told Oppy that they had to have the first date because of things beyond his control." At 5:45 P.M. Tolman called the general from Los Alamos. "GG said since he had talked to Y the upper crust wanted it as soon as possible," and though he was not in agreement there was nothing he could do about it. Oppenheimer then got on the line, and Groves stressed to him "the importance of trying to arrange for the 14th and . . . to tell his people that it wasn't his fault but came from higher authority."

The following day Oppenheimer called with bad news that the early date was not possible because of problems with the high-explosive castings; another date was proposed. Over the next week Groves continued to urge Oppenheimer to have the gadget ready as soon as possible. Finally the date of July 16 became fixed, with a shot time of 4:00 A.M.

In addition to Groves's direct concerns over scheduling the test there were other matters that needed attention, including safety precautions against the possibility of radioactive fallout. To handle medical questions, he approached one of the few doctors in the country working in the new field of radiology.

Stafford Leak Warren was born in 1896 in New Mexico and graduated from the University of California, Berkeley, in 1918. After medical school he taught at the University of Rochester from 1925 to 1943. In March 1943 Albert K. Chapman, an Eastman Kodak vice president, arranged a luncheon meeting at the Rochester Club with Groves and Col. James C. Marshall. During lunch on March 19 Dr. Warren was asked about his work with radiation and isotopes. After lunch Chapman left, advising Warren to do whatever the officers asked of him.

> The officers then took Dr. Warren to a private room, where one of them closed and locked the door and closed the transom while the other looked into the closet. Both of them then looked out of the closed window, and, when they sat down, there was a moment of quiet, during which they seemed to be listening.[14]

With the room secure Groves asked Warren if he would be interested in working on an important medical program for the government. When Warren replied that he was busy, Groves countered that replacements could be found for his other work and research funds supplied. Groves stressed the importance of the work but said it was so confidential that nothing could be explained until Warren accepted the offer. The fifteen-minute meeting concluded with Warren being given a telephone number of another doctor already on the staff in New York to whom he could talk. Shortly thereafter Warren joined the project as a consultant and soon after was chief medical adviser to Groves and head of the MED medical office.

As an author of a book about radiation safety during the Manhattan Project has put it, "Safety never commanded topmost concern at Los Alamos. Getting the job done came first. In testing the bomb, however, safety may have ranked even lower than normal."[15] Groves was concerned about radiation and also the prospect of earth shock. If there were problems in either of these areas, angry citizens could bring lawsuits against the government for health or property damage.

So some safety precautions were taken at the site. Three observation shelters located ten thousand yards from ground zero were built of wood with walls reinforced with concrete under layers of earth. Each shelter was under the supervision of a scientist until the shot was fired, when a medical doctor was in command.[16] If the radiation instruments indicated that one or more of the observation posts needed to be evacuated, vehicles were on standby with drivers familiar with the desert roads.

Fallout spreading beyond the site was a bigger concern. Shortly before the test Joseph O. Hirschfelder and John Magee of the Theoretical Division revised their conclusions about fallout, introducing some troubling concerns.[17] An off-site contingent of 160 enlisted men, under the command of Maj. T. O. Palmer, was stationed north of the test site with enough vehicles to evacuate the ranchers, farmers, and townspeople if necessary. Palmer had enough jeeps and trucks to remove up to 450 people and enough tents, food, and other supplies to last for two days.[18]

Earlier Groves had called Gov. John J. Dempsey and warned him that he might be asked to declare martial law in southwest New Mexico.

Weather was a vital factor in determining when to test. Accurate long-range and short-range forecasting of the weather was essential to picking the best date and time. Among Groves's least favorite people during his Manhattan years was Jack M. Hubbard, a meteorologist who joined the project in April. The choice of Hubbard was arrived at in a rather haphazard way, one that Groves later acknowledged as one of his mistakes. So as not to further exacerbate the military-civilian tensions at Los Alamos, Groves let Oppenheimer seek an academic candidate for the job. Oppenheimer asked Cal Tech president Robert A. Milliken for a meteorologist, and Milliken asked one of his faculty, a Professor Krick, to recommend someone for a special assignment. Krick did not know what the assignment would be and rather casually chose one of his lesser-qualified students, Jack Hubbard.

As the test date approached, Groves saw that Hubbard was not the person for the job and arranged for an air force meteorologist to be brought in to give him advice. The person sent was Col. Ben G. Holzman, one of the meteorologists who had selected the date for the Normandy landing. In a letter to Groves some years later Holzman corroborated Groves's opinion of Hubbard and the poor quality of the meteorological advice. "[I]t was quite a black spot on the science of meteorology for one of the most important events of our time. . . . I consider the meteorological operations for this most important historical event to be a travesty."[19]

Finally it was time for the test.[20] In the six months leading up to it Groves visited Los Alamos five times.[21] He spoke by phone with Oppenheimer and the others almost daily about the test preparations and other matters. As with almost everything else he did, Groves calculated his arrival with a larger purpose in mind. He did not want his presence to interfere with the hectic preparations at Trinity, so he planned to arrive at the site only a few hours before the time of the test.

On Wednesday morning, July 11, Groves left Bolling Field with Vannevar Bush and 1st Lt. John J. O'Connell (son of the chief inspector of the New York City Police Department), using the general's plane, piloted by Lieutenant Neal, to tour some of the facilities on their way to New Mexico. After a short stop in Omaha, they spent the night in Ogden, Utah, and on Thursday morning flew to Hanford, arriving at 10 A.M. They spent the rest of the day and the next morning touring Hanford and flew to Oakland, arriving on Friday afternoon. They stayed at Groves's favorite San Francisco hotel that night, the St. Francis on Union Square. They picked up Lawrence in Berkeley, flying on Saturday morning to Inyokern to check on the production rate of the lenses,[22] and then to Pasadena in the afternoon, where they spent the night. On Sunday morning they flew from Pasadena to Albuquerque, arriving at 12:40 P.M., according to Groves's travel orders. James Conant had flown to New Mexico separately and met up with Groves and Bush. From there they drove down Route 85, which paralleled the Rio Grande, turning east after San Antonio to the Trinity Base Camp, arriving in the late afternoon.

According to Groves the three reached the base camp at around 5:00 P.M.[23] At about 8:00 P.M. thunder and lightning began over the area, along with winds up to thirty miles per hour. After dinner at the mess Groves found Oppenheimer in the center of a large group of scientists who were urging him to postpone the test. Groves quit listening to Hubbard fairly early in the evening when he saw he was "obviously confused and badly rattled." Holzman said Hubbard was "frantic and incoherent." "From that time on I [Groves] made my own weather predictions with some advice from an Air Force meteorologist [Holzman] whom I had obtained only a few days before as I did not want to depend on a single man."[24]

Groves led Oppenheimer away from the group and went to his office, "where matters could be discussed calmly and without the unasked advice of so many young people totally unaccustomed to responsibility."[25] Groves was determined that the test was going to happen, and he listed the reasons to Oppenheimer. First, the electrical connections would get thoroughly soaked. It would take days for them to dry out, and even then the chances of a misfire would increase. Second, everyone was so keyed up that a test had to happen to break the tension. A postponement would be unendurable. Finally, there was the situation in Potsdam, which made delay impossible. At this point, at around eleven o'clock, Groves urged Oppy to go to bed; he left to get some sleep as well. Bush and Conant shared a tent with the general. With the wind and the rain, the tension, and a flapping tent canvas Bush and Conant got little or no sleep. None of this bothered Groves. As usual he fell asleep amid all of the commotion and strain, and slept soundly for about an hour.[26]

Up until the last minute Groves feared that the bomb might be sabotaged. He had Bainbridge, Kistiakowsky, and Lieutenant Bush spend the night under the bomb tower.

At 1:00 A.M. Groves arose and went to South 10,000, which served as the control point, and spent the next four hours with Oppenheimer, calming and reassuring him. By two o'clock the weather began to look better. After a discussion with his advisers on what the weather might be like over the next few hours, Groves decided that the test would occur at 5:30 A.M., rather than at the original time, 4:00 A.M. The decision was Groves's, and his alone.[27]

At 5:10 A.M. Groves returned the four miles to base camp to prepare for the blast.[28] With Bush and Conant on either side of the general the three lay facedown on a tarpaulin with their feet toward ground zero. They closed and covered their eyes.

> As we approached the final minute, the quiet grew more intense. I myself, was on the ground between Bush and Conant. As I lay there, in the final seconds, I thought only of what I would do if, when the countdown got to zero, nothing happened.[29]

Through the nearby loudspeaker they heard Samuel Allison count down the final seconds. Then came the burst of white light that filled the sky — "like the end of the world," wrote Conant. Conant had expected a relatively quick and bright flash, and the "enormity of the light and its length stunned" him. His "instantaneous reaction was that something had gone wrong and that the thermal nuclear transformation of the atmosphere, once discussed as a possibility and jokingly referred to a few minutes earlier, had actually occurred." The three turned around, sat up, and used the welder's thick, dark glass to view the tremendous light.[30] Shortly they lowered the glass and saw the purplish mushroom cloud form and rise. For Groves, "[T]he first reactions of the three of us were expressed in a silent exchange of handclasps."[31]

Groves also reflected on his precarious position:

> I personally thought of Blondin crossing Niagara Falls on his
> tightrope, only to me this tightrope had lasted for almost three years,
> and of my repeated, confident-appearing assurances that such a thing
> was possible and that we would do it.[32]

Enrico Fermi was with them at the base camp, and Groves described
Fermi's procedure of dropping pieces of paper as the shock wave hit to
measure the yield.[33] Groves had planned to discuss upcoming plans with
Oppenheimer and the others, but the event that they had just witnessed
left no one in a frame of mind to discuss anything.

Shortly after the test, General Farrell and Oppenheimer returned to the
camp from the control point. As Farrell congratulated Groves, his first
words were, "The war is over." Groves's reply was, "Yes, after we drop two
bombs on Japan."[34] Groves seems to have been persuaded by Admiral
Purnell that as few as two bombs would be sufficient. As he later
recounted:

> [I]t was not until December of 1944 that I came to the opinion that
> two bombs would end the war. Before that we had always considered
> more as being more likely. Then I was convinced in a series of dis-
> cussions I had with Admiral Purnell. Purnell had been Chief of Staff
> to Hart; he had spent a lot of time in the Philippines and also I think
> he had been on the China Patrol. He had much more background in
> the workings of the oriental mind than I had. In discussing it with
> him he said that he thought that two bombs would be sufficient. He
> said, "One is necessary to show what it can do, and the second one
> will convince them that we can produce them — that they are in
> production and that will give them an excuse (to surrender)."[35]

General Marshall also seems to have had the idea that two bombs
would be enough. During a break at the Potsdam conference he went to
Berchtesgaden to see Gens. Maxwell D. Taylor and George S. Patton. On
July 28, after lunch, General Marshall told them of the Trinity test and the
development of the Manhattan Project under General Groves (Taylor's old
tennis partner), saying, "Gentlemen, on the first moonlight night in
August, we will drop one of these bombs on the Japanese. I don't think
we will need more than two."[36] As we shall see, Groves, ever the prudent

military officer who planned for every contingency, was preparing several more bombs if two were not enough.[37]

Less than an hour and a half after the blast Groves called Mrs. O'Leary in Washington.[38] A prearranged code was used to pass on the results of the test. As soon as Mrs. O'Leary finished decoding the general's message, she went to George Harrison's office in the Pentagon to help draft the first cable to Secretary Stimson in Potsdam.

> Operated on this morning. Diagnosis not yet complete but results seen satisfactory and already exceed expectations. Local press release necessary as interest extends a great distance. Dr. Groves pleased. He returns tomorrow. I will keep you posted.[39]

Stimson received the message at 7:30 P.M. (1:30 P.M. Washington time). As he noted in his diary, ". . . I took it at once to the President's house and showed it to Truman and Byrnes who of course were greatly interested, although the information was still in very general terms."[40]

Back at Trinity there were other matters that concerned Groves that Monday morning. There was the question of radioactivity and whether it would be necessary to evacuate anyone.

There was also the matter of an inquisitive press. Obviously a huge explosion and bright light were going to arouse curiosity. At about 11:00 A.M. Groves heard from the officer he had stationed in the Associated Press office in Albuquerque, to guard against any revealing or alarming dispatches from being sent out, that the AP was about to write its own account if the army did not respond promptly. As usual Groves had anticipated the problem. He had planned for such an eventuality weeks before, and was ready with a cover story. He had chosen 1st Lt. Walter A. Parish Jr., "a very smooth young Texas lawyer" who was executive officer of the Intelligence and Security Division in the Washington office, to arrange the details. Parish first went to see Col. William O. Eareckson, commanding officer of the Alamogordo Air Base, with a letter from Gen. R. B. Williams, commanding general of the Second Air Forces at Colorado Springs, instructing him to carry out any orders that Parish might give him. It is always a delicate situation for a colonel to take orders from a lieutenant, and Groves noted Parish's "calm unfailing courtesy and great firmness," qualities that caused him to pick him for the job in the first place. The colonel turned over the responsibilities to his executive officer to work with Parish. The B-29s at Alamogordo were grounded for a few hours while the test took place.

Groves made a final edit to the press release and told Parish to give it out at once. It read:

> Alamogordo, N.M., July 16
>
> The commanding officer of the Alamogordo Army Air Base made the following statement today:
>
> "Several inquires have been received concerning a heavy explosion which occurred on the Alamogordo Air Base reservation this morning.
>
> "A remotely located ammunition magazine containing a considerable amount of high explosives and pyrotechnics exploded.
>
> "There was no loss of life or injury to anyone, and the property damage outside of the explosives magazine was negligible.
>
> "Weather conditions affecting the content of gas shells exploded by the blast may make it desirable for the Army to evacuate temporarily a few civilians from their homes."

Groves, Conant, and Bush left the base camp at about 3:00 P.M. and flew from Albuquerque at 4:00 P.M., along with Lawrence and Tolman. While they talked of what they had just seen, Groves was clearly focused on his next goal. "As for me, my thoughts were now completely wrapped up with the preparations for the coming climax in Japan."[41]

They arrived at Nashville after midnight and stayed the night. They departed Nashville at 8:00 A.M. and arrived in Washington at approximately 1:10 P.M. Groves walked in his office at about 2:00 P.M., cleaned up, and immediately went to the Pentagon to see Harrison.[42] That afternoon another cable was sent to Stimson prepared by Groves and Harrison.

> Doctor has just returned most enthusiastic and confident that the Little Boy is as husky as his big brother. The light in his eyes discernable from here to Highhold [Stimson's country estate on the north shore of Long Island, a distance of 250 miles] and I could hear his screams from here to my [Harrison's] farm [in Upperville, Virginia, a distance of 50 miles]."[43]

Harrison suggested that a longer memo be prepared for Stimson. Upon returning to his office Groves began the report and then recruited Farrell to help, as Groves had an appointment with the British ambassador he could not break. When he returned to the office around 7:00 P.M., they worked on it until well past midnight, using Mrs. O'Leary and Farrell's secretary, Miss Elliot, to type and retype draft after draft. Finally the report was finished and taken straight to the airplane, which had been held for them, for delivery by a special courier to Stimson in Potsdam.

Dated July 18, 1945, the memo has become one of the more widely quoted documents about the atomic bomb.[44] Groves provided the secretary with the basic facts as they were known at that point and with a graphic account of what had transpired. The energy generated was in excess of fifteen thousand to twenty thousand tons, a conservative estimate, he told the secretary. In addition to the tremendous blast effects there was, for a brief period, a lighting effect "equal to several suns in midday," after which "a huge ball of fire . . . formed which lasted for several seconds." The light was seen clearly in Albuquerque and other points to about 180 miles away; the sound was heard to about 100 miles. A crater was formed twelve hundred feet in diameter, and the steel tower evaporated. The structure holding Jumbo was twisted and ripped apart. "I no longer consider the Pentagon a safe shelter from such a bomb," Groves informed the current senior occupant. He told Stimson about the weather delay and the press release and included some impressions of his deputy, General Farrell. Farrell, a Catholic, used some religious imagery and phrases to describe his thoughts and emotions "at the birth of a new age." After the air blast came "the strong, sustained awesome roar which warned of doomsday and made us feel that we puny things were blasphemous to dare tamper with the forces heretofore reserved to The Almighty."

The courier arrived in Potsdam Saturday morning and delivered the report. Stimson read it, and wrote in his diary later that day:

> It was an immensely powerful document, clearly and well written and with supporting documents of the highest importance. It gave a pretty full and eloquent report of the tremendous success of the test and revealed a far greater destructive power than we expected in S-1. . . .

Stimson took the report to Marshall and had him read it. They conferred about it, and then Stimson went to the Little White House to see Truman. Stimson read the report in its entirety to the president and to

Byrnes, and later informed Churchill. "The President was tremendously pepped up by it and spoke to me of it again and again when I saw him."

For Groves the days following Trinity were even busier and more pressure filled than those that had come before. His final "preparations for the coming climax in Japan," as he termed it, involved the prompt transport of all of the components of the bombs to Tinian, the drafting of orders from Washington to the field authorizing the missions, the successful delivery of the bombs on their targets, and the preparation of public statements, press releases, and a lengthy report to inform the world of this new weapon. He was, as he said, "the prime mover in all the planning and the control of operations."[45]

The uranium for the Little Boy bomb was transported to Tinian in two "Bronx" shipments, one by ship and one by air.[46] The reason was, in part, the short time span between completing the enrichment at Oak Ridge and the demanding timetable Groves had set for combat use. There was not time to fabricate two solid pieces in the United States and then ship them to the Pacific. Rather, the projectile and the target insertion were composed of rings of HEU, a more modular design that allowed pieces to be shipped when they were finished to be fully assembled at Tinian.

The rings themselves were not of a standard size, nor were the target insert and the projectile of equal weight. Approximately thirty-eight kilograms of HEU, or 60 percent of the total HEU, were in the projectile, and made up of nine large washers bound together, about seven inches long, six and one-quarter inches in diameter, with a four-inch hole in the middle.[47] This portion was completed and shipped first.

To escort this most precious cargo General Groves chose one of his closest aides, Maj. Robert F. Furman, who was just back from Europe after helping round up the German scientists and seeing that they were settled in at Farm Hall. Groves told Furman to proceed to Los Alamos. There Capt. James F. Nolan from the Health Group joined him.[48] As a doctor, he would serve as a radiological officer as well. The two were ordered to wear the branch insignia of artillery officers and to hide the true purpose of their mission from everyone. After a dry run of transporting the uranium from Los Alamos to Santa Fe on July 13, a convoy consisting of a closed black truck preceded by four cars full of security men and followed by three more departed Los Alamos on Saturday morning, July 14, two days prior to Trinity. Inside the truck was the uranium projectile encased in a cylinder about eighteen inches in diameter and two feet long. The convoy proceeded to Albuquerque, passing through Santa Fe without incident except for a tire blowout on the car occupied by Furman and Nolan.

At Albuquerque three DC-3s were standing by at Kirtland Field. Two of the planes were for the security men. The uranium was carefully strapped to a parachute and put aboard the third plane, along with Furman and Nolan. The planes took off and landed at Hamilton Field, near San Rafael, twenty miles north of San Francisco, where even more security men met the group. The convoy wound its way through the streets of the city to Hunters Point Naval Shipyard. The uranium spent Saturday night and all day Sunday, the fifteenth, in the commandant's office.

Groves had Admiral Purnell and Captain Parsons arrange with the navy to have a speedy ship available to transport the uranium. Of course, as few as possible were told the purpose; even the captain was not to be informed of the secret cargo.[49] The ship chosen was the USS *Indianapolis* (CA-35), a fast heavy cruiser commissioned in 1932. Purnell went to San Francisco and told the captain that if the ship ran into trouble, the small cylindrical container was to be saved at all cost. The ship was to be tracked carefully as it made its way to Tinian; if anything happened to it, it would have been known within hours.

At about 3:00 A.M. on Monday, July 16, the *Indianapolis* was ordered to prepare to get under way. An hour later two army trucks came alongside, one carrying a large crate and the other a small metal cylinder. A large gantry crane quickly put both aboard. The crate was secured to the deck and surrounded by a marine guard. The cylinder went to the flag lieutenant's cabin, which was occupied by Furman and Nolan. A series of rings was welded to the floor with steel straps on hinges. The cylinder was put in the middle; the straps secured it in place. Furman affixed a padlock to the makeshift cage and put the key in his pocket.

As this activity was taking place on the dock at Hunters Point, the atomic age was being ushered in several hundred miles away at Trinity. At 8:00 A.M. the *Indianapolis* cast off and half an hour later passed under the Golden Gate Bridge, westward bound. Shortly after daybreak on Thursday, July 26, the *Indianapolis* reached Tinian. Without an adequate harbor the ship stopped a thousand yards from shore and an LCT (landing craft tank) came alongside. The crate and the cylinder were put aboard by crane. Captain Parsons met Furman and Nolan and informed them of the success at Alamogordo ten days earlier.

The rest of the uranium that made up the other portion of the bomb arrived at Tinian two days after the *Indianapolis* delivery. The target insert was made of six rings, smaller in diameter to fit into the hole of the projectile. The total weight of it was approximately twenty-six kilograms. They were cast later than the projectile rings and flown, two each, as the sole cargo aboard three C-54 Green Hornet flights. The planes departed from Kirtland Field the morning of July 26, about the time the *Indianapolis* was unloading its cargo off Tinian. The planes flew to Fairfield Sunsun airfield, outside San Francisco, and then across the Pacific to Tinian, arriving on July 28.[50] Escorting the shipments were intelligence officers Lt. Col. Peer de Silva and, from Groves's office, Maj. Claude C. Pierce.[51] Navy Lt. Cmdr. Francis A. Birch, in charge of assembly, also escorted the shipment.

With all of the components now in hand on Tinian, the final assembly of the Little Boy took place.[52] On July 30 the projectile and the target insert were positioned inside the casing (Unit L-11). When brought together by the force of the cordite high explosive to critical assembly, the two uranium pieces would be about seven inches long and 6¼ inch diameter. The following day the other components — the eight timers, the four radar devices for altitude determination ("Archies"), and the six barometric switches — were installed and checked. The completed Little Boy bomb was 120 inches long, 28 inches in diameter, and weighed about ninety-seven hundred pounds. A few minor last-minute adjustments needed to be made, but those could wait until there was a favorable weather forecast. Basically the bomb was ready for delivery on July 31.[53] Now it was up to the weather.

Groves of course knew of the plan to issue an ultimatum to Japan. It had been discussed during the Interim Committee meetings. He also knew that there would have to be a certain period of time between its issuance and the dropping of the bomb. The Potsdam Declaration was issued on July 26, and the bomb was basically ready on July 31. Groves later said that had it been ready much earlier, he would have had to contact Secretary Stimson for permission to drop it earlier. The weather would make the issue moot.

Getting all of the components of the Fat Man bomb to Tinian went on simultaneously with preparing Little Boy. The plutonium, the "active material for the hot FM," as it was referred to, had arrived on July 28, brought from Los Alamos by couriers Peer de Silva and scientist Raemer Schreiber.

Three B-29s departed from Mather Field in Sacramento on July 29 carrying Fat Man bomb cases, preassembly units F31, F32, and F33. The plane, the *Laggin' Dragon,* the last to take off, carried the F31 unit — which would soon be dropped on Nagasaki — in the front bomb bay; in the rear bomb bay was a ten-foot statue of Christ that was being taken to Tinian for one of the chaplains.[54]

Soon after clearing the ground, one of the panels on the right side of the plane came off. The life raft and other emergency gear flew out and wrapped itself around the tail, causing the plane to shake and vibrate. The pilot managed to turn the plane around and make an emergency landing. After making repairs, they left again around midnight for Hickam Field, Hawaii, and then arrived at Tinian at 12:30 P.M., August 2.[55]

Ever since arriving on Tinian, B-29 crews of the 509th had conducted daily solo runs, dropping gun- and implosion-assembly-type practice bombs. In order to season the pilots, real Japanese targets were chosen. The practice bombs contained only high explosives. The reports of these flights indicated that there was little interest in whether or not the bombs hit their targets (typically aircraft manufacturing plants); their only purpose was to give the pilots the experience of the sudden loss of a five-ton bomb and the opportunity to practice the power bank and turn required to get as far away as possible after release of the bomb before detonation. Since each of these attacks was carried out by a single unaccompanied plane, the Japanese made no effort to counterattack.

On July 22 General Marshall, from Potsdam, directed Gen. Thomas T. Handy to prepare a tentative directive for submission to the secretary of war and himself.[56] In response Handy sent back a directive, drafted by Groves, that became the formal order for use of the atomic bomb. Marshall cabled back to Handy on July 25 that the secretary had approved the directive.[57]

Acting Chief of Staff Gen. Thomas T. Handy issued the directive, to Gen. Carl "Tooey" Spaatz, commanding general, and U.S. Army Strategic Air Forces.[58] The directive called for the 509th to "deliver its first special bomb as soon as weather will permit visual bombing after about 3 August 1945 on one of the targets: Hiroshima, Kokura, Niigata and Nagasaki."[59] The second paragraph ensured that no further orders would be needed to continue dropping them on the target cities after the first one. "Additional bombs will be delivered on the above targets as soon as made ready by the project staff."

In actual fact authorization for the operation had been approved by Marshall much earlier. Handy's order was not essential, but he apparently preferred not to take responsibility and sent it to Marshall in Potsdam for final approval by the secretary and the chief of staff.

In effect, these orders made the final operational decision to use the bombs the responsibility of the commander in the field. This meant that Groves, through his representative on Tinian, General Farrell, was granted the authority to drop atomic bombs on Japan as they became available — predelegation as it would be known later, during the Cold War. At about the same time Groves transmitted to Handy a schedule that called for the use of atomic bombs at the rate of approximately one every ten days through the month of October.

In mid-July the U.S. Army Strategic Air Forces (USASTAF) was established in the Pacific under the command of General Spaatz with headquarters on Guam. Among other units, USASTAF included XXI Bomber Command headed by Maj. Gen. Curtis E. LeMay. The XXI Bomber Command included the 58th, 73rd, 313th, 314th, and 315th Bomb Wings, all B-29 units. All of these units, and others, were part of the Twentieth Air Force, under the personal command of General Arnold. The Twentieth Air Force had been established in April 1944 with a special command arrangement. Rather than having it subordinate to a theater commander, it operated directly under the JCS with the commanding general of the Army Air Forces, General Arnold — also a member of the JCS — as its executive agent.[60]

On August 1, 1945, major organizational and personnel changes took place. LeMay became Spaatz's chief of staff in USASTAF and XXI Bomber Command became the Twentieth Air Force under the command of Lt. Gen. Nathan F. Twining.

In an official air force history of the Silverplate program the following remarkable sentences describe the chain of command in these two periods.

> From 15 July until 1 August the 509th chain of command *seems to have been* from the President through Stimson to either Marshall or Groves to Arnold as CG AAF, and CG Twentieth Air Force and thence either to the 509th direct or through LeMay as CO XXI Bomber Command.
>
> Beginning 1 August, however, the chain *appears to have been* from the President, through Stimson to Marshall or Groves, to Arnold and thence either to the 509th direct or through Spaatz and Twining.[61]

That the author, an official air force historian, would use the phrases *seems to have been* and *appears to have been* to describe the command channels for the dropping of the atomic bomb acknowledges some degree of uncertainty as to how the orders were given and carried out. Although the air force historian does not explore why this ambiguity exists, it is clear that it is Groves's role in the procedures and orders that is the cause. As can be seen in the diagram that accompanied the discussion, Groves occupied the dominant position in ordering the use of the bomb, though he was not a part of the formal combat command structure.

Some years later, when doing research for his own memoir, Groves read the air force history and commented on the passage to his son.

> I am struck, in fact flabbergasted by the Air Force historian stating that I was above Arnold. Actually, of course, in a way, it was true in the sense that if Arnold had not done what I wanted I could have asked for Marshall to order him to do so. . . . Apparently this wording was not an accident because later on . . . it says, "that directive (the other enclosure) was issued to Spaatz under circumstances described above, after an exchange of views between Stimson, Marshall and Arnold at Potsdam and Groves, Spaatz and Gen Thomas T. Handy at Washington." Until I read this other page, I had not thought there was any significance in the order in which the names were listed but now I realize that there was and that it is the author's intent to indicate that on these matters I was controlling the situation, which was certainly a fact.[62]

Whether or not Groves honestly thought that just two bombs would end the war, he nevertheless proceeded in late July and early August, even up to August 13, with vigorous plans to produce, prepare, and use many more, if circumstances dictated. By the last two weeks of July the future availability of plutonium from Hanford, and the HEU from Oak Ridge, could be calculated fairly precisely, and thus the number and timing of ready bombs could be predicted with some certainty.

On July 19, with Trinity a success three days earlier, Oppenheimer cabled Groves about which path to pursue in bomb production over the next three and a half months. There was serious consideration by some at Los Alamos, even including Oppenheimer, of using the HEU from Little Boy to combine with the available plutonium to make composite cores of both materials, so that more bombs could be built.[63]

Groves replied the same day:

> Factors beyond our control prevent us from considering any decision, other than to proceed according to the existing schedule for the time being. It is necessary to drop the first Little Boy and the first Fat Man and probably a second one in accordance with our original plans. It may be that as many as three of the latter in their best present form may have to be dropped to conform to planned strategic operations.[64]

Groves told Oppenheimer that he would come out to see him to discuss the schedule and that by then the technical people would have figured out the advantages and delays of the various changes under consideration.

Rather than go all the way to Los Alamos Groves arranged to meet Oppenheimer in Chicago on July 24 to discuss it. During the period from his return from Trinity on July 17 until the Japanese surrender on August 14, Groves left Washington only twice, a quick trip to New York to see Sengier for a few hours on July 23, and the Chicago meeting with Oppenheimer on July 24. The matter was important, and it required they meet face to face.

Unfortunately, the details of what was discussed among Groves, Oppenheimer, and Tolman, who had just arrived by train from New Mexico, remain classified.[65] Groves was in Chicago for only a few hours. He used his private plane and pilot to fly from National Airport at 10:45 A.M., arriving in Chicago at 2:00 P.M. He had arranged to have Oppenheimer meet him at the airfield so they could talk on the drive into town. The meeting took place in the MED's Chicago Area Office. After the meeting Groves flew back, taking Tolman with him, and arrived in Washington at 11:30 P.M.

Stimson had cabled Harrison on the afternoon of July 23. The secretary wanted to know when the first Fat Man "will be ready for use and give approximate time when each additional weapon of this kind will be ready."[66] Harrison responded that the first one should be ready about August 6 and a second about August 24, with "Additional ones ready at accelerating rate from possibly three in September to we hope seven or more in December." Harrison informed the secretary that Groves was going to Chicago the next day to discuss future plans with Oppenheimer, and that further details would follow.[67]

From what transpired at the Chicago meeting it is evident that the production schedule was settled once and for all. During the next few days Groves prepared a memo for the chief of staff with the details.

The July 30 memo to the chief of staff provided General Marshall with additional information about the Trinity test, with what to expect in the way of damaging effects from the combat bomb, and with his plans for producing bombs for the rest of the year.[68] Little Boy was already at Tinian, and the final components for Fat Man were proceeding by plane that day from San Francisco. Groves informed Marshall that he saw "no reason to change our previous readiness predictions on the first three bombs." By "the first three bombs" Groves presumably meant Little Boy, Fat Man, and another plutonium bomb. At this juncture the third bomb would be ready for delivery on August 24 (this date was later advanced by six or seven days).

Groves's memo went on to say that three or four bombs should be ready in September, one of which would be an implosion design using U-235. The bomb would have an effectiveness of about two-thirds of the Trinity explosion, but that would improve and should be equivalent in power by sometime in November. For October, plans called for another three or four bombs, one of which would use U-235. The rate would rise to at least five bombs in November, to seven in December, and "increase decidedly in early 1946."

Groves's earlier memo to General Marshall on July 24 had provided approximately the same schedule of operations after dropping the first two bombs.

> The second implosion bomb should be ready 24 August. . . . Additional bombs will be ready for delivery at an accelerating rate, increasing from about three in September to possibly seven in December, with a sharp increase in production expected early in 1946.[69]

Thus, under Groves's plan at the end of July, there would have been a stockpile of approximately twenty bombs by the end of 1945, several of which would have used an all-HEU implosion-assembly core. There is no mention of additional Little Boy, gun-type bombs. By mid-October, if both materials were used in a composite core of HEU and plutonium, the number of bombs could increase slightly, but with an initial ten-day production setback that could be made up in about a month's time. No time need be lost if a decision were taken now to use the composite core design, but Groves counseled:

> From what I know of the world situation, it would seem wiser not to
> make this change until the effects of the present bomb are determined.

The weather over Japan during the first few days of August was over-
cast and rainy. One of the conditions on which the Interim Committee
had insisted was that the bombardier of the attacking aircraft have visual
acquisition of the target. Radar was not to be used. On August 4 Parsons
was informed that the weather was improving over Japan. For the first
time the crews were briefed and learned of their special mission. There
would be seven aircraft on Special Bombing Mission No. 13, three weather
observation planes to survey each target city, an instrumentation aircraft
(The Great Artiste), a photo-reconnaissance plane *(Necessary Evil),* a backup
plane on Iwo Jima, and the "Queen Bee," as it was called, the plane carrying the bomb. The pilot of the Queen Bee, Paul Tibbets named the
plane for his mother, the *Enola Gay*.[70]

On Sunday morning, August 5, with the weather prediction good for
the following day, General LeMay gave the order for an attack the following day. Final preparations got under way. At 2:00 P.M. Little Boy was
wheeled out of the assembly building, loaded onto a trailer, and taken to
the loading pit. After it was lowered into the pit, the *Enola Gay* backed
over the bomb and it was raised into the bomb bay. By six o'clock everything was ready. At 11:00 P.M. a special briefing was held for the aircrews
of the *Enola Gay, The Great Artiste,* and the *Necessary Evil* given by Tibbets,
Parsons, and Norman Ramsey. The Rev. Bill Downey, the Lutheran chaplain, offered a prayer.

Farrell informed Groves by cable on Saturday that the weather forecast
was favorable and that if it remained so the plane would take off at approximately noon on Sunday, August 5 (Washington time). Groves notified
General Marshall and Secretary Stimson through Harrison of the schedule.

Not many who served under General Groves ever countermanded his
orders. As the time approached to take off, Deak Parsons decided that he
would have to. The agreed-upon procedure was to have the plane take off
with the Little Boy bomb already loaded with its cordite high explosive.
The simplicity of the design made the bomb highly dangerous if there
were an accident. If the *Enola Gay* crashed and caught fire, the cordite
could detonate and drive the projectile into the target. The likelihood
would then be high that a nuclear explosion would result.

Parsons had witnessed many fiery B-29 crashes at the end of Tinian
runways and decided to change the procedure and insert the cordite bags

only after takeoff. The day before the mission Parsons practiced loading the bags through the small breech opening. After the bomb was loaded into the plane, he practiced for several more hours in the cramped bomb bay. Fearing that Groves would probably veto his decision, given a chance, Parsons decided not to tell him, though he did inform Farrell. With time short and communication difficult Farrell did not contact Groves either and permitted Parsons to proceed.

Groves rarely balked at taking risks. It was an essential part of being in command, and it went with the job. But in this instance he probably would not have allowed the different procedure, which had risks of its own. It would be difficult to load the powder in the dark and cramped bomb bay of the drafty airplane. If the powder were not placed correctly, the bomb might not function. Furthermore, there were not going to be any other combat missions using Little Boy–type bombs. Later, when Groves was reminded about Parsons's action and the crashes at the end of the North Field, he dismissed the danger: "[S]ure, sure, they just didn't have the nerve that was required that was all. There had been quite a few crashes but after all we had probably the best pilot in the Air Force, Colonel Tibbets. . . . If I had known about it in advance they would have had a very positive order over there."[71]

The time in Tinian was thirteen hours later than Washington. Early Sunday morning Groves went to the office to await word that the plane had taken off, or that the mission had been aborted. The plane had taken off at 2:45 A.M. on Monday, August 6, Tinian time —11:45 A.M. on Sunday, August 5, in Washington. Groves expected to hear by about 1:30 or 2:00 P.M. that the plane had taken off. When nothing was heard Groves decided that rather than making himself and everyone else nervous he would leave the office and play an hour or two of tennis. He left instructions with Major Derry to contact him if the cable came through. Groves took an officer with him to stay at courtside to receive Derry's call and to check with the office every fifteen minutes to see if the message had arrived. After tennis Groves returned to the office, but there was still no word.

Groves had arranged to have dinner with Mrs. Groves, Gwen, Dick and Patsy, and George Harrison at the Army-Navy Club on Farragut Square in downtown Washington that evening. If all had gone according to plan, he would have already known that the plane had taken off. General Handy was also there that evening and came by to ask Groves if he had heard anything. At about 6:45 P.M. Groves was called to the phone to take a call from Derry, who informed him that the plane had left on schedule. In fact, the bomb had already been dropped on Hiroshima about half an hour earlier, but word of this would take more time to reach Washington. Upon returning to the dining room, Groves informed Harrison and Handy of the news, gathered up his family, and left for the office. As they drove him across town, he told them he would spend the night at the office. Though this was the first time he had done so during the war, they did not ask why. Groves later wrote that "their lifetime in the Army had conditioned them well."[72]

Upon returning to the office, he tried to find out why there had been a communication problem, calling his friend the chief signal officer, Maj. Gen. Harry C. Ingles, along with Col. Frank McCarthy, secretary of the General Staff and Marshall's closest aide. The initial message concerning the plane's departure from Tinian was six hours late in reaching Washington. By early evening word should have arrived about the attack, but there was nothing. It proved to be a long night for Groves and his aides. The hours dragged on until finally at 11:30 P.M. a message arrived from Parsons.

> Results clearcut, successful in all respects. Visible effects greater than New Mexico tests. Conditions normal in airplane following delivery.
>
> Target at Hiroshima attacked visually. One-tenth cloud at 052315Z [11:15 P.M. August 5, Greenwich Mean Time]. No fighters and no flak.[73]

Groves was the first person in Washington to learn of the attack. He called McCarthy, who then contacted Marshall.

The *Enola Gay* and the other two airplanes landed at Tinian at 2:58 P.M. With much fanfare General Spaatz pinned the Distinguished Service Cross on Tibbets's coveralls. Deak Parsons went unrecognized at that moment, but later was given a Silver Star. "When Groves learned that, he snorted, 'There was never any question by anyone but that Parsons was running the show. Apparently, the only person who did not get that right was General Spaatz.'"[74]

On August 7 General Farrell was instructed by the War Department to engage in a propaganda campaign by dropping leaflets on Japanese cities and by frequent broadcasts over the radio.[75] A third element was to drop copies of Japanese-language newspapers with stories and pictures of the Hiroshima bombing. The purpose was to encourage the populace to pressure the military and the emperor to end the war, under the threat of further atomic bomb attacks. Farrell used navy and air force officers at Saipan and Guam to compose the text. Three young Japanese officers in the prisoners' stockade at Guam were detailed to translate the message into Japanese and draw the calligraphic characters.

In English the leaflet said:

> To the Japanese People:
> America asks that you take immediate heed of what we say on this leaflet.
>
> We are in possession of the most destructive explosive ever devised by man. A single one of our newly developed atomic bombs is actually the equivalent in explosive power to what 2000 of our giant B-29's can carry on a single mission. This awful fact is one for you to ponder and we solemnly assure you it is grimly accurate.
>
> We have just begun to use this weapon against your homeland. If you still have any doubt, make inquiry as to what happened to Hiroshima when just one atomic bomb fell on that city.
>
> Before using this bomb to destroy every resource of the military by which they are prolonging this useless war, we ask that you now petition the Emperor to end the war. Our President has outlined for you the thirteen consequences of an honorable surrender: We urge that you accept these consequences and begin the work of building a new, better, and peace-loving Japan.
>
> You should take steps now to cease military resistance. Otherwise, we shall resolutely employ this bomb and all our other superior weapons to promptly and forcefully end the war.

The newspapers that were dropped contained stories and photographs of the atomic bomb strike. Forty-seven Japanese cities were chosen, and the Seventy-third Wing of the USASTAF dropped some six million leaflets.[76] Because of a coordination failure, the leaflets were dropped on Nagasaki after the city was destroyed.

The question of uncertainty, ever present throughout the Manhattan Project, underscored the second combat mission, using the Fat Man implosion-type bomb. While the Trinity device had worked, there were significant differences between it and a deliverable bomb. The Trinity device had been detonated from a tower under strict conditions. It had neither a fusing system nor a ballistic case. A combat mission dropping an oddly shaped bomb over enemy territory from thirty thousand feet was quite a different thing. There were many things that could go wrong.

The original schedule called for dropping the second bomb on August 11, but a forecast of poor weather to begin on August 10 accelerated the date to August 9. Six B-29s were part of Special Bombing Mission No. 16, one less than the Hiroshima mission because there were only two targets instead of three. The specific B-29 that was chosen to carry the bomb was named the *Bockscar,* commanded by Maj. Charles W. Sweeney with weaponeer Cmdr. Frederick L. Ashworth, USN, in charge of dropping the bomb on the designated target.[77] Each of the eleven other members of the crew had their specific responsibilities. As for the other B-29s, two were for instruments and photos, two planes flew ahead to the primary and secondary targets to observe and report the weather, and a standby plane went to Iwo Jima in case of trouble with the strike aircraft.[78]

Just as they were about to take off it was discovered that the fuel transfer pump for two of *Bockscar*'s auxiliary tanks did not function, limiting how much gasoline would be available for the mission. There was not enough time to replace the pump, to empty the 640 gallons of trapped gasoline, or to transfer the bomb to another B-29. Bad weather was forecast for the next day. A discussion ensued among Farrell, Parsons, Tibbets, and Sweeney, who concluded that the mission had to go forward. The *Bockscar* took off at 3:47 A.M. (local time) and proceeded to the rendezvous point over the island of Yaku-shima, arriving there a little after 9:00 A.M. to wait for the other two B-29s, the instrument plane, *The Great Artiste,* and the photo plane, *Big Stink.* Five minutes later one of the planes arrived.

Sweeney identified the aircraft as the instrument plane but for some reason did not tell Ashworth. Ashworth later said that had he known that it was instrument plane, they would have proceeded on immediately.

This failure by Sweeney to inform Ashworth which plane had arrived could have jeopardized the mission. It has led to some hard feelings among members of the 509th after the war, and has contributed to a general criticism of Sweeney's performance during the mission by Tibbets and by others.[79] In another point of contention, Sweeney had been instructed to wait at the rendezvous point for only fifteen minutes, but circled for forty-five, using up precious fuel. Finally Ashworth suggested that they proceed on to Japan without the third plane.

In yet another mishap the pilot of the third plane, Maj. James I. Hopkins, broke radio silence and asked in the clear, "Has Sweeney aborted?" The message was slightly garbled; at Tinian all that was heard was "Sweeney aborted." When General Farrell heard this he ran out of the tent and threw up. Understandably, this caused near panic with fears about what might have happened. Had the *Bockscar* crashed or been shot down, did it jettison its bomb, and where was the crew? As a result of Hopkins's negligence, the rescue ships and planes were canceled. It would be more than two hours before it was learned that the mission had actually succeeded.[80]

En route they received a report from the weather observation plane, *Up an' Atom,* saying that the primary target, Kokura, was generally clear with 30 percent low cloud cover, which was expected to dissipate. Shortly thereafter they received a report from the Nagasaki weather plane, *Laggin' Dragon,* stating that the secondary target was clear and was expected to stay that way. Ashworth made the decision to attempt to attack Kokura.[81]

The precise aim point was the Kokura Arsenal, an enormous complex of about one square mile, surrounded by extensive minor industries and a residential area primarily of workers for the arsenal. As they approached Kokura, they discovered that it was partially obscured by heavy haze and smoke that had drifted from nearby Yawata — a city of a quarter of a million and the center of Japan's steel industry — which had been bombed the day before by more than two hundred of Twining's B-29s carrying incendiaries.[82] This made it extremely difficult to carry out the order to locate the aim point and bomb visually rather than using radar. Three bomb runs were attempted over Kokura, each from a different direction, all without being able to locate the target visually.[83] The runs took forty-five minutes, and with the fuel getting ever lower, they decided to proceed on to Nagasaki. A discussion among Sweeney, the navigator, and the flight engineer concluded that there was enough fuel for a single pass over Nagasaki with perhaps enough left over to reach Okinawa. Due to the heavy load the *Bockscar* would not be able to reach Okinawa with the bomb on board. It was therefore mandatory that either Nagasaki be bombed or, if that proved impossible, some other target be attacked (since no third target was assigned), after which the plane would have to ditch in the ocean.

To minimize the distance they flew directly across the island of Kyushu, a more hazardous route, rather than proceeding over water as the original flight plan had specified. As they approached Nagasaki they broke radio silence to alert the air-sea rescue organization that they might have to ditch. To their dismay Nagasaki was closed in as well, with low-lying clouds, between eight thousand and ten thousand feet, obscuring the city. A visual run did not seem possible. Ashworth was faced with a monumental decision. Though the orders were not to bomb using radar, Ashworth decided that it would have to be used. From their direction of approach, Nagasaki's mountains and bay showed prominently on the radarscope. The radar operator, Staff Sgt. Edward R. Buckley, and the navigator, Capt. James F. Van Pelt, were experienced at their jobs and were familiar with Nagasaki's profile, as was the entire crew. Ashworth instructed the bombardier, Capt. Kermit K. Beahan, twenty-seven years old that day, to follow as closely as possible what Van Pelt and Buckley were doing and be prepared to take over and make a visual run, if conditions changed. If conditions did not change, Ashworth told Beahan, then release the bomb by radar; he would take responsibility.

As the plane flew on its radar-bombing run, the city of Nagasaki and the aiming point were fast approaching. With about forty seconds to go until release, a hole in the clouds opened up. Beahan had only about twenty seconds in which to visually select a target.[84] He yelled to Sweeney, "I've got it! I've got it!" and was given control of the plane. Beahan released the bomb, which exploded forty-eight seconds later, at 11:02 A.M. (local time), at an altitude of approximately 1,650 feet, almost midway between the Mitsubishi Steel and Arms Works in the south and the Mitsubishi-Urakami Ordnance Plant in the north, along the Urakami River Valley.[85]

Sweeney made the sharp-angled turn to the left to escape the blast, because *The Great Artiste,* less than two hundred yards behind, had already released its three instrument packages and made a turn to the right. As the mushroom cloud billowed up, three shock waves hit the planes, more intense than those at Hiroshima.

After the drop the fuel gauge showed only 300 gallons of fuel remaining (of the 7,250 gallons that the plane had started with), barely enough, as it turned out, to reach Yontan Field on Okinawa, 350 miles southwest of Nagasaki. The almost two-hour flight was touch and go. In approaching the airfield Major Sweeney was unable to reach the control tower to notify them that they were about to make an emergency landing. The right outboard engine quit. With no alternative Sweeney chose to land in the midst of heavy air traffic, firing off all of the plane's flares to signal an extreme emergency. To get down quickly Sweeney brought the plane in faster than normal, touching down well over halfway up the seven-thousand-foot runway. It would have surely crashed if it had been equipped with conventional propellers, but the *Bockscar,* like the other B-29s of the 509th, had reversible pitch propellers, which helped stop the plane fifty feet from the end of the runway. Two other engines had quit. Some accounts say there were only thirty-five gallons of usable fuel remaining in the tanks, while others say there were only seven left — less than one minute of flight time.

Ashworth had sent a brief coded message while in the air that the bomb had been dropped on Nagasaki. With the plane now on the ground and the crew safe, it was essential for a more detailed report to be sent to Farrell and Parsons on Tinian. Sweeney and Ashworth asked to be taken to the senior ranking commander, which turned out to be Gen. Jimmy Doolittle of the Eighth Air Force. Greeted by Doolittle's chief of staff, General Partridge, Ashworth informed him that they had just dropped the second atomic bomb, on Nagasaki, and had to inform Strategic Air Forces on Guam. Ashworth then talked to Doolittle, and with his assistance the message was sent.

After refueling and a meal the crew took off, arriving at Tinian at about 10:30 P.M., some twenty hours after taking off. Unlike the return of the *Enola Gay,* there were no klieg lights, film crews, or ceremonies, only Tibbets, Purnell, and the ground crew. After debriefing the crew, Farrell sent a more detailed message to Groves.[86]

On the day after the Nagasaki bombing, August 10, Groves informed Marshall of an improved timetable for the next bomb.

> The next bomb of the implosion type had been scheduled to be ready for delivery on the target on the first good weather after 24 August 1945. We have gained 4 days in manufacture and expect to ship from New Mexico on 12 or 13 August the final components. Providing there are no unforeseen difficulties in manufacture, in transportation to the theatre or after arrival in the theatre, the bomb should be ready for delivery on the first suitable weather after 17 or 18 August.[87]

Marshall promptly returned Groves's memo with a handwritten directive written on the bottom stating, "It is not to be released over Japan without express authority from the President." The predelegation authority that had been granted to Groves to drop bombs "as made ready" was being revoked. The president was inserting himself back into the chain of command, becoming more fully aware of what was transpiring and engaged in the process. There was good reason to do so, for the first indications of a Japanese surrender offer reached Washington that day via the

"Magic" decrypts.[88] The offer had a single condition, that the emperor be maintained.

Henry Wallace reported in his diary that day that the president said to the cabinet that

> he had given orders to stop atomic bombing. He said the thought of wiping out another 100,000 people was too horrible. He didn't like the idea of killing, as he said, "all those kids."[89]

What would have been the next target city and how would the attack have been conducted?[90] It is impossible to know because the process never proceeded far enough to provide an answer. There were still two remaining targets from the original list, Kokura and Niigata, but other cities were possibilities. The same day, August 10, Groves sent a memo to General Arnold:

> General Farrell has recommended in APCOM 5313 DTC 07/06/28 that Tokyo be added to the list of approved targets which originally consisted of Hiroshima, Nagasaki, Kokura and Niigata. Additional targets must be selected and made available to COMGENUSASTAF before additional strikes can be made by the 509th Group.
>
> You may wish to give consideration to easing the requirements for visual bombing of targets so that the weather difficulties encountered in the last strike, which might well have resulted in failure, may be avoided on future strikes.[91]

In an August 11 letter to Farrell, Groves said,

> I saw General Marshall this morning and it was decided that no further shipments of material should be made to the Theater until the question of the Japanese surrender was decided. We are, of course, accumulating supplies here so that they can be shipped promptly in the event there is no change in the situation.[92]

An active sphere of plutonium for unit F32 was scheduled to leave Kirtland the evening of August 12.[93]

> My opposition to rushing over material for a third strike was that we should not be put in a poor position historically by showing that we

were anxious to punish Japan unnecessarily. This was in accord with the philosophy of Stimson about the importance of the future historical position of the United States. If I had not wanted to delay shipment I would not have gone to see Marshall.[94]

Groves never seemed troubled by any moral questions presented by the use of the atomic bomb, either at the time or over the following twenty-five years until his death. He never expressed any doubt that using the bomb was the correct decision. From the moment he took charge his goal was to have a bomb ready as soon as possible to use against America's enemies, to shorten the war and ultimately to save American lives.

Shortly after the war he said,

> I have no qualms of conscience about the making or using of it. It has been responsible for saving perhaps thousands of lives. If the bomb had not been used the Japs would have held out for 60 to 90 days longer. We know what that would mean in the sacrifice of human lives. . . . I had staked my reputation and long service in the Army on the successful construction of this bomb, as I believed it would do what it has done — save thousands of lives. From an official standpoint I knew its success would be greatly to our advantage and from a personal standpoint it might save my own son.[95]

Groves's beliefs were widely shared by his civilian and military colleagues. In the minds of those conducting the war the distinction between the military and civilians had eroded under the pressure of the war. All Japanese were combatants in some fashion and therefore targets. Civilian-area bombing had become routine. The cities that were chosen as targets for the atomic bombs were considered legitimate military targets. There was no such thing as a pure military target isolated from the rest of the society. Cities housed industries that supported the war effort. Some were ports; some had arsenals, bases, and military headquarters. If there had been no atomic bomb, General LeMay would have destroyed them as he did the dozens of other Japanese cities.

But even with this consensus, several of Groves's military and civilian counterparts recognized that some actions by the United States were slipping over the line, especially the relentless bombing of cities. They clearly recognized that these targets were mainly convenient, not strategic, and they sometimes hinted that they were troubled by it. But the momentum of the war and the psychological difficulty of speaking out allowed them to overcome doubts and go on as they were.

Key air force leaders such as Arnold, Spaatz, Twining, LeMay, and Norstad all defended strategic bombing as morally just.[96] It was better than having ground forces slug it out in bloody land battles. Strategic bombing ended wars faster, reducing American casualties. In their view the atomic bomb was merely a more effective way of achieving these goals.

Leaders cannot appear sentimental. They must be resolute and tough. To carry out such decisions hardens men to suffering and death. General Marshall put it succinctly:

> It is a grim fact that there is not an easy, bloodless way to victory in war and it is the thankless task of the leaders to maintain their firm outward front which holds the resolution of their subordinates. Any irresolution in the leaders may result in costly weakening and indecision in the subordinates.[97]

Stimson did not shirk the responsibility that goes with conducting warfare. Looking back on his tenure, he said that there were

> too many stern and heartrending decisions to be willing to pretend that war is anything else than what it is. The face of war is the face of death; death is an inevitable part of every order that a wartime leader gives. The decision to use the atomic bomb was a decision that brought death to over a hundred thousand Japanese. No explanation can change that fact and I do not wish to gloss it over. But this deliberate premeditated destruction was our least abhorrent choice. The destruction of Hiroshima and Nagasaki put an end to the Japanese war. It stopped the fire raids, and the strangling blockade; it ended the ghastly specter of a clash of great land armies."[98]

CHAPTER TWENTY-ONE

War Hero for a Day

• *Mid-August–December 1945* •

With the successful use of the bombs and the Japanese surrender that quickly followed, the extraordinary pressure that Groves had imposed on himself and on everyone around him was off. But almost immediately a whole new set of issues and problems surrounding the bomb came to the fore and had to be dealt with.

For Groves, in those first few days of the atomic era, providing the story of the bomb to the public headed the list. It was now time that the existence, purpose, and activities of the Manhattan Project become known, at least in part. There was great curiosity about this new superweapon, about the secret cities where it had been developed and built, and about the interesting personalities who were responsible for its creation. In an instant Groves went from being a secret figure to a public one.

Groves was aware of what might be said about the bomb, and he attempted to control and shape the news stories that were published about it, to the degree that he could. Always security conscious, Groves now faced new challenges to prevent certain secrets from becoming known.

There were other responsibilities as well. It was time to recognize the hard work of the past three years by publicly honoring key individuals and corporations. While many were leaving the project, eager to return to their careers, Groves sought to maintain the facilities and recruit new personnel until the basic questions about the future of atomic energy and the bomb were decided upon. One central question — who was going to be in charge, the civilians or the military? — began to be publicly addressed within days of the end of the war, with the scientists taking a prominent role in swaying public opinion. Groves unavoidably found himself in the middle of the debate and at the receiving end of many scientists' animosity. By the end of the year his reputation and public image had suffered badly, never really to recover.

As the bomb had neared completion Groves set in motion elaborate preparations to inform the world about this new weapon once it had been used.[1] In effect he set up his own public relations firm within the MED. Characteristically, he tried to anticipate what would be needed and oversaw the drafting of virtually all of the initial public statements, press releases, and reports. Also essential were photographs of the key sites and key individuals. Groves prohibited any extemporaneous radio interviews given by Manhattan personnel. Only previously cleared scripts could be used. Like all of the other aspects of the Manhattan Project, it was conducted under Groves's strict security procedures and guidelines.

Prior to the spring of 1945 anything to do with public relations and the press had been the responsibility of the MED's counterintelligence organization and of the Office of Censorship, which Groves worked with closely. The goal was preventive in nature — to keep things as secret as possible, to black out completely any news about the bomb that might reveal its existence.

When it came time to prepare for a positive program, a program that would divulge information rather than limit it, Groves put Lt. Col. William A. Consodine in charge of the MED's Public Relations Program, with Maj. John F. Moynahan as his assistant. Consodine, as we have seen, was one of Groves's closest aides and worked on security and intelligence matters, as well as on the highly secret Combined Development Trust. Moynahan was a staff reporter for the *Newark Evening News.* In May 1942 he had gone on active duty and served with the Army Air Forces and OSS before joining the MED, perhaps at Consodine's behest. At the end of July 1945 Moynahan accompanied General Farrell to Tinian to oversee the release of information in the Pacific.[2] The stories about the aircrews and bombing missions were clearly going be dramatic ones, and the press corps covering the Pacific war would have to be taken care of. By the same token, Consodine saw the potential of too much focus on that aspect of the project, and tried to avert it. "The Air Forces should be restricted by military order from any publicity. While they will be given due credit, they will not be allowed to steal the show," he told Groves.[3]

Groves's Washington office was the central point in handling all phases of the project's public and press relations, with support provided at the major sites. It also coordinated the public relations with the Office of Scientific Research and Development, the Office of Censorship, and the War Department Bureau of Public Relations. A long list was drawn up of things that could not be talked about, including the characteristics of the "gadget," future plans for employment, and the quantity and quality of the active materials. Groves even arranged to send letters from himself and from Byron Price of the Office of Censorship to the editors of the local newspapers in Tennessee, Washington, and New Mexico, thanking them for their cooperation and observance of the security regulations.

The release plan was to begin with the president's announcement of the successful bombing of Japan several hours after it occurred. A primary purpose of this initial information was to impress upon the Japanese that the overwhelming power of the bomb left them no choice but to surrender. The first public announcement by President Truman came at 11:00 A.M. on Monday, August 6, almost sixteen hours after the attack. Truman himself was still aboard the USS *Augusta,* returning from the Potsdam conference, which had adjourned on August 2. He would not arrive back at the White House until the late evening of the seventh, some thirty-six hours later. He was informed of the Hiroshima attack while on board.

Groves and Consodine had overseen preparation of Truman's statement in collaboration with Stimson's aides, chief among them Arthur Page. Stimson left Germany on July 27 and was back in the United States the next day, going first to Highhold to rest from the long trip. He returned to Washington on July 30 and worked with Bundy, Harrison, Page, and Groves on the president's statement.

> We made some changes in it which were induced by the difference of psychology which now exists since the successful test. I did not realize until I went over these papers now what a great change that had produced in my own psychology. We put some more pep into the paper and made it a little more dramatic and then sent a telegram to the President telling him what we had done and asking for his authority.[4]

Some of the pep and drama was purposely menacing, a clear warning to the Japanese that they should surrender. The first paragraph described the one bomb that was dropped on Hiroshima as having the power of more than twenty thousand tons of TNT, more than two thousand times larger than the British "Grand Slam." The second paragraph denounced the Japanese for starting the war. For this they "have been repaid many fold" and were warned, "the end is not yet."

After providing some background on atomic energy and the enormous

American achievement, Truman described the next steps that would follow.

> We are now prepared to obliterate more rapidly and completely every productive enterprise the Japanese have above ground in any city. We shall destroy their docks, their factories, and their communications. Let there be no mistake; we shall completely destroy Japan's power to make war.

Japan had promptly rejected the Potsdam ultimatum of July 26. "If they do not now accept our terms," Truman warned, "they may expect a rain of ruin from the air, the like of which has never been seen on this earth."

A secondary purpose of the releases was to inform the American people of what had been accomplished by their government. Stimson was at Highhold, his home on Long Island, when he learned "that the S-1 operation was successful," as he recorded it in his diary. He spent the day on the phone talking to, among others, Marshall, Groves, Harrison, Lovett, and his aide Col. William H. Kyle.

The release of Stimson's own statement soon followed the president's and provided greater detail about the "Herculean effort on the part of science and industry working in cooperation with the military authorities." To keep "the people of the nation as completely informed as is consistent with national security," Stimson provided the broad outlines of "the story behind this tremendous weapon." It was in his statement that the American people first learned of the secret sites in Tennessee, Washington, and New Mexico, the major companies that built and operated the plants there, the contribution of American science, the expenditures of almost two billion dollars, and the names of key figures in the effort, including several from Great Britain and Canada. Stimson singled out Groves for special tribute.

> His record of performance in securing the effective development of this weapon for our armed forces in so short a period of time has been truly outstanding and merits the very highest commendation.

Stimson concluded by detailing some of the efforts the government had made thus far to deal with the future of atomic energy, its potential peaceful applications, as well as "the grave problems that arise concerning the control of the weapon." Members of an Interim Committee had been charged with formulating recommendations to the president concerning the future course of the United States in this field.

In the era before television most people got most of their information about the day's events from newspapers. On Friday, April 13, 1945, Groves had gone to the offices of the *New York Times* to meet with the managing editor, Edwin L. James,[5] about hiring the paper's science writer, William L. Laurence, to write about the secret project for future publication. Laurence had written several articles for the *New York Times* on uranium fission in 1939 and 1940.[6] In 1943 and 1944 Laurence submitted articles to Price and Lockhart at the Office of Censorship but was denied permission to have them published. After speaking with James, Groves met with Laurence and invited him to write about a secret project. He promised Laurence full access but insisted that Laurence would be able to publish nothing until he, Groves, granted permission. He also told Laurence that the story was a big one — the biggest that Laurence would ever have a chance at. Whether or not Laurence guessed what Groves was hinting at, he wasted no time in agreeing to Groves's terms. Plans were quickly made with the *Times* administrative offices to grant Laurence a leave of absence to work under contract for the Manhattan Engineer District, an unusual journalistic arrangement.[7]

Groves was as good as his word. For the next three months he gave Laurence almost complete access to every phase of the project. The only matters held back were such things as the existence of the Combined Development Trust and its operations, and Alsos. Laurence spent time at Oak Ridge, Hanford, Los Alamos, and the university laboratories at Columbia, Chicago, and Berkeley. He witnessed the Trinity test from Compania Hill and wrote a fictitious account to be used if the explosion unexpectedly killed all of the scientists and observers. Fortunately, it did not have to be used.

When the time came to deliver the bombs on Japan, Laurence was flown to Tinian. He was scheduled to accompany the mission to Hiroshima but arrived too late. He did attend the briefings that evening and watched the *Enola Gay* take off at 2:45 A.M. and return. For the Nagasaki mission he was on board *The Great Artiste,* the instrument aircraft that accompanied the *Bockscar.*[8]

Groves did not simply leave Laurence to write these stories as he saw fit. On May 1, 1945, at the initial meeting of the Interim Committee, one of the first items brought up was the question of how to inform the public about the Manhattan Project. Groves informed the committee that he had already engaged Laurence, who was at their service.

Laurence's first assignment was to draft a statement that would be used by the president when he announced the dropping of the first bomb on Japan. The draft announcement was presented to the Interim Committee on May 15. It did not meet with their approval. As we have seen, the statement was worked upon by many hands before it was released on August 6. Such was the case with all of Laurence's stories. Everything he wrote before Hiroshima was submitted to Colonel Consodine. Consodine in turn gave them to his staff, Moynahan, Lt. Col. Clyde H. Mathews, and Capt. Kilburn R. Brown, who rewrote and edited them, as many as half a dozen times, before Groves finally found them satisfactory. The drafts of Laurence's articles are preserved with the Manhattan records in the National Archives. Groves's handwriting is evident on many of the pages, making additions, deletions, and corrections.

In all Laurence produced a series of ten stories on various aspects of the project, and an eyewitness account of the mission to Nagasaki.[9] The series ran in the *Times* between September 26 and October 9 and was awarded a Pulitzer Prize in 1946.

Groves was introduced to the public as head of the atomic bomb program in an August 6 press release. It provided much detail about his career and his supervision of the Manhattan Project.

> A soft-spoken Major-General with a flair for the "Impossible" emerged today from the shadows of Army-imposed anonymity to be revealed as the driving force behind a $2,000,000,000 "calculated risk" which he directed to successful completion in three years as one of the world's greatest scientific and engineering achievements, the large-scale tapping of the energy within atoms to produce a new weapon of war.

One should not unhesitatingly accept a press release about a person, especially one written by that person. But one should not totally dismiss it either. In Groves's case what he was actually like and what he wanted to be like are both revealed.

> A pleasant-mannered, gracious officer who outwardly never shows the strain and worry of his job, General Groves is a constant source of amazement to his associates and subordinates because of his ability

to handle the variety of complex details attendant to the project with a minimum of confusion and the smoothness with which he can treat multitudinous technical and administrative problems during a day's work. Firm and blunt when the occasion demands, but withal considerate and fair-minded, and interested in the welfare of his associates, he has the deep respect and admiration of his staff for his ability to organize and get things done. He has a prodigious memory and often confounds his staff by recalling names, dates and incidents long past buried in files.

Soon an article profiling Groves appeared in *Collier's* magazine, with some quotes from the general.

My emotional graph is a straight line, . . . I never worried. This job would never have been done if I had. I never had any doubts. Not having any doubts, I could not feel very surprised or elated by our success.[10]

As Groves well knew, the initial statements by the president and secretary of war, and a few press releases, were not going to be enough to satisfy the curiosity of the public about this revolutionary event. Something more extensive and substantial was going to be needed.

In the spring of 1944 Groves asked Professor Henry DeWolf Smyth, chairman of the Princeton Department of Physics and, at the time, an associate director of the Metallurgical Laboratory at the University of Chicago, to prepare a report that would describe the atomic bomb program, with eventual public release in mind. Smyth was a logical choice. Princeton had been associated with several early milestones of atomic energy research, and Smyth had served as a member of the S-1 Section and a consultant to the S-1 Executive Committee.[11]

Smyth recalled suggesting the idea of a report to Arthur Compton in early 1944, and to James Conant during one of his visits to Chicago. In early April Smyth went to Washington to see Groves and Conant, and the matter was settled. Groves arranged for Smyth to have access to all of the laboratories and plants. Over the next year Smyth drafted the report; by June 1945, with help from Richard Tolman and extensive editing by his technical aides, William A. Shurcliff and Paul Fine, it took final form.

There were two principal reasons for issuing the report. The first was to tell the American people what their government had been doing regarding this new weapon and source of energy, and to prepare them for the future. To quote the preface, "The ultimate responsibility for our nation's policy rests on its citizens and they can discharge such responsibilities wisely only if they are informed." Up until now those responsibilities had been shared by a very few, in isolation and in secret, during wartime. Now with peace not far away the responsibility was going to have to be borne by the entire society. And to be able to make wise choices, the public would at least have to be knowledgeable about the pertinent scientific and technical information and the scope and scale of the bomb project.

The second reason was probably more important to the security-conscious Groves. It was to establish a baseline of information beyond which those who had worked on the project could not go. Without a clear line of what could and could not be said, the thousands of people who were returning to normal life might divulge too much.[12] As Professor Smyth said, "I have always found it curious that two lines of reasoning quite opposite in the abstract led in practice to the same conclusion."[13] A corollary reason was that the report could help discipline the scientists and control the temper and tone of the discussion concerning the bomb.[14]

When the report was released on August 12, it was much in demand. Journalists used it for their stories; inquisitive congressmen called Groves's office for copies. It is an extraordinary document and remains in print today.[15] It is not a popular history; rather, the "report is intended to be intelligible to scientists and engineers generally and to other college graduates with a good grounding in physics and chemistry." Several of the thirteen chapters are quite heavy going. The first two chapters trace the course of twentieth-century physics and describe the problem of trying to control fission, complete with formulas, graphs, and abstract concepts and principles. The chapters that follow provide extensive details about how to produce plutonium, separate isotopes, and design and construct a bomb. At the time some felt that the report provided too much detail, with useful clues and shortcuts about what worked and what did not. Groves relied heavily upon his scientific advisers to review draft upon draft to decide what could be released. Among the first readers were the Russians, who undoubtedly found in it certain things that assisted their efforts.[16]

Groves at a later date drafted a lengthy memo explaining its background, intent, and why it was published.[17] The memo, which was never published, provides some interesting insights into Groves's attitudes toward security and secrecy. Contrary to what many might think, at least in theory he had a realistic and rather nuanced view of security and especially of secrecy. The determined agent, he felt, will go to any length to get the information, and thus:

> Maintaining security is always a losing battle in the end. Just as a young athlete loses his strength and skill after a comparatively few years, just so Time, the inevitable, conquers security.

On secrecy, his attitude can only be termed progressive:

> No one can predict exactly the scientific developments of the next decade or two, but it can be assumed that most of them will come from the minds of young men working untrammeled and undirected, with full access to information, in an atmosphere of freedom. During the war just past, it was demonstrated most conclusively that America's capacity to win wars with new weapons — of which the atomic bomb is by far the most distinguished example — depends on the general scientific, technical, and industrial strength of the country, not on secret researches in either private or government laboratories. There is good reason to believe that any rival country, even if given complete blueprints of the bomb would not be able for many years to construct a duplicate. Therefore we should put our trust in continued scientific progress rather than solely in the keeping of a secret already attained.[18]

There was enormous interest in what had happened to the two cities that had been bombed, and Groves was determined that he was going to be among the first to find out. He would be called upon to answer endless questions; thus he needed accurate and comprehensive information, and that meant getting some of his people to Hiroshima and Nagasaki as quickly as possible.

With indications that the Japanese were likely to surrender soon, Groves wasted no time in ordering teams to Japan to survey the damage. On August 11 he directed Colonel Nichols to select appropriate personnel for a mission and to procure the special equipment they would need. At the same time he instructed General Farrell on Tinian that he would lead the teams going into Japan and to assemble whatever available medical, scientific, and intelligence personnel were already there. With the forthcoming occupation by U.S. troops in mind, on August 12 Marshall informed General MacArthur that Groves had ordered Farrell to organize three groups to survey Hiroshima and Nagasaki and to secure information about Japanese activities in the field of atomic weapons. The Hiroshima and Nagasaki groups

> should enter those cities with the first American troops in order that these troops will not be subjected to any possible toxic effects although we have no reason to believe that such effects actually exist.[19]

Nichols quickly assembled a group of fifteen officers and twelve enlisted men, chiefly medical doctors, and had them at Hamilton Field, California, on August 12. The following morning they departed for Tinian, arriving there on the fifteenth to join Farrell. Out of the available personnel General Farrell formed three teams. He included himself, Brig. Gen. James B. Newman Jr., and a number of medical and intelligence officers in an advance party.[20] The second group was to go to Nagasaki and was headed by Col. Stafford Warren, chief of the MED's Medical Section.[21] Warren was ordered to take measures to ensure the safety of troops who would occupy the two cities; investigate and record any radioactivity that might be present; and report on the amount of blast and other damage caused by the bomb.[22] Warren's deputy Lt. Col. Hymer L. Friedell led the Hiroshima team.

In a letter to Farrell on August 20 Groves told him that the effort to learn what had happened in Hiroshima and Nagasaki was of the utmost importance, especially the effects having to do with radioactivity.[23]

Farrell and his advance party finally entered Japan on September 5. On the eighth and ninth they made a preliminary inspection of Hiroshima, and of Nagasaki on the thirteenth and fourteenth; he then cabled Groves with his first impressions. Using Geiger counters and other instruments, they determined that no significant amounts of radioactivity persisted.[24] Members of the signal corps took pictures of the damage wrought by the bomb.

Farrell flew over the city at three hundred feet and examined it in detail on the ground. Speaking in the third person he reported his preliminary inspection to Groves.[25]

> The airplane photographs taken after the strike did not give an adequate picture of the complete leveling of Hiroshima and did not pre-

pare Farrell for the scene of utter devastation. The city is substantially destroyed and gave General Farrell awful proof of the terrible power of the atomic bomb. . . .

Everything is blasted with some evidence of burning up to a radius of two miles from the point of detonation. . . . All of the staff of an army headquarters located in the city were casualties. Of the nine thousand military personnel in Hiroshima four thousand were killed outright, three thousand wounded and only two thousand escaped. The report of a Japanese official indicates from seventy five to one hundred twenty thousand dead and from seventy-five to two hundred thousand wounded. Nearly ninety percent of the city or a total of sixty eight thousand buildings were destroyed or damaged.

The first Western journalist, perhaps the first Westerner, to enter Hiroshima was the Australian Wilfred Burchett.[26] After covering Okinawa, Burchett had reached Yokosuka in late August with an advance party of marines. He took a twenty-one-hour train trip from Tokyo to Hiroshima and arrived in the early-morning hours of September 3. He walked the streets and visited the hospital, where the doctor told him of a terrible sickness that was killing his patients. Burchett wrote his story and managed to get it out to London's *Daily Express,* where it was published on September 6. Burchett's story told of the devastation and of people dying of an "atomic plague."

> For no apparent reason their health began to fail. They lost appetite. Their hair fell out. Bluish spots appeared on their bodies. And then bleeding began from the ears, nose and mouth. At first, doctors told me, they thought these were the symptoms of general debility. They gave their patients Vitamin A injections. The results were horrible. The flesh started rotting from the hole caused by the injection of the needle. And in every case the victim died.

On the morning of September 7 Burchett returned to Tokyo just in time to attend a press conference at the Imperial Hotel where General Thomas Farrell explained that the bomb had exploded at a height intended to reduce the risk of radiation. American officials were skeptical of the reports of radiation, believing that anyone close enough to the hypocenter to be irradiated could not have survived the blast. Burchett's questions about what he had seen fell on deaf ears.

Rumors spread of lingering radiation at Hiroshima and Nagasaki that, according to some press reports, would make the cities uninhabitable for many years. By early September these rumors became serious enough that Groves organized a press tour of the Trinity test site. Groves contacted Dr. Louis H. Hempelmann in Santa Fe on August 25, and was told the Trinity site would be safe enough for a visit by newspaper reporters in a few days.

Two weeks later Groves led a group of twenty journalists, plus several scientists and doctors, to the Trinity crater.[27] Among the pictures taken that day was a famous one, reproduced many times since, of Groves and Oppenheimer standing next to the twisted wreckage of the tower that had been obliterated on July 16. Groves's purpose in undertaking the trip and staging the photo opportunity was to reassure the public. If the two leaders could stand in the bomb crater safely, at ground zero, then all of this talk of radiation sickness must be exaggerated. But a close look at those photographs shows that everyone present was wearing special disposable booties on their feet to prevent contamination from the site. The irony was that Trinity site was probably more contaminated with lingering radiation than Hiroshima and Nagasaki. The reason was simple. The bomb had been detonated on top of a hundred-foot tower and the fireball — a quarter of a mile across — had hit the ground, leaving behind large quantities of radioactive isotopes. At Hiroshima and Nagasaki the bombs were detonated at altitudes of 1,950 feet and 1,650 feet to maximize blast, and thus avoided such lingering radiation.

The attitude and approach that Groves took to the stories about radiation would be repeated over and over by a succession of government officials in the decades that followed. The official line always sought to project an image that authority knew what it was doing, even when it did not. The government did not want to admit that it was ignorant or confused about the effects of the bomb, but had to project an image of certainty and competence. Groves was an early exemplar of the government authority figure seeking to soothe the public's fears. In fact, however, Groves had little real understanding of the effects of radiation in the bombed cities — or at Oak Ridge or Hanford, for that matter.

Part of Groves's ability to control the story lay in his use of secrecy and security classification. The underlying assumption is that the figure of authority has more knowledge than the public, by virtue of having access to classified information. Unfortunately, Groves implied as the argument unfolded, certain important facts could not be shared because they would aid our enemies. The public would just have to trust its leaders and hope that they made wise choices when it came to the military and civilian uses of atomic energy.

Throughout the fall and winter of 1945 Groves made a grand tour visiting the university laboratories and industrial contractors that had participated in the Manhattan Project. It was time to recognize the hard work that so many had engaged in. At each stop he would make a speech reviewing the broad achievements and then focus on the specific work that had been done at the location where he was speaking.

The first three weeks of October were busy as Groves — accompanied by General Farrell and Colonel Nichols, and Commodore Parsons and Commander Ashworth representing the navy — awarded the Army-Navy "E" Award flag (for high achievement in the production of war equipment) to a dozen industrial firms and universities.[28]

The speeches followed a common pattern, with the emphasis on how the bomb shortened the war and saved American lives. "Millions of American parents are in your debt." Another theme was the teamwork of American industry, labor, science, and the military, a unique partnership that got the job done.

In remarks at Oak Ridge on August 29 he told his audience, "The war ended sooner because of us." A great many thousands of American lives were saved. "The job was well done." "You will never have to be apologetic of what you did during this war." The element of surprise was central, and that was possible because of the secrecy that was maintained. "It is not an inhuman weapon. I have no apologies or excuses for its use. I think our best answer to anyone who doubts this is that we did not start the war and if they don't like the way that we ended it, to remember that they started it."

In letters of commendation to scientists and corporate leaders he thanked them individually for their contributions. In a letter to Hans Bethe he also encapsulated his view of the role of the bomb in ending the war.

> The major final factor which determined the surrender of Japan, as thoughtful people everywhere must agree, was the atomic bomb. That surrender was of course an ultimate certainty anyway, but the war would surely have continued for weeks and perhaps months longer had it not been for the timely knockout blow provided by the completion

and use of the atomic bomb. That weapon, therefore, saved friend and foe alike, from further death, destruction and misery.[29]

Groves oversaw awards to a group of army officers that ran the MED, and to members of his staff. On October 30 Jean O' Leary was awarded the Exceptional Civilian Service Award.[30] At the end of November fifty-four officers were given awards.[31]

Groves himself was honored several times in these early months. The army can often be a petty place, where envy over the recognition of fellow officers grates on some. How Groves's accomplishments were recognized became a source of controversy within the army and was an early indicator that his military career might not continue to flourish.

On August 9, 1945, the day Nagasaki was bombed, Secretary Stimson recommended to General Marshall that Groves be awarded the Distinguished Service Medal at the earliest possible date. In making the recommendation, Stimson recognized that such awards were normally not given until the officer in question had completed the assignment for which the award was made, but in this case Stimson felt that it was appropriate that an exception be made. General Marshall responded immediately and agreed with Stimson, but issued two caveats:

> . . . I feel that decorating Groves immediately would tend to add fuel to the flames of our present difficulties with respect to public reaction to the atomic bomb, and I believe we should wait until this has died down somewhat. I recommend, therefore, that this matter be held in abeyance and rechecked with you in a week or ten days.
>
> I assume that you have considered the effect of singling out Groves and omitting the small group of leading scientists and similar group of heads of plants producing the explosive.

Stimson concurred "that the matter be held in abeyance for a week or so until the present furor has died down somewhat." He recognized the need to honor the scientists and industrialists who had played leading roles in the project. "But I feel that Groves' work stands head and shoulders above even the invaluable work of some of the others because he was the responsible agent of the government and shouldered more of the ultimate responsibility than did any of the others." In an afternoon ceremony on September 12 Secretary Stimson awarded the Distinguished Service Medal to Groves. The citation read:

Major General Leslie Richard Groves, as Commanding General, Manhattan Engineer District, Army Service Forces, from June 1942 to August 1945 coordinated, administered and controlled a project of unprecedented, world-wide significance — the development of the Atomic Bomb. His was the responsibility for procuring materiel and personnel, marshalling the forces of government and industry, erecting huge plants, blending the scientific efforts of the United States and foreign countries, and maintaining completely secret the search for a key to release atomic energy. He accomplished his task with such outstanding success that in an amazingly short time the Manhattan Engineer District solved this problem of staggering complexity, defeating the Axis powers in the race to produce an instrument whose peacetime potentialities are no less marvelous than its wartime application is awesome. The achievement of General Groves is of unfathomable importance to the future of the nation and the world.[32]

Stimson recorded in his diary that day, "The citation itself was a strong one and I added that the case was rare indeed where a single individual had the fortune to be as effective as Groves had in the winding up of a great war."

But these well-deserved commendations carried a sting as well. Three years earlier, in July 1942, General Somervell had recommended Groves for a DSM for the construction work he accomplished from November 1940 to February 1942 during the mobilization. But to avoid drawing attention to Groves's new secret position as head of the Manhattan Project, the DSM was not awarded at the time. Now with the end of the war there was the problem of what to do about this previous DSM. The Decorations Board felt it should be changed to a Legion of Merit, while some of his supporters, including Maj. Gen. Thomas M. Robins, acting chief of engineers, and Lt. Gen. Leroy Lutes, who had replaced Somervell as head of Army Service Forces, felt the DSM was completely justified and asked the board to reconsider.[33] In early January 1946 the board recommended disapproval of the DSM. General Lutes made one more try, writing the board on March 1 to remind members that some officers working for Groves had received the DSM.[34] But the board refused to be swayed.

At the same time President Truman suggested that Groves be promoted to the permanent grade of major general. At that point Groves was a temporary major general with a permanent rank of colonel. Marshall replied on September 10 that the law required that Groves be placed on the list only by a board of officers.[35] With the war ended the army was preparing to significantly reduce the number of officers holding temporary appointments, setting a lower number for each rank. Some officers were going to have to be "busted back" (reduced in rank) to meet the lower levels. Any proposal to promote certain officers in this environment was thus a delicate one. The solution in Groves's case was to not use any of the army's allotted number of brigadier general slots. The position of assistant chief of engineers came by statute with the rank of permanent brigadier general, and thus did not count against the army's quota. And so on October 16 Groves was nominated for the position of assistant chief of engineers, while retaining his temporary rank of major general. But some felt this was the same old Groves, riding roughshod over others to advance himself.

On September 21 Groves, his wife, and daughter traveled to New York City. He was greeted by a crowd of five thousand at City Hall, where Mayor Fiorello H. La Guardia presented him with a scroll. This ceremony was followed by a luncheon given in his honor by Thomas J. Watson Sr., president of IBM, at the Waldorf-Astoria Hotel. Groves sat on the dais flanked by Watson and Mayor La Guardia. Some 250 dignitaries attended from the corporate, military, and political worlds. In Watson's introductory remarks he extended sincere thanks and appreciation for Groves's "leadership in developing the organization and carrying through the work to completion which ended in the atomic bomb, which shortened the war. . . . saving the lives of an enormous number of American boys."[36]

After a standing ovation and prolonged applause, Groves spoke. He reviewed his experiences in organizing the project, in recruiting Du Pont, Tennessee Eastman, and Union Carbide, and praising them for their willingness to commit themselves to the job. American labor was another essential partner. The third member of the team was the scientists, with the army overseeing all the parts. In this speech, and in others, Groves presented the building of the bomb as a case study of American exceptionalism. Certain things were absolutely essential. There had to be teamwork, ambition to get the job done and get it done right, and confidence and faith in one another. There were special qualities about America that made it uniquely fit for such a project, according to Groves. When other nations had these qualities they too might achieve what we had done, but Groves did not see any others at the time.

Groves received several honorary doctorates from colleges and universities as well. On October 21 he received a doctorate of laws from the University of California and five days later gave the featured address at the inauguration of Ralph C. Hutchinson as president of Lafayette College in Easton, Pennsylvania, where he was awarded a doctorate of science. His speech reviewed the history of warfare from the Greeks to the Middle Ages, then through the ages of gunpowder, steam, and oil, to the present atomic age.[37]

This recognition of Groves culminated in an elaborate banquet on February 26, 1946, with the Chemical and Metallurgical Industry Award given to Groves; he accepted in the name of the Manhattan Engineer District. The affair was held at the Waldorf-Astoria Hotel. Photographs of the event show the head table as a tier of daises seating more than two hundred Manhattan Project prime contractors.

But these busy weeks were not only filled with celebratory speeches and award ceremonies. A fundamental issue loomed. What should be done with the colossal empire that Groves had created? It was not at all clear. The vast number of decisions did not end with the completion and use of the bomb. What of the personnel at the major and minor sites? What about the contracts with the universities and corporations? Though the intense pressure that all had worked under was now lifted, the office remained busy with telephone calls, visits, and visitors. Groves's short-term goal was to keep the Manhattan District intact and operating until the situation clarified itself. Groves expected that Congress would decide in early fall, a month or two away, the future policy with respect to the atomic bomb and relieve the army and himself of responsibilities.

In late August Groves submitted a memo to the chief of staff and the secretary of war detailing his short-term plans to curtail production.[38] Up until this point, he noted, the "atomic bomb project has been carried forward without any lessening of effort in order that we would have a maximum number of bombs available in the remote event that Japan should fail to go through with her surrender." Under the proposed plan production of material at Oak Ridge and Hanford would be reduced by about 15 percent, resulting in a cut of operations costs by more than 30 percent.

In July Groves had approved plans to build about twenty Nagasaki-type bombs by the end of 1945. The production of the necessary parts for the final assembly of the bombs would continue, and Los Alamos would work on improving the current bombs, although on a reduced scale. In an interesting comment, he referred to the hydrogen bomb. "No efforts will be directed toward the design of a super bomb. Such developments would require several years and must be conducted under the highest order of secrecy."

A key question entailed the future of Los Alamos and especially its leadership. Oppenheimer was anxious to turn over the management of the laboratory to a successor, and Groves wanted a change as well.

> I did not want to keep Dr. Oppenheimer as the Director for two reasons, first, anything he did as Director from that time on would be an overwhelming anticlimax; second, I was trying to get out of the project everyone about whose security background there could ever be a question.[39]

At Oppenheimer's suggestion Groves eventually chose Norris Bradbury.[40] This would prove to be a sound decision, though Bradbury had to be urged to take the job. A Berkeley Ph.D. in physics, Bradbury had taught at Stanford before the war. As a member of the Naval Reserve, he had been called to active duty at Pearl Harbor. He spent several years doing research on high explosives, and had been brought to Los Alamos by Captain Parsons to head the X-1 Group in the Explosives Division, which was responsible for field-testing the explosive lens of the implosion bomb. Like Groves, Bradbury believed that it was important that Los Alamos be kept going. To close it would be to lose the unique skills and experience that had accumulated during the war. Ongoing development of nuclear weapons must be continued at least until the questions of domestic and/or international control were settled. As an interim move Groves chose Col. Lyle Seeman to be assistant director of Los Alamos in early September.

In addition, Groves took up Bradbury's suggestion of starting an educational institute at the laboratory, which would allow employees to take advanced undergraduate and graduate courses for credit. By the end of 1945 the situation at Los Alamos had stabilized, at least for the short term.

A related problem was the MED's contract with the University of California, which wanted to withdraw as the contractor. Groves convinced the university to continue by granting it a new concession. The university did not want the manufacture and assembly of bombs to take place at Los Alamos, feeling it was inappropriate for a university to be associated with such work. Groves acquiesced to this request and transferred the production, assembly, and storage of fission cores to Sandia Base at Kirtland Field in Albuquerque.

In addition to the exodus of scientists, who left to return to their normal lives at universities and elsewhere, an even heavier attrition was taking place among the MED's military officers. Many of those who had run the project during the war had been noncareer reservists. With the war over these temporary officers were eligible for discharge and were anxious to return to civilian life. For Groves the problem of replacing them was formidable.

Groves needed about fifty regular officers to run the various elements, men young enough to break into the atomic field but senior enough to have demonstrated an ability to accept responsibility.

The most obvious place to find talented officers was among West Point graduates. Those who ranked among the first five or ten in their class were preferred, with no one below the first 10 percent considered. A successful athlete demonstrating above-average determination and a will to win would also be considered.

Groves's insistence that only the most highly qualified officers meeting the strictest selection standards be assigned to the MED was not greeted with enthusiasm by the War Department General Staff, which saw no reason why the Manhattan Project should get first choice of the best officers in the army. Groves negotiated with Generals Handy and McNarney, who objected that he was asking for too many good men. Groves next went to General Eisenhower, now chief of staff, to seek his support, but he sided with the WDGS. Typically, Groves did not give up. Soon after being rebuffed, Groves brought the matter up with Secretary of War Robert P. Patterson; a "quite forceful, and occasionally even heated discussion" ensued among Eisenhower, Handy, and Groves over the matter of the choice personnel. After about five minutes, Patterson said, "I agree with Groves," and, "I want him to have as many officers as he decides he needs and of the quality he thinks he needs, and I want him to have complete freedom of choice."[41]

Another of Groves's responsibilities as MED "trustee" was to decide upon the future of the research and development facilities. Groves formed an Advisory Committee on Research and Development to help with guidance and to make recommendations. The committee consisted of seven scientists representing a broad spectrum of experience within the Manhattan Project. From this committee came a series of recommendations, the most profound of which was the idea of designating some facilities as national laboratories where unclassified nuclear energy research would be performed. These laboratories would serve as places to funnel federal funds into promising research areas. Groves lost no time acting on the idea. As originally envisioned, national laboratories would be located in all parts of the country. The first two were Argonne National Laboratory, to be administered by the University of Chicago, and a new facility, Brookhaven National Laboratory, located at Patchogue, Long Island, to be administered by a new organization, Associated Universities, Inc., representing a consortium of twenty-nine universities.

Groves entered into similar negotiations with President Hutchins of the University of Chicago. There was some tension between them. Groves took umbrage at Hutchins's outspoken views on the importance of placing the development and control of nuclear technologies under international control. But he also recognized that the Met Lab, which had been housed at the university, had gained a wealth of experience in nuclear technologies. The university was also successful in recruiting first-rate talent from other MED facilities for its postwar faculty. As had been anticipated prior to the end of the war, all MED activities at Chicago were moved to the Argonne forest, about fifty miles south of Chicago, where the nuclear pile that Fermi had first constructed in the squash court under the stands at Stagg Field had been moved in 1943. The initial postwar contract with Chicago was also a short-term one. Groves was aware that the situation would remain extremely fluid until Congress legislated a structure within which postwar nuclear research could be pursued.

The immediate postwar situations at Lawrence's Radiation Laboratory and at Oak Ridge were a little clearer. Almost immediately after the war, Groves ordered the shutdown of most of the electromagnetic separation facilities at Y-12. It was now clear that the process, while marginally effective in the short run, was too wasteful and inefficient to serve as the primary technique for isotope separation. Much better at producing highly enriched uranium was the gaseous diffusion plant. A portion of the Y-12 facility was converted to produce stable isotopes of elements other than uranium. The air-cooled reactor at X-10 continued to produce small quantities of plutonium and other rare radioactive isotopes for research purposes.

With Groves's encouragement, Ernest Lawrence returned to the activity that had been at the core of his professional life prior to the war: cyclotron development and research, a field that would soon be known as high-energy physics, or particle physics. At the end of December Groves agreed to provide Lawrence with $170,000 to finish construction of the

184-inch cyclotron whose completion had been interrupted at the beginning of the war. He also provided money out of MED operating funds for construction of a new kind of accelerator, the synchrotron, a device that Edwin McMillan had conceived of while working at Los Alamos. As with the case of Chicago and Los Alamos, more aggressive support would have to await action by Congress.

The question of secrecy, of what information should be declassified, not only had national security implications but was directly relevant to non-military research on atomic energy as well. Not all of the issues were resolved by the publication of the Smyth Report. What was needed was a more systematic approach to the declassification of information. At the end of October 1945 Groves chose his close adviser Richard Tolman to form a declassification committee to recommend a detailed program and a process to accomplish the release of scientific information, without endangering national security. The committee consisted of Robert F. Bacher, Arthur H. Compton, Ernest O. Lawrence, J. Robert Oppenheimer, F. H. Spedding, and Harold C. Urey, with Lt. Col. John R. Ruhoff, the Madison Square Area engineer, as the nonvoting secretary.

The committee members grappled with the problem, which continues to this day, of distinguishing between harmful and harmless information. Their problem, as now, was how to strike a balance between restricting information that protects national security and releasing information that enhances scientific research and informs a democratic society. Their report was approved in March 1946, and a Declassification Guide was prepared. A special organization was set up under the direction of Lt. Col. Ruhoff and began operating on April 1, 1946.

> Our fundamental belief is that the release of basic scientific and technical information obtained during the development of the bomb would, over a sufficiently long term, not only enhance our national welfare but actually conduce to our national safety. . . . We believe that nearly everyone will agree that there is much that can be disclosed at the present time without danger to our military security. We are convinced that practicable and sound principles can be formulated which will make it appropriate to release such information at once provided the release is carried out with circumspection and discernment under competent and informed guidance. The needless withholding of new developments is bound to delay progress in technical

fields, and hence to have serious consequences for our national welfare and security, while disclosure of a great store of new and useful
information will stimulate the growth and development of science
and industry.[42]

A contributing factor to Groves's declining reputation in the final months
of 1945 was the destruction of the Japanese cyclotrons in late November.[43]
The incident also brought embarrassment upon the army as a whole and
was used by the scientists as an example of how the military would be
expected to act if it were left in charge of atomic energy.

Concurrent with the teams that were sent to Hiroshima and Nagasaki
to investigate the effects of the bombs was a third mission to survey
Japanese activities in the area of nuclear science. Unlike the wartime fears
aroused by the possibility of a vigorous German bomb program, assessments of Japanese atomic research concluded that it was never much of a
threat. All along Groves had been quite certain that Japan posed no danger.
It did not have enough scientists trained in nuclear physics, it did not have
enough uranium or uranium ore, and the industrial capability needed to
produce fissionable materials exceeded its abilities. As a leading authority
later concluded,

> The Japanese endeavor was badly fragmented, inadequately staffed,
> indifferently pursued, and plagued by doubt and ambivalence at the
> individual level.[44]

A few Japanese scientists did carry on modest research programs using
whatever experimental equipment they had on hand, including a few
cyclotrons. By late 1945 there were two cyclotrons at the Nishina
Laboratory of the Institute for Physical and Chemical Research in Tokyo,
two at Osaka Imperial University, and one at Kyoto Imperial University.
Four of them were inoperable; the only one of any value had been made
in the United States, erected at the Nishina Laboratory in 1938, and made
obsolete by the end of the war.

In October 1945 General MacArthur gave permission for research with
cyclotrons to aid studies in biology and medicine.[45] Any work applied to
warfare or the separation of isotopes was forbidden. The laboratories were
monitored and had to submit regular reports. Of course, cyclotrons are
laboratory instruments and are not capable of producing atomic bombs.

The cause of the destruction of the cyclotrons was a series of confused orders that never came to the attention of higher authorities, who might have stopped it in time. On September 5 the General Staff issued instructions directing destruction of all enemy war equipment, except new or unique items, which were to be saved for examination. On October 30 the JCS cabled the commanders in the Pacific area and China, instructing them to seize all facilities for research in atomic energy and to take into custody all persons engaged in such research. When the cable came to Groves's office he instructed one of his officers on its contents, but as later events would show, his intentions were not as clear as they might have been. He later said he was anxious that the cyclotrons be secured and not be destroyed, but something quite different was conveyed to Maj. Amos Britt. On November 7 a message to General MacArthur was prepared in Groves's office to go out in the secretary of war's name. It explicitly ordered the destruction of the five cyclotrons. The next day the message was cleared through Mr. Patterson's office as a routine matter without being brought to his attention and was sent over his special channels. On November 24 General MacArthur's headquarters in Tokyo reported that the cyclotrons had been seized on November 20 and that destruction had begun on November 24. The cable was sent to nine people in authority, including General Groves, but none of the intended recipients saw it, because it was handled and filed by subordinate staff officers. And so the destruction of the cyclotrons continued.

MacArthur's headquarters first became aware that there might have been a mistake when a November 28 cable directed staff to send one of the cyclotrons (already destroyed) back to the United States for study. The discrepancy was brought to the attention of MacArthur, who cabled Eisenhower about the conflicting instructions emanating from the War Department. Apparently, Eisenhower never saw the cable, for no action was taken at that time.

The matter came to everyone's attention when a United Press news article was published on November 30 reporting that the American occupying forces in Japan had destroyed the cyclotrons. Over the next few days an explanation of where the order originated was eventually determined. The problem had originated in Washington, and to try to stem the public relations disaster a statement from the secretary of war was felt to be in order. According to Groves, Patterson first proposed a tough no-nonsense explanation that actually justified the action. "In order to ensure peace for generations to come we desire to eliminate to the maximum extent possible, the Japanese war-making potential." Groves said that this did not appeal to him. After some discussion a far different statement was issued on December 15 in which Patterson admitted that a mistake had been made, regretted the hasty action that had been taken, and took full responsibility.

Groves's account in his memoir does not quite tie up all of the loose ends about this episode. He admits that the lesson he learned from the affair was that a commander must always make his intentions unmistakably clear to his subordinates. This he says he did not do — but then the issue was not terribly complicated. Either the cyclotrons were to be secured, with some possibly sent back for study, or they were to be destroyed. Groves makes it seem that his real intention was to save the cyclotrons. In fact, the message that he had his assistant prepare was clear as could be that they were to be destroyed:

> After all technical and experimental data has been secured it is requested that the cyclotrons at the Institute for Physics and Chemical Research (Riken), the Imperial University of Kyoto and the Imperial University of Osaka be destroyed.[46]

It is hard to see how Groves's subordinate, especially one who was new to the MED, could arrive at such a definitive conclusion from the supposedly confused message of his superior.[47]

With Patterson's admission of error the story quickly died down, but as far as the American scientists were concerned another black mark had been attached to the name of Gen. Leslie Groves.

In just a few short months Groves had gone from a celebrated war hero to a figure that represented in many people's minds the dangers of military control. This quick descent in reputation turned out to be permanent: Groves was relegated to the sidelines in most of the accounts of the Manhattan Project that were written over the next forty years.

PART SIX

Final Battles

• 1946–1948 •

CHAPTER TWENTY-TWO

Caught in the Middle:
Fights Over Domestic Control
· 1946 ·

While Groves was being feted for building the bomb and helping to bring the war to a swift and sudden end, there were ominous signs of a growing opposition to giving him a significant role in the peacetime future of atomic energy. In getting the job done, Groves had stepped on a lot of toes inside and outside the army. He had amassed a large and potentially powerful group of enemies, and with the war now over Groves was soon to find out just how powerful they were.

Groves was slow to realize that new forces had quickly emerged in the aftermath of the war and the use of the bomb. Though excellent at anticipating some things, there were others that he either did not see coming or was unable to adapt to. During the war Groves had the absolute authority to take virtually any action he deemed necessary to get the job done, all with the highest security under the greatest secrecy. He flourished in this unique environment. Literally overnight circumstances were transformed. The war was over, the bomb worked, his major patrons Stimson and Marshall quickly left their positions, and there was a clamor by almost everyone to have a say in what this new terrifying weapon meant. New skills were needed to survive in this postwar world, and unfortunately for Groves he did not possess them.

Since early 1944 several government officials had been giving thought to the problem of the domestic control of atomic energy in the postwar period. As we have seen, one of the major tasks of the Interim Committee was to consider this issue and to prepare draft legislation to deal with it. Shortly after V-J Day the committee sent to President Truman the draft of a bill that it had prepared with considerable input from the War Department and the Manhattan Project. The Royall-Marbury draft legislation was the forerunner of the May-Johnson Bill that had been worked out with the Interim Committee and reflected its members' views. In late August 1945 Groves had meetings and phone calls with Royall and Marbury and helped craft the language.

The president, in his October 3 message to Congress, had emphasized the need for prompt action on the bill. On the same day Rep. Andrew J. May of Kentucky, chairman of the House Military Affairs Committee, introduced the bill (H.R. 4280) and scheduled one day of hearings the following week. Sen. Edwin M. Johnson of Colorado, a member of the Senate Military Affairs Committee, introduced the bill in the Senate (S. 1463) and thus it became known as the May-Johnson Bill. Unbeknownst to its sponsors, and to Groves, the legislation was about to run into a buzz saw.

Immediately after Hiroshima and Nagasaki some of the Manhattan scientists began to organize themselves to address the many new and complicated questions surrounding the bomb.[1] For three years they had been silenced by General Groves, but now that the bomb was no longer a secret it was time to speak out about the dangers it portended and how it might be controlled. At each of the Manhattan sites scientists' associations were formed to discuss these matters. On November 1, 1945, the groups from Chicago, Oak Ridge, Los Alamos, and New York joined together to launch the Federation of Atomic Scientists, reorganized the following month as the Federation of American Scientists. Also in December the first issue of the *Bulletin of the Atomic Scientists* was published. By the end of the year an active and vocal scientists' movement was well in place.

High on the scientists' agenda of crucial issues was the question of civilian versus military control of atomic energy. When the text of the May-Johnson Bill was made public in early October, the FAS leadership was horrified. The bill called for control of all research, production, and exploitation of nuclear energy by a part-time nine-member commission headed by a chairman who had to answer only to the president. The commission was given absolute authority in determining which information about nuclear energy was to be kept secret. The bill also allowed for the chairman, and any or all of the commissioners, to be military officers. It also provided for a full-time administrator and deputy administrator, either or both of whom could be military officers. This suggested to the scientists the probability that military control of nuclear policy would continue, and that General Groves would likely remain in charge, with his characteristic tight control over the distribution of information.

These fears were reinforced by the manner in which Chairman May conducted its single scheduled day of hearings on the bill. On October 9 the witnesses included Groves, Conant, Bush, and Patterson, now secretary of war after Stimson's retirement two weeks earlier. The four were enthusiastic supporters of the bill. Secretary Patterson told the committee that:

> The War Department has taken the initiative in proposing that it be divested of the great authority that goes with the control of atomic energy, because it recognizes that the problems we face go far beyond the pure military sphere. . . . The wisest minds in our Nation will be required to administer this discovery for the benefit of all of us.

Groves professed to be anxious for the bill to pass so that he might be relieved of the heavy responsibilities that he had shouldered for the past three years. "In coming before your committee today we are appealing for an opportunity to give you our existing powers," he told them.[2]

Groves explained that there would be a commission of nine members with revolving membership, appointed by the president. The commission would have broad powers, but they were justified so as to meet its broad range of responsibilities. Bush and Conant concurred. Many assumed that the bill would have smooth sailing through the House, but adverse reaction in the Senate, in the press, and among the public suddenly erupted.

The scientists questioned the bill and spoke out vigorously against it. With their new commitment to educating the public about nuclear energy and their rapid organizing effort, they soon became effective lobbyists against the bill. To their satisfaction, they found that senators and congressmen, the press, and the public listened to them when they identified themselves as scientists who had worked on the bomb.

While President Truman had originally supported passage of the May-Johnson Bill, he and his allies in Congress were surprised and a bit unnerved by the furor surrounding the one day of hearings. Press reports appeared suggesting that May was railroading the bill through the House. Reluctantly, May scheduled a second day of hearings for October 18, which would allow the scientists an opportunity to express their opinions on the bill. Among the scientists appearing before the committee that day were Leo Szilard, Arthur Compton, Robert Oppenheimer, and Harold Urey.

The hostility by the committee to the scientists who opposed the bill was palpable. For their part, with one notable exception, the scientists attacked the bill's provisions for making the commission and its chairman powers unto themselves, for needlessly preventing the free flow of ideas, for unnecessarily denying freedom of action in the realm of laboratory research, and for opening the door to continued military control of nuclear energy research and development. Only Oppenheimer endorsed the bill on all counts, much to the consternation of those scientists who had accepted his earlier, wartime assurances that they could trust him to represent their interests in Washington. With the Senate also critical, it was clear that the May-Johnson Bill was in serious trouble.

Arthur H. Vandenberg, the ranking member of the Senate Military Affairs Committee, challenged the bill on the grounds that it dealt with a subject beyond the competence of a standing committee. He proposed a resolution that a special joint committee of both houses be created. While his resolution was voted down, it set the stage for creating a special Senate committee, one that would have a profound impact on the issue.

On October 10, the day after the House Military Affairs Committee's hearings, the freshman senator from Connecticut, Brien McMahon, introduced a resolution calling for a special Senate committee to study the nuclear energy issue, to consider all proposed legislation, and to make recommendations to the Senate. On October 26 the resolution passed and, fending off a challenge from Senator Johnson, McMahon was named chairman of the new committee, quite a feat for a senator with less than a year in office. The committee consisted of six Democrats — McMahon, Richard B. Russell, Edwin C. Johnson, Tom Connally, Harry F. Byrd, and Millard E. Tydings — and five Republicans — Vandenberg, Warren R. Austin, Eugene D. Milliken, Bourke B. Hickenlooper, and Thomas C. Hart.

McMahon named James R. Newman as the committee's special counsel and physicist Edward Condon as its scientific adviser. From November 27 to December 20 the committee held a series of public hearings. Twenty-two witnesses were called, principally scientists and technical experts, military officers, and corporate officials.[3] They discussed the "appalling destructiveness" of the atomic bomb, the difficulties other countries would find in trying to make their own, the limited prospects of defense, the kinds of "secrets" about the bomb and the limited chances of retaining them, and possible peacetime benefits. The press and public followed all of these issues closely.

The choice of Condon sent a strong signal to Groves. The two had crossed swords before — first during Condon's brief stay at Los Alamos, and more recently in June 1945, when Groves learned that Condon had accepted an invitation to attend the celebration of the 220th anniversary of the Russian Academy of Science, to be held in Moscow and Leningrad from June 15 to June 18. Groves argued to the War Department that Condon knew too much and might inadvertently give the Soviets information about the Manhattan Project. Groves even expressed fears that the Soviets might kidnap Condon.[4]

Twice Groves succeeded in delaying Condon's departure from La Guardia Airport, where a Russian plane was preparing to fly to Moscow. The first time he had help from the New York City police. One of Groves's security aides, Lt. John J. O'Connell, was the son of the chief inspector of the New York City Police Department, a good man to have on one's staff. The Russian plane that Condon was preparing to board had mechanical problems that took thirty-six hours to correct. Apparently Inspector O'Connell had arranged to have the mechanics delay and delay finishing their work. After dark a truck ran into the plane and damaged one wing so that it was unable to take off for an additional several days. O'Connell apparently arranged this "accident" as well. In a letter to the chief inspector, Groves thanked him "for pulling me out of what could have been a very embarrassing position." Condon did not give up and continued preparations to leave. As he protested against the efforts to prevent him from going, Groves had the head of the passport division revoke his passport.[5]

Needless to say, Groves and Condon did not like one another. Just before he was to testify before the McMahon committee, on November 27, Groves sent the following memo to Patterson:

> I want to express for your eyes only my belief as to what is behind the detailed questions that have been submitted by Dr. Condon for the Senate Committee. It is obvious to me from the nature of the questions asked that they were propounded by individuals who had considerable information concerning the various sections of our work. I have known for some time that Dr. Condon and Dr. Leo Szilard, as well as many other curious individuals, have been seeking the type of information that is contained in the questionnaire. I feel certain that if such information is given it will soon be public property.[6]

Throughout November and December the defects of the May-Johnson Bill and its unlikely adoption were driven home to President Truman, who confided to his close advisers that he was withdrawing his support of the bill. On December 20 Brien McMahon introduced S.1717, his bill to control and regulate atomic energy.[7] As Groves was well aware, McMahon, Newman, and his staff had relied heavily on the advice of the scientists to draft it.

On January 22 the special committee began a second round of hearings that lasted until April 8. Forty-nine witnesses appeared, including military officers, scientists, cabinet members, educators, and businessmen. President Truman sent a letter to McMahon on February 1 indicating five points he considered desirable in atomic energy legislation: (1) a commission of three full-time civilian members, (2) exclusive production and ownership by the government of fissionable material, (3) compulsory nonexclusive licensing of private patents on devices utilizing atomic energy, (4) guarantees of free scientific research, (5) provisions to facilitate the establishment and enforcement of international agreements. The committee received more than seventy-five thousand letters from the public. On April 19 the committee reported the McMahon Bill to the Senate.[8]

Groves testified on February 27, and it was soon apparent that deep antipathy existed between the general and the chairman. When ordered to supply the committee with certain information, he flatly refused.[9] This exacerbated the civil-military issue even further: Groves's high-handed manner foreshadowed for some how the military might act. He also objected to the exclusion of active military from the commission in the McMahon Bill, favored the provision of a single executive that had been in the May-Johnson Bill, and criticized the inadequate security provisions. He expressed his preference for the part-time commissioners of the May-Johnson Bill, believing that they would be more capable men than the full-time commissioners called for in the McMahon Bill. In the course of the hearings Groves and McMahon became bitter enemies. McMahon showed his outright contempt for Groves. He belittled Groves's accomplishments during the war and even implied that he was an underachiever and an opportunist because of his erratic promotion course.

Throughout the first six months of 1946 the battle was fought on a broad front: in committee hearings, in the halls of Congress, in the offices of the War Department, and in the press. Truman eventually backed the McMahon Bill, and the War Department was forced to fall into line. Groves found himself almost alone, without the direct and open channels to the secretary and the undersecretary of war, the chief of staff, and the president to which he had become accustomed throughout the war. While he had friends in Congress and the press, his word no longer carried the weight it once had and he no longer made those who opposed him tremble.

From the start, as we have seen, a dominating issue became precisely what role the military should play in the field of atomic energy.[10] The extreme position was *none,* and unfortunately for Groves Chairman McMahon held it. McMahon had seen an opportunity to make the issue of atomic energy his own and had moved quickly to stake his claim. He was ambitious and combative and Groves was a perfect foil to help him in his rise.

The fact was that no one was suggesting that the military should remain in control. McMahon's position of totally excluding the military resulted in some wild rhetoric.[11] More moderate voices felt there should be shared responsibilities and consultation.

Groves's position on this issue was repeatedly misrepresented and distorted at the time, and solidified into myth in retrospective accounts of these battles. The view that Groves was a proponent of military control is false, as is the allegation that he wanted to run the program after the war.[12] As he said, "There was nothing I wanted more than to be rid of my responsibilities." His goal was to put the MED's operations on an orderly and sound footing, and he had hoped that this could occur quickly in the fall of 1945.

Groves was not opposed to civilian control of atomic energy. His criticism of the McMahon Bill was twofold. The bill contained what he felt was an ineffective executive structure, and it *totally* excluded the military. The bill proposed that the five commission members would have equal power, with one, appointed chairman by the president, to act as their spokesman. This arrangement could hardly be more different from the way he had operated during the war. How could a group of five lead an organization? As he pointed out in his memoir, "ever since the tribunes of Rome no executive group has ever functioned well."[13]

On the second issue Groves felt that the military, the ultimate user, must have some role in the many responsibilities associated with the new weapon. All of his actions in criticizing the McMahon Bill and, as we shall see, in creating and running the Armed Forces Special Weapons Project (AFSWP) were directed at ensuring that the military would be prepared to carry out basic responsibilities with regard to nuclear weapons.

Nevertheless, by early 1946 the military-versus-civilian controversy had turned into an extremely divisive issue. The United States was in one of its periodic mood swings over how it viewed the military, sometimes as the savior of democracy, sometimes as its destroyer. Marquis Childs wrote a series of articles in the *Washington Post* about the undesirability, even the dangers, of military control and painted Groves as power hungry and autocratic.[14] The portrait drawn was of a military filled with bunglers and warmongers, antidemocratic and incompetent, and the purest expression of these traits was General Groves. The caricature of Groves was taking shape, and his reputation was sinking further into eclipse.

Before the bill became law, Groves's position of guaranteeing a role for the military found an advocate in Senator Vandenberg.[15] Vandenberg, stimulated by Groves's testimony on the dangers of totally excluding the military, proposed an amendment establishing a "military liaison board" consisting of representatives from the War and Navy Departments, which the commission should "advise and consult with on all atomic energy matters which the board deems to relate to the common defense and security." Although the Senate committee approved Vandenberg's amendment by a vote of ten to one (with McMahon dissenting), it was intensely attacked for giving too much control to the military. In response Vandenberg softened a second version. The reworded amendment had the members of a Military Liaison Committee chosen by the president. As to their activities: The committee should advise and consult with the commission on all matters that the committee considered to relate to military applications of atomic energy; the commission and the committee should keep each other fully informed on all matters concerning atomic energy; the committee was authorized to make written recommendations to the commission; and in the event that the committee concluded that the commission was taking actions adverse to the military departments, it could refer the matter to one of the service secretaries, who could, if he concurred, appeal to the president for a final decision. Thus revised, the amendment was found acceptable and unanimously approved by the Senate committee.

For Groves the Vandenberg amendment that created the MLC was the only bright spot in an otherwise bleak situation. The MLC could serve as the vehicle to keep the Atomic Energy Commission (AEC) informed of the views of the military departments in regard to weapon development and manufacture and also be a monitoring body, representing the military's interests in the operations of the AEC.

The six members were chosen in July, August, and November of 1946. Lt. Gen. Lewis H. Brereton, USAF, was designated chairman. Maj. Gen. Lunsford E. Oliver and Col. John H. Hinds were the army members and Rear Adms. Thorvald A. Solberg, Ralph A. Ofstie, and William

S. Parsons were the navy members. General Groves replaced Oliver on January 31, 1947.

The first MLC meeting was held on October 12, 1946, and dealt with organization and function. Capt. Frederick L. Ashworth, weaponeer on the Nagasaki mission, was selected acting secretary. At the second meeting on October 16 the committee met with General Groves. He expressed the opinion that the AEC would not be a strong organization, that it would be under pressure to divert fissionable material for power, and that the MLC must be a watchdog for the military. He urged them to consider themselves a totally unified organization whose first goal was national security.

Throughout October officers of the MED provided orientation conferences on the topics the MLC would soon be responsible for: raw materials, foreign intelligence, security, legal issues, and military operations.

The bill specified that all of the assets of, and the commitments entered into by, the Manhattan Engineer District would become the responsibility of the AEC. After three years of work requiring the most intense concentration, and after a further year of virtually nonstop political maneuvering over the fate of the program he had created, Groves was worn out. He later recalled that this period from August through December 1946 was one of the most difficult of his tenure, because "everyone knew that I was in a caretaker position, and they had no assurance that my views would be those of the Commission. After the commissioners were finally appointed, it was quite evident that my views would not be accepted without a long-drawn delay."[16]

The atomic energy bill came to the Senate floor on June 1, and after three hours of debate and a few minor amendments it passed unanimously. The bill then went to the House Committee on Military Affairs on June 5; after hearings on June 11 through June 13, it went to the full House. The House took up the bill on July 17 and passed it with major changes on July 20. The House added seventy-one amendments, the great majority of which were accepted by the Senate with little question. Three were of importance to the military. One of these required that the director of military application be a member of the military services. Another provided that the president might direct the commission to deliver fissionable material to the armed forces in such quantity and for such use as he considered necessary in the national defense. The third stated that the president might direct the commission to authorize the armed forces to manufacture, produce, or acquire any equipment or device utilizing fissionable material or atomic energy as a military weapon. The president signed the Atomic Energy Act of 1946 (Public Law 585, Seventy-ninth Congress) into law on August 1, 1946.

Under terms of the act responsibility for the direction and control of atomic energy was now in a civilian commission of five members appointed by the president. Truman's first choice for position of chairman of the commission was James B. Conant, but opposition to him within the scientific community, led by the Federation of American Scientists, was fierce. Conant had been a supporter of the May-Johnson Bill, and thus it was politically impossible for him to accept. Over the next several weeks Truman received a great deal of advice with regard to the appointment of a chairman.

There was considerable support for naming David E. Lilienthal chairman, but Truman kept his own council. Finally, on September 20 Truman offered him the job. Truman had already selected two commissioners, Sumner T. Pike, formerly chairman of the Securities and Exchange Commission, and Lewis L. Strauss, an aide to Herbert Hoover and a Navy Reservist who had served in the Bureau of Ordnance and as an assistant to the secretary of the navy during the war. Lilienthal recruited the final two members in October. William W. Waymack, an editor of the *Des Moines Register and Tribune* and a Pulitzer Prize winner, said yes on the eighteenth, and Robert F. Bacher, a physicist at Cornell University and a Los Alamos group leader, agreed a week later. The formal announcement of the five appointees to the AEC was made on October 28, 1946.

Within a few days Lilienthal named three temporary staff members: Carroll L. Wilson, an assistant to Vannevar Bush at OSRD; Herbert S. Marks, an assistant to Dean Acheson; and Joseph Volpe. Groves had known Wilson and Marks at a distance, and Volpe had been on Groves's legal staff during the war. They had worked together closely, and Groves thought highly of him. When Groves heard of this appointment, he branded Volpe a "traitor." Loyalty was high on Groves's list of virtues, and Volpe's action effectively ended their relationship.

Groves was irreconcilably bitter at the overall turn of events. It was going to be hard enough to turn the organization he had built and nurtured over to anyone, but to give it to David Lilienthal was going to be extraordinarily difficult. Groves had neither respect for nor trust in Lilienthal. Over the next three months Groves did his best to impede the efforts of the newly appointed chairman, to make his job as difficult as possible, and to ensure a place for the military.

According to a scientist who observed Groves at close range, he was an unhappy and lonely figure in Washington during this period.

> Driving back from military bases with the General, I would listen to his lament as he detailed his woes. He felt that he was not given proper credit for his work during the war. . . . To top matters off, Groves found that the new AEC chief never once asked his advice on anything. General Groves later stated, "Mr. Lilienthal had made it very plain that he wanted no advice of any kind from me. He wanted nothing what- soever to do with me. He thought that I was the lowest kind of human being, and he was not going to get anything from me."[17]

As is the case with almost any piece of enacted legislation, details of the execution of the act's provisions are promulgated in one or more execu- tive orders drafted by those responsible for carrying out the terms of the act and signed by the president. In the case of the Atomic Energy Act of 1946, the most pressing issues facing the commission were appointing a staff and establishing the ground rules of how to accomplish the MED's transition to the AEC. There were further problems of finding office space, providing for funding, and working out the details of the transfer. The AEC staff, Wilson, Marks, and Volpe, and Groves's representative, Colonel Vanden Bulck, chief of the MED's Administrative Division, quickly agreed upon temporary office space in the New War Building. Lilienthal and the commissioners had their offices on the sixth floor, and the rest of the staff occupied offices on floors two, three, and four, while permanent office space was being arranged elsewhere.

On November 4 Lilienthal, Bacher, and Pike visited district headquar- ters in Oak Ridge. From November 13 through November 20 all five commissioners visited the major facilities at Oak Ridge, Los Alamos, Berkeley, and Chicago, escorted by Colonel Nichols.[18]

There was agreement on a mechanism whereby start-up funds for the commission would come out of MED-appropriated money. In a "hard ses- sion" with General Groves on December 3 the commissioners "reached a kind of tacit understanding that we would try to transfer the responsibil- ities on December 31, 1946."[19] This gave the principals less than a month to work out the details.

Negotiations to decide on what was to be transferred proved extremely difficult. There were the outsized egos of the two principals — Groves and

Lilienthal — and there was also Groves's disparaging assessment of his suc-cessor's experience and ability. He felt the commissioners were amateurs, not up to the task of handling the extensive facilities of the MED. Finally, even after all was said and done, even after passage of the Atomic Energy Act, there remained — in Groves's mind at least — the fundamental issue of balancing civilian and military control.

The Atomic Energy Act stated that "all property in the custody and control of the Manhattan Engineer District" would be transferred to the commission. On December 9 Marks drafted language for the executive order that reflected this part of the act. The AEC position was that all property and functions would be transferred without exception. Subsequently, the AEC would transfer back such property and functions as seemed appropriate for the military to control. Groves saw things dif-ferently. There was not going to be any temporary transfer and then retransfer back.

Acting on Groves's orders, Nichols informed the commission that the weapons storage facilities, Sandia base at Albuquerque, the heavy-water installations at several army ordnance locations, and several other sites would be exempt from transfer. He also informed them that the district's Murray Hill Area Office in New York City (which managed ore procurement) and the MED intelligence records and operations would remain with the army.

Groves's position was that the excluded sites were not under the exclu-sive control of the MED but jointly operated with other military units. Groves was unwilling to yield the Murray Hill office unless the commission assumed his responsibilities as chairman of the Combined Development Trust. The commission had thus far been able to get so little information out of Groves on the operations of the trust that it was unwilling to assume such responsibilities until it knew more. And as to the foreign intelligence files and functions, Groves argued, they should go to the recently formed Central Intelligence Group (CIG), and not to the AEC.[20]

The CIG was established on January 22, 1946, with Capt. (soon to be Rear Adm.) Sidney W. Souers appointed the first director of Central Intelligence (DCI). Lt. Gen. Hoyt S. Vandenberg succeeded him in June and soon after issued a memorandum to the National Intelligence Authority (NIA) concerning the collection and evaluation of intelligence related to foreign atomic energy developments. Up until then, Vandenberg noted, this activity had been performed by the MED's Foreign Intelligence Branch under General Groves's personal direction. The DCI recom-mended that "the personnel and working files of the Foreign Intelligence

Branch operated by General Groves should be transferred to the Central Intelligence Group."[21]

At the meeting of the NIA on August 21 the issue of transferring General Groves's foreign intelligence files and personnel was the sole topic discussed. Secretary of War Patterson strongly urged that they be transferred to the CIG without delay. To concerns about sharing the information with the AEC, General Vandenberg said that if the NIA authorized the CIG to furnish it information, he would do so. A directive was prepared for the president's approval, calling for the transfer of the personnel and working files "at the earliest practicable date."[22] The matter of the "practicable date" was not fully resolved for another eight months.

The commission was totally mystified at this turn of events, but stuck to its position of total transfer and retransfer. Nichols, in a series of meeting with Carroll Wilson and the commissioners, was unable to convince them otherwise. For the rest of the month, as the deadline loomed, Lilienthal and the commission forced Groves to concede his property a bit at a time in an ongoing test of wills.

A final maneuver by Groves and the army was the attempt to have Nichols nominated as director of military application. Four days before Christmas this idea was presented by Secretary Patterson to Lilienthal, who firmly resisted. Here was another indication to some, if any more were needed, that Groves was still out to control things. Later that day Lilienthal met, and continued to argue against the idea, with Secretary of the Navy Forrestal, Admiral Nimitz, Admiral King, and Vannevar Bush, looking for suggestions of someone else who was not so closely aligned with Groves.[23]

And so it went right down to the wire. On December 23 Marks completed the executive order on the transfer. On December 30, with less than thirty-six hours left, Lilienthal met with Secretary of War Patterson to deal with some final issues. Lilienthal reluctantly agreed to the commission assuming responsibility for the Combined Development Trust, something they had learned about only recently.

On the issue of the MED's intelligence files, Patterson (who shared Groves's doubt that Lilienthal and his staff could handle them responsibly) refused absolutely their transfer. It was agreed that they would temporarily remain in the custody of the War Department, but that three designates of the commission would be allowed to examine them.

The transfer of the files and personnel in the MED's intelligence division to the Central Intelligence Group was completed on February 18, 1947. Those who were transferred from the MED became the Nuclear Energy

Group in the Scientific Branch of the Office of Reports and Estimates on March 28. They were to prepare estimates of the capabilities and intentions of foreign countries in the field of nuclear energy and to represent the DCI in dealings with the AEC. The directive, which authorized the DCI to coordinate all intelligence related to foreign developments of atomic energy, was not, however, issued until April 18, 1947.[24]

On the issue of Sandia and the storage sites the commission finessed things by agreeing that while nominally these would go to the AEC, there would be no actual transfer of properties or of military personnel.

Finally in a ceremony at the White House on the afternoon of December 31, four of the five commissioners and Carroll Wilson joined Groves, Nichols, and Patterson to witness President Truman sign the executive order that legally ended the army's stewardship of the atomic energy program.

Going on simultaneously with the deliberations over the future control of atomic energy were preparations for a series of atomic bomb tests in the Pacific Ocean. Groves's role in what became known as Operation Crossroads was one of supplying the bombs, the scientists, and the technicians, and providing information.

On August 28, 1945, Fleet Adm. Chester Nimitz suggested that the surviving ships of the Japanese navy be sunk.[25] Army Air Force Generals LeMay, Arnold, and Barney Giles suggested using atomic bombs to sink them. On October 16 Chief of Naval Operations Adm. Ernest King told the JCS that the "threat of atomic explosives is producing radical thinking about future warfare, which will influence the size and nature of the Army and the Navy."[26] He recommended that surplus U.S. ships also be used as atomic targets, and that two tests be conducted at a location in or near the Caroline Islands in the Pacific. A subcommittee of the Joint Staff Planners (JPS) was formed on November 9 under General LeMay's direction[27] and its report, completed on December 21, proposed three types of tests: an airdrop (Able), a second at the water's surface or slightly below (Baker), and a third several thousand feet underwater (Charlie). The JCS approved these recommendations on December 28. General LeMay suggested that General Groves be chosen to head the operation, but since it was largely a naval operation the JCS appointed Vice Adm. William H. P. Blandy to be commander of the Joint Task Force. For Groves the choice of Blandy was one more sign that the conditions under which he operated during the war were over. His absolute power over all things atomic was coming to an end. Many new players were staking their claims about what to do with this new weapon.

On January 5, 1946, Admiral Blandy submitted his plan, based on the JPS conclusions, recommending that the tests be held in the Marshall Islands, specifically in the lagoon at Bikini Atoll, with nearby Kwajalein and Eniwetak used as staging and support bases. He also named the operation, Crossroads. As time went on, additional tasks were added. What began as a modest test of the effects of an atomic bomb on a fleet of naval vessels soon grew as more experiments were added. It was eventually said that Crossroads was the "greatest scientific test ever held up to that time." When the plan was shown to Groves and Oppenheimer they agreed, as Oppenheimer said, that "it would be difficult to get it more complicated."[28]

Crossroads can be partially understood as one of the initial contests of an intense interservice rivalry between the Army Air Forces — struggling to free itself from the army — and the navy.[29] For the air force the atomic bomb was a mixed blessing. If just one of these weapons could destroy a city, or sink a naval fleet, why did the United States need a large force of bombers and other aircraft? The navy was faced with similar dilemmas. Were naval fleets obsolete, defenseless against this formidable new weapon? It was hoped that the upcoming tests would prove otherwise. The rivalry among the services, with the army a participant as well, would be unceasing throughout the Cold War, and characterized much military behavior over the next decades.

As noted, Groves supplied the bombs, scientists, technicians, and information to Operation Crossroads. Los Alamos scientists headed three of the nine technical groups, and Col. Stafford Warren headed a fourth, Radiological Safety. The Crossroads staff had its share of differences and disputes with Groves, who started a few and was drawn into others.

Groves's refusal to provide highly secret information about the bomb to some of the participants was in keeping with his strict standards of security.[30] One matter concerned technical details and data about Fat Man, information an aircrew needed to know to be able to carry out its task. The AAF had decided to have a competition among five pilot-bombardier teams to select the one that would drop the Able bomb on the ships. In order to conduct their practice drops they needed specific information about the ballistics, shape, and geometry of the bomb, information that Groves was not about to give out to anyone. Only after General LeMay went directly to Groves to argue his case did Groves release some, but not all, of the data.[31]

A second matter had to do with the training of weaponeers. The First Ordnance Squadron had disbanded shortly after Nagasaki, and the AAF had no one who knew how to prepare and arm the bomb. Again Groves refused to supply this sensitive information. Only after General Arnold presented the air force's case directly to Groves did he agree that Manhattan would train six AAF officers as bomb commanders and five junior officers as weaponeers.[33] This arrangement for technical training was an important milestone, as it opened new roles for the AAF to perform. As we shall see, Groves would soon greatly expand this training for the military during the final months of the MED, and more extensively when he became chief, AFSWP.

The first test, Able, an airdrop from a B-29, was conducted at approximately 9:00 A.M. on July 1 (Bikini time). The bomb was off target, eighteen hundred feet short and to the left of its aim point the USS *Nevada,* more than three times the prescribed distance. The explosion sank only five ships. This poor showing proved embarrassing to the fledgling air force, which never conclusively discovered the reason for its targeting error.[32] The second test, Baker, held on July 25 (Bikini time) was detonated ninety feet under water, sending a shaft of water twenty-five hundred feet in diameter a mile into the air. The radioactive water fell on the target vessels and exposed many of the men who later boarded the ships.[34]

There was controversy over whether to conduct the third test. Groves was firm in urging that it be canceled. He offered several reasons in an August memo to the JCS. The director of Los Alamos, Norris Bradbury, had informed Groves that nothing more would be learned in a third test, and Groves recounted this to the JCS. A second reason was the impact on Los Alamos. To conduct a third test in 1947 or 1948 would divert scarce laboratory personnel and interfere with the research and development program. Finally, there was the fact that there were only a few bombs in the stockpile.

> I wish to call to the attention of the Joint Chiefs of Staff that even a single atomic bomb can be an extremely important factor in any military emergency. It is imperative that nothing interfere with our concentration of effort on the atomic weapons stockpile which constitutes such an important element in our present national defense.[35]

Only Groves, and maybe one or two others, knew precisely how many bombs there were in the stockpile. There was fissionable material enough for about thirty, but of actual bombs — that was a number Groves kept to himself.[36]

CHAPTER TWENTY-THREE

"The Best, the Biggest and the Most": Fights Over International Control

Groves never took the prospect of effective international control of atomic energy seriously. It was one potential path out of the dilemma that mankind now found itself in, but for Groves the approach was too idealistic. He believed the only true way to ensure the national security of the United States was through its own initiatives, not by trusting some world body, no matter how well intentioned.

Groves was an early believer in how fundamentally the atomic bomb was going to affect military thinking and warfare. He was called upon in these early months to articulate those thoughts to a military struggling to find meaning and answers about this new weapon.[1]

In a confidential memo to John M. Hancock, a member of the U.S. delegation to the United Nations Atomic Energy Commission, dated January 2, 1946, Groves laid out his thoughts about the future of the U.S. Army as it was influenced by atomic weapons.[2] Groves saw two possibilities, though it is abundantly clear which he felt was the more likely. The first was that "satisfactory world agreements with respect to atomic energy . . . ensure that atomic bombs will not be used under any circumstances." Such agreements, Groves felt, must provide for complete information at all times about activities of any and all nations in the atomic field. Representatives must have full access to a suspect nation and be able to observe and raise questions about any activity. In short, this meant the abandonment of all rights of privacy, in the home, laboratory, and plants, in the United States and around the world. It also meant that the United States would have to give up its present supply of bombs, though it might be possible to arrange that a small number be retained under the auspices of an international agency or some other body. If the agreement were broken, there probably would be "an atomic weapons armament race."

A second and more likely course from Groves's point of view was that no world agreement would be reached, and each of the three major nations would have atomic bombs within "15 or 20 years or even 5 or 10." In a prescient forecast Groves stated,

> Should there be an armament race in atomic weapons, — and the world could not long survive such a race — then the United States must for all time maintain absolute supremacy in atomic weapons, including number, size and power, efficiency, means for immediate offensive use and defense against atomic attack. We must also have a worldwide intelligence service which will keep us at all times completely informed of any activities of other nations in the atomic field and of their military intentions.

In this memo, written only a few months after Hiroshima, Groves became one of the first to advocate a "preventive war." The idea of preventing potential adversaries from obtaining the bomb in the name of peace became, over the next few years, a position a widely shared among a variety of military leaders and civilian officials.[3]

> If we were truly realistic, instead of idealistic, as we appear to be, we would not permit any foreign power with which we are not firmly allied, and in which we do not have absolute confidence to make or possess atomic weapons. If such a country started to make atomic weapons we would destroy its capacity to make them before it had progressed far enough to threaten us.

Groves sounded the clarion call:

> If there was only some way to make America sense now its true peril some 15 to 20 years hence in a world of unrestricted atomic bombs, the nation would rise up and demand one of the two alternatives essential to its very existence. Either we must have a hard-boiled, realistic, enforceable, world-agreement ensuring the outlawing of atomic weapons or we and our dependable allies must have an exclusive supremacy in the field, which means that no other nation can be permitted to have atomic weapons.

Groves's suggestions of what must be done to build a future army around atomic weapons were an extremely accurate blueprint of what, in the main, was done over the next half century. Many of his conclusions were repeated ad infinitum throughout the Cold War:

- Special air units with modern aircraft will need to be trained for instant attack, capable of delivering atomic bombs any-where in the world from domestic and overseas bases.
- "All possible methods of delivery of atomic weapons including aircraft, guided missiles, rockets and submarines should be studied and developed."
- "Our intelligence forces must be strengthened many-fold, made world-wide and be competent to always know and to give prompt, accurate and complete answers to the question: 'What are other nations doing in the atomic weapon field?'"
- "Many government and military installations must be arranged by construction, concealment, dispersal and other means so as to continue to function during an enemy attack with atomic weapons."
- "Defense against the atomic bomb will always be inade-quate. . . . There must, nevertheless, be continued research of the highest quality and urgency in the defensive field."
- "With atomic weapons, a nation must be ready to strike the first blow if needed. The first blow or series of first blows may be the last."
- In sum, "[i]f there are to be atomic weapons in the world, we must have the best, the biggest and the most."

While Groves predicted many of the things that did happen, he also got many things wrong. For example, he did not think that the United States or any other nation — presumably Russia, when it eventually got the bomb — would build very many. His logic was that all one had to do was draw, with a compass, three-mile circles on a map to see that a few bombs could destroy many cities. As he told one military audience: Take the Pacific Coast, for example; fourteen bombs would be enough to wipe out all the cities on the coast.[4] Why build more? Groves probably would have been astounded if he had been told that over the next four decades the United States would build approximately seventy thousand nuclear weapons and have at one time more than thirty-two thousand in the active stockpile.

His prediction about deploying atomic weapons overseas was also wrong. He thought the United States would hesitate to store atomic bombs abroad because they would be "too highly susceptible to sudden attack or capture by specially trained Commandos or Ranger troops."[5] Just a few years later the first U.S. nuclear weapons began to be deployed abroad. At the peak in the 1970s some fifteen thousand were outside U.S. borders — in foreign lands and on or under the high seas.[6]

On the matter of defending against atomic bombs, especially those delivered by missiles, Groves was not optimistic.

> Interception by guided missiles is conceivable but such defense is effective only if all atomic bombs are prevented from reaching the target. Our prior experience with interception of aircraft or buzz bombs shows us that this is an impossible requirement to satisfy.[7]

Another approach to defense was dispersal of strategic targets. "This would necessitate in effect, moving our cities, factories, army and navy camps, and governmental centers underground. The complications to such a scheme are obvious." Groves concluded that the "true defense against the atomic bomb lies in international control of atomic energy and fundamentally, the control of war itself." While international cooperation might be hoped for, the army needed to prepare as if there would be no control.

Well before the end of the war several Manhattan Project officials began to consider the many implications of nuclear energy in the postwar world. On September 30, 1944, Vannevar Bush and James Conant sent two papers to the secretary of war for his consideration.

They argued that trying to keep secrets about the bomb was impossible. Physics respected no geographical boundaries, and other industrial nations would take approximately the same amount of time as the United States to develop their own nuclear technologies. Clearly the nation they had in mind was the Soviet Union. The attempt to control the raw materials to prevent others from building the bomb would not work either, especially if heavy hydrogen were used as a material. Furthermore, if hydrogen were used, then bombs a thousand times larger were likely in the future. To avoid "an extremely dangerous" arms race, they advocated "complete disclosure of the history of the development and all but the manufacturing and military details of the bombs as soon as the first bomb had been demonstrated." They also proposed "free interchange of all scientific information on this subject be established under the auspices of an international office deriving its power from whatever association of nations is developed at the close of the present war."

How long the American atomic monopoly would endure was an extremely important question to Western policy makers during the first few postwar years. It was equally difficult to know Stalin's intentions as well as his technological capabilities to build a bomb. Policy makers and the public alike listened to General Groves's opinions on these matters. His position as head of the Manhattan Project gave his judgment special weight.

Why Groves believed that it would take "ten, twenty or even sixty years"[8] for the Soviets to develop their first atomic bomb, rather than the four years it actually did take, needs some explanation.[9] The most basic reason was derived from the common belief of the day, that high-grade uranium ores — those containing significant amounts of uranium oxide — were scarce.

Groves had overseen two global surveys of uranium and mistakenly concluded that Russia did not have access to high-grade ores. Some accepted the comforting belief that Russia might not have any uranium at all.[10] Exactly how and why the general could have come to this conclusion is unclear. With the Soviet Union occupying one-sixth of the earth's landmass it was likely that major uranium deposits would be found somewhere, and after Hiroshima, they were. Prior to 1940 some exploration had taken place to discover uranium reserves in the Soviet Union, but the scale and scope of the effort remained limited. The heightened interest in atomic physics after 1940 naturally provoked Russian scientists to want uranium for their experiments, and modest amounts were obtained. After Stalin ordered a crash program to build the bomb in August 1945, major deposits were discovered within a short time.[11]

A secondary reason was Groves's assessment of Soviet technological abilities. Knowing better than anyone what it took to build the bomb, he could not imagine that the Soviet Union could duplicate such a feat in the near future, especially after being devastated by war.

The writer Merle Miller has Groves telling one audience

> that the United States didn't need to worry about the Russians ever making a bomb, "Why," he said, smiling, "those people can't even make a jeep." You should have heard the applause; thunderous is the only way to describe it; a great many people stood and cheered."[12]

A little more than seven weeks before the first Soviet detonation, Groves wrote to a journalist to say, "In fact I would be very much surprised if

Russia is not further down the atomic road than we were in December of 1942."[13] This presumptuous attitude about the capabilities and aspirations of a lesser-developed nation would be repeated many times over the decades that followed as similar assessments were made about China, India, Pakistan, Iraq, and North Korea, to name a few.

Groves believed that there were certain secrets — mainly technical in nature — about the bomb, and that they should be kept from others. In 1948 Groves argued that even if the Russians were given all of the thousands of blueprints for the plants, it would still take them until 1955 to produce successful atomic bombs in quantity. Groves believed this because of his low opinion of Russian industry and technical skills. They could not come close to "duplicating the magnificent achievement of the American industrialists, skilled labor, engineers and scientists who made the Manhattan Project a success."[14]

He not only convinced himself, but he convinced others as well. In a letter to Groves seven months before the explosion of the first Soviet bomb in August 1949 Conant said,

> By the way, I am more and more inclined to think that history will record that you had the best of the guesses when you gave the twenty-year end of the target rather than the five years which I put as the short end.[15]

To help answer the crucial question of what other nations might be doing in the atomic energy field, Groves was instrumental in establishing a system to detect atomic explosions at long distances. The fears of what others were doing ran deep; after all, it had been the fear of a German bomb that helped initiate the American program. As we have seen, in the fall of 1943 General Groves had been asked by General Marshall to take the responsibility for collecting information about the atomic energy activities of other nations. In carrying out this responsibility Groves introduced two innovative approaches to gathering atomic intelligence: first, the Alsos concept, and second, the notion of radiological surveillance.[16]

During the war Groves initiated radiological surveillance methods in an effort to learn whether Germany was operating an atomic reactor as part of a bomb program. The Alsos team had supplied Maj. Robert F. Furman with water samples from the lower Rhine River, a likely coolant, to see if there were traces of radioactivity. A second method was xenon detection using air-sampling apparatus aboard A-26 aircraft that flew over suspected sites in Germany. While the fears about the German bomb subsided, the techniques of atomic intelligence were soon refashioned and directed toward Russia.

Knowledge of another secret program to detect atomic explosions at long distances, this one using high-altitude weather balloons, was only recently declassified. In what was known as Project Mogul, New York University received an Army Air Forces contract in November 1946 to develop "constant-level" balloons capable of floating for as long as forty-eight hours at altitudes between ten and twenty kilometers.[17] The instrument payload would detect the sound waves from an atomic explosion from great distances. During the first week of June 1947 members of the research team launched three balloon trains from the Alamogordo Army Air Field.

It would be one of these balloon clusters that crashed on a ranch in the New Mexico desert and started the Roswell myth about flying saucers, the confiscation of the extraterrestrial crew, and the government cover-up that, according to the true believers, continues to this day.[18]

Groves probably did not know about the highly secret Mogul program, but he had one small brush with the Roswell incident. Reporter Bob Considine had asked Groves about the reports from New Mexico about flying discs. Skeptically, Groves responded,

> I know nothing about flying discs and I know no one who does. Before even a real clue to a theory can be developed you will have to catch one or get movies of it. The top of Pikes Peak in the early morning hours should be a better point of observation than the Broadmoor [Hotel, where Considine was staying]. The following must not be attributed to me: "Anyone who lived in Washington as many years as you did must have seen stranger things than even flying discs."[19]

On October 3, 1945, President Truman in a message to Congress publicly committed the United States to seeking an agreement on the international control of nuclear technology.

> The release of atomic energy constitutes a new force too revolutionary to consider in the framework of old ideas. . . . The hope of

civilization lies in international arrangements looking, if possible, to the renunciation of the use and development of the atomic bomb.[20]

Truman gave no details. Byrnes was not enthusiastic. His undersecretary, Dean Acheson, had urged that a substantive set of proposals be drafted, and had helped write Truman's speech to Congress. Secretary of War Patterson and Vannevar Bush also made direct appeals to Byrnes on November 1. Byrnes countered by asking Bush to set down his views in writing. There was some urgency in this, since the British and Canadian prime ministers were due to arrive in Washington on November 11 to begin negotiations on, among other things, the question of international control of nuclear energy.

Bush proposed a three-step process beginning with the formation of a United Nations agency to oversee scientific research, including nuclear research. Member states would commit themselves to opening their laboratories to foreign scientists. Once this was accomplished, the exchange of information would be broadened to include industrial applications and a formal agreement on international inspection under United Nations control. In the final stage member states would renounce the use of fissionable materials for the purpose of building weapons, an agreement that would have to be safeguarded by regular and stringent inspections.

On November 15 the three leaders announced their conclusions on atomic energy. The Truman-Attlee-King Declaration stated that the three countries agreed that the free exchange of nuclear information was desirable, but that this exchange must be controlled and limited to peaceful purposes.

A second outcome of their discussions was new arrangements for Anglo-American collaboration. Groves and George Harrison assisted Secretary Patterson in preparing two memorandums, which were issued on November 16. The first directive called for continued cooperation among the three states, as well as the continuation of the CPC and CDT. The second document set forth guidelines to be followed by the CPC for a new document to replace the Quebec Agreement. Over the next two months a subcommittee of the CPC, composed of Roger Makins of the British Embassy, Lester B. Pearson, the Canadian ambassador to the United States, and General Groves, drafted a report outlining the main points to be included in a new document. Groves warned that many of the suggested provisions were in possible violation of Article 102 of the United Nations Charter.[21] On February 15, 1946, the CPC took up the subcommittee's proposals.

On January 7, 1946, Secretary of State Byrnes announced his appointment of a five-member committee "to study the subject of controls and safeguards necessary to protect this government." He asked Undersecretary of State Dean Acheson to serve as chairman. The other members were Groves, John J. McCloy, Vannevar Bush, and James Conant. Byrnes expected the committee to report to him, to the U.S. representative to the United Nations Atomic Energy Commission, and to the appropriate congressional committees. Groves, Bush, and Conant had been supporters of the May-Johnson Bill, while Acheson and McCloy had not. A week later at the initial meeting of the committee Acheson proposed that they appoint a panel of scientific experts to educate and advise them about nuclear energy.

Groves disagreed, noting that at least three on the committee (himself, Bush, and Conant) already knew more about aspects of the problem than any panel that might be assembled.[22] In this instance, and in many more to come, the committee outvoted Groves. Acheson named David E. Lilienthal, chairman of the Tennessee Valley Authority, as head of the advisory group.[23] From the point of view of winning Groves's cooperation for the ensuing process, Acheson could not have made a worse choice.

Part of Groves's dislike of Lilienthal was due to the undeclared state of war that existed between the TVA and the Army Corps of Engineers. There were long-standing battles over jurisdiction and influence with Congress, along with philosophical differences about water management and flood control. Groves called the TVA "essentially a new-dealer do-gooder [organization] with no regard for the American taxpayer."[24] Given Groves's fierce sense of loyalty to the Corps of Engineers and its traditions, this was reason enough to dislike and distrust Lilienthal, but he soon had several others.

Acheson gave Lilienthal the authority to name his own board of consultants. Lilienthal chose J. Robert Oppenheimer; Charles A. Thomas, the vice president of Monsanto Chemical Company who had served as a consultant during the war on the chemistry of plutonium; Chester I. Barnard, president of the New Jersey Bell Telephone Company, who had considerable experience during the war in negotiating relationships between government and industry; and, upon Groves's recommendation, Harry A. Winne, a General Electric vice president and engineer who had worked at Oak Ridge.

Under Oppenheimer's guidance, the Lilienthal board undertook an intensive study of the basics of nuclear physics. This allowed them to examine the question of secrecy and the atomic bomb independent of Groves's views. The Lilienthal board was much more successful than the McMahon committee in getting access to information that Groves considered sensitive. After weeks of discussion, deliberation, and hard labor throughout February, the board prepared a four-volume draft report for consideration by the secretary of state's committee that would be "a place to begin, a foundation on which to build."[25] It was elegant and simple. The members read sections of the draft to Acheson on March 7.[26]

An international agency, the Atomic Development Authority, would mine and refine uranium, maintain a monopoly of fissionable material, and distribute a "denatured" form of it to nations for peaceful purposes. In its denatured form the material would not be able to be used for weapons purposes. The plan had some features of Groves's idea of controlling the ore, but the intent of the two could not be more different. For Groves it was a way to keep an American monopoly for as long as possible. The Lilienthal board's proposal required all nations, including the United States, to renounce and forsake ownership of nuclear weapons, something Groves found intolerable.

After hearing the presentation, Groves doubted the practicality of trying to control raw materials. Nations might cheat by exploiting low-grade ores to obtain sufficient fissionable material and make a nuclear weapon clandestinely. Earlier Groves had rejected this position out of hand. He believed that it would be possible for the United States to monopolize the world's supply of high-grade ores to prevent another nation from becoming a nuclear power.

The other strident objection to the Lilienthal panel's plan came from Vannevar Bush, who saw that possessing atomic bombs equalized the military might of the United States against the much larger armies of the Soviet Union. Rather than giving up its nuclear weapons immediately, Bush urged a process in which the United States gave up its nuclear weapons only after a series of steps aimed at moderating and liberalizing Soviet society. The Acheson committee remained split along these lines for several weeks until Lilienthal, in an act of desperation, finally insisted that the plan his panel had laid on the table be either accepted or rejected. To his surprise the Acheson committee voted its acceptance, but as was so often the case in such situations, privately Groves informed Patterson that the whole notion of denaturing uranium was a pipe dream that would never work. He also predicted that Byrnes would never accept the plan, now dubbed the Acheson-Lilienthal plan. But just to make sure, Groves called together a group of leading Manhattan Project physicists, including Oppenheimer, to consider the question of whether or not it was really possible to denature uranium in a way that rendered it permanently unavailable for use as bomb material. The group concluded that there was no such technique possible. Any process that might be used to make uranium incapable of being burned in a bomb but still usable as fuel for a reactor was capable of being reversed. No one seemed to pay much attention to the ensuing press release. Groves wrote it off to the fact that the public, the press included, was too ignorant about nuclear energy to understand the significance of the finding. For that, Groves was as much to blame as anyone.

Byrnes did not reject the Acheson-Lilienthal plan, at least not in any obvious way. In fact it was probably he who saw to it that the report was prematurely released to the press, before the committee had authorized it. He may have welcomed the plan as a tool for prying the State Department out of the shadow of the War Department with regard to matters of international control of nuclear energy. But his next step could not have been more designed to imperil the plan. He and Truman decided to appoint Bernard Baruch as representative to the United Nations Atomic Energy Commission. It is hard to imagine an appointment more calculated to discourage those who had seen in the plan a chance to prevent an arms race with the Soviet Union.

The popular image of Baruch was of a down-to-earth, no-nonsense, practical philosopher; an adviser to presidents whose office was a park bench in Lafayette Square. Baruch had made his fortune on the stock market, and among insiders in and out of government he was known to be extremely conservative and extremely vain. At the time Byrnes announced his appointment to the United Nations Atomic Energy Commission, Baruch was seventy-five years old and, though spry, was also quite well known for his distrust of and hostility toward the Soviet economic and social experiment. Members of the Acheson-Lilienthal committee were quite dismayed at the appointment. In his journal Lilienthal said that when he read the news he was "quite sick."[27]

Baruch's first step was to surround himself with an inner circle of advisers comprised of people with whom he had worked quite closely, individuals who shared his views on politics and economics and who were as wary of alliances with the Soviets as he. Save for Fred Searls, a mining expert who had advised Groves with regard to the Manhattan Project's uranium ore survey, none of them had any scientific training. Baruch requested that members of the Acheson and Lilienthal panels stay on as informal advisers, but — sensing the drift — none was willing to do so. When he approached Oppenheimer to serve as scientific consultant, Baruch spoke of "preparing the American people for a refusal by Russia." It was quite clear to Oppenheimer that Baruch was primarily interested in using Oppenheimer's reputation, not his advice. As Acheson and Lilienthal had expected, Baruch was unwilling to accept the Acheson-Lilienthal plan. He made several major modifications, which they saw as calculated to force a refusal of cooperation from the Soviets. The Acheson-Lilienthal plan had called for nuclear disarmament. Baruch and his trusted aides changed the demand to total disarmament. The Baruch scheme also emphasized severe punishments for those who violated the terms of the agreement, and it substantially modified the concept of the international oversight authority, which in the Acheson-Lilienthal plan had responsibility for controlling the mining of uranium ores and the distribution of uranium metal. These kinds of activities were now to be the responsibility of private enterprise. The most controversial of Baruch's changes was the insistence on abolishing veto power among the United Nations Security Council members.

Most of the nation's military leaders were unenthusiastic about the emphasis in the Baruch plan on total disarmament and the stress on severe punishment for those that violated its provisions. The Joint Chiefs of Staff's endorsement of the plan was slow in coming and at best lukewarm. The military establishment, in general, was no more conversant with the details of the manufacture or the working characteristics of nuclear weapons than any other agency. Like it or not, they had to rely on Groves.

Baruch turned to Groves for help. Not surprisingly, Groves had been supportive of the changes that Baruch had made in the Acheson-Lilienthal plan, and he responded enthusiastically to Baruch's call for help and advice. Groves detailed many MED personnel to provide assistance to Baruch's office. Groves recommended that Richard C. Tolman serve as his scientific adviser. Through Tolman Groves in effect decided which information about the atomic bomb and nuclear technology could be provided to Baruch, to his assistants, and, in turn, to the United Nations. A series of seven volumes titled *Scientific Information Transmitted to the United Nations Atomic Energy Commission by the United States Representative* was prepared under Tolman's supervision.

Groves also suggested his former deputy Maj. Gen. Thomas F. Farrell, now retired, as the best person to help assess Soviet prospects for developing nuclear weapons. Farrell's estimate of twenty years coincided with his former boss's prognostication. Groves also suggested that Edgar Sengier be appointed as a consultant to Baruch's office, to join Fred Searls as an expert in uranium ore. Sengier, the Belgian mining magnate who had worked closely with Groves during the war in supplying the Manhattan Project with ore from the Belgian Congo, was now reaping even richer rewards in Manhattan Project contracts. It was hardly surprising that Sengier was as opposed as Baruch to the idea of placing control of uranium mining under an international authority.

Though negotiations with the Soviets on international control of atomic energy dragged on beyond Groves's retirement at the end of February 1948, it had been obvious from the opening exchanges between Baruch and his Soviet counterparts that there was never any chance of an agreement. Groves had been the odd man out during the Acheson-Lilienthal deliberations, but it was his view in the end that prevailed. America would go it alone.

CHAPTER TWENTY-FOUR

Chief of Special Weapons

· *1947–1948* ·

Groves's imprint in establishing practices and procedures during the Manhattan Project have lasted more than half a century. He had a similar influence through his short tenure as chief of the Armed Forces Special Weapons Project (AFSWP). The little-known AFSWP was basically created by Groves as his final redoubt. During the first few years after the war it was clear that the bomb was here to stay, though it was unclear what role it would play in America's security policy. How many would be built, where would they be stored, who would have custody of the bomb? These questions and many others lay in the future, but while they were being deliberated the bomb had to be taken care of. It would be AFSWP's duty to store, transport, maintain, and assemble the bomb. Groves's impact on these procedures and practices would be profound in the decades that followed.

No longer responsible for the nation's nuclear resources, Groves was finally forced to recognize how hard he had been driving himself over the past four years and how exhausted he was.[1] He was ordered to report to the army's Pratt General Hospital in Coral Gables, Florida, for an extended rest. On January 18, 1947, he left for Florida, stopping en route at Hobcaw Barony, south of Myrtle Beach, South Carolina, to visit Bernard Baruch at his winter residence. He arrived at Pratt on January 23 and until March 10 was provided with a suite of rooms reserved for VIP patients. He was joined by his wife for the first vacation they had had since 1940.[2] Nevertheless, Groves kept in touch with his office on a daily basis. He even heard about news reports in the *Washington Post* and the *New York Times* that he was being mentioned for the position of commissioner of the All-American Football Conference.

Among the things he received from the office was testimony of Bernard Baruch before McMahon's special committee, testimony in which Baruch had roundly praised Groves's leadership and vision as head of the MED. In a long letter of thanks to Baruch, Groves informed Baruch that he, Groves, had recently been "ordered to duty" as a member of the Military Liaison Committee, which, Groves said, was a distinct surprise. Groves went on to say that his appointment came after "Senatorial criticism of the function of the Atomic Energy Commission with respect to Military matters." Groves informed Baruch that

> The assignment places me in the position of being responsible in the eyes of many for safeguarding of the national security as it relates to atomic energy.

But unfortunately, unlike the days of the MED when he had the authority to carry out his responsibilities,

> [I am] without at the same time given any authority to carry it out. I would be at the mercy of a commission, which has already point-edly avoided using the men who are most experienced in the field of atomic energy. This has been particularly emphasized by the appoint-ment [to the AEC] of a young officer from the General Staff as the Director of Military Applications, instead of Colonel Nichols. . . .[3]

Groves was not as surprised as he pretended to be. He knew well before January 1, 1947, that he was going to be appointed to the Military Liaison Committee. On December 23 Secretary of War Patterson, one of Groves's ardent admirers, told him that he should forget any thought of retirement. The army still needed him.

Three days later Gen. Lewis Brereton, the first chairman of the MLC, had told Groves that this position

> will be so closely tied in with the work of the Military Liaison Committee . . . that serious difficulties to the point of unworkability will arise unless he is a member of the Military Liaison Committee.

While appointment to the MLC came as no surprise to Groves, it most assuredly did to David Lilienthal. The news — delivered by General Brereton on January 30 — rattled Lilienthal to the point that he fired off a completely and atypically unguarded memo to Patterson stating his total opposition to such an appointment and detailing the many ways in which

Groves had, over and over again, attempted to undermine and otherwise impede the process of transfer of resources and responsibilities from the MED to the AEC. For his trouble Patterson informed Lilienthal that the appointment not only had been approved but had been already announced to the press as well. It is fairly obvious that among them, Groves, Brereton, and Patterson had finessed Lilienthal.

Groves checked out of Pratt on March 10, after having lost approximately forty pounds, and returned to Washington. Throughout March and April he attended MLC meetings, gave speeches, had appointments, and worked on AFSWP activities. On April 21 he checked into Walter Reed General Hospital and was on sick leave from May 9 until June 8. The problem was arthritis in both knees and numbness in his right leg, which had been noticeable since the previous August. His blood pressure was 134/90, and an electrocardiogram showed no evidence of heart disease. He kept several personal appointments, even traveling to New York for a weekend celebrity golf tournament at the Columbia Country Club on May 17–18, 1947. He gave several speeches (National Industrial Conference Board at the Waldorf-Astoria on May 28), attended the MIT class of 1917 reunion in Portsmouth, New Hampshire, on June 6, and visited the Officers Reserve Association on June 10.

Many of his staff had gone to the new organization. The Washington offices of the AEC in February 1947 were well represented with Groves's people. Colonel Kirkpatrick was assistant to General Manager Carroll L. Wilson. His former executive officer James B. Lampert, and John A. Derry, were in Wilson's office as well.[4] Capt. Robert A. Lavender continued to advise on patent questions. Joseph Volpe was deputy general counsel (and later general counsel); Maj. Fred B. Rhodes was head of Security and Intelligence. Col. Kenneth E. Fields was in the Research Division, along with Capt. George Rebh. Maj. Robert Coakley was in the Office of Information. Jean O'Leary had a new job, as did some of the other secretaries. Gerry Elliott worked in Jannarone's office, and Patricia Cox Owen supported the MLC staff.

Groves's closest aide at the time was Maj. Robert P. Young. Young graduated eighth in the West Point class of 1942 and chose the Corps of Engineers. After serving in the Mediterranean theater during the war and a short stint on the General Staff in 1945, he joined the MED in 1946 and was the general's aide for the next year and a half before returning to Harvard for a master's degree in civil engineering. The relationship between the fifty-year-old general and the twenty-six-year-old major was close. The two kept in contact, exchanged letters, and saw one another occasionally for many years afterward.

With Young we see a more relaxed Groves, not always evident in earlier tension-filled years or with many other people. With more time on his hands and not as nimble on the tennis courts as he once was, General Groves took up golf. Often his partner in 1946 and 1947 was Major Young. At the first tee, as the game was about to begin, they would banter and negotiate, with Groves trying to get strokes and an edge over his younger opponent. Groves, according to Young, was a gamesman, always finagling on how to win without cheating, using psychological warfare and intimidation to gain an advantage against an opponent. Groves gave Young a book, *The Theory and Practice of Gamesmanship or The Art of Winning Games Without Actually Cheating.* Competition drove him on the golf course, as it had at West Point and up the army career ladder.

In a letter to Young, now at Harvard, Groves recounted two unpleasant experiences he recently had at the Army-Navy Country Club playing golf with his aide Capt. Kenneth Cooper and his friend Brig. Gen. Willard A. Holbrook.

While playing the ninth hole his second shot landed in a snake hole. "Being a gentleman of honor, I brought the situation to the attention of my opponents, requesting a decision as to whether the ball could be moved out of the snake hole without penalty." He was told no, to play as it lay.

A second instance occurred a week later, this time on the fifth hole. Just as he was swinging his club, a low-flying plane, approaching to land at National Airport, cast its shadow directly over the general, causing him to flub the shot into the rough. His plea to take a mulligan was to no avail.[5]

Later Cooper recounted the episode, using it as an insight into Groves's character. Cooper at the time was a twenty-three-year-old captain (class of 1944) who was being asked by his boss, a two-star general, if he could play the shot over. Mustering up his courage, Cooper told him no, he must play it as it lies. The lesson was that if you stood up to Groves and your position was sound and correct, he would respect you. Had Cooper allowed Groves to replay the shot, that may not have been the case.[6]

Groves had proposed that Nichols be appointed director of military application. Throughout 1946 Nichols had continued to operate the plants at Oak Ridge and to turn out highly enriched uranium. He was involved as Groves's representative in many of the issues involving the transition. He led the commissioners on a tour of the facilities in November, and had strong views on who should have custody of the weapons themselves.[7] These views, which reflected Groves's as well, torpedoed any chance that he would be chosen. Lilienthal selected Col. James McCormack to be director.

Groves's behavior and his positions on substantive issues led to greater and greater isolation. Everything that had once worked now backfired. Doors that had been open during the war were now closed. Though he had Patterson's support, many of his peers now brushed him off with impunity. But Groves continued to fight to carve out a role for the War Department, and himself, to control certain aspects of the military application of atomic energy.

Groves had arranged in the summer of 1945 for the MED to take over what was then known as Oxnard Field, at the base of the Manzano Mountains, adjoining Kirtland Air Field, in Albuquerque.[8] In September 1945, to provide ongoing ordnance engineering as well as to alleviate crowding at Los Alamos, Z Division was established.[9] While Z Division's transfer to the AEC on January 1, 1947, was yet another prize that Groves lost to the civilians, the ordnance activities that went on at Kirtland would eventually bear fruit.

As the transfer date approached, negotiations continued over what facilities and personnel would go to the AEC and what would remain with the military. The AEC took a firm position while the War Department fought a rear-guard action, still hoping as late as December to retain control of Los Alamos. Groves saw to it that exempt from transfer was the key installation where the AFSWP would carry out its activities — Sandia Base. His nationwide empire had shrunk to a former army airfield in Albuquerque.

The idea of an interservice organization to continue the military functions of the Manhattan District originated with Groves shortly after passage of the Atomic Energy Act in August 1946.[10] In a letter to Secretary of War Patterson on September 18 he proposed the establishment of "a War Department Atomic Energy Committee," to which army members of the MLC would also be assigned. Groves appeared before the November 7 MLC meeting and argued for a joint command to handle all atomic matters pertaining to the military, including training, storage, and combat delivery. On November 15 Groves further proposed that the War Department join with the navy in establishing an atomic energy command, by presidential directive, if necessary. Groves's plan was examined, and largely supported, by the Plans and Operations Division of the General Staff, then under Lt. Gen. Lauris Norstad. On January 29, 1947, Secretary of War Patterson and Secretary of the Navy Forrestal issued a directive, made retroactive to January 1, 1947, creating the Armed Forces Special Weapons Project (AFSWP). The directive limited the new agency to "all military service functions of the Manhattan Project as are retained under the control of the Armed Forces." Three were mentioned: (1) the training of personnel required by the project and of bomb commanders and weaponeers, (2) military participation in the development of atomic weapons of all types, and (3) "developing and effecting joint radiological safety measures in coordination with established agencies." The chief was to be selected by the two service chiefs, with a deputy similarly selected from the opposite service. Both were to be members of the MLC.

Groves was appointed chief — on February 28, 1947 — and Rear Admiral Parsons was appointed deputy chief. Groves chose Col. Sherman V. Hasbrouck as his chief of staff and Col. Harry M. Roper as deputy chief of staff, both having joined the MED the previous April. In the first months the numbers of AFSWP personnel at headquarters were few, only about twenty officers and a dozen civilians. On April 15 the AFSWP headquarters office moved from the New War Building across the river to the fifth floor of the Pentagon.[11]

In a memorandum to the Joint Chiefs of Staff on April 4, 1947, Groves proposed ten basic functions for the AFSWP:[12]

- Technical training of special personnel in the military use of atomic energy, including bomb commanders and weaponeers.
- Military participation in the development of atomic weapons of all types.
- Development of joint radiological safety measures.
- Storage and surveillance of atomic weapons in the custody of the armed forces.
- The proposal of recommendations to assure uniform policy on security measures within the armed forces.
- Assisting the services in preparing courses and training instructors on atomic energy.
- Assisting in the development of war plans and on the technical considerations affecting the employment of atomic weapons.

- Preparation of materials to educate the public on atomic weapons.
- The command of military units assigned to the storage, surveillance, and assembly of atomic weapons.
- Appropriate staff assistance to the MLC.

Groves's proposed charter was under consideration by the military departments for three months. By the summer Groves had organized the project into six divisions: Operations and Training, Fiscal and Logistics, Radiological Defense, Development, Security and Personnel, and Administration. Several revisions took place, including recognition in September of the newly independent status of the air force.

After his initial effort to commandeer the best officers to run the MED at war's end, he took another personnel action that would have lasting impact. Groves created an engineer battalion to take over many of the tasks that were performed by civilian scientists during the Manhattan years. In a January 1946 memo he argued that the military could not rely on civilian scientists when it came to preparing bombs for possible use.[13] A group of officers would have to be trained in the technical aspects of the bomb.

The bomb was not going to disappear. As the nation deliberated about atomic energy's role in its military and diplomatic policies, Groves moved forward to develop a trained cadre of military officers with knowledge of how to assemble and maintain the bomb, how to transport and store it, and how to prepare it for use.

The operational unit charged with these tasks was the 2761st Engineer Battalion (Special), which was activated on August 19, 1946, at Sandia Base.[14] To fill the officer positions Groves, and his executive officer Col. James B. Lampert, scrutinized the lists of recent West Point graduates and ordered the top-ranking graduates, overwhelmingly engineers, to report to Sandia Base.

By any standard the original sixty Regular Army officers whom he chose comprised an impressive group. The first ones began reporting to Sandia Base on September 15, 1946; by December 1 the battalion was sixty strong. Fifty-two of its members were engineers.

In July 1946, Groves selected Col. Gilbert M. Dorland, a thirty-three-year-old West Point graduate and engineer working at Oak Ridge, to command the special battalion.[15] In a meeting with Groves in Washington Dorland was informed of the general problem of the steady loss of civilian scientists, and how it was necessary for the military to assume their responsibilities. Groves told him to first go to Los Alamos and Albuquerque, look around, and report back with an assessment. When Dorland returned and told Groves what he thought should be done, the general gave him the job, saying, "Go do it."

While the government deliberated what to do about the bomb, Groves set about to build storage bases for the tiny number of them in the stockpile, and the components.[16] In early 1946, with the war over for only a few months, Groves, anticipating a future need, began to prepare for the more permanent storage of the nation's atomic bombs. The code name of the highly secret storage project was Water Supply. Groves selected three sites, each in a separate area of an existing military installation.[17]

Site A was Sandia Base, a separate area of about twenty-seven hundred acres at Kirtland Air Force Base in Albuquerque, New Mexico. Site B was Killeen Base, a separate area of about fifteen hundred acres at Fort Hood, Texas. Site C was Clarksville Base, a separate area of about twenty-five hundred acres at Fort Campbell, Kentucky. Groves made a weeklong tour from August 19 to August 25, visiting his new storage sites.

During AFSWP's first two years of operation, 1947 and 1948, no complete weapons were turned over to it, but the sole type of bomb in the stockpile at the time, the Mark III, was not designed to be fully assembled.[18] It was basically a reengineered Fat Man bomb, but it remained difficult to quickly assemble and maintain in a ready state.

On July 13, 1945, at the Trinity site Groves's deputy, Gen. Thomas F. Farrell, signed a receipt for the active material and handed it to Dr. Louis Slotin, who was in charge of the nuclear assembly. Acceptance of the receipt signaled transfer of the plutonium from the civilian scientists to the military to be expended in a test.[19] The issue of the custody of the active material and the weapon would become a difficult and contentious one in the aftermath of the war and would not be fully resolved for many years.[20]

Groves, and his military colleagues, had strong feelings about the issue. He did his utmost to ensure that those who were responsible for employment of the weapons, the military, be given as much access to and control over them as possible. This basic issue was part of the reason that Groves created the AFSWP.

The language in the Atomic Energy Act of 1946, Section 6 (a) (2), allowed differing interpretations.

> The President from time to time may direct the Commission (1) to deliver such quantities of fissionable materials or weapons to the armed forces for such use as he deems necessary in the interest of national defense or (2) to authorize the armed forces to manufacture, produce, or acquire any equipment or device utilizing fissionable material or atomic energy as a military weapon.[21]

Some saw this as the logical division of authority between the AEC and the military. The AEC would manufacture the weapons, and the military would store and use them. The military position was that this newest weapon in the nation's arsenal should be under its control and custody. If they were to carry out their responsibilities to safeguard and defend the United States, the armed forces must have the weapons in strategic locations available for instant use. Others were not so sure. The principle and practice of civilian control that had been fought for so persistently and enshrined in the atomic energy legislation must include the weapons themselves.

The issue was unresolved when the transfer of the Manhattan District to the AEC took place on January 1, 1947. Final resolution would take place in phases over the next twenty years as the military eventually gained full control and custody of more and more of the arsenal. The stockpile grew in that time from a handful of bombs at a few sites in the United States to more than thirty-two thousand weapons at hundreds of domestic and foreign locations.

But legalities and legislation aside, during the two-and-a-half-year period from the fall of 1945 to the beginning of 1948 the fledgling arsenal was basically under the control of General Groves. The AEC held theoretical custody, but the AFSWP had practical custody of both the nuclear and non-nuclear components.[22] As MED commanding general until the end of 1946, and as chief of special weapons until his retirement in early 1948, Groves basically retained custody of the weapons.[23]

There were few greater secrets in these early years than how many bombs there were in the U.S. stockpile. Groves, and perhaps two or three others, knew this number precisely. For his first two years as president, Truman did not know how many there were. Nor did Truman know such details as where they were located, how they would be transported to forward bases, or how they might be used.

Though much had happened to Groves's compartmentalization structure, at least with regard to the size of the stockpile it was still intact.

Dorland has recounted how it was done. "The stockpile, which had been at Los Alamos, was moved to Sandia base, and became my personal responsibility. I was the custodian of the components of the bombs that we had."[24] The bombs of the day — it must be emphasized again — were not fully assembled. Several of Dorland's assistants were responsible for counting different components of the bomb. One would count the number of cores, another the number of non-nuclear assemblies, and a third the number of initiators. They would give their tallies to Dorland, who was then able to determine the number of potential operable bombs. He then "typed out the inventory . . . put one copy in the safe that I, alone, had the combination to, and in the early days put the other copy in a belt that I wore on me, a body belt, and stuck a gun in the shoulder holster under my arm and went to Washington and handed the copy to Groves personally."[25]

David Lilienthal was amazed to find that when he took over as head of the AEC in January 1947, there was only one bomb that was "probably operable" in the stockpile.[26] In the spring the AEC submitted its first official inventory; by this time there were roughly a dozen cores, but apparently not all of them were fully operable. On April 3, at a White House briefing, Lilienthal reported the number. He observed that the news came to Truman as "quite a shock."[27]

This hypersecrecy — fostered by Groves — over important details about the stockpile made for difficulties with the military services and the combatant commands in their war planning.

For example, in November and December 1945 the Joint Intelligence Committee (JIC) of the Joint Chiefs of Staff prepared a very imprecise target list without knowing how many bombs there were in the arsenal or much about what there was to hit in the Soviet Union. The JIC presumed that twenty to thirty bombs were available,[28] and listed twenty urban centers that contained important targets. In fact, there were far fewer bombs; at the end of 1945 perhaps only three or four existed.[29]

PART SEVEN

Slowing Down

· *1948–1970* ·

CHAPTER TWENTY-FIVE

Retirement and a New Career

• 1948–1961 •

Leslie Groves was never terribly concerned with material things. Of course he wanted his family to be comfortable, and he did his best on a modest army salary to provide for them, but what drove him was not the making of money or what it could buy. He was content to remain in uniform, knowing full well that he could have commanded a salary in civilian life many times that which he received from the army. During the 1930s, and especially when he was in charge of army construction, he regularly received offers of employment from civilian contractors who recognized just how good he was at getting a job completed on time and under budget. Groves never had any trouble turning down such offers. He loved the army. He loved the Corps of Engineers, and his plan was to serve until the statutory retirement age of sixty-four. In 1946 that meant another fourteen years. But this was a new army, one shaped by the largest war in history and now confronted with a new adversary, and thus bound to be far different from the army of the 1920s and 1930s. There could be no returning to that. There were new responsibilities, new personalities, and new issues, and these many factors caused Groves to rethink his life plan.

Groves had difficulties in what would be his last two years in the army. He no longer had the direct and easy access to the highest levels of government and the military, and his authoritarian style, which had proven so effective during the war, was no longer respected in the rough-and-tumble postwar environment. The rules had changed, and Groves did not adapt to them. He had been a master bureaucratic warrior in the wartime setting. The circumstances under which he operated were unique. He had enjoyed full backing from the president, the secretary of war, and the chief of staff. He had operated in great secrecy with an unlimited budget and no congressional oversight. None of these conditions could be sustained into the postwar period. Now there were many players all vying to shape and direct atomic energy policy. He was just one voice among many.

Groves played the hand that was dealt him, but often his cards were weak and on several occasions he misplayed them, contributing to negative opinion about the army and himself. As the specter of possible retirement from the army loomed in the spring and summer of 1947, other paths opened up.

Groves still had connections inside the Atomic Energy Commission; after all, many of the staff were Manhattan veterans. He had chosen them, trained them, and worked with them. These included the directors of several of the laboratories as well as many in key positions at AEC headquarters and elsewhere.

In Washington information is a coin of the realm; to have good sources with solid, interesting material means you can play the game and influence policy or events. Several of his old team kept him informed of what was going on inside the AEC, but others did not, like Joseph Volpe, whom he considered a "traitor." Groves had similar pipelines to the Joint Committee on Atomic Energy (JCAE). The committee's executive director was Fred B. ("Dusty") Rhodes, who had served as head of the MED's Washington Liaison Office counterintelligence operation. Another staffer was David Teeple, who had conducted counterintelligence operations out of the New York Area Office. His expertise included tapping telephones — a service that Groves sometimes had availed himself of during the war.

From time to time Groves used the information he did have to hinder the AEC and stymie his opponents. Sometimes he shared information with the press. His favorite recipients were Hearst syndicated columnist Bob Considine and the New York Sun's George Sokolsky. If information were especially damaging — of the sort that might lead to a congressional investigation — he would pass it to Sens. Arthur H. Vandenberg (R-MI) or Bourke B. Hickenlooper (R-IA) of the JCAE, or to Rep. J. Parnell Thomas (R-NJ), chairman of the House Committee on Un-American Activities. Groves had testified before Thomas's committee on several occasions on Soviet espionage activities in connection to the atomic bomb and the shipment of uranium to Russia during the war.

In early July 1947 a story broke in the newspapers that certain classified MED documents were in the possession of former army men and had disappeared. Groves tried to use the news to stir up trouble for the AEC. He arranged with Senator Hickenlooper to be called to testify on the matter before an executive session of the JCAE. On July 13, 1947, Groves drafted a memo to David Lilienthal, in the name of General Brereton, chairman of the Military Liaison Committee, asking that the AEC "meet with General Groves and certain of his staff" to evaluate the situation.[1]

A few days later, on July 17, Arthur Sylvester, a reporter for the *Newark Evening News,* the city's largest newspaper, published a story evaluating the MED's security record during the war.[2] Sylvester concluded that Groves's performance was not very good. He pointed out that at the time that AEC took over for the MED, sixty-seven important top-secret documents were unaccounted for at Los Alamos; there were also thirty-five missing at Oak Ridge. Sylvester's implication was that Groves and his staff were poor at their jobs. He noted that the AEC had been able to immediately locate all the missing documents, and that the MED had not even been aware, until then, that the documents had been missing.

On July 19 Groves authored a statement for release to the press in which he flatly denied having been the source of any leak to Congressman Thomas or anyone else or that he had "encouraged a campaign to discredit the administration of the Atomic Energy Commission."

He then protested, at length, any suggestion that at war's end he had favored continued military control of atomic energy; he also denied that there had been any lapses of security within MED during the war. But in a manner that was typical of him, Groves did not let the matter rest there. The following morning the editor of the *News,* Lloyd M. Felmly, received a call from William A. Consodine, now a Newark lawyer and formerly one of Groves's closest and most loyal aides during the war. Consodine informed Felmly that "an enraged Major-General is on his way here from Washington to see you about Sylvester's story." Consodine asked Felmly if he would see Groves, and when Felmly agreed, Consodine suggested that they meet for lunch at the exclusive Down Town Club. When he arrived, to the editor's amazement and consternation, he found that Groves was not alone. He had with him a Mr. Sam McKee, the agent in charge of the FBI's Newark office. Reluctantly, Felmly heard Groves out. Groves objected to the story, claimed it was unfair and inaccurate, and demanded a retraction. For his part Felmly stood behind his reporter. He told Groves that he had confidence in Sylvester's accuracy and integrity and that unless Groves could document his charges, not a word would be retracted. During the entire exchange, Agent McKee never uttered a word.

Felmly was beside himself with rage, feeling that the presence of an FBI agent was an attempt at intimidation and thereby a threat to his newspaper's freedom. He fired off a complaint to FBI headquarters and received, in short order, a full apology from Director J. Edgar Hoover.

Hoover also denied that the FBI had authorized its Newark agent to be present at the luncheon. At this point Consodine lamely informed the editor that the FBI agent had been present at the luncheon as Consodine's personal guest. To add insult to injury, the entire tale became the focus of a Martin Agronsky radio commentary on July 26, in which the moral drawn from the story was that such actions by Groves, Consodine, and Thomas were abusive attempts at intimidation, and, had they been successful, would have represented

> a dangerous breach of the freedom of the press. . . . This seems an abuse by an important army officer of his authority and for the purpose of endangering a fundamental civil liberty. It seems, too, an excellent subject for investigation by the House Un-American Activities Committee whose Chairman, Mr. Thomas, has shown himself so consistently zealous of threats to American security.[3]

Such incidents — and this was not an isolated one — could not help but contribute to negative opinion about the army. Still, Groves's standing with some of the more conservative members of the House and Senate remained high. Just how high was demonstrated in early December when the presidential nominations for military promotions were sent to the Senate. Certain senators were displeased that Groves's name was not among those on the list to become a permanent major general of the Regular Army. Republican Sen. J. Chandler Gurney from South Dakota, chairman of the Military Affairs Committee, telephoned General Eisenhower and Secretary Royall to inform them that fifty-three senators had objected to the omission of Groves from the list. Furthermore, he added, Senator Hickenlooper would object to unanimous consent in approving the nominees and would ask each nominee how much greater were his responsibilities, and how much better he had performed, than Groves. In short, the none-too-subtle message was that unless Groves was added to the list, nobody would be confirmed.[4]

Groves went to see Eisenhower and told him he felt discriminated against and that such action made retirement more attractive. Eisenhower's excuse, that a smaller army required some officers to be reduced in rank, fell on deaf ears, and the five-minute meeting was over. Groves's Anglophobic antennae must have been extra sensitive that day, for he later noted in his private papers,

> I believe this was my final break with Eisenhower, although I knew
> that he had been previously displeased with my belief that we did not
> owe the British any great debt of gratitude. His feeling was that of a
> typical Britisher — that the British had carried the burden of the War
> and that we should forever be in their debt.[5]

Groves also went to see Royall, who was more sympathetic and assured
him that he would be on the next list. Still, the Senate refused to take any
action on the nominees until Groves's name was among them. On
December 13 Secretary Royall had sent a memo to President Truman recommending that the army member of the MLC be given the rank of lieutenant general and that Groves — already occupying the position as a
major general — be given the appointment. On December 17 the president sent his name to the Senate. Both Eisenhower and Royall asked for
Groves's help with the Senate to push through all of the nominations.
Groves made some telephone calls; when the new list was submitted, it
was confirmed almost immediately. These bold maneuvers did not endear
him to his growing band of critics inside and outside of the military.

Throughout this time Groves's archenemy Lilienthal was trying to have
him removed from his positions as chief of AFSWP and as the army representative on the MLC. In an October 15 meeting with Secretary of the
Army Kenneth Royall, Lilienthal described Groves's failure to accept the
civilian commission. General Groves, he said,

> disagreed with the law and he had no confidence in the men named
> to administer the law, and furthermore conducted himself in a way
> that carried out his fundamental disagreement and opposition to the
> Commission.[6]

Royall had helped draft the May-Johnson Bill and was one of Groves's
strongest supporters. He told Lilienthal that the previous day the three
chiefs of staff had met in Eisenhower's office to discuss this matter and had
concluded that Groves should continue as head of AFSWP.

Groves still had strong allies and was considered in the public mind the
defender of atomic secrecy and military might. Nevertheless, even some
of the people who had worked closely with Groves during the war, and
who had the highest respect for his oversight of the Manhattan Project,
now realized that his animosity toward Lilienthal and the AEC was a

serious impediment to the smooth workings of the commission and its related organizations.

After repeated failures by others to dislodge Groves, James B. Conant, a member of the General Advisory Committee, volunteered to attempt the deed. Conant's foremost concern was how to create an effective and smooth-running weapon bureaucracy. In late December 1947 he, Oppenheimer, Bush, and Air Force Brig. Gen. James McCormack, director of military application, began quietly plotting to remove Groves.[7]

More than a year earlier Bush had had doubts about Groves's future in the army. In a letter to Conant, criticizing the film *The Beginning or the End,* he said:

> Of course the whole thing [the film] blows up Groves, and this did
> not happen by accident, and I believe that you and I both take the
> point of view that this is his affair and not ours. As a matter of fact,
> it may be the last straw that breaks the camel's back in the Army, but
> I believe he has no career there anyway.[8]

In an early-January letter to Oppenheimer, chairman of the GAC, Conant proposed major changes within the weapon bureaucracy, with no role for Groves. Conant's "Personal and Confidential" letter asked for anonymity for everyone concerned, considering "the drastic nature of the proposals in so far as personalities are concerned." With regard to filling a single army slot on the Military Liaison Committee, Conant proposed "the bringing back from West Point of our friend [that is, Kenneth D. Nichols] to replace Groves who should be eliminated by all means from this picture."[9]

In a four-hour meeting on January 17, 1948, Conant, Bush, and Oppenheimer discussed the situation with Secretary of Defense Forrestal. Three weeks later Oppenheimer told Lilienthal about the meeting and how the three had "insisted that Groves must get out."[10] While they were discussing it, Royall joined them and agreed. It was no longer possible for the military leadership to ignore the clamor. The problem, as it turned out, never had to be confronted head-on but was dealt with in a final meeting with the outgoing chief of staff.

On January 30, 1948, General Eisenhower met with Groves to evaluate his performance during the previous year. It was Eisenhower's last week as chief of staff before Gen. Omar Bradley would replace him on February 8.

Groves and Eisenhower had never particularly liked one another. From Groves's perspective, Eisenhower was too much the bureaucrat. Rules and protocol bound him; he slavishly observed the custom of seniority even at the cost of competence. Groves considered him devoid of imagination, an army commander who sought consensus rather than simply taking charge, an ineffective administrator without a sense of loyalty to the army traditions that Groves held to be fundamental and sacred. At their meeting Eisenhower made it clear to Groves that he would never again have the kind of influence on policy he had enjoyed during World War II, and that he would never be chief of engineers.

Eisenhower had a long list of complaints. Groves was not just rude to his fellow officers, whom he perceived as standing in his way, but he was also ruthless to the point of cruelty. He was arrogant and insensitive, and treated peers and subordinates with callousness in his single-minded drive toward his goals. Once he set his mind to accomplishing a task, Groves did not abide by the rules. Moreover, Groves had never paid his time-in-grade dues; rather, he had shamelessly maneuvered advancement out of turn. Most egregiously, in the last round of promotions Groves had used his influence with members of Congress to blackmail the army leadership.

The handwriting had been on the wall for some time, and Groves knew it. Besides all else that confronted him, the most basic problem was that there was no place for him in this army. At fifty-one he was too young to become chief of engineers, a logical next step and a position that he would have loved to hold. He had risen a little too fast, and now there was a problem in finding the right slot. It would not be in the larger army, which rarely let the aloof, semi-independent engineers serve in high staff and command positions. When the supreme Allied commander in Europe had moved up to become the chief of staff, it became clear that those marked for advancement were going to be the generals who had served in that theater, leading armies, corps or divisions. Their counterparts in the Pacific theater would be next in line, and then would come all the rest.

Recognizing that his hand was weak, Groves folded his cards and left the table for good. On Monday, February 2, 1948, Lt. Gen. Leslie R. Groves announced his retirement, effective at the close of business, February 29, 1948, after twenty-nine years and four months of active service as a commissioned officer in the Regular Army. His last Efficiency Report, written by Deputy Chief of Staff Gen. J. Lawton Collins, noted that Groves was

An intelligent, aggressive, positive type of man with a fine, analytical mind and great executive ability. His effectiveness is unfortunately lessened somewhat by the fact that he often irritates his associates, but he has extraordinary capacity to get things done.[11]

As final recognition of Groves's accomplishments, and to soften his retirement somewhat, Senators Hickenlooper and Vandenberg cosponsored a bill, dated February 25, 1948, that authorized the promotion of Lieutenant General Groves, without confirmation by the Senate, to the permanent grade of major general, effective the day prior to his retirement.[12] Section Two of the bill further authorized placement of Groves on the retired list with rank and grade of lieutenant general with honorary date of rank July 16, 1945 — the date of the Trinity explosion.

Secretary of the Army Royall advised sponsors of the Senate and House bills that the army generally frowned upon special legislation promoting individual military officers, but, given the exceptional service rendered by Groves to the nation, the Department of the Army, with the concurrence of the Departments of the Navy and the Air Force, endorsed the bill.[13] Congress did not finalize the action until passage on June 24, but for Groves it was worth the wait.[14]

As it became clearer to Groves that his army career might be cut short, it was only prudent to explore other possibilities. He had always been as frugal as possible and had invested for the long term in the stock market, but his economic situation could hardly be considered prosperous. He worried about having enough money put away for a comfortable retirement, or for Grace if he were not there.

In June 1947 James Rand, president of the Remington Rand Corporation, approached Groves. They met at the Waldorf-Astoria and discussed the possibility of Groves joining Remington Rand as director of research. Groves showed some interest but did not commit himself at that time.

Immediately after his conversation with Eisenhower at the end of January 1948, Groves contacted Rand and accepted the offer, and a few days later agreed to Rand's generous salary terms. Remington Rand issued a press release announcing that Groves was to become vice president in charge of research and development of office equipment at the company's Laboratory of Advanced Research on Wilson Avenue in South Norwalk,

C1. Hanford, on the banks of the Columbia River, June 1944.

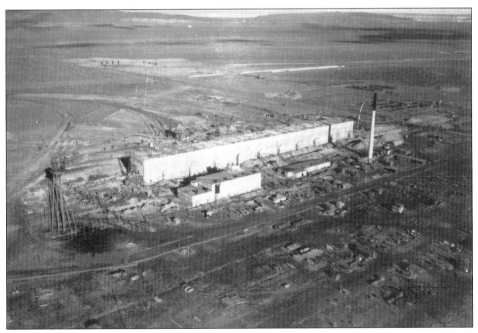

C2. The Hanford B-Plant (221-B building or B-canyon) chemical separation complex, under construction in June 1944. The main building was 810 feet long, 102 feet wide, and 85 feet high.

C3. On Tinian, July–August 1945. From left, Dr. Norman F. Ramsey, Capt. William S. "Deak" Parsons, and Brig. Gen. Thomas F. Farrell.

C4. The crew of the B-29 bomber *Enola Gay*, which dropped the bomb on Hiroshima on August 6, 1945. The plane was named after the mother of the pilot, Col. Paul W. Tibbets Jr., standing fourth from left.

C5. Maj. Gen. Thomas Francis Farrell, chosen by Groves to be his deputy in early 1945. Involved in target selection, the Trinity test, and in late July 1945 went to Tinian to oversee delivery of the atomic bombs against Japan.

C6. Daughter Gwen and wife Grace discover what the head of the family has been doing for the past three years.

C7. Secretary Stimson awarding Groves the Distinguished Service Medal, September 4, 1945.

C8. Groves awarding Edgar Sengier the Presidential Medal of Merit in a private ceremony in 1946. Sengier and the Belgian firm Union Minière were the source of 3,700 tons of uranium ore.

C9. General and Mrs. Groves relaxing on the boardwalk in Atlantic City.

C10. After retiring from the army in 1948 Groves worked
for Remington Rand and moved to Darien, Connecticut.
Photo dates c. 1960.

C11. Groves at the Chevy Chase Club, Chevy Chase, Maryland, July 12, 1970, the day before he died.

Connecticut. His appointment was to be effective March 1, 1948, the day after his retirement from the army.

In a letter to his son at the time, Groves said:

> If you should happen to meet anyone who knows me and be asked as to why I am retiring, you can merely state that you don't know but you imagine that I felt the long-continued strain was not conducive to a long life — the strain was continuing instead of stopping and that I had always believed that youth should lead the Army and that I saw no reason to change my mind now just because I was becoming an old man in the military sense. You might add if you feel it desirable that I assured you I felt I could be of more value to the security of the United States outside the Service than in.[15]

There was much to be done. As usual, whenever there was much to do, Groves did it with dispatch. Professionally, he turned over the responsibilities for the Armed Forces Special Weapons Project to his successor and former subordinate, Kenneth D. Nichols. Privately, he began to look for a house near Norwalk and quickly settled on neighboring Darien as the place to live. Darien, Connecticut, is a pleasant suburb on Long Island Sound about an hour's train ride from New York City. By February 24 he had narrowed his search for a house down to two. Grace was vacationing in Mexico with the Gregorys and was not scheduled to return until the second week in March. His son Richard was stationed in West Germany, where his wife, Patsy, had joined him with their first child. Groves met his daughter Gwen, now a sophomore at Bryn Mawr, in New York, and they proceeded to Darien to look at the houses under consideration. The final choice was a just-completed eight-room house on a wooded acre lot at 9 Dellwood Road, a few minutes from the Boston Post Road. Groves decided to purchase the adjoining acre lot as well for more space, and paid approximately thirty-five thousand dollars.[16]

James Henry Rand Jr. liked to collect famous people, to be among them, and he attracted several to positions in his company.[17] Rand was ten years older than Groves. Born in North Tonawanda, New York, his father, James Sr., had founded the Rand Kardex Bureau in 1886. His son graduated from Harvard in 1908 and joined his father's firm as a sales manager. Displaying what would become a lifelong pattern for the independent entrepreneur, he broke with his father in 1915 and founded the American Kardex Company. For ten years the two were business rivals, after which they reconciled their differences and merged their companies. Two years later, in January 1927, the Rand Kardex Company, now under James Jr.'s control, merged with the Remington Typewriter Company to form Remington Rand. Throughout the 1930s and early 1940s Remington Rand was a leading manufacturer of adding machines, punch card tabulators, and office equipment.[18] In 1937 the company added a revolutionary new product, the electric shaver. In 1943 Rand bought Rock Ledge Estate in Rowayton, Connecticut, for his company headquarters.

Rand was an autocratic chief executive. After the war, he envisioned his company doing pioneering research work and established a laboratory in South Norwalk in 1946, near Rowayton. Rand saw Groves's role in the company as the able administrator of scientific research, someone who had good contacts with the military and the government. But these goals would never be fully realized, as Groves's skills and abilities were largely underutilized. Groves's name and reputation turned out to be the most valuable asset for Rand. His mere association with Remington Rand was good advertising for the company, and that seemed enough for James Rand.

For Groves's part he had been on the go, nonstop, for more than seven years, working under enormous pressure. A rest was welcome. He often mentioned to his wife that he did not want to run anything again. The salary was excellent, and not the least of inducements was the prospect of retiring at age sixty-five with a handsome pension. Somewhat uncharacteristically, he seemed content to accept the situation.

Throughout his time with Rand Groves remained a public figure, much in demand as a speaker and commentator on politics and on domestic and international events. In the thirteen years he spent with the company he gave countless speeches, most having little to do with Remington Rand. In almost all of them he recounted his experiences in the Manhattan Project and the lessons that could be derived from them. While the structure and content of these speeches were largely the same, from time to time Groves would introduce a new twist. For example, in response to public interest in the question of whether it had been necessary to use the atomic bomb to end the war, Groves did not simply declare this to be the case. Instead he noted that the war was costing $250 million a day. Thus, if the use of the atomic bomb had shortened the war by as few as eight days, it had paid for itself. Periodically Groves attended trade shows and industry conferences and gave speeches, usually written by the company's public relations department, on the new Remington Rand electric typewriter or the wonders of the amazing new computer, the UNIVAC.

Groves is deserving of a page or two in the early history of computers. James Rand, always on the lookout for new business opportunities, saw the new field of computers as a promising one after the war. Groves was involved in two Remington Rand acquisitions that figure significantly in the evolution of the computer industry.

Though there are several candidates for the title of inventor of the computer, many agree that J. Presper Eckert and John Mauchly deserve it. They built the first digital, general-purpose, electronic computer, known as ENIAC.[19] This was done at the Moore School of Electrical Engineering at the University of Pennsylvania under a contract by the War Department, specifically to help compute the ballistic trajectories of artillery shells.

The ENIAC computer made its formal debut to the public on Valentine's Day 1946. It was an enormous contraption, eighty feet long, eight feet high, weighing thirty tons and filled with seventeen thousand vacuum tubes and seventy thousand resistors. A few months earlier Los Alamos scientists had used it to run calculations, possibly on hydrogen bomb designs. Though Groves never commented upon these uses of the computer while he was still in charge of the MED, it may have been where he first became acquainted with, and intrigued by, these new calculating marvels. From the moment of their births, the two infant technologies — nuclear weaponry and computers — were entwined in a dependent relationship that continues unbroken to this day.[20]

Others in the burgeoning field quickly saw the value of Eckert and Mauchly's work — which spelled trouble for the naive pair, who had little or no sense of how to transform their innovative ideas into a solid business. After conflicts with Penn over patents, they struck out on their own, founded the Electronic Control Company in October 1946, and started work on UNIVAC, short for "universal automatic computer." Over the next three years their sales efforts and ability to raise capital were lackluster, and they teetered on the edge of bankruptcy.

Remington Rand had been looking into the new field of computers and on February 18, 1950, agreed to pay off the pair's debts and acquire the company. A contract was worked out whereby Eckert and Mauchly would have a free hand but report to Groves, the vice president in charge of research and development. The first UNIVAC was sold to the U.S. Census

Bureau; it became operational in March 1951. The following year the UNIVAC made headlines by predicting the results of the presidential election. With only 7 percent of the vote tabulated UNIVAC predicted a landslide victory for Eisenhower. In the early 1950s *UNIVAC* became a household name and was synonymous with the word *computer.*

The second Rand acquisition that involved Groves came in 1952, when Remington Rand bought Engineering Research Associates (ERA) of St. Paul, Minnesota.[21] William Norris, a cryptology analyst for the navy during the war, founded ERA. To preserve the expertise developed in this arcane field, Norris and several colleagues were encouraged to form a private company to continue to do cryptological work for the navy.[22] The company, ERA, was established in 1946 with funding by John Parker, an investment banker. It was located in St. Paul and did highly secret work, mainly building cryptoanalytic drum memory devices. The company moved into the computer field and delivered the first ATLAS computer to the National Security Agency's navy component, Communications Supplementary Activity — Washington (CSAW), in December 1950.

In the fall of 1951 Rand approached John Parker with an offer to buy ERA. In December the terms of the sale were reached; after obtaining government approval, the sale became official in May 1952.

For a brief time in 1952 and 1953 Remington Rand was the only computer vendor in the world. With the delivery of the IBM 701 in April 1953, however, Remington Rand began to be eclipsed by that company's more aggressive marketing techniques. In 1955 the number of orders for IBM computers overtook Remington Rand. In 1957 William Norris left, along with Seymour Cray, to start Control Data Corporation. Remington Rand's involvement in the burgeoning computer industry was effectively over. By 1965 IBM's share of the market had grown to more than 65 percent, while Sperry Rand's had shrunk to just 12 percent.

In 1953 Groves became a company director with general responsibilities for corporate administration and the formulation of policy. In 1955 he was named a vice president of Remington Rand when it became a division of Sperry Rand, Inc. Along the way Rand had collected another famous general, Douglas MacArthur. With the merger MacArthur was made chairman, James Rand was vice chairman, and Harry Vickers from Sperry was CEO of the new company.[23]

During his time at Remington Rand, Groves kept in touch with Manhattan Project veterans who had positions of responsibility within the

Armed Forces Special Weapons Project, the AEC, or the Joint Committee on Atomic Energy. Every six months, at first by General Nichols and later by others, he was sent the AEC's semiannual report, with a summary of the contents in which the Manhattan Engineer District was mentioned. He was ever alert to the least hint of a suggestion that there had been mismanagement, waste, or lack of foresight in the operation of the Manhattan Project. As time went on, such references to the Manhattan Engineer District grew less and less frequent, but Groves was no less zealous in keeping track of them. He also was attentive to what he considered misleading accounts of the building of the Pentagon, especially when it came to the issue of how much the building had cost.

It was often the case that questions of Groves's competence in running the Manhattan Project were entwined with issues of national security. For example, when the Rosenberg and Fuchs spy cases broke in 1950, Groves testified several times before congressional committees and made his opinions known in radio, television, and newspaper interviews. With regard to the Rosenberg case he pointed out that had he had his way with regard to compartmentalization at Los Alamos, David Greenglass, Ethel Rosenberg's brother, would never have known that the molds he was making were for an implosive lens, let alone what they were to be used for. But in fact, Greenglass should not have known anything. It was the scientists who talked in his presence who made his betrayal of secrets possible.

In the case of Klaus Fuchs, compartmentalization would not have been effective; Fuchs's work required that he know almost everything there was to know about the bomb and how it worked. When he was first presented with the British list of scientists proposed for work on the Manhattan Project, Groves pointed out, he rejected it on the grounds that there was no evidence that these people had been subjected to careful security checks. The British protested that they themselves had carefully checked each of these people and vouched for their loyalty. Groves was finally forced, grudgingly, to give way. As it turned out, even the most superficial check of Klaus Fuchs's background should have set off alarm bells. After the revelations of Fuchs's treachery, Groves was quite bitter, not that the event had happened but that none of the British who had treated his suspicions with such contempt ever bothered to send a note of apology.

Compared to all that had come before, life for Groves throughout the 1950s was fairly easy going. He joined the Wee Burn Country Club and the Tokeneke Club near his Darien home and had plenty of time to play golf and tennis. When in Manhattan he frequently used the University Club of New York, which he also joined. He and Grace traveled extensively to western Europe, to the Caribbean, and to South America.[24]

Frequently — such as when they visited Spain in 1957 — he was feted and honored for his accomplishments in the Manhattan Project. He met Francisco Franco and made headlines in the local newspapers: EL PADRE DE LA BOMBA ATOMICA, EN MADRID. Often they were accompanied on such trips by one or more of their grandchildren, who remember Groves as a stern but loving man who always made sure that they were supplied with sufficient quantities of sweets. In some ways Grandfather Groves was mellowing.

There was time for volunteer work. In early 1950 he accepted the position of chairman of the New York campaign at the urging of John D. Rockefeller Jr. to raise funds for the United Negro College Fund.

On December 18, 1954, daughter Gwen married John Alan Robinson at St. Luke's Episcopal Church in Noroton, followed by a reception at the Wee Burn Country Club. Robinson was from Yorkshire, England, and was studying for a Ph.D. at Princeton. At the time Gwen was a research assistant at the Institute for Advanced Study.

While in the army Groves never allowed any political opinions to interfere with his job, though of course he had political views and strong opinions. When it came to politics and serving military officers, George Marshall was the standard. There were not too many people whom Groves held in the highest esteem, but the chief of staff was one of them. He was the model to be followed. Under this rule the office was always to be respected, no matter what you thought of the man.

Groves had had mixed ideas about President Roosevelt. Like many others, he was often impatient with the president's political maneuvering, which he saw as vacillation rather than strategies to build consensus. Groves was suspicious of the social welfare programs and the prolabor positions of the New Deal and the Democrats. He had little use for Mrs. Roosevelt. He thought her a busybody who usurped power and interfered in certain government departments.

Groves's political differences with Vice Pres. Henry Wallace became quite heated. During the war, Roosevelt had designated Wallace as his representative on the Top Policy Group. Groves was obliged to periodically review for Wallace, and the other members of the committee (Bush, Conant, Stimson, and General Marshall), the status of the program. Early in 1944 Wallace told Groves there really was no longer any need to keep him informed, so the practice stopped.

After the war during the battle over the May-Johnson Bill, Wallace, who was then Truman's secretary of commerce, spoke out forcefully against allowing the military to serve on the proposed Atomic Energy Commission. Later Wallace broke publicly with the Truman administration, and was forced to resign, over the manner in which Bernard Baruch was representing the United States in the United Nations Atomic Energy Commission. It was Wallace's contention that the cause of world peace would be strengthened greatly if the United States followed the position originally outlined in the Acheson-Lilienthal plan.

Once Groves retired from the army he no longer had to be circumspect in voicing his opinions of liberal politicians such as Wallace. In October 1948 Groves gave a speech at the American Legion National Committee banquet in which he accused Wallace, who was then a third-party candidate for the presidency, of having been influenced by the Soviets. Groves implied that Wallace's positions with regard both to the May-Johnson Bill and to the Baruch plan had been shaped by this influence.[25]

After the November election, Groves became somewhat bolder, implying publicly that Wallace was implicated in security leaks to the Soviet Union. Such charges reached a climax when right-wing Republicans accused Wallace of having used the wartime lend-lease program to smuggle heavy water and uranium to the Soviets without Groves's permission. Groves made no public statements on the matter, but in an interview with a reporter from the Associated Press published in December 1949 he suggested that during the war he had withheld secret reports from Wallace because he distrusted him and believed he was a security risk. Such reports prompted the House Un-American Activities Committee to call both Wallace and Groves to testify concerning Soviet wartime lend-lease acquisitions.

Finally, Wallace had enough. In March 1951 he sent a long letter to Groves and pointed out that it had been he, Wallace, and not Groves, who had asked that he no longer be included in the information loop concerning the Manhattan Project. He then accused Groves of having publicly charged Wallace with having been disloyal to the United States. Wallace demanded an immediate apology. Should one not be forthcoming, Groves would hear from Wallace's lawyers. Wallace waited almost four months. When Groves did not respond, Wallace wrote again and informed Groves in no uncertain terms that should Wallace not hear from him immediately he would begin legal proceedings. After consulting with his attorneys Groves capitulated completely. He informed Wallace that, unfortunately, his own remarks had been taken out of context; he had never believed and never said that Wallace was disloyal, nor had he, as reported in the press, compared Wallace to Aaron Burr or Klaus Fuchs.[26]

In public speeches, radio interviews, and writings during the 1950s a clearer vision of Groves's political views comes into focus. Certain themes are articulated and repeated that place Groves in the conservative wing of the Republican Party, closer to the views of Senator Taft than the more moderate, centrist Eisenhower.

Groves was critical of most U.S. foreign aid, seeing it as an attempt to buy friendship abroad. He advocated getting out of the United Nations.[27] The national debt, he felt, was too large, and it was time to get our economic house in order. Groves was vehemently antiunion, a position held in extremis by his employer James Rand. The labor unions, he felt, had become powerful largely through Roosevelt's policies, aided and abetted by Harry Hopkins.

The moral condition of America was in trouble as well, in Groves's view. Too many privileges and too much permissiveness were eroding the character of the American people.

In general Groves was suspicious of the press, especially the "liberal" newspapers headed by the *New York Times*. He was distrustful of most reporters, believing that they had their own agendas. More than once Groves charged that they took his words out of context and distorted what he had said.

As far as the Russians were concerned, Groves did not have much use for them. He distrusted them. They never kept agreements, and the hope of foolproof inspection plans to verify nuclear weapon treaties was unrealistic, he thought. Domestically there were dangers, too. "There's the deep-dyed Red, there's the Pink Red, then there are little shades of Pink. Every one of those is a source of danger," Groves said in a radio broadcast.[28]

In 1949 there was a flurry of talk that Groves intended to become the Republican candidate to try to defeat the Democratic senator from Connecticut, Brien McMahon.[29] He was approached to run but had no real interest in the job.

By the 1952 elections Groves was becoming move involved politically, though usually behind the scenes. In the campaign leading up to the presidential nominating convention he favored Robert Taft, the choice of the conservative wing of the Republican Party. But when Eisenhower got the nomination, Groves was ready to support him in any way he could. In a letter he said that the key to a Republican victory was for Ike to unify the party by reaching out and taking positive action.[30] Groves provided a list of more than two dozen prominent Republicans and suggested that Ike write personal letters to them requesting their views on certain matters.

Throughout their entire twenty-five-year relationship General Groves and Robert Oppenheimer each held the other in high regard, respected each other's skills, and carried on a cordial correspondence until Oppenheimer's death in 1967. This was true even through that most tumultuous of episodes, the Oppenheimer security case. In the late 1940s and early 1950s concerns and fears about Communism, the Soviet Union, and atomic espionage percolated into a highly combustible mix. Groves played only a marginal role in the Oppenheimer case, a travesty of justice whose wounds have still not healed almost half a century later.

Groves had sent a memo to Secretary of War Patterson, dated March 24, 1947, in which he pointed out that during the war Oppenheimer, and some of the other scientists working on the atomic bomb project, had been cleared under extraordinary circumstances. Groves went on to note that at the time the army's security organization, which was not under his control, did not want to grant Oppenheimer clearance primarily because of his past associations. But, Groves continued, "my careful study made me feel that, in spite of that record, he was fundamentally a loyal American citizen and that, in view of his potential overall value to the project, he should be employed." Groves emphasized that he had never had any reason to question that decision. Oppenheimer did a superb job and proved himself to be a loyal American. At the time, Groves noted, he did not have to satisfy the demands of the Atomic Energy Act's provisions concerning security.

> As I have long since informed the Atomic Energy Commission, I do not consider that all persons cleared for employment by the Manhattan District, while under my command, should be automatically cleared by the Atomic Energy Commission but that the Commission should exercise its own independent judgement based on present circumstances.[31]

Later that year Groves wrote to Lilienthal with a similar warning:

> I desire to bring to your attention that in the past I have considered
> it in the best interests of the United States to clear certain individuals
> for work on the Manhattan Project despite evidence indicating con-
> siderable doubt as to their character, associations and absolute loyalty.

With the commission about to assume responsibility, Groves recommended that "those persons on whom derogatory information exists
[should] be eliminated."[32]

The Atomic Energy Commission never acted on Groves's recommendation. Over the next few years the issue of Oppenheimer's associations
with left-wing organizations prior to the war became the subject of
scrutiny by some members and staff of the Atomic Energy Commission
and of the House Un-American Activities Committee. Though no official
action was taken, we now know that the FBI kept Oppenheimer under
continuous surveillance and that J. Edgar Hoover was watching the situation closely. How much of this Groves actually knew and how much of it
he only surmised is not known. However, on May 18, 1950, as a result of
some speculation about Oppenheimer's loyalty in the press, Groves wrote
Oppenheimer a warm, supportive letter:

> If at any time you should feel that it were wise, I would be pleased to
> have you make a statement of the general tenor of that which fol-
> lows: "General Groves has informed me that shortly after he took
> over the responsibility for the development of the atomic bomb, he
> reviewed personally the entire file and all known information con-
> cerning me and immediately ordered that I be cleared for all atomic
> information in order that I might participate in the development of
> the atomic bomb. General Groves has also informed me that he per-
> sonally went over all information concerning me which came to
> light during the course of operations of the atomic project and that
> at no time did he regret his decision."[33]

Oppenheimer was extremely grateful for this vote of confidence, and
said so in a note responding on May 24.

The story of the Oppenheimer security case is well known, even if some of
the details and certain individuals' motives remain unclear.[34] It started on

November 7, 1953, when William L. Borden, former executive director of the Joint Committee on Atomic Energy, sent a letter to FBI director Hoover charging that Oppenheimer "more probably than not" had been, and continued to be, an "agent of the Soviet Union." Upon learning of the charges in early December President Eisenhower ordered a "blank wall" between Oppenheimer and atomic secrets and directed the AEC to study and resolve the matter. For many years after the war Oppenheimer had served in various prominent positions, but his influence was waning. He had not been reappointed to the AEC's General Advisory Committee in 1952, and his only official connection to the government was as a consultant. His security clearance was scheduled to expire at the end of June 1954.

The AEC presented Oppenheimer with a list of charges derived in part from Borden's letter. Oppenheimer decided to request a hearing, and from April 12 through May 6, 1954, closed hearings were held in Washington, taking testimony from more than forty witnesses, one of whom was General Groves. The hearing could better be termed a trial whose unspoken purpose was to destroy Oppenheimer, which it did. A three-man special board found Oppenheimer loyal but by a two-to-one margin voted him a security risk. Oppenheimer's appeal to the five AEC commissioners fared no better. They handed down their four-to-one decision on June 29, 1954, the day before his clearance would have expired.

On the eve of Oppenheimer's AEC security hearing, Chairman Lewis Strauss, the person largely responsible for orchestrating the hearings and the judgment that the hearing board eventually rendered against Oppenheimer, questioned Groves about the significance of his 1950 remarks. Groves's only response to Strauss was that he had thought about it for a long time before writing to Oppenheimer and that the letter should be read with great care.

In testifying before the board of inquiry on April 15, 1954, Groves repeated the circumstances under which Oppenheimer was originally granted clearance and emphatically defended the correctness of that decision.

In response to questioning:

> I think he did a magnificent job as far as the war effort was concerned.
> Q. Based on your total acquaintance with him and your experience with him and your knowledge of him, would you say that in your opinion he would ever consciously commit a disloyal act?
> A. I would be amazed if he did.

He was asked if he had complete confidence in Oppenheimer's integrity.

> I know of no evidence of any kind that was ever given that would refute in any way O.'s statement, "I never accepted Communist dogma or theory."
>
> I would be astounded if Oppenheimer had ever consciously committed a disloyal act.

The lawyer presenting the case against Oppenheimer, Roger Robb, then asked him if, under current AEC regulations, he could clear Oppenheimer. The answer that Robb knew would be forthcoming was that were Groves a commissioner, he would not be able to grant Oppenheimer security clearance.[35]

Groves's answer was unavoidable and the only one he could have given. "I would not clear Dr. Oppenheimer today if I were a member of the Commission on the basis of this interpretation." This had been Groves's position all along. He had repeatedly made the point that under the provisions of the Atomic Energy Act of 1946 it was illegal for Oppenheimer to be given access to atomic information, but this had been violated for more than seven years. In the end he was perhaps more cautious than he might have been. He could have supported Oppenheimer to a greater degree, more along the lines of John Lansdale. Nevertheless Oppenheimer always expressed his gratitude to Groves for what he had done, and they remained warm personal friends until Oppenheimer's death in 1967.

Groves's assessment of who had organized the effort to bring down Oppenheimer centered on two people, J. Edgar Hoover and Lewis Strauss. Hoover, according to Groves, had always resented the fact that Oppenheimer was permitted to work on the project. He had many Communist friends, he had attended pro-Communist gatherings, he had contributed to left-wing causes, his wife's first husband had been a Communist, his wife was sympathetic to the party, and his brother Frank's wife was outspoken in her support. Hoover probably resented the fact Groves had been able to decide questions of security by his own book and ignore the FBI's security investigators.

Groves, on the other hand, believed Hoover never understood what disastrous results might have occurred if Oppenheimer had been barred. Oppenheimer already knew everything, and to reject such a person might provoke him into committing treason. In situations of these sort Groves followed another rule.

In all of my decisions on similar security problems, and there were others, I was always mindful of the fact that treasonable actions often result from mistreatment (in their own minds) of individuals. I was fully aware of Benedict Arnold, and had concluded that he was encouraged to become a traitor because he had been unfairly treated after the battle of Saratoga.[36]

Groves later speculated on the relationship between Strauss and Oppenheimer. They had originally been friends; Groves did not know what caused the break or when it happened. Strauss was a trustee of the Institute for Advanced Study and he had been instrumental in getting Oppenheimer appointed as head. Perhaps Strauss felt that Oppenheimer had not shown a sufficient degree of gratitude to him for his effort. When Strauss and Lilienthal had a falling-out, perhaps Strauss expected Oppenheimer to side with him instead of sympathizing with Lilienthal.

With regard to the Borden letter, Groves was unable to explain why he wrote it. Borden had long had the information and never did anything about it. "I have never been able to lose entirely the feeling that he was induced by the FBI to undertake this effort to destroy O. But also I have never heard anything to do with such a scheme."

As an alumni organization, the Association of Graduates (AOG) of the U.S. Military Academy has few equals. Normally more than 90 percent of the graduates are members, and they actively involve themselves in the affairs of West Point. The AOG revises and publishes the *Register of Graduates of the U.S. Military Academy* each year, with detailed career summaries of its graduates.[37] It also publishes the magazine *Assembly,* which covers news about West Point, articles, class notes, and obituaries.

In 1958 Groves was nominated to be a member of the board of trustees of the AOG, for a three-year term, after which he became the president, serving four terms from June 1961 to June 1965. Old classmates, colleagues, and Manhattan veterans served with him on the board or as his vice presidents.[38]

A major responsibility of the president is to preside over the annual meeting. These meetings are a time for the alumni to gather, to recognize distinguished graduates in attendance, to report on gifts and endowments, and to present various awards, the most prestigious of which is the Sylvanus Thayer Medal, awarded annually to an outstanding citizen whose service to country best exemplifies the ideals of the West Point motto,

"Duty, Honor, Country."[39] During Groves's tenure as president he oversaw the selection of three recipients. At the annual meeting on May 12, 1962, Groves introduced that year's winner, General of the Army Douglas MacArthur, and read the citation.

The audience rose in a standing ovation as General Groves presented the medal, after which MacArthur delivered his address. At MacArthur's request the speech he gave that day, titled "Duty — Honor — Country," was delivered before the entire Corps of Cadets. It was a virtuoso performance, and to enhance the effect MacArthur spoke without notes, his listeners assuming that he was speaking extemporaneously. In fact he had memorized and rehearsed every word and inflection, knowing that these would probably be his last public remarks. The eighty-two-year-old icon held his audience spellbound as he inspired the cadets to fulfill the academy's motto, and closed with the words, "Today marks my final roll call with you. But I want you to know that when I cross the river my last conscious thoughts will be of the Corps and the Corps; and the Corps. I bid you farewell."

In Groves's report as president he emphasized "the reaction of the audience to General MacArthur's emphasis on what graduates are supposed to do, which is to fight for the country and not worry too much about all the things that are talked about so much in the liberal press."[40]

CHAPTER TWENTY-SIX

Last Years

• *1962–1970* •

By the early 1960s Groves had distanced himself from the public stage, and from the controversies that often resulted from his speaking out on the issues of the day. His final years were more private ones. As soon as Groves reached his sixty-fifth birthday in August 1961, he retired from his position at Sperry Rand. In his retirement he finished his book on the Manhattan Project and closely monitored what others said about it, quick to correct their mistakes and record his strong opinions in memos and letters. He enjoyed visiting his son and daughter and being a grandfather to their children. There was more time to see friends and keep up an extensive correspondence, to travel abroad, and, as we have seen, to serve as president of the West Point alumni association. A final duty, as he saw it, was to help establish a retirement home for widowed army wives.

Groves had begun work on an autobiographical account of the Manhattan Project several years before he retired. It was published in 1962 under the title *Now It Can Be Told: The Story of the Manhattan Project*.[1] Part of the reason for writing the book was his dissatisfaction with most of the secondary histories of the Manhattan Project, especially those purporting to explain his motivations and his contribution.

When Groves began his research, the records of the Manhattan Project had been transferred to the National Archives but were not available to the public. Most of the documents, especially those necessary to understand his role in the project, were still classified "Top Secret."

Groves prided himself on keeping the amount of paperwork necessary to run the project to a minimum. Nevertheless, the Manhattan Project records, then stored in the National Archives facility (known as the Torpedo Factory) on Picket Street in Alexandria, Virginia, occupied thousands of feet.[2] Beyond these holdings are other documents at Argonne, Los Alamos, the Lawrence Radiation Laboratory, Oak Ridge, Hanford, university libraries, and the Library of Congress. One shudders to think how much paper there would have been had Groves taken the opposite approach.

Soon after the war Groves told his close assistant Gavin Hadden, and others, that he intended to write three documents: a book for publication, a second containing secret information that could not be published until after certain material was declassified, and a series of profiles of the interesting personalities that he had known. Everything and everyone would be dealt with frankly; the material would be published only after their deaths.[3]

Groves had Hadden assemble, write, and edit the massive *Manhattan District History*, a body of material that continues to assist anyone delving into the history of the bomb. Hadden continued to work for the general during Groves's yearlong tenure as chief of the AFSWP. With his assistant Frank Fallowfield, Hadden largely completed the project by January 1948.[4] Of course Groves's intentions were not only, or even primarily, scholarly. Rather, he felt that a fully documented record of his administration of the project was the best defense against any future congressional investigation.

Groves arranged with Eisenhower, soon after he became chief of staff in November 1945, to keep certain files apart from the rest.

> The papers themselves were to be under my complete control subject only to the written orders of the Chief of Staff or the Secretary of War. The physical location of the papers was to be in special safes in the War Plans Division of the War Department General Staff.[5]

During the last few months before his retirement, Groves separated the papers into various files and removed those that were personal and not secret. Part of his intent was not to let certain records go to the AEC, where they might be used in future attacks on him or the MED.

> These files are designed primarily to preserve those papers which pertained to strictly military operations and those which it would be difficult to disprove unjust attacks on the reputation of the War Department and of its responsible officers, particularly Secretary Stimson, yourself [Nichols] and me.[6]

He also established a personal file, which contained references to secret matters, with plans to remove these papers to his personal control as they were declassified. On the day of his retirement, February 29, 1948, he authorized his successor as chief of AFSWP, Kenneth Nichols, to be the caretaker of the papers. The records remained under Nichols's custody until he retired in 1953, after which they were transferred to the adjutant general. In 1958 they were transferred, along with many other army records, to the custody of the National Archives facility in Alexandria. Physical transfer to the National Archives Building in downtown Washington was made in 1968. The records were transferred once more upon completion of the Archives II building in College Park, Maryland, in the mid-1990s.

The selected records that Groves felt would be most useful in writing his history comprise a separate entry within Record Group 77, Records of the Office of the Chief of Engineers. The entry is titled "Correspondence ('Top Secret') of the Manhattan Engineer District, 1942–1946," sometimes referred to as General Groves's Top Secret Files, and has been so heavily used that the National Archives has placed the documents on five rolls of microfilm for easier access to scholars. Groves was provided with a special safe to store the papers he was working on. For Groves, maintaining the secrecy of these papers was very important. According to Edward Reese, archivist of modern military history at the time, when he handed Groves a slip of paper containing the combination of the safe, Groves studied it for a moment, crumpled it up, popped it into his mouth, and solemnly chewed and swallowed it. Groves was once again in his element!

But the documents alone were not always sufficient to jog Groves's memory on many aspects of the project. Groves therefore wrote many individuals who had played key roles in various phases of the project to ask specific questions or to invite more personal and general recollections of how some particular phase of the project started and how it carried out its mission. Once word got around to Manhattan Project veterans that Groves had embarked upon writing a book, he got many unsolicited comments and anecdotes, some of which are now the only record we have of some of the project's activities.

Groves had always worked best when he had someone else write a first draft of a letter, memo, or report. He would then rewrite it and revise it until he was satisfied. Though *Now It Can Be Told* contains no acknowledgments, Groves worked closely with his son Richard in drafting and redrafting the chapters of the book. It would be not be an overstatement to say that the book was really coauthored. Richard admits that he wrote significant sections of the book and edited others. The Japanese edition lists them as coauthors.

As with all such enterprises, *Now It Can Be Told* reflects the interests and perspectives of its author. It is also unsurprising that some controversial issues are sidestepped or that certain persons whom he did not get along with are ignored. Groves makes no mention, for example, of the fact that most telephones at the various sites were tapped. There is only passing reference to Leo Szilard. There is no mention of the fact that it was the radio chemist Martin Kamen who had first suggested that the electromagnetic separation process would work more effectively if it were divided into two stages — the so-called Alpha and Beta tracks. Groves had Kamen removed from the project as a possible spy and after the war made sure that the FBI made Kamen's life difficult. Groves discussed the Combined Development Trust and the fact that money for the trust, in the amount of $37.5 million, was obtained in the form of personal checks to himself. He explained that this was a way of meeting long-term commitments within a budgeting system, which required year-to-year authorization from Congress. But this is only part of the story. Nowhere is it mentioned that Secretary of War Stimson and Secretary of the Treasury Morgenthau almost came to blows over the matter. Nor does Groves acknowledge that Congress knew nothing about the trust until the middle of 1947 — and when members did find out they were not pleased. There is no discussion of his restructuring of the Columbia University laboratory, primarily to remove Harold Urey as the administrator.

The book is tightly focused on the period from September 1942 through 1946. It is an account of Groves's running of the Manhattan Project and not an autobiography that describes his entire life. In the foreword he stated his goals: to fill in the gaps in the American public's understanding of the project and to record the lessons he learned while in charge. The MED was unique up until then in terms of its size and administration. One factor that Groves emphasized that contributed to its success was its cohesiveness, where each person understood the job he had to do, did it, and did nothing more. The book is dedicated "To the men and women of the Manhattan Project, and to all those who aided them in their yet unparalleled accomplishment." Groves hoped that whatever knowledge was gained from his account — the successes and the failures — would help others lead similar public or private projects in the future.

Groves tells us how he did the things he did, but there is little explanation of where he learned these methods and techniques. There is nothing about his upbringing, his education, or his experience in the interwar army; no mention of his parents, wife, or children.

On what it does include the book has several interesting accounts not found anywhere else. There are the revelations about the German scientists at Farm Hall, mention of Moe Berg, and a discussion of Groves's disagreement with Secretary Stimson about targeting Kyoto. Fundamentally the book is Groves's contribution to a category of writings about running large organizations and the leadership lessons necessary to do it. Groves comes through as a recognizable American type, a can-do, no-nonsense master CEO who gets the job done.

Almost forty years later the *Engineering News-Record*, the major American magazine covering the construction industry, acknowledged his accomplishment by selecting the Manhattan complex as one of the "Top Projects" of past 125 years and Groves as one of the "Landmark Project Managers."[7]

The book's text reflects Groves's philosophical commitments and attitudes, not only about the project but also on bureaucracy and the proper relationship between the public and private sectors. He is frank concerning his general distrust of academic scientists, especially theoreticians, though he lavishes praise on Ernest Lawrence, Robert Oppenheimer, Norman Ramsey, and Luis Alvarez. *Now It Can Be Told* is a valuable historical document in itself because of what it reveals about its author and his understanding of the history and significance of the Manhattan Project and its consequences.

The book came out in early 1962 and was reviewed widely in the major newspapers and magazines.

The *New York Herald Tribune* reviewer saw better than most the importance of the book. Thus far the story had been largely told from the scientists' point of view. With this book

> we see this mammoth venture through the eyes of the tough, blunt bear of a man who was its top administrator . . . the chief value of the book lies in the picture it gives of the controversial General himself and of the magnitude and complexity of the tasks which he handled with such assurance and audacity. The long chapters dealing with the construction and operation of the atomic plants at Oak Ridge and Hanford may not be everybody's cup of tea, yet they may well be the most valuable part of the book.[8]

David Hawkins, the historian at Los Alamos, reviewed it for *Saturday Review*, noting what he saw as a blind spot in the general:

> Groves is motivated by a simple and all-sufficing patriotism that is untroubled by what others see in the atom. He does not probe for any new vision of national interest in the age he helped create.[9]

Groves sent copies of the book to family, friends, and associates with appropriate inscriptions. To Oppenheimer: "Could anyone else have run Los Alamos as well." To Douglas MacArthur: "With a salute to the greatest military commander of our time, if not of all history." To his daughter: "To Gwen, a daughter who was never curious about my work, and hence was never a security worry." To President Truman: "With admiration for the courage and wisdom you displayed in approving the use of the bomb to end the War and thus to save American lives, and with gratitude for your unfailing courtesy to me." To Mrs. O'Leary: "Without your super-competent assistance my task would have been immeasurably harder." To Gilbert Church: "The man who built Hanford."

But *Now It Can Be Told* is not the only written account of Groves's views on people, issues, and events. From the moment that he completed his responsibilities with the Manhattan Project Groves adopted a defensive and protective stance against what others wrote about his handiwork. He read their articles and books, and in some cases took voluminous notes on what was right and what was wrong with their accounts. Sometimes he contacted the writers to tell them what he thought. He was frequently interviewed for oral history projects and television and radio programs, and wrote several articles for publication. Another ongoing exercise over the years was the writing of memoranda to the files about incidents or persons. This substantial collection of notes, commentary, interviews, and articles makes an excellent supplement to *Now It Can Be Told* and provides, in less varnished form, his opinions about the events, the politics, and the personalities connected to the bomb. Always fairly direct in his views, in these largely unpublished writings he held nothing back.

In a typical example we can closely follow Groves's reading of Lansing Lamont's *Day of Trinity*, published in 1965.[10] Next to his citations of page and line are his notes and comments. He points out errors of fact (Einstein was never at Los Alamos) and things that he felt Lamont exaggerated. This tendency to embellish met with his disapproval: "I doubt all this. Typical TIME writing." He also noted his opinions about scientists and others ("If Condon is meant, he left because he was unstable and always had been").

Volume 1 of the history of the Atomic Energy Commission, *The New*

World, 1939–1946, by Richard Hewlett and Oscar Anderson, published in 1962, is generally considered to be a model official history, well written and authoritative. Every scholar since has drawn upon it repeatedly as a basic source. Groves found it wanting. His chief criticism was that the authors did not interview those who knew the most about the project, namely himself and other senior officials.[11] For Groves the book relied too heavily on documents and memoranda. He is critical of academic historians who "almost invariably accept as authentic anything that has appeared in print. . . . The historian's preoccupation with pieces of paper and disregard of the characteristics of the men involved lead to excessive use of such source material and thus distort history."[12]

Furthermore, the authors, according to Groves, did not have an appreciation of how the Manhattan Project operated.

> They were fully aware that [a] policy of committing as little information as possible to writing had been in force throughout the project, but despite this they persisted in depending upon such documents as were available and upon their interpretations of the reasons for major decisions, most of which they pulled out of thin air.

Groves read the book carefully and made hundreds of comments, with references to page and line, pointing out inaccuracies and where he thought the authors misunderstood something or exaggerated a point.

> It could have been a great historical document. As it is, it is completely unreliable and cannot be trusted. Unfortunately, I am afraid it will be considered a source book by historians forever for apparently that is the way that historians operate. They accept anything that has appeared in print as true.[13]

In almost two hundred pages of detailed commentary we get his clear-cut opinions about the men he came in contact with during the war and afterward.

On Lilienthal and the first AEC commissioners Groves had this to say: "Lilienthal, of course, was completely unprepared for the position technically. He had been a self seeker and quite radical in his viewpoint. He was always most sanctimonious in his expressed attitudes" (page 4). Pike was a poker-playing buddy of Truman's. Waymack had "no personal qualifications whatsoever as far as I could discover." Strauss "had a sound layman's view point on science"; Bacher was "a thoroughly competent physicist but with no experience in management." "The general manager, Carroll L. Wilson, was a complete misfit. He had had no experience whatsoever in management and he displayed these weakness[es] throughout his term of office" (page 4).

On certain scientists and their impracticality: "Tuve was a very competent operator as well as a distinguished scientist. Briggs was old and ineffectual. . . . Gunn was extremely difficult to get along with . . . Pegram was an elderly dean at Columbia. He did not have any real backbone or courage. . . . Urey was completely ineffectual. . . . He was not a doer himself and could never make decisions. At heart he was a coward" (page 20). "Murphree was a seasoned executive but he lacked drive. A wonderful example of scientific impracticality" (page 24). "Dunning was a great talker who could always use five minutes and a thousand words to say yes or no. No one had much confidence in his operational capacity" (page 39). "In the early stages it was the scientists who moved slowly. They were simply not accustomed to moving with courage and rapidity. While Szilard urged them on, he did not really have their respect or confidence. . . . None of them [Urey, Pegram, Beams, Gunn] were go-getters; they preferred to move at a pipe-smoking academic pace"(page 18).

As always, there is a special word for Szilard and the other troublemakers: "There was no question, but that he had no moral standards of any kind" (page 69). "Actually after initiating the idea of the first Einstein letter he was a positive detriment to the Project" (page 120). "He fancied himself as a genius, and he never was one. He can best be described as an opportunist" (page 143). Zay Jeffries "was a constant trouble maker in our operations. He was always full of ideas which disrupted sound procedures" (page 104). "The Jeffries report, like everything else he did, was not of great value" (page 105). "I was never sure of how much value Franck was to the laboratory, but he was a constant source of trouble. He apparently thought that because he was a Nobel Prize winner that he knew more about everything than a normal person. . . . He was always stepping out of his field and trying to enter that of national and international affairs where he was a babe in the woods"(page 104). As for certain foreigners: "Both Simon and Peierls were Germans and professors as well. Their superior attitude stemmed from that as well as from their recent experiences in Britain" (page 82).

But there were other scientists with much to admire. Groves thought highly of Bush and Conant. "Conant could be relied upon, he had considerable experience in industrial consulting work . . . was shrewd and able" (page 32). Laughlin M. Currie "was a top notch chemist and a very smooth operator. He was able to carry out the most difficult assignments with skill and drive. He had to show all proper outward deference to Urey and permit Urey to think, or rather pretend, he was running everything, while at the same time he, Currie, was fully responsible for the complete operation"(pages 43–44).[14]

George Harrison "was not a strong advisor. He was always seeking to protect Mr. Stimson's future position"(page 130). On Averill Harriman:"I know of no one in public life who, in my opinion has been as overrated as Harriman"(page 137). He was "always a 'weak sister'" (page 124). Joseph Grew, on the other hand, "was a most able Under-Secretary" (page 134).

On Carroll Wilson and Herbert Marks: "Both of them were completely inexperienced. . . . Neither of them ever displayed any grasp of international affairs." They "were typical of what has been a festering sore spot in Washington for many years now, the self styled able young man. The only part of the definition which is correct is the use of the word young" (pages 167–168).

Groves had opinions about some of the high political figures as well. "Churchill was always informed on the matters which he wished to discuss. Roosevelt never was. Churchill received the advice of people who were fully aware of all the details, such as John Anderson. Whereas, Roosevelt listened to Harry Hopkins who knew nothing" (page 88). "Morgenthau was not qualified in any way to be a cabinet officer and in addition, he was not very bright. He apparently was the connecting link between a highly intelligent father and a very intelligent son, but he himself never displayed any signs of real intelligence, and he undoubtedly was responsible for untold war killed and wounded on all sides by his proposal of unconditional surrender" (page 108). "The actions of Congressman Engel were, in my opinion, most reprehensible. Like many Congressmen of today (Fulbright, Javits, etc.) he had no real regard for security or the national welfare" (page 116).

Groves repeatedly takes the authors to task for some infraction. "This paragraph is generally wrong" (page 94). "The authors have over-simplified the problem"(page 93). "The authors again have failed to understand how the Manhattan Project was operated" (page 98). "It is a mis-statement to say that the first bombs were expected to yield less than 1,000 tons equivalent"(page 102). "Another case of over-dramatics" (page 113). "This whole treatment of the formation of the interim committee is off-base" (page

115). "The authors do not seem to understand just how things are handled in a properly run organization, whether it is military or non-military" (page 129). "The text is misleading" (page 131). "The authors are apparently unfamiliar with the speed of sound"(page 133). "The authors' failure to talk to participants continues to lead to fictional instead of factual writing"(page 133a). "More sloppy writing" (page 154).

Occasionally there is a good word. "This is a first class recital of events as they actually occurred" (page 131). General "Arnold did not have too good a memory but his advice was always sound and his manner was so pleasant that he was universally popular except for a few old enemies. I found him an ideal officer to work with"(page 139).

Groves was much more favorably disposed towards the first volume of the British official history, *Britain and Atomic Energy 1939–1945* by Margaret Gowing. Though first published in London in 1964, it took some time for Groves to get around to reading it; his fifty-seven pages of comments and annotations are dated December 15, 1969. In them we learn further details not found elsewhere about Groves's opinions. While Mrs. Gowing did not interview the general, he is much less critical of her than he was of Hewlett and Anderson.

Many of his comments underscore how tentative everything was at the time. Even the best historians cannot escape the knowledge of how things actually turned out. Groves always emphasized how uncertain were the technologies, how unknown was the amount of material needed, or to what levels uranium had to be enriched. Other themes recur, such as the impracticality of scientists when it came to taking action and getting things done.

Obviously, there is much more commentary about the British and the British role. Groves felt that if it had truly been a joint effort, it would not have succeeded, for too many officials would have slowed things down. Akers was a poor choice to head the scientific effort. He should have leaned over backward to avoid the appearance of being interested in postwar commercial activities. On the contrary, he was always interested in these possibilities. Oliphant was the British Szilard, in Groves's view, "always overly impressed with his ability in areas outside of pure atomic physics" (page 9).

Groves's lifelong Anglophobia comes through loud and clear, once again. "[T]he general British laxity on security" (page 12). ". . . the general air of superiority which seems to be inherent in the English. Only the English could be so annoying towards everyone not of their nationality"(page 12). Groves had a special place reserved for Dr. W. L. Webster of the British Central Scientific Office in Washington. It was Webster, we recall, who resisted Groves's efforts to have security checks done on Fuchs and the others of the British mission. He was the one Englishman whom Groves "cordially disliked." He was obnoxious to Groves and everyone he dealt with in the MED. "He was thoroughly imbued with all the traits of the Englishman who regarded all Americans and colonials as well, as being inferior beings."

In his comments Groves often takes ample credit for certain events. In the well-known story of removing Bohr and his family from Denmark to England via Sweden because he was in danger of arrest by the Germans, Groves says, "Also I had been asking for him. I particularly did not want him to fall in German hands. I also thought he might be useful at Los Alamos. Not so much for any useful ideas he might have but mostly for the inspirational and possibly catalytic effect he would have on the scientists there. To most of them he was on an almost superhuman plane" (page 30).

And again, "If there was ever anyone who needed a wet nurse it was Bohr. He was prone to do the most idiotic things imaginable and it was impossible without being obnoxious to protect him from his own follies. On one occasion in the height of the morning rush hour he crossed Massachusetts Avenue on foot right in front of the Naval Observatory where there was no crosswalk for several hundred yards in each direction. One of my security agents in reporting it to me said 'He was too far away to stop him and all I could do was to follow across saying my prayers every foot of the way'" (pages 30–31).

The American scientists are ranked again, Lawrence "head and shoulders above the others. . . . Murphree was a complete flop" and "Urey was a washout, he inspired lack of confidence, he was unstable administratively and scientifically, and worst of all was a coward."

There were some British whom Groves got along with and admired: Chadwick, Sir Charles Hambro, F. M. Maitland Wilson, and Roger Makins. "On the whole a very good book and a very accurate one for general understanding" (page 57).

In 1968 Groves purchased a volume in the American Heritage Junior Library series on *The History of the Atomic Bomb* for his grandson.[15] Before sending it Groves read it and was shocked at the number of inaccuracies he found. In characteristic fashion he fired off a lengthy memo to the chairman of the editorial committee with his detailed comments about what he thought of the work.[16] He also noted that it was evident to him that the author had a great deal of animosity toward him, but that is a "common tendency on the part of the smart young writers to downgrade everyone who carried senior military responsibilities successfully." As we have see with his other annotations and comments, he supplies details not always captured elsewhere.

Always a bit sensitive as to his weight and appearance, he apparently took affront at the author's depiction of him:

> "Greasy" Groves, as he had been called at West Point, where he graduated fourth in his class in 1918, would have looked out of place as a combat officer: a stout two hundred pounds, he had a weakness for double portions of dessert and was known to keep candy in his office safe.

He responded with characteristic understatement.

> This is a little far-fetched as I was in top-notch physical condition, much more so because of my tennis playing, than almost any other officer of my experience. At that time I was still able to play 5 sets of singles and as many more of doubles in an afternoon as there was time for.

He noted the many factual inaccuracies. "Hanford was not an abandoned railroad town." "Pontecorvo was never employed on the Manhattan Project." The reason Bohr did not wear the oxygen mask was that he could not fit his enormous head into it. The *Indianapolis* did not carry the full amount of U-235 for the bomb because it had not all been produced by the time it left. "Mr. Truman had approved my plans for the use of the bombs, as far as I knew, and he knew I planned to continue dropping bombs until the Japanese surrendered." "It had always been intended by me that the use of the bomb would provide the Japanese with an excuse, which they so badly needed, to surrender."

Groves kept a keen eye trained on the arguments and controversies about the necessity of using the bomb to end the war. After Hiroshima and Nagasaki questions were raised about whether the bomb was really necessary, whether Japan was on the verge of surrendering, and whether it should have been used.

Groves watched the debate unfold. He played a modest role in the preparation of Stimson's article "The Decision to Use the Atomic Bomb," in the February 1947 issue of *Harper's* magazine, answering questions to refresh the secretary's memory and reviewing it just before publication.[17]

While he fully supported Stimson's account, he was suspicious of others who strayed from the historical record, portraying themselves with stronger views about the bomb than they had held at the time. One such example of this tendency that caught Groves's attention was a passage in Eisenhower's memoir *Mandate for Change,* which came out in November 1963.

The passage in question concerned Eisenhower's account (on pages 312–313) of a meeting he had with Stimson in Potsdam in July 1945, soon after the Trinity test.

> The Secretary, upon giving me the news of the successful bomb test in New Mexico and of the plan for using it, asked for my reaction, apparently expecting a vigorous assent. During his recitation of the relevant facts, I had been conscious of a feeling of depression and so I voiced to him my grave misgivings, first on the basis of my belief that Japan was already defeated and that dropping the bomb was completely unnecessary and secondly because I thought that our country should avoid shocking world opinion by the use of a weapon whose employment was, I thought, no longer mandatory as a measure to save American lives. It was my belief that Japan was, at the very moment seeking to surrender with a minimum of loss of "face." The Secretary was deeply perturbed by my attitude, almost angrily refuting the reasons I gave for my quick conclusions.

Groves was skeptical and decided to look into it. He sent for copies of the relevant pages of Stimson's diary, which had recently become available. The reference librarian wrote to Groves and supplied him with Stimson's diary entries for July 20 and July 27, 1945, the days that Stimson and Eisenhower were together. Groves wrote back thanking him and shared his doubts about Eisenhower's account.[18] Groves had considerable contact with Eisenhower after he became chief of staff in November 1945. For the next two years he kept Ike apprised of basic matters concerning the nation's atomic weapons. Never once had he heard such views expressed; nor had he ever witnessed Eisenhower take a strong stand on any issue with his civilian superiors. Groves considered Eisenhower to be the always cautious soldier who would not want to jeopardize his position or career by taking unpopular stands, and in July 1945 that is exactly what misgivings about using the bomb would have been. Eisenhower would have been all alone in holding such views, and he was not the kind to man such barricades. It would have been totally out of character for Eisenhower to talk that strongly to Stimson. Groves suspected that Ike had made the entire incident up to put himself in a better moral and historical light.

Groves concluded his letter by suggesting that perhaps the best way to handle it would be to recruit a historian to write a piece titled "The Potsdam Conference as Viewed by Stimson." Close aides could be interviewed to shed light on the subject. Groves wanted the record corrected for future writers. The article, if written, would not find any hint of Eisenhower's account in Stimson's diary, with the obvious implication that Eisenhower had offered no such opinion. Librarian Gottleib replied that perhaps Elting Morison, a Stimson biographer, might do it and that he would contact him. Gottleib reported back that Morison felt it would not be the best article.[19]

Groves continued his research, writing to Col. William H. Kyle, Stimson's military aide at the time. Kyle was with the secretary in Germany on the dates in question. Throughout 1945 Kyle and Groves had dealt with one another on many occasions as the bomb neared completion.[20] He asked Kyle to recall any mention of Eisenhower objecting to the use of the bomb. He listed the reasons why he thought it unlikely. If Eisenhower had objected as strongly as he said he had, Stimson would surely have made an entry of it in his diary and mentioned it to Kyle. Groves told Kyle about a review by Richard Rovere in *The New Yorker* that questioned Eisenhower's accuracy. Groves said that Rovere was "no favorite of mine," but, as he wrote in a memo to the file, "while I have always thought that [he] was too biased in everything I believe in I think in this case he makes a very good point."[21]

Kyle replied recalling the events of July 20 and 27. He said he had "no recollection of the Secretary ever making the statement that anyone was against using the atom bomb" and agreed with Groves that "Eisenhower would have been entirely out of character to have expressed any such opinion to the Secretary."[22] In a follow-up letter Kyle offered to have a friend of his write to Ike to clear up the matter, but Groves thought better of it. He wrote in the margin of Kyle's letter, "Drop it." He also wrote in the margin, "I did not believe that Eisenhower ever spoke to Stimson in the way he detailed in his book, LRG." For Groves, at least, the historical record had been corrected.[23]

Groves was beginning to slow down. He was no longer able to play tennis on a regular basis. In 1964 he and Grace sold their house in Darien and moved back to Washington. Besides being in closer contact with old acquaintances and friends, Groves, like so many retired military personnel who settle in the metropolitan area, wanted to be near Walter Reed Hospital. They rented a spacious apartment on the second floor at 2101 Connecticut Avenue, about half a mile north of DuPont Circle, near the south end of the Taft Bridge. Thirty years earlier he had lived near the north end of the bridge and walked over it hundreds of times to meet Pot Graves as they made their way to the chief's office on the Mall. Much had happened since then.

After graduation from West Point, his son Richard had followed in his father's footsteps and chosen to serve in the Corps of Engineers. In October 1945, four months after graduation, he married Patricia Hook, the daughter of a navy doctor, and their first child, Carolyn, was born the following July. Over the next decade three more children would arrive — Patricia, Richard, and Ann — supplying the grandparents with much to look after.[24] In July 1968 General Groves was extremely pleased when his son was promoted to brigadier general.

Groves participated in several commemorations of the Manhattan Project. In October 1962 and September 1967 Groves visited Oak Ridge to participate in the twentieth and twenty-fifth anniversaries. In his 1967 visit he took his grandson Richard, then fourteen, and was proud to show the facility to him.

But Groves was not quite finished with getting things done. For several years he had been on the executive committee of the Army Distaff Foundation, an organization whose mission was to provide for the widows of career army officers. The main project was a three-hundred-unit residence on a fourteen-acre tract of land across from Rock Creek Park at Oregon and Nebraska Avenues in Northwest Washington. Groves was named president on August 26, 1960. When Groves joined the board, the organization was in considerable disarray. The fund-raising efforts had been badly managed; there was no clear concept of what kind of building to build, how big it should be, and how much money would be required. Groves took charge. He organized the building committee and oversaw the development of the criteria required to define the nature and size of the building. He was elected chairman of the board, reorganized the fund-raising effort, and appointed people who proved to be extraordinarily effective, a skill he apparently never lost. After having raised the required funds, he negotiated for the board with the architects and then with the contractors. While not quite the Manhattan Project or army construction, he was back in his element, getting things done. He was in part motivated by his concern for Grace's welfare if he should die before her. But as was the case all of his life, he was also driven by a deep commitment to the army and to his vision of the things that made the United States great. The wives of army officers who serve their country deserve to be taken care of should their husbands die before them.

In 1966 and 1967 Groves's health noticeably deteriorated. His breathing became labored; he could not walk stairs easily. He once fainted without warning while walking through Union Station, causing considerable consternation and turmoil. His aortic valve had become badly calcified. His activities were largely restricted to answering the mail he still received from scholars and journalists writing about the Manhattan Project or requesting interviews. He continued to write detailed critiques of works about the Manhattan Project and of the public events of the day.

The 1960s were a tumultuous decade, and General Groves observed many of the events with skepticism and disfavor. There were riots in the ghettos and protests against the Vietnam War — evidence that the society had lost its direction and respect for the law. An enthusiast of the Age of Aquarius, or of President Lyndon Johnson's "Great Society," he was not. He wrote many memos to the file about the social ills he was witnessing, and through them we see his conservative opinions in full bloom.

The ghetto riots of the late 1960s were troubling, and Groves believed it was easy to see why they occurred.

> The trouble really started soon after Roosevelt was elected President and he and his wife started to curry favor of the negro population, and encouraged the philosophy that it was alright to be lawless if you thought your cause was a just one.

This led to sitdown strikes in the automotive factories in 1936 and the Michigan governor's refusal to enforce the laws by calling out the National Guard. Further contributing to the current state of affairs was the Wagner Labor Act, which gave overwhelming power to the unions. Passage of various minimum wage laws and certain decisions by the Warren Supreme

Court eroded respect for the law even further, in Groves's view. "This has reached a climax in the Johnson Administration and was well illustrated in Washington when the rioters and looters knew they could behave as they did with complete impunity." The police were ordered not to use force. Only much later were federal troops brought in, and then in driblets, "apparently in accord with the Kennedy-Johnson-MacNamara operations in Viet Nam of gradual escalation and, like in Viet Nam, it was a complete failure." There have been times when troops were ordered to shoot looters on sight and to shoot to kill, such as the Baltimore fire of 1904 or the San Francisco earthquake of 1906, Groves noted.

> The Johnson Administration has no one to thank but themselves and their Democratic predecessors for the mess to which they have brought the country. And, in turn, the majority of American citizens share the responsibility for having elected and re-elected to high office and control of our Government men who were so lacking in moral fiber and principles.

In comments on a *New York Times* op-ed piece about national service as an alternative to the draft Groves dismissed the young writer as a hypocrite in comments bitter with scorn.[25] The real reason for draft avoidance, Groves believed, is that young men fear getting hurt during actual fighting. Some avoid military service by going into elementary school teaching, but few will remain when they are no longer subject to the draft. His Anglophobia had not diminished, for he took a swipe at New York City: "Its history of patriotism has never been up to that of the rest of the country from the very first days when it was a Tory center during the Revolution."

There was always time to devote to honoring friends and colleagues. On February 25, 1967, he went to Princeton to pay his final respects to Oppenheimer, who had died three weeks earlier. A service was held at Alexander Hall with I. I. Rabi, Philip Morrison, Robert Serber, John Lansdale, and many others in attendance. The new director of the institute, Carl Kaysen, introduced the three main speakers, Hans Bethe, Henry Smyth, and George Kennan, who spoke eloquently about Oppenheimer.

There was also time to honor or assist West Point classmates. At a ceremony in May 1969 at Bolling Air Force Base the ashes of Eugene Vidal (father of the writer Gore Vidal) were to be spread by helicopter over the

Virginia countryside. General Groves came with a dozen others to pay their respects.[26]

Groves got involved in trying to help another classmate, Linson Dzau.[27] Dzau came from a wealthy Chinese family and had been appointed to West Point by the Chinese government. After graduating with Groves in the class of November 1918 he worked at the Chinese Embassies in Washington and London and served with the Chinese delegation at the Paris peace conference. Over the next two decades he held many other positions. Caught up in the civil war, the Communists stripped him and his family of all their possessions; eventually he landed in Macau as a refugee in 1951. He taught at various Catholic schools and in 1965 founded Linson College. Dzau and his school were in serious financial difficulties, and when this came to the attention of his West Point classmates, they rallied and contributed some twenty thousand dollars. The bonds of West Point are strong: If a classmate is in trouble, you try to help him. Groves was part of this effort and with several others wrote to the Henry Luce Foundation to request a modest grant of fifty thousand dollars to save Dzau's school.

On the morning of February 27, 1970, in a ceremony at the White House, President Richard M. Nixon presented James B. Conant, Vannevar Bush, and Leslie R. Groves with the Atomic Pioneer Award. AEC Chairman Glenn T. Seaborg, who felt that special recognition should be given to these three men for their contributions in the development of nuclear weapons during World War II, had initiated the occasion. Furthermore, each had taken unpopular stands with regard to postwar nuclear energy policies and, as a result, had never received the kind of public recognition his work deserved. Groves's citation read,

> For his exceptional contributions to the national security as Commanding General of the Manhattan Engineer District, United States Army, in developing the world's first nuclear weapons during World War II, and for his pioneering efforts in establishing administrative patterns adopted by the Atomic Energy Commission in effecting the use of atomic energy for military and peaceful purposes.

The press release noted that the Atomic Pioneer Award "is a special award which will not be bestowed on any other recipients in the future." After the ceremony the medalists and their guests went to the State

Department for a luncheon. Groves's guests included Major General Nichols, Mrs. O'Leary, Joe Volpe, and the chief of engineers, Lt. Gen. Frederick J. Clarke and his wife. Also with him were Mrs. Groves, his son and daughter-in-law, and granddaughter Carolyn.

It would be Groves's last public appearance. His health declined rapidly. Previously the doctors at Walter Reed had told him that he had calcification of the heart exit valve, or aortic valve.[28] They said that if they operated to repair or replace the valve he would have a 50 percent chance of surviving the operation and a 50 percent chance of its being successful. He decided not to take the odds. At the beginning of April he applied to the Veterans Administration for total disability. In order to receive this he had to prove that his heart condition began before he retired. In support of his application Dr. Charles E. Rea, who was in charge of the Oak Ridge Hospital during the war, noted Groves's medical history. He had come to the hospital on several occasions complaining of chest pains and pains radiating down both arms. Occasionally he had spent a day or two in the hospital resting from the enormous strain he was under, a fact known to few. Though repeated EKGs produced indeterminate and inconsistent results, one of the diagnoses at the time suggested the beginnings of arteriosclerosis around the openings of the coronary arteries and the aortic valve. Now, Rea noted, General Groves was unable to walk at a normal pace or walk up stairs or an incline without possible fainting spells.

A few days before he died he was awarded 100 percent disability — though he would not live to collect the first payment. Nevertheless, the pension did provide income to his widow, Grace, who would live for another sixteen years.

On Monday evening, July 13, Groves suffered a severe heart attack at his home on Connecticut Avenue. He was rushed to Walter Reed Army Medical Center about three miles away. Hospital records show that he was admitted at 2200, or 10:00 P.M. According to the notes of the attending physician, "[He] expired 2315 hrs 13 July 1970," as a result of "acute pulmonary edema calcific aortic stenosis. . . . Wife present at time of death."[29]

AEC Chairman Glenn Seaborg issued a statement the next day.

> He was a constant source of amazement to all of us in the Manhattan Project because of his ability to handle a variety of complex details with a minimum of confusion. He had the deep respect and admiration of his staff for his ability to organize and to get things done. And he had a prodigious memory and often confounded his staff by

recalling names, dates, and incidents buried in the files.... Twenty five years ago this coming Thursday, the world's first nuclear explosion took place at Alamogordo, New Mexico, and the anniversary serves to remind all of us of the great debt which the Free World owes to the work of General Groves.[30]

Over the next few days obituaries and editorials appeared in dozens of newspapers extolling "The Man Who Built the Bomb," the "A-bomb Quarterback," concluding that he was the right man, in the right job, at the right time and that "[m]any thousands of American men are alive today because Gen Groves' team won the bomb race."[31]

The *Washington Post* noted that

> The talents for organization and administration, the daring and imagination he brought to his mammoth assignment were peculiarly American, reflective in a sense of the genius of America."[32]

The *Chicago Daily News* said,

> Gen Groves believed passionately in what he was doing. He believed that wars were to be won, and that weapons were to win them with. His soldier's mind brooked no moral doubts. He figured that if the United States did not get the bomb first it could become a decisive weapon in the hands of the enemy (and he was right). Facing that challenge, he drove his team to the limits of human endurance. And he won and held their respect in the process."[33]

The funeral service was held at the old Fort Myer Chapel at 11:00 A.M. on July 17. Retired Chief of Chaplains Maj. Gen. Luther D. Miller, a friend of the general's, who read as his text most of the verses of the forty-fourth chapter of Ecclesiasticus from the Book of Common Prayer, conducted the service.

> Let us now praise famous men, and our fathers that beget us. The Lord hath wrought great glory by them through his great power from the beginning.... Their bodies are buried in peace; but their name liveth for evermore. The people will tell of their wisdom, and the congregation will show forth their praise.

For reasons not altogether clear Chaplain Miller omitted verses nine and ten, though they better describe what had happened to the man they had assembled to mourn:

> And some there be, which have no memorial; who are perished as though they had never been; and are become as though they had never been born, and their children after them. But these were merciful men, whose righteousness hath not been forgotten.[34]

Following the service his body was interred at Arlington Cemetery in a grave next to his brother Allen on a steep hillside east of the Custis-Lee Mansion, almost within sight of the Pentagon.

AFTERWORD

Leslie Richard Groves operated in the world personally and professionally according to well-defined principles of conduct. These attitudes were shaped initially by growing up in the military culture of army posts at the beginning of the twentieth century. The dynamics within his family fostered competition and a striving for excellence. His energy, drive, and determination were prodigious; from an early age he was confident that if he set himself a goal he could attain it.

The two most formative experiences of Groves's early life were attendance at West Point and becoming an army engineer. Determined to get into the Military Academy, once there he excelled, graduating fourth in the class of November 1918. Over the next three decades Groves would rise farther and faster in his army career than any of his classmates. At graduation he chose the Corps of Engineers, by consensus the elite branch of the army with the smartest and ablest officers, almost a caste by itself with its own canon and precepts. Throughout the 1920s, with steady ambition and hard work, he performed a variety of assignments well, gaining experience and learning from senior engineers.

During the 1930s Groves carried out a series of difficult and demanding tasks, which would position him to be chosen to run the Manhattan Project in 1942. Ability explained part of his success, but equally important was his association with Col. Ernest "Pot" Graves, perhaps the best-known engineer within the corps at the time. Graves embodied certain qualities that his protégés sought to emulate. He was physically tough, liked hard assignments in the field, was very smart and politically savvy, and influenced the careers of a group of younger officers who would go on to play key roles in World War II.

To those who listened — and Groves was one of his prime pupils — Graves delivered lessons about how to be an exceptional engineer and army officer. Intelligence headed the list as a necessary attribute. Graves also stressed the importance of having mental discipline and the ability to analyze problems and devise solutions. Equally important was judging people. The ability to size someone up quickly and identify his strengths and weaknesses was fundamental to success. An officer has to delegate, Groves learned, but he should delegate wisely to those he feels confident can do their jobs. He will drown in detail and fail if he tries to do everything himself. Groves had extraordinary instincts when it came to taking the measure of others, and his success depended on it.

With his mentor's lessons well learned, Groves continued to gain confidence as he took on larger and larger projects. After attending the Command and General Staff School and the Army War College — essential to higher rank and advancement — he eventually rose to be deputy chief of construction within the Corps of Engineers. In this position Groves oversaw more than eight billion dollars' worth of army projects during the mobilization period from 1940 to 1942 — a vast expenditure, four times that of the Manhattan Project. His schedule was extraordinary: countless telephone calls and meetings, and a grueling travel routine. Groves took it all in stride as part of the job. In the process he amassed a long list of contacts: contractors, architects, scientists, corporate officials, civilian and army engineers, craftsmen, and artisans. Many would soon be recruited to assist him in building the bomb.

The military is predicated on hierarchy and rank, and learning how to make decisions to lead. Its culture and ethos, rituals and traditions, are intended to bring out the qualities that it needs to fight and win a war, or to prepare for one.

It is instructive to compare and contrast Groves with the other American generals of World War II about whom much has been written. Here a distinction might be made between *leadership* and *command*. While sometimes the terms are used synonymously, highlighting their differences may offer insight into Groves and the others.

Leadership is composed of a combination of qualities that inspire and motivate others to follow and to achieve goals. The qualities of leadership can include instilling fear in subordinates, appealing to their self-interest, or projecting that most alluring character trait, charisma. Military leaders must be able to persuade those under them to follow unswervingly, to risk danger, to kill, and possibly even to die. Probably the purest American examples in World War II of this type of leader are Patton, MacArthur, and Bradley.

Command, on the other hand, is a managerial function — the coordinating of military forces. The skills required are somewhat different, though they may share qualities with leadership. Intelligence and sound judgment about others are shared characteristics. But command also requires attending to detail, steadiness of application, the ability to pursue many tasks simultaneously, trust in subordinates, and a talent for spotting and solving problems before they arise. Leaders may not be effective commanders, nor commanders necessarily effective leaders. The American military in the mid–twentieth century underwent a managerial revolution as it grew enormously during World War II and then remained substantial throughout the Cold War. Groves fits very well as the archetype of the commander as manager.

By Groves's measure of character he probably judged his own father as someone who was a bit weak and vacillating, though he no doubt loved him and learned from him. Toughness was prized. The ability to carry on under enormous pressure and not break was the measure of a man. We will never know if he ever had any fundamental misgivings or doubts about his own capabilities to perform the Manhattan mission. Probably he did not. As an officer who held integrity above all else, if he had felt that he was not up to the job the only proper course would have been to ask Secretary Stimson and General Marshall to find another man.

Though the record is limited, his mother was probably the more influential figure in his early development, urging all of her sons to excel. Her early death, when he was only sixteen, and the scarce resources that had always been a fact of family life forced Groves to conclude that if he was to make his way in the world it would have to be by his own wits, intelligence, and drive.

Though Groves was the son of an army chaplain, he hardly ever attended church and did not seem to draw upon organized religion for any reason. It did not seem to matter, as his sense of personal integrity was forged early and did not need continual renewal through churchgoing. From his father he gained a pronounced sense of what was right and wrong and was scrupulously honest. Every penny of the two billion dollars the Manhattan Project spent was accounted for, and every other penny before and after that as well.

No one can ever remember hearing him swear. Neither did he erupt in volcanic outbursts or routinely dress down subordinates. His mere presence was normally enough to deliver the message loud and clear. Those who entered his orbit could feel the emanations that declared, in no uncertain terms, that you had better do your job, do it well, and do it on time. One indicator of a disciplined mind is how well one expresses oneself orally and in writing. Groves's memos and letters are well written and focused, as is his memoir.

Material things meant little to him. Not until he retired from the army and worked for a corporation did he receive a decent salary and begin to enjoy some of the better things of life.

He was highly intelligent though not an intellectual. He had little exposure to or appreciation of the cultural side of life. There were few visits to the art museums or the concert hall, except occasionally as a companion to his wife, Grace, whose interests centered on music and the arts.

Yet many of the qualities that made Groves effective as a military administrator and manager caused him difficulties in other areas of life. In some respects he was his own worst enemy. As one of his aides later said of him, "there was not a humble bone in his body." He thought highly of himself and did not conceal it. He had a tendency to talk down to people, and not surprisingly, many resented it. He was used to giving orders and having others follow them. He was brusque and highly opinionated. His strong personality rubbed some people the wrong way. For him many, if not most, things *were* black and white.

He prospered in the highly secret, authoritarian military world of the Manhattan Project where he had, basically, absolute power and control. Soon after the end of the war everything changed, and changed very quickly, and Groves was not able to adjust to it. The many questions surrounding the military and civilian applications of atomic energy were now being discussed openly. The debate involved the executive and legislative branches of government, scientists, the press, and the public in questions of how control of nuclear power should be allocated among the various agencies, departments, committees, and services. Groves fought against the prevailing winds and lost. Even after the Atomic Energy Commission took over from the army Groves kept up a guerrilla campaign to maintain a vigorous role for the military. But a long list of enemies was waiting for revenge and eventually they got their chance. Equally important was the fact that there was no real place for him — an engineer, still young, with no combat experience — in an army undergoing extensive demobilization. Amid the new rules of the postwar world he was unceremoniously forced into retirement.

Groves did not turn out to be a bureaucratic survivor, that Washington personality type who conforms and adapts so as to fight another day. With a new adversary and the establishment of a large military to fight the Cold War, a whole new stage was set. Military officers and their civilian leaders adopted new skills to advance and prosper in Washington's corridors of power. Groves would not be among them. He observed the drama from afar, remaining a semipublic figure called upon periodically to comment on this or that development, but with little influence. He spent a great deal of his time after retiring from the military preserving and protecting the legacy of the Manhattan Project against what he considered were ill-informed journalists and historians, or anyone else who did not see the story his way.

From all indications Groves was not a very complicated person. He did not seem to have any obvious complexes or secrets that might be keys to his personality. If there were any bothersome afflictions of that sort, his iron will and determination pushed them aside. What you saw was pretty much what you got. Everything was essentially on the surface. His supreme self-confidence, for example, did not develop to compensate for feelings of doubt or inadequacy, as is sometimes the case in others. He was confident of what he could do and with brute force got it done.

Several years after his death Groves's son Richard, himself by then a major general in the Army Corps of Engineers, paid tribute to his father:

> He was determined to succeed in everything he undertook; he played to win; he had no use for losers; moral victories did not exist for him. He took great pride in accomplishing those things that he considered worthwhile; to his way of thinking they were the funda-mental virtues. He expected others to feel the same way about them and to strive to excel in them, too. He sought his own qualities in others and, when he found them, he put them to work for him, regardless of whatever niceties of seniority might be involved.

He was intensely proud of his American heritage, proud of being in the U.S. Army, proud of being a West Pointer. Of the latter he said, when his life was nearly done, "To be a graduate of West Point is an honor that comes to few and one that is never forgotten by any former cadet."

There have been few people in this world who equaled the genius of Leslie Richard Groves for getting things done. The world needs such people from time to time, and it is clear in Groves's case that he was the right man at the right place at the right time. Controversies about the bomb will continue, though from the vantage point of half a century cer-tain things now appear clear. It seems to this author that the bomb's sudden and dramatic use ended one war and in the process had a profound influence in preventing another. As I have demonstrated throughout, Groves was the dominant force in the early completion and use of the bomb. By influencing all that came in the bomb's wake he thus deserves a more central role in twentieth-century history than has previously been acknowledged.

The dual promise of the bomb was recognized early on. Niels Bohr noted that while the bomb was a menace to civilization, it "would offer quite unique opportunities to bridge international divergencies." Albert Einstein had a similar thought: Perhaps the bomb "may intimidate the human race into bringing order into its international affairs, which without the pressures of fear, it would not do."

The grand schemes and goals of the movement for international control of atomic energy were only partially achieved, and in ways admittedly incomplete and imperfect. The nuclear arms race between the United States and the Soviet Union was inherently dangerous, and at times threatened to spin out of control. Nevertheless, the two main protagonists did manage to create a semblance of order and control built around fear of the bomb as its centerpiece.

For Groves the challenge of mastering atomic energy was evidence, as he said, "that when man is willing to make the effort, he is capable of accomplishing virtually anything." His success has left us with numerous challenges of our own: threats of terrorist A-bombs, environmental problems seemingly without solutions, new aspirants, and new arms races. Mastering these will require special skills. It would be good to have the general here to help.

PERSONAL INTERVIEWS

Interviews indicated with an asterisk were conducted by telephone; all others were conducted in person.

ROBERT S. NORRIS

    Agnew, Harold, April 15, 1998*

    Camm, Lt. Gen. Frank A. (USA, retired), February 16, 1998

    Clarke, Lt. Gen. Frederick (USA, retired), July 23, 1999*

    Cooper, Lt. Gen. Kenneth (USA, retired), November 24, 1997

    Cushman, Lt. Gen. John H. (USA, retired), January 9, 1998

    Derry, John, February 10, 1999*

    Dorland, Gilbert M., August 17, 1999*

    Finley, Luke W., March 25, 1998*

    Furman, Robert F., March 28, 1998

    Grayson, Cary T., April 29, 1998*

    Graves, Lt. Gen. Ernest (USA, retired), December 29, 1997

    Groves, Richard H., December 15, 1998; January 25, 1999; June 1, 1999

    Hadden, John L., January 9, 1999*

    Hammel, Edward F., February 17, 1999*

    Hawkins, David, June 6, 2000*

    Jannarone, Ann, April 29, 2000*

    Johnson, Allan C., April 2, 9, 1998*

    Johnson, Lyall, March 23, 1999; December 9, 2000*

    Keyes, Geoffrey, April 14, 1998*

    Kolodney, Morris, February 12, 1999*

    Martin Elliott, Gerry, March 23, 1998*

    Marks, Anne Wilson, October 24, 2000

    Nichols, Maj. Gen. Kenneth D. (USA, retired), May 16, 1998

    Olsson, Virginia J., April 30, 1998*

    Owen, Patricia Cox, April 11, 1998

    Rehl, Donald C., November 22, 1999*

    Rhodes, Fred B., December 7, 1998*

    Shapiro, Maurice M., February 25, 1998*

    Shivers, Rufus, May 14, 1998*

    Volpe, Joseph, June 13, 2001

Watson, Coni O'Leary, April 4, 1998
Westcott, James Edward, March 2, 1998★
Young, Maj. Gen. Robert P. (USA, retired), January 19, 1998★

STANLEY GOLDBERG

Bacher, Robert, April 12–14, 1990
Bainbridge, Kenneth T., February 13, 1990
Campbell, Patsy Groves, June 23, 1993
Chorpening, Maj. Gen. Claude (USA, retired), August 21, 1992
Derry, John, November 20, 1991
Furman, Robert, February 10, 1990
Lansdale, John, February 7, 1990; March 19, 1991
Larmee, Charlotte Groves, August 14, 1991
Marks, Anne Wilson, September 24, 1991
McGowan, Louise Jackson, July 23, 1991
Martin, Gerry Elliott, August 30, 1991
Morrison, Philip, February 12, 1989
Mountain, Thomas, August 6, 1991
Pash, Boris, April 18, 1990
Robinson, Gwen Groves, May 3, 1991
Robinson, Alan, May 2, 1991
Rotblat, Joseph, July 9, 1993
Rhodes, Fred B., September 17, 1991
Shurcliff, William S., July 13, 1989
Seaborg, Glenn, April 17, 1990
Volpe, Joseph, September 6, 1991
Watson, Coni O'Leary, February 9, 1989

ABBREVIATIONS

In citing works in the notes, shortened citations have been used where possible. Works frequently cited have been identified by the following abbreviations:

Alperovitz, *Decision.*   Gar Alperovitz. *The Decision to Use the Atomic Bomb and the Architecture of an American Myth.* New York: Alfred A. Knopf, 1995.

Ambrose, *Duty.*   Ambrose, Stephen E. *Duty, Honor, Country: A History of West Point.* Baltimore: The Johns Hopkins Press, 1966.

AMG   Allen Morton Groves (brother)

AMG, Appreciation.   *Allen Morton Groves: An Appreciation Together with Selections from His Orations and Essays.* Privately printed, n.d.

Badash, et al., *Los Alamos.*   Badash, Lawrence, Joseph O. Hirschfelder, and Herbert P. Broida, eds. *Reminiscences of Los Alamos, 1943–1945.* Dordrecht, Holland: D. Reidel Publishing Company, 1980.

Ball, *Command.*   Ball, Harry P. *Of Responsible Command: A History of the U.S. Army War College.* Carlisle Barracks, Penn.: Alumni Association of the U.S. Army War College, 1984.

Bernstein, "Seizing."   Bernstein, Barton J. "Seizing the Contested Terrain of Early Nuclear History: Stimson, Conant, and Their Allies Explain the Decision to Use the Atomic Bomb," *Diplomatic History* 17, winter 1993: 35–72.

Bernstein, "Struggle."   Bernstein, Barton J. "The Struggle Over History: Defining the Hiroshima Narrative." In Philip Nobile, ed., *Judgment at the Smithsonian.* New York: Marlowe & Company, 1995, 127–256.

Bernstein, "Understanding."   Bernstein, Barton J. "Understanding the Atomic Bomb and the Japanese Surrender: Missed Opportunities, Little-Known Near Disasters, and Modern Memory," *Diplomatic History* 19, spring 1995: 227–273.

Bernstein, *Atomic Bomb.*   Bernstein, Barton J., ed. *The Atomic Bomb: The Critical Issues.* Boston and Toronto: Little, Brown and Co., 1976.

Bowen, *Silverplate.*   Bowen, Lee. *Project Silverplate, 1943–1946.* Vol. 1 in Lee Bowen and Robert D. Little, eds., *The History of Air Force Participation in the Atomic Energy Program, 1943–1953.* Washington, D. C.: USAF Historical Division, 1959.

Brown, MacDonald, *Secret History.* Brown, Anthony Cave, and Charles B. MacDonald, eds. *The Secret History of the Atomic Bomb.* New York: Dell, 1977.

Brown, *Neutron.* Brown, Andrew. *The Neutron and the Bomb: A Biography of Sir James Chadwick.* Oxford: Oxford University Press, 1997.

Cameron, *Barbarians.* Cameron, Nigel. *Barbarians and Mandarins: Thirteen Centuries of Western Travelers in China.* Chicago: University of Chicago Press, 1970.

Carlisle, *Supplying.* Carlisle, Rodney P., with Joan M. Zenzen. *Supplying the Nuclear Arsenal: American Production Reactors, 1942–1992.* Baltimore: Johns Hopkins University Press, 1996.

Chaplain Groves, *Papers.* Groves, Chaplain Leslie Richard, Sr. *Papers of Chaplain Leslie Richard Groves.* 2 vols. Arranged by Lt. Gen. Richard Hulbert Groves. Unpublished manuscript, December, 1988.

Chappel, *Before.* Chappel, John D. *Before the Bomb: How America Approached the End of the Pacific War.* Lexington: University Press of Kentucky, 1997.

Christman, *Target.* Christman, Al. *Target Hiroshima: Deak Parsons and the Creation of the Atomic Bomb.* Annapolis, Md.: Naval Institute Press, 1998.

Cohen, *Keys.* Cohen, Paul A. *History in Three Keys: The Boxers as Event, Experience and Myth.* New York: Columbia University Press, 1997.

Compton, *Quest.* Compton, Arthur H. *Atomic Quest: A Personal Narrative.* New York: Oxford University Press, 1956.

Conant, "History." Conant, James B. "A History of the Development of an Atomic Bomb," spring 1943, Bush-Conant Correspondence, S-1 Files, Office of Scientific Research and Development (OSRD), Box 3, Folders 1 and 4, RG 227, NARA.

Cosmas, *Empire.* Cosmas, Graham. *An Army for Empire: The United States Army in the Spanish American War.* Columbia: University of Missouri Press, 1971.

Coster-Mullen, *Atom Bombs.* Coster-Mullen, John. *Atom Bombs: The Top Secret Inside Story of Little Boy and Fat Man.* Privately published, 2001.

Dawidoff, *Catcher.* Dawidoff, Nicolas. *The Catcher Was a Spy: The Mysterious Life of Moe Berg.* New York: Pantheon Books, 1994.

Dickson, *Chronicle.* Dickson, Col. B. A., AUS, retired. *A Chronicle of the Class of November, 1918, United States Military Academy, West Point, New York.* 1968.

Ermenc, *Scientists.* Ermenc, Joseph J., ed. *Atomic Bomb Scientists: Memoirs, 1939–1945.* Westport, Conn.: Meckler, 1989.

Fine, Remington, *Corps.* Fine, Lenore, and Jesse A. Remington. *The*

Corps of Engineers: Construction in the United States. Washington, D.C.: Center of Military History, U.S. Army, 1989; first published 1972.

Fleming, *West Point*. Fleming, Thomas J. *West Point: The Men and Times of the Military Academy.* New York: William Morrow and Co., 1969.

FOIA Freedom of Information Act

Frank, *Downfall*. Frank, Richard B. *Downfall: The End of the Imperial Japanese Empire.* New York: Random House, 1999.

***FRUS: Conference of Berlin*.** U.S. Department of State. *Foreign Relations of the United States: Conference of Berlin (Potsdam), 1945.* 2 vols. Washington, D.C.: U.S. Government Printing Office, 1960.

***FRUS: Intelligence*.** U.S. Department of State. *Foreign Relations of the United States: 1945–1950, Emergence of the Intelligence Establishment.* Washington, D.C.: U.S. Government Printing Office, 1996.

Gerrard-Gough, Christman, *Experiment*. Gerrard-Gough, J. D., and Albert B. Christman. *The Grand Experiment at Inyokern.* Washington, D.C.: Naval History Division, 1978.

GGG Gwen Griffith Groves (mother).

GGR Gwen Groves Robinson (daughter).

Giovannitti, Freed, *Decision*. Giovannitti, Len, and Fred Freed. *The Decision to Drop the Bomb.* New York: Coward-McCann, 1965.

Goldberg, "Compartmentalization." Goldberg, Stanley. "Groves and the Scientists: Compartmentalization and the Building of the Bomb," *Physics Today,* August 1995: 38–43.

Goldberg, "Hanford." Goldberg, Stanley. "General Groves and the Atomic West: The Making and the Meaning of Hanford." In *The Atomic West,* edited by Bruce Hevly and John M. Findlay (Seattle: University of Washington Press, 1998), 38–89.

Goldberg, "Inventing." Goldberg, Stanley. "Inventing a Climate of Opinion: Vannevar Bush and the Decision to Build the Bomb," *ISIS* 83 (1992): 429–452.

Goldberg, "Oppenheimer." Goldberg, Stanley. "Groves and Oppenheimer: The Story of a Partnership," *Antioch Review* 53 (fall 1995): 482–493.

Goldberg, "Racing." Goldberg, Stanley. "Racing to the Finish: The Decision to Bomb Hiroshima and Nagasaki," *The Journal of American–East Asian Relations,* summer 1995: 117–128.

Goldberg, "Reins." Goldberg, Stanley. "Groves Takes the Reins," *Bulletin of the Atomic Scientists,* December 1992: 32–39.

Goldberg, "Words." Goldberg, Stanley. "A Few Words About This Picture," *Invention & Technology,* fall 1991: 49–54.

Goldberg, *Pentagon.* Goldberg, Alfred. *The Pentagon: The First Fifty Years.* Washington, D.C.: Historical Office, Office of the Secretary of Defense, 1992.

Goodchild, *Oppenheimer.* Goodchild, Peter. *J. Robert Oppenheimer: Shatterer of Worlds.* Boston: Houghton Mifflin Co., 1981.

Goudsmit, *Alsos.* Goudsmit, Samuel A. *Alsos.* Los Angeles: Tomash Publishers and the American Institute of Physics, 1983; originally published 1947.

Gowing, *Britain.* Gowing, Margaret. *Britain and Atomic Energy 1939–1946.* London: Macmillan, 1964.

Groueff, *Manhattan.* Groueff, Stephane. *Manhattan Project: The Untold Story of the Making of the Atomic Bomb.* Boston: Little, Brown and Co., 1967.

GWG Grace Wilson Groves (wife)

Hacker, *Dragon.* Hacker, Barton C. *The Dragon's Tail: Radiation Safety in the Manhattan Project, 1942–1946.* Berkeley: University of California Press, 1987.

Hales, *Spaces.* Hales, Peter Bacon. *Atomic Spaces: Living on the Manhattan Project.* Urbana: University of Illinois Press, 1970.

Harrison-Bundy Files. Records of George L. Harrison and Harvey H. Bundy from Record Group 77. Reproduced as a NARA microfilm, nine rolls (M1108).

Hawkins, *Trinity.* Hawkins, David. *Toward Trinity.* Vol. 1 in *Project Y, the Los Alamos Story.* Los Angeles: Tomash Publishers and the American Institute of Physics, 1983.

Helmreich, *Ores.* Helmreich, Jonathan E. *Gathering Rare Ores: The Diplomacy of Uranium Acquisition, 1943–1954.* Princeton, N.J.: Princeton University Press, 1986.

Herken, *Winning.* Herken, Gregg. *The Winning Weapon: The Atomic Bomb in the Cold War, 1945–1950.* New York: Random House, 1981.

Hershberg, *Conant.* Hershberg, James. *James B. Conant: Harvard to Hiroshima and the Making of the Nuclear Age.* New York: Alfred A. Knopf, 1993.

Hewes, *Root.* Hewes, James E. , Jr. *From Root to McNamara: Army Organization and Administration, 1900–1963.* Washington, D.C.: Center of Military History, U.S. Army, 1975.

Hewlett, Anderson, *New World.* Hewlett, Richard C., and Oscar E. Anderson Jr. *The New World, 1939–1946,* Vol. 1 in *A History of the United States Atomic Energy Commission.* University Park: Pennsylvania State University, 1962.

Hewlett, Duncan, *Shield*. Hewlett, Richard C., and Francis Duncan. *Atomic Shield, 1947–52,* Vol. 2 in *A History of the United States Atomic Energy Commission.* University Park: Pennsylvania State University, 1969.

Hoddeson, et al., *Critical*. Hoddeson, Lillian, Paul W. Henriksen, Roger A. Meade, and Catherine Westfall. *Critical Assembly: A Technical History of Los Alamos During the Oppenheimer Years, 1943–1945.* New York: Cambridge University Press, 1991.

Hoddeson, "Change." Hoddeson, Lillian. "Mission Change in the Large Laboratory: The Los Alamos Implosion Program, 1943–1945." In Peter Galison and Bruce Hevly, eds., *Big Science: The Growth of Large-Scale Research* (Stanford, Calif.: Stanford University Press, 1992): 265–289.

Hodgson, *Colonel*. Hodgson, Godfrey. *The Colonel: The Life and Wars of Henry Stimson, 1867–1950.* New York: Alfred A. Knopf, 1990.

Hounshell, "Du Pont." Hounshell, David A. "Du Pont and the Management of Large-Scale Research and Development." In Peter Galison and Bruce Hevly, eds., *Big Science: The Growth of Large-Scale Research* (Stanford, Calif.: Stanford University Press, 1992).

Irving, *German*. Irving, David. *The German Atomic Bomb.* New York: Simon & Schuster, 1967.

Jones, *Wizard War*. Jones, R.V. *The Wizard War.* New York: Coward, McCann and Geoghegan, 1978.

Jones, *Manhattan*. Jones, Vincent C. *Manhattan: The Army and the Atomic Bomb.* Washington, D.C.: Center of Military History, U.S. Army, 1985.

Keown-Boyd, *Fists*. Keown-Boyd, Henry. *The Fists of Righteous Harmony: A History of the Boxer Uprising in China in the Year 1900.* London: Leo Cooper, 1991.

Kirkpatrick, "Gaps." Kirkpatrick, Charles E. "Filling the Gaps: Reevaluating Officer Professional Education in the Inter-War Army, 1920–1940." Paper presented at the 1989 American Military Institute Annual Conference, April 14–15, 1989.

Kurzman, *Blood*. Kurzman, Dan. *Blood and Water: Sabotaging Hitler's Bomb.* New York: Henry Holt and Co., 1997.

Kurzman, *Day of Bomb*. Kurzman, Dan. *Day of the Bomb: Countdown to Hiroshima.* New York: McGraw-Hill Book Company, 1986.

Lanouette, *Genius*. Lanouette, William, with Bela Silard. *Genius in the Shadows: A Biography of Leo Szilard, The Man Behind the Bomb.* New York: Charles Scribner's Sons, 1992.

Lansdale, Manuscript. Lansdale, John. *John Lansdale, Jr. Military Service.* Unpublished manuscript, 1987.

Lawren, *General*.　Lawren, William. *The General and the Bomb: A Biography of General Leslie R. Groves, Director of the Manhattan Project.* New York: Dodd, Mead and Co., 1988.

LC　Larmee Collection. Letters retained by Charlotte Larmee, daughter of Owen Griffith Groves.

Lilienthal, *Journals*.　Lilienthal, David E. *The Journals of David E. Lilienthal,* Vol. 2: *The Atomic Energy Years, 1945–1950.* New York: Harper & Row, 1964.

Little, *Foundations*.　Little, R. D. *Foundations of an Atomic Air Force and Operation Sandstone, 1946–1948.* Vol. 2 in *The History of Air Force Participation in the Atomic Energy Program, 1943–1953.* Washington, D.C.: Air University Historical Liaison Office, 1959.

LOC　Library of Congress

LRG　Leslie Richard Groves

LRG Appointment Book.　Groves Diary, Boxes 1–4, Entry 7530G, Papers of LRG, RG 200, NARA.

LRG, Diary Notes.　Folder, Letter Pocket, Box 5, Entry 7530C, Papers of LRG, RG 200, NARA.

LRG Official Records.　Richard H. Groves compiled his father's official records (normally referred to as a 201 file) into this 725-page bound volume.

LRG, "Atom General."　Groves, Lt. Gen. Leslie R., USA, retired. "The Atom General Answers His Critics," *Saturday Evening Post,* June 19, 1948.

LRG, "The A-Bomb."　Groves, Lt. Gen. Leslie R. "The A-Bomb Program." In Fremont E. Kast and James E. Rosenzweig, eds., *Science, Technology and Management* (New York: McGraw-Hill, 1963).

LRG, *Grandchildren*.　Groves, Leslie R. *For My Grandchildren.* Entry 7530N, Papers of LRG, RG 200, NARA.

LRG, *Now*.　Groves, Leslie R. *Now It Can Be Told: The Story of the Manhattan Project.* New York: Harper and Brothers, 1962. Reprinted in 1983 by DaCapo Press, New York.

MacEachin, *Final Months*.　MacEachin, Douglas J. *The Final Months of the War with Japan: Signals Intelligence, U.S. Invasion Planning, and the A-Bomb Decision.* Washington, D.C.: CIA Center for the Study of Intelligence, 1998.

Marshall Diary.　Marshall, Brig. Gen. James C. *Chronology of District "X" (17 Jun 42–28 Oct 42),* USACE, Office of History.

Meigs, *Optimism*.　Meigs, Mark. *Optimism at Armageddon: Voices of American Participants in the First World War.* New York: New York University Press, 1996.

Millett, *Organization.* Millett, John D. *The Organization and Role of the Army Service Forces.* Washington, D.C.: Office of the Chief of Military History, Department of the Army, 1954.

MNG Marion Nash Groves (sister-in-law)

Montgomery, "Dust." Montgomery, M. R. "Impalpable Dust," *The New Yorker,* March 27, 1989.

NARA National Archives and Records Administration, College Park, Md.

Nenninger, "Leavenworth and Critics." Nenninger, Timothy K. "Leavenworth and Its Critics: The U.S. Army Command and General Staff School, 1920–1940," *Journal of Military History,* April 1994.

Nichols, *Trinity.* Nichols, Maj. Gen. K. D., USA, retired. *The Road to Trinity: A Personal Account of How America's Nuclear Policies Were Made.* New York: William Morrow and Co., 1987.

Nye, *USMA.* Nye, Roger H. *The United States Military Academy in an Era of Educational Reform.* Unpublished Ph.D. dissertation, Columbia University, 1968.

OGG Owen Griffith Groves (brother)

Ohl, *Supplying.* Ohl, John Kennedy. *Supplying the Troops: General Somervell and American Logistics in WWII.* DeKalb: Northern Illinois Press, 1994.

Papers of LRG, RG 200, NARA. Papers of General Leslie R. Groves, Record Group 200, National Archives and Records Administration, College Park, Md.

Pash, *Alsos.* Pash, Col. Boris T., AUS, retired. *The Alsos Mission.* New York: Award House, 1969.

Paul, *Rivals.* Paul, Septimus H. *Nuclear Rivals: Anglo-American Atomic Relations, 1941–1952.* Columbus: Ohio State University Press, 2000.

Powers, *Heisenberg.* Powers, Thomas. *Heisenberg's War: The Secret History of the German Bomb.* New York: Alfred A. Knopf, 1993.

PRO Public Records Office

Rearden, *Formative.* Rearden, Steven L. *The Formative Years, 1947–1950.* Vol. 1 in *History of the Office of the Secretary of Defense.* Washington, D.C.: Historical Office, Office of the Secretary of Defense, 1984.

RG Record Group

RHG Richard Hulbert Groves (son)

RHG, *Ancestors.* Groves, Richard Hulbert. *Our American Ancestors: Richard Hulbert Groves and His Direct-Line Antecedents.* Unpublished manuscript, n.d.

RHG, *Cuba.* Groves, Richard Hulbert. *Off to Cuba with the Regulars: A Narrative of Events Surrounding the US Army's Operations in the Santiago Campaign (June–August 1898.)* Unpublished manuscript, n.d.

RHG, *DNO.* Groves, Richard Hulbert, *DNO Chronology.* Unpublished manuscript, n.d.

RHG, *War Stories.* Groves, Lt. Gen. Richard H. *War Stories, Jump Stories and Other Lies and Half-Truths: A Melange of Recollections, Stray Thoughts and Artifacts, Arranged More or Less Chronologically.* Vol. 1, part A (1923–1936), part B (1936–1945), unpublished manuscript, n.d.

Rhodes, *Making.* Rhodes, Richard. *The Making of the Atomic Bomb.* New York: Simon & Schuster, 1986.

Robertson, *Sly and Able.* Robertson, David. *Sly and Able: A Political Biography of James F. Byrnes.* New York: W. W. Norton and Co., 1994.

Sanger, *Working.* Sanger, S. L. *Working on the Bomb: An Oral History of WWII Hanford.* Portland, Ore.: Continuing Education Press, Portland State University, 1995.

Schaffer, *Wings.* Schaffer, Ronald. *Wings of Judgment: American Bombing in World War II.* New York: Oxford University Press, 1985.

Serber, *Peace.* Serber, Robert, with Robert P. Crease. *Peace & War: Reminiscences of a Life on the Frontiers of Science.* New York: Columbia University Press, 1998.

Sherwin, *Destroyed.* Sherwin, Martin J. *A World Destroyed: The Atomic Bomb and the Grand Alliance.* New York: Alfred A. Knopf, 1975.

Sigal, *Fighting.* Sigal, Leon V. *Fighting to a Finish: The Politics of War Termination in the United States and Japan, 1945.* Ithaca, N.Y.: Cornell University Press, 1988.

Smith, *Peril.* Smith, Alice Kimball. *A Peril and a Hope: The Scientists' Movement in America, 1945–47.* Cambridge, Mass.: MIT Press, 1965.

Smith, Weiner, *Oppenheimer.* Smith, Alice Kimball, and Charles Weiner, eds. *Robert Oppenheimer: Letters and Recollections.* Cambridge, Mass.: Harvard University Press, 1980.

Smyth Report. Smyth, Henry DeWolf. *Atomic Energy for Military Purposes.* Stanford, Calif.: Stanford University Press, 1989; first published in 1945.

Spence, *China.* Spence, Jonathan D. *The Search for Modern China.* New York: W. W. Norton and Co., 1990.

Stewart, *Organizing.* Stewart, Irvin. *Organizing Scientific Research for War: The Administrative History of the Office of Scientific Research and Development.* Boston: Little, Brown and Co., 1948.

Stimson, Diary. Henry L. Stimson Diary, Sterling Library, Manuscript and Archives, Yale University.

Stimson, Bundy, *Service.* Stimson, Henry L., and McGeorge Bundy. *On Active Service in Peace and War.* New York: Harper & Brothers, 1947.

Stover, *Handymen*. Stover, Earl F. *Up from Handymen: The United States Army Chaplaincy, 1865–1920,* Vol. 3. Washington, D.C.: Office of the Chief of Chaplains, Department of the Army, 1977.

Szasz, *British*. Szasz, Ferenc Morton. *British Scientists and the Manhattan Project: The Los Alamos Years.* New York: St. Martin's Press, 1992.

Szasz, *Day*. Szasz, Ferenc Morton. *The Day the Sun Rose Twice: The Story of the Trinity Site Nuclear Explosion July 16, 1945.* Albuquerque: University of New Mexico Press, 1984.

Thayer, *Management*. Thayer, Harry. *Management of the Hanford Engineer Works in World War II.* New York: ASCE Press, 1996.

Top Secret. Correspondence from Record Group 77 of special interest to LRG. Reproduced as a NARA microfilm, five rolls (M1109).

Tyler, *Leavenworth*. Tyler, Orville Z., Jr. *The History of Fort Leavenworth, 1937–1951.* Fort Leavenworth, Kans.: Command and General Staff College, 1951.

USACE, *History*. U.S. Army Corps of Engineers. *The History of the US Army Corps of Engineers,* 2nd ed. Alexandria, Va.: Office of History, Headquarters, U.S. Army Corps of Engineers, 1998.

USAEC, *Oppenheimer*. U.S. Atomic Energy Commission. *In the Matter of J. Robert Oppenheimer: Transcript of Hearing Before the Personnel Security Board, Washington, D.C., April 12, 1954, through May 6, 1954.* Washington, D.C.: U.S. Government Printing Office, 1954; Cambridge, Mass.: MIT Press, 1971.

Washburn, "Attempt." Washburn, Patrick S. "The Office of Censorship's Attempt to Control Press Coverage of the Atomic Bomb During World War II," *Journalism Monographs* 120 (April 1990).

Wyden, *Day One*. Wyden, Peter. *Day One: Before Hiroshima and After.* New York: Simon & Schuster, 1984.

Zachary, *Frontier*. Zachary, G. Pascal. *Endless Frontier: Vannevar Bush, Engineer of the American Century*. New York: The Free Press, 1997.

NOTES

PREFACE

1. The atomic bombing of Japan was number 1 in a list of the 100 top news stories of the 20th century, according to a ranking by 67 prominent journalists and scholars, sponsored by the Freedom Forum's Newseum. The atomic bomb test of July 1945 was number 48, and the Manhattan Project's work on the bomb was number 64. Associated Press, February 24, 1999. In a wider survey conducted by the Newseum and *USA Weekend* 36,151 Americans were asked to choose the top 10 news stories from the list of 100, chosen by the journalists in the earlier survey. Among men the dropping of the atomic bomb ranked number one, and for women it tied for fourth. Eric Newton, "Story of the Century," *USA Weekend,* December 24–26, 1999, 6–10. New York University's journalism department sponsored another 100-best-of-the-century list. The 37 judges found John Hersey's *Hiroshima,* first published in *The New Yorker* in August 1946, to be the best work of 20th-century journalism. Felicity Barringer, "Journalism's Greatest Hits: Two Lists of a Century's Top Stories," *New York Times,* March 1, 1999, B1. There is one inadequate biography. William Lawren, *The General and the Bomb: A Biography of General Leslie R. Groves, Director of the Manhattan Project* (New York: Dodd Mead and Co., 1988). Lawren focuses on the war years, providing no background on Groves's earlier life and little after the war. The book is riddled with errors and lacks any critical analysis.

2. The Manhattan Project is often depicted as one of the transforming events in the creation of what has become known as Big Science, the heavily funded, interlocking government-science relationship that characterized Cold War research and development as well as serving as a precursor to the "military-industrial complex" that emerged after World War II.

3. John C. Fredriksen, "Groves, Leslie Richard, Jr.," *American National Biography* (New York: Oxford University Press, 1999), vol. 9, 673–674. There are also problems in David M. Kennedy, *Freedom from Fear: The American People in Depression and War, 1929–1945* (New York: Oxford University Press, 1999), 657–665.

4. The author carries over several of these mistakes, and adds others, in a new work. John C. Fredriksen, *American Military Leaders: From Colonial Times to the Present* (Santa Barbara, Calif.: ABC-CLIO, 1999), vol. 1, 306–308. Other authors describe him as: the "shadowy director of the MED" (Hales, *Spaces*); "a testy Army general," "hard-working but difficult," "efficient and demanding," "corpulent in double starched uniforms that constrained his bulk" (James W. Kunetka, *City of Fire: Los Alamos and the Birth of the Atomic Age, 1943–1945,* Englewood Cliffs, N.J.: Prentice Hall, 1978, 4, 5); "Rough, simple men like Leslie Groves" (Robert Jungk, *Brighter Than a Thousand Suns,* New York: Harcourt, Brace and Co., 1958, 143); "The chunky, often gruff general" (Frank Carey, AP Science Writer, "Leading A-Bomb Figure Is Dead," various newspapers, July 1970). Remarkably, the cover of the Da Capo paperback edition of *Now It Can Be Told* lists the author as General Leslie M. Groves.

5. Wesley G. Jones, "A Conversation with an Engineer Legend: Retired Major General Claude Henry Chorpening — Class of November 1918," *Assembly,* September 1992, 20.

INTRODUCTION At the Top of His Game

1. The classic description of spring in Washington is by Louis J. Halle. Conveniently, he describes the spring of 1945. *Spring in Washington* (New York: Atheneum, 1963).

2. Letters by Marion Nash Groves, Owen's wife, suggest that Grace moved to 36th Street on September 8 after the lease on the Oliver Street house ran out. Dick left on August 9 to go to Nicaragua and did not return until November 6. Letter, MNG to OCG, August 31, 1939, LC; Letter, MNG, September 2, 1939, LC.

3. The description of the office, and some other details in this introduction, are taken from a 17-page paper written by Gwen Groves Robinson about 15 years after the events she describes. It is contained in Papers of LRG, RG 200, NARA, a valuable compilation of reminiscences, photos, and documents that General Groves intended for his grandchildren. A bound copy, titled *For My Grandchildren,* is in the National Archives. Today the building is part of the State Department, and the offices on the fifth floor are occupied by the Arms Control and Disarmament Agency. Through the years major alterations have taken place; the original offices no longer exist.

4. In an April 23, 1945, memo to Secretary Stimson Groves estimates that by June 30, 1945, the cost will be $1.95 billion, 21. It is notable that he says, "If the war ends before June 1946, appreciable savings will result. Otherwise the total to be expended by 30 June 1946 will be 2842 millions of which 1660 millions will be for construction and 1182 millions for operations." LRG, Memo for the Secretary of War, Atomic Fission Bombs, April 23, 1945, Papers of LRG, RG 77, NARA.

5. LRG, *Now,* 28. In September 1942 the office consisted of two rooms, 5120 and 5121. A year later it had grown to seven rooms, and by the summer of 1945 a few more were added for the public information section.

6. April 26, 1945, telephone call, 10:28 A.M. LRG Appointment Book.

7. Army lost to the Hopkins Club, 10–7. Richard Groves scored two goals. "Hopkins Beats Army Stickmen," *Washington Post,* April 29, 1945, 6. On June 30 he was selected first-team All-American for the 1945 season.

8. Groves's Efficiency Reports and other official records have been collated by his son Richard and bound into a 725-page volume, cited throughout as LRG *Official Records.*

9. Private correspondence with author.

10. LRG, *Now,* 231.

11. James Les Rowe, *Project W-47* (Livermore, Calif.: JAARO Publishing, 1978); Harlow W. Russ, *Project Alberta: The Preparation of Atomic Bombs for Use in World War II* (Los Alamos, N.M.: Exceptional Books, 1990), 18.

12. Kenneth P. Werrell, *Blankets of Fire: U.S. Bombers Over Japan During World War II* (Washington, D.C.: Smithsonian Institution Press, 1996), 159–163. About one-third of the more than 300 bombers used in the attack flew from Tinian. The rest used Saipan and Guam.

13. Szasz, *Day.*

14. Groves, *Now,* 267. Notes on Initial Meeting of Target Committee, LANL Archives, A-84-019, 11–10.

15. Szasz, *British.* Szasz primarily focused on the prominent group at Los Alamos. A list titled "British Mission," apparently compiled in 1953, has 93 names (including a few French and Canadians), along with arrival and departure dates. Many worked at more than one place, such as Berkeley and Y-12 or New York and K-25. Some worked a few months, others a few years. Folder 7, Box 1, Entry 7530T, Papers of LRG, RG 200, NARA.

16. Szasz, *British,* 40; Brown, *Neutron.*

17. Hoddeson, et al., *Critical,* 360–362.

18. The National Academy of Sciences panel of November 1941 estimated a total cost of $133 million.

19. Groueff, *Manhattan,* 170–178.

20. Paper by GGR, *For My Grandchildren,* Papers of LRG, RG 200, NARA.

21. Paper by GGR. Within the family General Groves was called DNO (originally spelled — and pronounced — *Deno,* and later shortened to *DNO*) by his wife and children. The derivation of the name will be examined in chapter 7.

CHAPTER I Family Heritage: The Groveses in America

1. The date is not absolutely certain but generally accepted, and was so entered in his official records. I will distinguish between Groves and his father by adopting the family custom of referring to Leslie Richard Groves Jr. as Dick. Information on the Groves and Griffith families comes from an unpublished genealogy compiled by Richard Hulbert Groves, Dick's son, that he has kindly shared with me; I refer to it as RHG, *Ancestors.* See also William T. Groves, *History and Genealogy of the Groves Family in America: Descendants of Nicholas La Groves of Beverly Mass* (Ann Arbor, 1915). As cited in the preface, there is an unpublished manuscript written by Dick late in life, and edited by his son, titled Papers of Lieutenant General Leslie R. Groves. A copy is among his papers at the National Archives, RG 200, Entry 7530N. The original purpose of the exercise was to write down his memories for the benefit of his grandchildren. I refer to it as *Grandchildren.* The drafts have material that was not included in the final version. It covers Dick's life through 1931 and has several important appendices.

2. General Groves found 26 by the name of Le Gros in the Jersey telephone book during a three-day visit there in the summer of 1965.

3. The surname was spelled variously as *Legros, Le Grove, Lagro, Lagroe, La Grove, Grove, La Groves,* with the eventual transformation into *Groves.*

4. The DAR and SAR were not organized until the 1890s. It appears that Groves family eligibility was not established until 1969, the year before Dick's death.

5. Groves family legend has it that as a young girl Jemima saw George Washington. Richard H. Groves, through diligent research, has tracked this down and concluded that it is probably true. In the fall of 1789 President Washington toured New England, in the fashion of an old Roman Triumph. On October 23 his itinerary took him through Palmer and Brookfield to Spencer, which meant he passed close by Brimfield. Throughout his tour the local populace lined the route; periodically President Washington would stop and greet his fellow citizens. Thus it is entirely possible that Jemima, then 15, saw, and perhaps met, the new president on that day.

6. The eldest son was Eugene Allen (b. June 14, 1852), followed by Adelbert Dexter (July 18, 1853–April 17, 1896), Lavello Jason (b. October 19, 1854), Leslie Richard, and daughter Leonie Nancy (January 21, 1859–March 10, 1894).

7. *Hamilton Literary Magazine,* November 1900, 123.

8. RHG, *Ancestors,* 273–277; LRG, *Grandchildren,* 7–8.

9. Letter, Owen Griffith to William M. Owen, March 8, 1856, LRG, *Grandchildren,* 187.

10. Letter, Owen Griffith to William M. Owen, March 19, 1863, LRG, *Grandchildren,* 189–190.

11. Letter, Owen Griffith to "Brother & Sister," March 20, 1963, LRG, *Grandchildren,* 192.

12. The phrase used was, "Constant disaffection among the Officers of the Regiment." LRG, *Grandchildren,* 7.

13. *Utica Observer Dispatch,* June 27, 1892.

14. Whitestown Seminary was founded in 1828 as the Oneida Institute of Science and Industry.

15. In 1793 the Rev. Samuel Kirkland founded the Hamilton Oneida Academy. Originally an Indian school, the academy was chartered as Hamilton College in 1812.

16. There are strong ties between Hamilton and the Groves family. A relative, William Groves, was one of the first two graduates of the college in 1814. Leslie Sr. was in the class of 1881; his two sons Allen (class of 1913) and Owen (class of 1916) attended; and Dick would receive an honorary doctor of laws degree on June 5, 1955. The Griffith side of the family had a strong affiliation as well. Gwen Griffith's brother William Morton Griffith was in the class of 1880 — a year before his future brother-in-law — and his adopted son W. M. Griffith Jr. was in the class of 1921. See *Hamilton College and Her Family Lines,* compiled by William de Loss Love III (1963), 178, 180. Hamilton was a male college and did not become coed until 1978, when it merged with its traditional "sister" and neighboring college, Kirkland.

17. The January 1881 issue of the *Hamilton Literary Monthly* has an essay by Groves titled "The Greek and Christian Theory of the State," 169–176.

18. In the annual circular for the fall 1881 term, which opened on August 29, he is listed as principal and teacher of classics, mathematics, and natural sciences. See also Andrew C. White, ed., *Decennial History of the Class of 1881 of Hamilton College: 1881–1891* (Ithaca, N.Y.: Andrus & Church, 1892), 12–13; Andrew C. White, ed., *Class of '81, Hamilton College: The Record of Forty Years, 1881–1921* (Clinton, N.Y.: Typescript, n.d.); LRG, *Grandchildren,* 4; *Hamilton Literary Monthly,* September 1881, 78.

19. He tried several cases before the Supreme Court of the State of New York. LRG, *Grandchildren,* 5.

20. LRG, *Grandchildren,* 5. Auburn Theological Seminary, in Auburn, New York, was founded in 1818. The last building left standing on the site of the seminary is the Willard Chapel.

21. Program of the Graduation Ceremony, Auburn Theological Seminary, May 9, 1889. Leslie, as one of 22 in the class, spoke on "Cyrus, the Shepherd of the Lord."

22. *Hamilton Literary Monthly,* May 1890, 358.

23. LRG, *Grandchildren,* 183. In a letter to Gwen from Manila on their 10th wedding anniversary he recalled that day and repledged his love. "If I can believe

your pictures you are even more beautiful than on that glad day, ten years ago. . . . It does not seem long ago. But we thought ten years ahead a long time to look forward to. And we did not think that, in ten years, we would be on opposite sides of the globe, and able to endure the separation with fortitude. And, yet, we can endure it. You are more to me now than then, but we have learned to do our duty more faithfully and, I assure you, you have the right to be nearer proud of me than ever before." Letter, Chaplain Groves to GGG, April 29, 1900, Chaplain Groves, *Papers,* vol. 1, 458.

24. LRG, *Grandchildren,* 197.

25. Daniel Scott Lamont was born on February 9, 1851, on his family's farm in McGrawville, New York. He held various positions in the New York State government and became private and military secretary to Gov. Grover Cleveland in 1883. During President Cleveland's first term he was his private secretary, after which he became involved in various business ventures. During Cleveland's second term he served as secretary of war for the entire four years. He was elected vice president of the Northern Pacific Railway Company in 1898, serving until 1904. He died on July 23, 1905, at his Dutchess County home in Millbrook, New York.

26. LRG, *Grandchildren,* 201.

27. LRG, "The Army as I Saw It," *Grandchildren,* 197.

28. *Hamilton Literary Magazine* 33 (December 1898), 177.

29. LRG, *Grandchildren,* 201.

30. Daniel S. Lamont to Chaplain Groves, postmarked October 20, 1896, author's collection.

31. Later it became the Third Presbyterian Church. By the 1960s the Booker T. Washington Community Center stood on the site of the manse. Howard D. Lewis, "Groves Calls Defense Agency Unwieldy," *Knickerbocker News* (Albany, N.Y.), February 7, 1963.

CHAPTER 2 Growing Up in the Army (1897–1913)

1. At the time the government did not pay transportation costs for the families of military personnel. Mr. Lamont, who had close connections with the Northern Pacific Railroad and would later be its vice president, may have arranged for railroad passes for Gwen and the children.

2. LRG, *Grandchildren,* 201. Originally Fort Vancouver, the log stockade on the north bank of the Columbia River was an important Hudson's Bay Company trading post. It was taken over by the army as Columbia Barracks in 1849.

3. The best overall history is David F. Trask, *The War with Spain in 1898* (New York: Macmillan Publishing Co., 1981). For the military aspects see Cosmas, *Empire.* For the diplomatic aspects see John L. Offner, *An Unwanted War: The Diplomacy of the United States and Spain Over Cuba, 1895–1898* (Chapel Hill: University of North Carolina Press, 1992).

4. A five-member naval court of inquiry was immediately appointed, and its report was made public on March 28. It concluded that the *Maine* had been sunk by a submerged mine. In 1976 Admiral Hyman G. Rickover published *How the Battleship Maine Was Destroyed* (Washington, D.C.: Naval History Division, Department of the Navy, 1976). His conclusion was that there had been spontaneous combustion of coal in the bunker adjacent to the reserve magazine, thus causing an accidental, internal explosion rather than a purposeful, external one.

The fullest treatment of the ongoing controversy is Peggy and Harold Samuels, *Remembering the Maine* (Washington, D.C.: Smithsonian Institution Press, 1995). A recent investigation by the *National Geographic* does not resolve the issue. Thomas B. Allen, "Remember the Maine?" *National Geographic,* February 1998, 92–111.

5. The authorized strength of the Regular Army was 2,143 officers and 26,040 enlisted men. Its principal units were 25 regiments of infantry and 10 of cavalry — all at minimum strength. By the end of the war U.S. military forces numbered almost 275,000 officers and men.

6. Dick's son Richard H. Groves (RHG) composed an extensive narrative of his two grandfathers' experiences taken from letters and memoirs. I refer to it as RHG, *Cuba.*

7. Letter, Chaplain Groves to GGG, June 4, 1898, Chaplain Groves, *Papers,* December 1988, vol. 1, 108. He received $128 for travel expenses for the eight-day trip and was pleased with the amount. Chaplain Groves sent part of every payment back to Gwen.

8. Letter, Chaplain Groves to GGG, May 28, 1898, Chaplain Groves, *Papers,* vol. 1, 92.

9. Letter, Chaplain Groves to GGG, May 30, 1898, Chaplain Groves, *Papers,* vol. 1, 95.

10. Letter, Chaplain Groves to GGG, May 31, 1898, Chaplain Groves, *Papers,* vol. 1, 98.

11. The chaplain may have volunteered but more likely was levied.

12. Unfortunately, Chaplain Groves lost most of the letters Gwen wrote to him during his time in Cuba. "Did I tell you I lost my photographs and all papers and records at the last." Letter, Chaplain Groves to GGG, July 30, 1898. RHG, *Cuba,* 241.

13. The two volumes of the *Papers of Chaplain Leslie Richard Groves* total 978 pages of letters written by Chaplain Groves from May 1898 to August 1901.

14. Most of the surviving letters from Chaplain Groves to his children are to Owen. They are part of a collection of letters that Owen Groves retained and which was passed on to his daughter Charlotte Groves Larmee; I refer to this collection as LC.

15. Letter, Chaplain Groves to OGG, June 5, 1898, LC.

16. Interview with RHG by Robert S. Norris, June 1, 1999.

17. Letter, Chaplain Groves to GGG, May 23, 1898, Chaplain Groves, *Papers,* vol. 1, 76.

18. "If the Lord's work can prosper, if we can lead some to Christ and help others, is it not worthwhile? We may think it hard, and it may be, but if it is for the glory of God, then we can endure it." Letter, Chaplain Groves to GGG, June 2, 1898, Chaplain Groves, *Papers,* vol. 1, 105.

19. His pay remained the same at the end of 1900, plus a 10 percent supplement for foreign service. He hoped that an army bill before Congress at the time would include a pay raise for chaplains. Letter, Chaplain Groves to presumably GGG, December 23, 1900, Chaplain Groves, *Papers,* vol. 2, 769. The bill passed and, as of the beginning of 1901, his pay was $150 a month. Letter, Chaplain Groves to GGG, March 31, 1901, Chaplain Groves, *Papers,* vol. 2, 866. Army officers had to pay for their meals, though they did get a ration allowance, and also an allowance for one or two horses, depending on rank and branch. By February 1913 his pay had risen to $332.90 a month, a sum that made the chaplain feel, according to Gwen "so pessimistic — over his financial affairs," though she believed "we are not so poverty stricken." Letter, GGG to OGG, February 9, 1913, LC.

20. RHG, *Cuba,* 291.

21. "For the first time since the Civil War . . . chaplains participated in combat operations." Stover, *Handymen*, 111. Most people in the army do not consider chaplains to be soldiers.

22. Letter, Chaplain Groves to GGG, June 13, 1898, Chaplain Groves, *Papers*, vol. 1, 150.

23. RHG, *Cuba*, 26; letter, Chaplain Groves to GGG, May 30, 1898, Chaplain Groves, *Papers*, vol. 1, 95.

24. "Tampa . . . presented throughout May and June an appalling spectacle of congestion and confusion in which the Army's haste and mismanagement made almost insurmountable the port's already formidable obstacles to rapid, orderly embarkation." Cosmas, *Empire*, 195. Captain Wilson was also aboard the *Seneca*.

25. Letter, Chaplain Groves to GGG, June 9, 1898, Chaplain Groves, *Papers*, vol. 1, 135. "I am just an insignificant little speck in this great ocean of humanity." Letter, Chaplain Groves to GGG, June 11 1898, Chaplain Groves, *Papers*, vol. 1, 144. "On this boat that seems so large, with the wickedness and God evidently forgotten, I seem like a speck only. My influence is nothing at all. Only two or a few have come within my range at all. It seems to me I do nothing." Letter, Chaplain Groves to GGG, June 18, 1898, Chaplain Groves, *Papers*, vol. 1, 161.

26. Letter, Chaplain Groves to GGG, June 13, 1898, Chaplain Groves, *Papers*, vol. 1, 152.

27. The 8th Regiment was part of the 5th Army Corps, under the command of Maj. Gen. William R. Shafter. Fifth Corps was made up of most of the Regular infantry, cavalry, and field artillery, the 1st U.S. Volunteer Cavalry (Rough Riders), and eight of the best-trained and best-equipped volunteer infantry regiments. Cosmas, *Empire*, 194. At almost 17,000 strong, the major components of 5th Corps were 1st Division (under Maj. Gen. Jacob F. Kent), 2nd Division (under Brig. Gen. Henry W. Lawton), and Cavalry Division (under Maj. Gen. Joseph W. Wheeler).

28. Cosmas, *Empire*, 205–230.

29. Captain Wilson began a narrative in the 1930s based on entries in his diary. RHG, *Cuba*, 125.

30. "Campaigning a La Hobo" (reflections, probably written around 1932), Chaplain Groves, *Papers*, vol. 1, 224.

31. Letter, Chaplain Groves to GGG, July 3, 1898, Chaplain Groves, *Papers*, vol. 1, 184.

32. "Campaigning a La Hobo" (reflections, probably written around 1932), Chaplain   Groves, *Papers*, vol. 1, 225. Also in RHG, *Cuba*, 130.

33. "Campaigning a La Hobo" (reflections, probably written around 1932), Chaplain Groves, *Papers*, vol. 1, 226. See also Stover, *Handymen*, 115–116.

34. On board the *Concho* the chaplain "was one of the most afflicted men on the entire ship of fever-stricken souls. With disease withering him and death looking him in the face, he would stagger out on deck and go down into the hold to perform the last rites and do the military honors above dead heroes ere we committed them to the sea." The Rev. Peter MacQueen, M.A., "Chaplains in the Army of Invasion and Occupation," *Congregationalist*, October 13, 1898, 481.

35. Many soldiers recovered at Camp Wikoff, at Montauk, Long Island (named for Charles A. Wikoff, an officer killed at San Juan Hill and buried by Chaplain Groves). The camp was hurriedly arranged in late July to receive the V Corps troops returning from Cuba. Troop shipments mainly took place throughout

August; by the end of October some 21,000 had passed through. A 500-bed hospital was built to care for the huge numbers of soldiers who had caught various diseases in Cuba. Overall around 2,500 officers and men perished of sickness during the short war, 10 times the number killed in action. Cosmas, *Empire,* 259–260, 275. Stover, *Handymen,* 117. Captain Wilson sailed from Cuba on August 13, disembarking at Montauk Point on the 20th. After a stay in Cheyenne he returned to Cuba at the end of the year for occupation duty until November 1899. While convalescing at his mother's home Chaplain Groves provided an account of his experiences. "The Yellow Fever Cure," *Utica Daily Press,* August 29, 1898.

36. RHG, *Cuba,* 46.

37. LRG, *Grandchildren,* 12.

38. LRG, *Grandchildren,* 213–226.

39. Capt. Sherman L. Fleek, "The Army's VIII Corps and the 'Splendid Little War,'" *Army,* November 1989, 52–58.

40. Brian McAllister Linn, *The U.S. Army and Counterinsurgency in the Philippine War, 1899–1902* (Chapel Hill: University of North Carolina Press, 1989).

41. Letter, Chaplain Groves to GGG, December 14, 1899, Chaplain Groves, *Papers,* vol. 1, 272.

42. Letter, Chaplain Groves to Jane, February 18, 1900, Chaplain Groves, *Papers,* vol. 1, 372. Six companies were at headquarters, Cuartel de Meisic, and the other six were located from half a mile to two miles away. Occasionally Chaplain Groves would venture farther into the countryside or to nearby islands.

43. Brig. Gen. A. S. Daggett, USA, Retired, to the Military Secretary, November 24, 1906. Folder, LRG, Commendations and Qualifications, Box 3, Entry 7530C, Papers of LRG, RG 200, NARA.

44. His schedule for Sunday, April 15, 1900: "Bible class 8:30 to 9:30. Service at San Fernando Police Station at 10. 3d Hospital at 2:30. Guardhouse at 5. Regimental service at 7:30." Letter, Chaplain Groves to GGG, April 15, 1900, Chaplain Groves, *Papers,* vol. 1, 436.

45. The war intruded into his letters from time to time. "There are attempts all around to puncture sentinels with bolos [machetes] and to burn property, but the attempts have been promptly met and the Insurrectos will soon get tired of the sure death, or capture, that seems to await them. Our regiment is as superior as ever in its work, and it is very unfortunate for the fellow that runs against them." Letter, Chaplain Groves to GGG, May 16, 1900, Chaplain Groves, *Papers,* vol. 2, 481. Combat casualties totaled: 777 killed in action; 227 mortally wounded; and 2,911 wounded but not mortally. In addition 589 died from accidents and 2,572 perished from diseases. Stover, *Handymen,* 136, footnote 8. The number of Filipinos who died in the war is a subject of some controversy and uncertainty. The most careful estimate is John M. Gates, "War-Related Deaths in the Philippines, 1898–1902," *Pacific Historical Review* 53, no. 3 (August 1984), 367–378. He concluded that direct war-related deaths, excluding those caused as a result of the cholera epidemic of 1902, are between 34,000 and 97,000.

46. "It would be a comfort to me some days to know if I am doing any good with all the effort. I cannot draw crowds anywhere, but I have the same old way of keeping at it and trying. I wish I could be more attractive to the crowd — I

wonder why I am not. I have the notion that I am rather an amiable chap, on the whole, for some people like me. But then, I am what I am and I guess I will keep on trying." Letter, Chaplain Groves to GGG, March 7, 1900, Chaplain Groves, *Papers,* vol. 1, 391.

47. "Prompt obedience only results from always expecting it and seeing that it is obtained. It takes time and care to break a colt and keep him obedient. So it does with boys." Letter, Chaplain Groves to GGG, June 14, 1900, Chaplain Groves, *Papers,* vol. 2, 506. In an apparent response to Gwen's question about sending the boys to dancing school, he had this to say: "As to the dancing business, you may say that I do not wish the boys to attend and, if your friends are impertinent enough to cross-examine you, you may add that we have no ambition but that they grow up men. If you want to convince your own mind, just count up the adult males of your acquaintance who attended dancing schools as children, and those who did not, and compare at your leisure. No arguments are needed. I would prefer boys trained in a mining camp." Letter, Chaplain Groves to GGG, August 21, 1900, Chaplain Groves, *Papers,* vol. 2, 614.

48. "I do not know yet whether any of the bills providing for the chaplains for all regiments have passed. One's indignation at the politicians increases mightily when there is such a need and they are trimming for political advantage and spoils, and give no heed. . . . Politicians, rather than earnest men. The world is very evil." Letter, Chaplain Groves to Jane, June 13, 1900, Chaplain Groves, *Papers,* vol. 2, 509.

49. Letter, Chaplain Groves to GGG, June 24, 1900, Chaplain Groves, *Papers,* vol. 2, 521.

50. Letter, Chaplain Groves to Jane, et al., July 14, 1900, Chaplain Groves, *Papers,* vol. 2, 549.

51. "Perhaps 70 percent were poor peasants, male and young. The rest were drawn from a broad mixture of itinerants and artisans: peddlers and rickshaw men, sedan-chair carriers, canal boatmen, leather workers, knife sharpeners, and barbers, some dismissed soldiers and salt smugglers. They were joined by female Boxer groups, the most important of which was named the Red Lanterns Shining, girls and women usually aged twelve to eighteen whose female powers were invoked to fight the 'pollution' of the Chinese Christian women, which was believed to erode the strength of Boxer men." Spence, *China,* 232–233. The literature on the Boxer Rebellion is extensive. Diana Preston, *The Boxer Rebellion: The Dramatic Story of China's War on Foreigners That Shook the World in the Summer of 1900* (New York: Walker & Company, 2000) and Cohen, *Keys,* are good places to start. To conform to the vast contemporary literature at the time, the Wade-Giles Romanization will be retained as much as possible.

52. Cameron, *Barbarians,* 371–399.

53. At the Empress Dowager Tzu Hsi's court anti-Boxer and pro-Boxer factions influenced her actions at different periods during the uprising.

54. Nat Brandt, *Massacre in Shansi* (Syracuse, N.Y.: Syracuse University Press, 1994). In the four-month uprising at least 32,000 Chinese Christians were slain, along with more than 185 Protestant missionaries and members of their families, and 47 Roman Catholic clerics and nuns.

55. Some 4,000 people were inside a limited area, of whom about 2,700 were Chinese Christians and 400 servants. Besides the 473 civilians — diplomats, their families, and their staffs — there was a garrison of 400 men. Cameron,

Barbarians, 378. The imperial army troops never pressed as hard as they could have. If they had, all of those inside might have been killed.

56. Spence, *China,* 234. There were also attacks on mission compounds and foreigners outside Peking. They were particularly vicious in Shansi, Chilli, and Henan. The worst atrocity occurred in Shansi. There the Manchu governor summoned the missionaries and their families to the provincial capital of Taiyuan, promising protection from the Boxers. Once they arrived he ordered all forty-four men, women, and children killed. *Ibid.* See also Cameron, *Barbarians,* 378–379.

57. The War Department also ordered 9th Infantry Regiment and Battery F, 5th Artillery, to China from the Philippines. The regiment set sail on June 27 and arrived off Taku on July 6. Maj. Gen. Adna R. Chaffee was notified in Cuba that he was to command American forces. He proceeded to San Francisco and sailed with the 6th Cavalry Regiment, arriving on July 30.

58. The battle of Tientsin began with the Chinese shelling foreign settlements on June 17. The next week was one of extreme danger for the 900 foreign civilians, of whom a 25-year-old American mining engineer named Herbert Hoover was one. The siege lasted until July 13, when foreign relief contingents — numbering between 5,000 and 6,000 — stormed the walled city and were victorious. Chinese casualties were estimated at more than 5,000, with 750 to 775 on the foreign side. The following days saw unrestrained killing, looting, and raping. Cohen, *Keys,* 53. This was the scene that Chaplain Groves witnessed a few days later. "As we came to Tientsin, everything was wrecked. Houses all burned out. . . . Everything was riddled with bullets . . . it was a burying grounds, with dead Chinamen and homeless dogs all about. The dogs and hogs acting as scavengers and eating up the bodies." Letter, Chaplain Groves to GGG, July 28, 1900, Chaplain Groves, *Papers,* vol. 2, 578. The 9th Infantry arrived in China on July 6.

59. The American forces that set out totaled some 2,200 men, mainly from the 9th and 14th Infantry Regiments, plus a battalion of marines. Keown-Boyd, *Fists,* 158.

60. "No one seems to have hazarded a guess as to how many Boxers there were still upon the scene. As the story of their 'uprising' unfolds they recede further and further into the background and by the beginning of August they are seldom mentioned except as mere miscreants to be hunted down and killed on the spot or executed after a perfunctory trail." Keown-Boyd, *Fists,* 159.

61. Letter, Chaplain Groves to GGG, August 8, 1900, Chaplain Groves, *Papers,* vol. 2, 599.

62. Brig. Gen. A. S. Daggett, Retired, to the Military Secretary, November 24, 1906, Folder, Groves, L. R., Commendations and Qualifications, Box 3, Entry 7530C, Papers of LRG, RG 200, NARA. The United States was the only country that had a policy of evacuating the remains. Eventually the remains of 138 soldiers and civilians who died in China were transported back to the United States. Dead allies were either buried or burned. Chinese were left in the field to be eaten by dogs or pigs. Maj. William C. Harlow, *Logistical Support of the China Relief Expedition* (Fort Leavenworth, Kans.: U.S. Army Command and General Staff College, 1991), 158–159.

63. Calvin P. Titus, "A Day in the Life of a Company Bugler," *Assembly,* fall 1964, 2–5; Letter, Chaplain Groves to GGG, August 16, 1900, Chaplain Groves, *Papers,*

vol. 2, 602; Letter, C. P. Titus to LRG, August 15, 1964, in LRG, *Grandchildren,*
229–230. For his efforts in the China Relief Expedition, Titus was awarded one
of of only 4 army Congressional Medals of Honor (as were 22 in the navy and 33
marines) and was offered a "mustang" promotion. Instead of the promotion, his
request to attend West Point was granted and he graduated in the class of 1905.
He was turned down for the chaplain corps because he was not ordained and
was a lieutenant colonel of the infantry when he retired in 1930. In 1900
American units (and veterans from the other seven nations) that served in the
rescue mission organized the Military Order of the Dragon to commemorate
the expedition; its motto: "Not to conquer but to save." See *Military Order of the
Dragon: 1900–1911* (Washington, D.C.: Press of Byron S. Adams, 1912).
Composed of enlisted men, the group at first held regular reunions, but as their
numbers dwindled they widened criteria of membership to include officers,
and then, in the 1940s, the sons of members. In 1946 Dick and Owen were
inducted as Hereditary Mandarins. See the correspondence on this matter in
Folder 201, Box 3, Entry 7530C, Papers of LRG, RG 200, NARA.

64. Inside the compound 66 Westerners had been killed and 150 wounded. "How
many Chinese within the legation were killed or wounded seems to be a
matter no one thought to record." Cameron, *Barbarians,* 381. In the aftermath
the city was extensively sacked by the foreign troops. The formal peace treaty,
known as the Boxer Protocol, signed in September 1901, was extremely harsh
and completed the humiliation of China. One of its many terms was payment
of an indemnity in gold for damages to foreign life and property in the amount
of $333 million, a staggering sum. Spence, *China,* 235. The treatment would be
a turning point. "Rarely in history has a single year marked as dramatic a
watershed as did 1900 in China. The weakness laid bare by the Allied pillage of
Peking in the wake of the Boxer Rebellion finally forced on China a polar
choice: national extinction or wholesale transformation not only of a state but
of a civilization. Almost overnight Chinese — imperial government, reformers,
and revolutionaries — accepted the challenge." Mary Clabaugh Wright, ed.,
China in Revolution: The First Phase, 1900–1913 (New Haven, Conn.: Yale
University Press, 1968), 1.

65. Richard A. Steel, *Through Peking's Sewer Gate: Relief of the Boxer Siege, 1900–1901,*
edited and with an introduction by George W. Carrington (New York: Vantage
Press, 1985), 16–17, 19–20, 24–25. Steel was aide-de-camp to General Gaselee.
See also Keowyn-Boyd, *Fists,* 169, 180–182.

66. Michael H. Hunt, "The Forgotten Occupation: Peking, 1900–1901," *Pacific
Historical Review* 48, no. 4 (November 1979), 501–529.

67. "Someone, last evening, promised to loot candles enough for evening services.
Just think of stealing lights to hold services with! But then these things are
being abandoned by their owners, they fall to someone and might as well fall
to a good use." Letter, Chaplain Groves to GGG, August 22, 1900, Chaplain
Groves, *Papers,* vol. 2, 610. Before leaving Peking Chaplain Groves got a sable
coat for Gwen. "It is not loot, but was purchased from the British looters, at a
regular sale." Letter, Chaplain Groves to GGG, October 9, 1900, Chaplain
Groves, *Papers,* vol. 2, 667. Afterward, the purchase from the "British thieves"
nagged at his conscience. He also sent a soapstone Confucius to Allen, along
with some silk, buttons, and jade to the others. Later, in recounting the looting
by the British, he told his sister-in-law, "If there is anything fills me with

shame, it is my English blood, but I hope six generations in America have evolved something besides bristles on my back." Letter, Chaplain Groves to Jane, November 17, 1900, Chaplain Groves, *Papers,* vol. 2, 741. As we shall see, Anglophobia would be strongly developed in Dick as well.

68. Letter, Chaplain Groves to GGG, August 19, 1900, Chaplain Groves, *Papers,* vol. 2, 607. According to the U.S. Army Military History Institute, the maximum strength of the American China Relief Expedition was 4,636 in September 1900. The total casualties suffered between May 1900 and May 1901 was 102: 33 killed in action; 18 mortally wounded; 47 from disease; and 4 from accidents. Only two chaplains accompanied American forces to suppress the Boxer Rebellion, Groves and Walter Marvine of the 9th Infantry Regiment. Marvine remained at the Tientsin hospital from July 1900 to January 1901, and thus Chaplain Groves was the sole chaplain in Peking. Stover, *Handymen,* 105, 133–135.

69. "It was not quite up to expectations. But it was worth doing. . . . The filth of the city continues in the sacred city." Letter, Chaplain Groves to presumably GGG, August 28, 1900, Chaplain Groves, *Papers,* vol. 2, 620.

70. Letter, Chaplain Groves to GGG, October 16, 1900, Chaplain Groves, *Papers,* vol. 2, 677. Chaplain Groves was extremely proud of the exploits of the 14th and in the aftermath monitored the U.S. newspapers that reached Manila to ensure that it was given proper credit. Later he said of the Japanese, "Mrs. Allison will enjoy Japan and will doubtless rave a bit over the beauties of Japanese life, and will probably never know the moral rottenness of a large portion of the people there." Letter, Chaplain Groves to GGG, June 14, 1901, Chaplain Groves, *Papers,* vol. 2, 923.

71. Letter, Chaplain Groves to Jane, November 17, 1900, Chaplain Groves, *Papers,* vol. 2, 740–744. The 9th Infantry Regiment, a squadron of cavalry, and an artillery battery remained as a legation guard. In May 1901 Chaffee marched the remaining troops out of Peking and the China Relief Expedition came to an end.

72. Letter, Chaplain Groves to GGG, November 16, 1900, Chaplain Groves, *Papers,* vol. 2, p. 732. It did not come as a great surprise, because she had long been ailing.

73. "Tonight, news of a battle, south about 60 miles, or so, that makes things look different. All day, I have had a feeling that we were not home yet. The news is serious. A good many killed and wounded. Capt. Wilhelm is reported seriously wounded [he later died]. If they are going to kick up this way, the war will never end." Letter, Chaplain Groves to GGG, June 10, 1901, Chaplain Groves, *Papers,* vol. 2, 920–921.

74. "I think it might also be of interest to some of your psychologist friends that I didn't talk until I was over 4 years of age. Then I distinguished myself by saying my first word, 'cheese.'" Ermenc, *Scientists,* 205.

75. The chaplain had no objection to cards. He approved of anything that sharpened their mental skills, but never on Sunday.

76. "I can tell you I am glad my boys are such good boys when I am so far away from them. I feel sure they are trying to do right all the time. My bitty bitty boy must grow to be a brave boy, brave enough to tell the truth always and do the truth too. I know he is a kind little chap that everyone likes. I like him too and would like to hug him now as hard as a big bear would hug." Chaplain Groves to OGG, n.d. but written from Manila, LC.

77. Interview with RHG by Robert S. Norris, January 25, 1999.
78. Chaplain Groves to OGG, n.d., LC. The letter can be dated from the fact that a poem on the envelope states, in part, "There is a little boy as good as gold, / who today is seven years old." If taken literally, this dates the letter as having been written on May 11, 1900.
79. Letter, Chaplain Groves to OGG, May 23, 1900, LC.
80. Letter, Chaplain Groves to OGG, July 22, 1900, LC.
81. LRG, *Grandchildren*, 19–94.
82. Named for General Anthony Wayne, the fort's barracks are now a museum and the site of Wayne State University in Detroit.
83. Allen informed his father, "I have not missed an example in algebra, am at the head of the class in English and probably know more about Latin and Greek history than anybody else in the class." Letter, Allen M. Groves to Chaplain Groves, September 10, 1904.
84. LRG, *Grandchildren*, 29.
85. LRG, *Grandchildren*, 32–33.
86. In a comment more than 50 years later General Groves reflected on a brief stopover he made in the spring of 1906. "Although the earthquake had occurred in April, San Francisco was still in the aftermath of disaster, and the people were cooking in the streets and in their yards. By that time most of them had cook-stoves and there were very few campfires. Most of the rubble had been cleared away and people were busily engaged in recreating their city. Unlike what would have happened today, they did it themselves and did not come to Washington and cry on the steps of the Capitol for a Federal handout. Just how much Federal assistance they got I do not recall, but I doubt if there was any." In the aftermath of the disaster General Funston sent troops into the city with orders to shoot looters on sight. "This provided quite a contrast to the mawkish handling of such scoundrels during the recent riots in this country because of the cowardly influence of the do-gooders." LRG, *Grandchildren*, 44.
87. It was financed with a first mortgage of $4,000 from Gwen's sister-in-law, Aunt Jo, the wife of William M. Griffith. LRG, *Grandchildren*, 48.
88. In another coincidence Lt. Richard Hulbert Wilson, Dick's future father-in-law, was stationed there with the 8th Infantry, just out of West Point, in 1877.
89. Dick spent the summer of either 1909 or 1910 in Pasadena with Aunt Jane. Apparently, for a short time in 1908 Dick went to the one-room post school, the sole member of the sixth grade. He recalls an excellent teacher, a corporal with only a year and a half of high school, who "kept us all working like dogs." When he left Dick was sent back to Pasadena. LRG, *Grandchildren*, 65.
90. RHG, unpublished manuscript, 103.
91. *Hamilton Literary Magazine*, February 1916, 262–263.
92. AMG, *Appreciation*, 8–9. This small privately printed book was composed by his Delta Upsilon fraternity brothers, a year after his death, with editorial work by Homer W. Davis (class of 1916) and R. B. Warren, his roommate for three years. *Hamilton Literary Magazine* 52 (December 1917), 120. Chaplain Groves, Uncle William Griffith (class of 1880), and Owen were also members of Delta Upsilon.
93. The practice continued in college. "Far from being a grind, he was often careless in his methods of work, as he was in his dress. However, he had that very valuable

ability of being able to work under pressure. Even if he did keep his History notes on the backs of old envelopes, if he did study his Greek or Latin on the way to breakfast in the morning or while getting undressed, if he did have to sit up all night before each day that a prize essay was due, typing it, correcting it on the way down the Hill and getting it into prex's hands just before the stroke of twelve — even if he did all these things, his work almost always excelled all those with whom he competed or worked. Not scholarly methods you say! Perhaps not, but the results were scholarly. *AMG, Appreciation,* 14.

94. "Although he was on the college debating team both his junior and senior years, his real capacity in this line was not fully appreciated; for a certain self-consciousness hindered his public speaking, and on the platform, he never did himself full justice." *AMG, Appreciation,* 9.

95. Owen was author of 11 pieces in the *Hamilton Literary Magazine* between October 1914 and April 1916, on such topics as "Defoe's Mastery of Verisimilitude" and "A Tenderfoot Goes Hunting" (about experiences at Fort Apache, Arizona), as well as the poems "The Vision of Mirzah," "The Indian Dance," and "Arizona."

96. Letter, GGG to OGG, June 24, 1912, LC. Maurice E. McLoughlin, the "California Comet," was a famous tennis player of the day, known for his serve. He won the U.S. Open singles title twice (1912, 1913), and the doubles title three times (1912, 1913, 1914). He was inducted into the National Tennis Hall of Fame in 1957.

97. Letter, GGG to OGG, October 13, 1912, LC.

98. Letter, GGG to OGG, October 27, 1912, LC.

99. Letter, GGG to OGG, November 10, 1912, LC.

100. Letter, GGG to OGG, November 3, 1912, LC.

101. Letter, GGG to OGG, November 10, 1912, LC. The combined popular vote of Theodore Roosevelt's Bull Moose Party (27.8 percent) and William Howard Taft's Republican Party (23.5 percent) was greater than that gained by Woodrow Wilson (42.4 percent). In electoral votes, though, Wilson received 435 of 531.

102. Letter, GGG to OGG, November 10, 1912, LC.

103. Letter, GGG to OGG, Easter Sunday 1913, LC.

104. Letter, GGG to OGG, March 27, 1913, LC.

105. Letter, GGG to OGG, April 6, 1913, LC.

106. Gwen, Owen, little Gwen, and Allen — who was in his third year at Hamilton College — did not join Dick and Chaplain Groves until the summer.

107. Edmund Bristol Gregory (class of 1904) was quartermaster general during World War II. As we shall see, Dick's friendship with Gregory would play an extremely important role in his career.

108. RHG, *DNO,* 119.

109. He was baptized Chauncey Hulbert Wilson. An elder brother named Richard Hulbert Wilson died at three years old shortly after Chauncey's birth. In 1872 or 1873 Chauncey took the name of his elder brother, as was often done in those days. His mother's name was Helen Maria Hulbert.

110. Obituary, Richard Hulbert Wilson, Cullum No. 2666, *Sixty-eighth Annual Report of the Association of Graduates of the United States Military Academy at West Point, New York,* June 11, 1937, 86–90.

111. For six years he served at Fort Gaston, California, a one-company post in the

mountains of northern California on the west bank of the Trinity River near the Klamath River, in the Hoopa Valley Indian Reservation. Wagons could not reach it; all supplies had to be brought in by pack mule. Lieutenant Wilson used his time to cultivate his linguistic abilities in Greek, Latin, Spanish, French, German, and Italian, some of which he would later pass on to his daughter. *Sixty-eighth Annual Report of the Association of Graduates of the United States Military Academy at West Point, New York,* June 11, 1937, 87. The Indian tribes he knew included the Apache, Modoc, Oglala Sioux, Arapaho, and Shoshone.

112. In the late nineteenth century army officers were detailed to the Interior Department's Bureau of Indian Affairs to serve as agents to the various tribes under the bureau's control.

113. On December 30, 1878, the post was officially designated Fort Washakie in honor of the Shoshone chief. A picture from about 1898 of the chief — then more than ninety years old — holding baby Grace survives.

114. LRG, *Grandchildren,* 307–337. Colonel Wilson remained in command of the 14th Infantry until he retired in June 1917. He died in Seattle on March 21, 1937, and was survived by his wife, Grace Chaffin Wilson, his daughter Mrs. Grace Wilson Groves, and a second daughter, Miss Mary Helen Wilson. *The Association of Graduates, USMA, Register,* autumn 1937.

115. During the summer of 1911 a second battalion of the 14th came down from Fort Missoula, Montana, with a third battalion stationed at Fort Lincoln, North Dakota. LRG, *Grandchildren,* 86.

116. LRG, *Grandchildren,* 324. "The New Year's receptions were festive, really, and full of good cheer. At one of these New Year's festivals a guest was the newly arrived Chaplain who was an old-timer with the regiment, sharing their experiences in war and their traditions. He had been absent in Arizona for his health and was now rejoining his organization and he was accompanied by a towering, black-haired youth whom he introduced as "'my little boy, Dick.'" Ibid., 337.

117. RHG, *DNO,* 126.

118. Letter, GWG to RHG, ca. 1981.

119. In 1884 the army recommended that a fort be built in Seattle. The chamber of commerce bought and presented 700 acres on Magnolia Bluff to the army in 1898. During the summer of 1901 the fort was assigned its permanent troops. In World War II more than one million troops passed through on their way to the Pacific, and over 200 new buildings were built quickly to accommodate them. In 1964 the army declared most of Fort Lawton property surplus; in 1972 the surplus lands were transferred to the city of Seattle. In 1973 the site was named Discovery Park, at 534 acres the city's largest.

120. Letter, OGG to LRG, May 18, 1966, LC.

121. Letter, GGG to OGG, February 20, 1913, LC.

122. Letter, GGG to OGG, February 20, 1913, LC.

123. Letter, GGG to OGG, February 23, 1913, LC.

124. Letter, GGG to AMG, April 4, 1913, LC.

125. Letter, GGG to OGG, March 27, 1913, LC; GGG to AMG, April 4, 1913, LC.

126. Letter, GGG to OGG, March 2, 1913, LC.

127. Letter, GGG to OGG, March 19, 1913, LC.

128. Letter, GGG to AMG, March 30, 1913, LC.

129. Letter, GGG to AMG, March 30, 1913, LC.

130. Letter, GGG to OGG, March 27, 1913, LC. Two days later in a letter from Gwen to Owen, Dick wrote within the body of the letter, "I am going to *West Point. Me*" (emphasis in original). His mother continued, "Fresh kid. He needs the discipline that West Point might give him and I for one say let him go and get all that's coming to him." Letter, GGG to OGG, March 29, 1913, LC.

131. LRG, *Grandchildren,* 93, 94.

132. Washington State Board of Health, Bureau of Vital Statistics, Certificate of Death, No. 1565, Gwen Griffith Groves, signed Robert M. Hardaway, July 2, 1913.

133. Letter, Jane Griffith to OGG, July 4, 1913, LC.

134. Letter, Jane Griffith to OGG, July 4, 1913, LC.

135. RHG, *Ancestors,* 310.

CHAPTER 3 Dick Defines His Future (Summer 1913–June 1916)

1. Some years later Dick's wife, Grace Wilson Groves, whose father was Chaplain Groves's regimental commander, noted that her father had felt that Dick had been "treated unfairly by the Chaplain who devoted his attention too greatly to the older boys in college." "Dick," Colonel Wilson, opined, "deserved better treatment than he received."

2. LRG, *Grandchildren,* 90. According to Dick's records, his father loaned or gave him $104.65 during the year. Ibid., 251.

3. LRG, *Grandchildren,* 91.

4. LRG, *Grandchildren,* 91.

5. LRG, *Grandchildren,* 92.

6. English Language and Literature I and II (B); Library Economy (B); Mathematics I, II, III, IV (C); French I and II (B); Military Science I and II (B and C); Hygiene (E). A = 96–100; B = 86–95; C = 76–85; E = failed. RHG, *DNO,* 140–141.

7. Dick entered Queen Anne on January 24, 1913, to begin the second semester. When the family arrived in Seattle there was only one week left in the first semester, and school authorities informed Dick he would have to wait for the beginning of the second semester to enroll. This annoyed Dick and his father. LRG, *Grandchildren,* 89.

8. Leslie Richard Groves, Permanent Record, Queen Anne High School. English Composition and Rhetoric (G); Latin (G); Economics (E); History (G); Geometry (E); Physiography (now Physical Geography) (G); Penmanship (P). P = passing, 70–80 percent: G = good, 80–90 percent; E = excellent, 90–100 percent. Queen Anne was built in three phases, in 1905, 1915, and 1951, and was located at 201 Galer Avenue. It closed in 1981 and is now an apartment building.

9. *Official Register of the Officers and Cadets, United States Military Academy for 1916* (West Point, N.Y.: USMA, 1916), 65. There were two appointments each for the territories of Alaska and Hawaii, two for Puerto Rico, and four for the District of Columbia. There were also appointments for qualified enlisted men from the National Guard and the Regular Army, and one Filipino per class.

10. This was as a result of an act of Congress approved May 4, 1916. Prior to this it had been 1 cadet from each congressional district, 1 from each territory, 1 from the District of Columbia, and 10 at large to be appointed by the president. By 1916 there were 435 Congressional seats for the 48 states. According to the new act, the maximum number of cadets was 1,332.

11. This was a few days shy of his 17th birthday, the minimum age limit.

12. Barry was born in 1855 and was a graduate of the class of 1877. In 1893 Secretary of War Daniel Lamont selected Barry for duty in his office. It was Barry who wrote to the commanding general at Vancouver Barracks to arrange for the chaplain's "good home." Barry served in the Philippines and in China when Chaplain Groves was there. He was the 27th superintendent of the USMA, from August 1910 to August 1912. Obituary, *Fifty-first Annual Report of the Association of Graduates of the United States Military Academy at West Point, New York,* June 14, 1920, 108–111.

13. Groves family lore has it that he was prepared by Lt. Edmund B. Gregory, then with the 14th Infantry and later to be quartermaster general during World War II. Groves said later that he had stayed at the house of Captain Castner, an old 14th Infantry friend, at Vancouver Barracks when he took his first examination for West Point. LRG, *Grandchildren,* 113. He also saw Castner in Washington while he was cramming for the March 1916 exam, and frequently several years later while he was at Engineer School. Examinations were given on the third Tuesday of March, which was the 17th. See also *Personal and School History Sheets,* March 1916, vol. 2, USMA Archives, where he wrote that he took the exam in March 1914 at Vancouver, Washington.

14. The move to the new Cambridge campus occurred in the summer and fall of 1916. MIT, *Catalog, 1916–17,* December 1916, vol. 52, 45.

15. LRG, *Grandchildren,* 97.

16. LRG, *Grandchildren,* 98. Tuition at MIT was $250 a year.

17. LRG, *Grandchildren,* 251.

18. The Braves had borrowed the Red Sox park (open since the 1912 season), because their own field, South End Grounds in Roxbury, held only 10,000, and the new concrete-and-steel Braves Field — a mile away to the west — was still under construction and would not be ready until the next season.

19. LRG, *Grandchildren,* 99.

20. MIT, *Catalog, 1913–14,* December 1913, vol. 49, 38 ff. MIT did offer a five-year option for students who wanted to complete work in two allied fields, or those who wanted, along with their professional studies, increased work in the humanities and general science, or those who wanted to redistribute the work in their field over a five-year period. See Ibid., 44 ff.

21. MIT, *Catalog, 1913–14,* December 1913, vol. 49, 80.

22. LRG, *Grandchildren,* 100.

23. MIT, *Catalog, 1914–1915,* December 1914, vol. 50, 556, 568.

24. LRG, *Grandchildren,* 101–102. The 700-acre summer surveying camp operated from 1911 until 1953.

25. MIT, Office of the Registrar, Transcript of Leslie Richard Groves, Jr. RHG, DNO, 153.

26. Freehand Drawing (L—barely passed), Mechanical Drawing (C—passed with credit), Descriptive Geometry (P—passed), English Literature (P+), Applied Mechanics (P), Topographical Drawing (P).

27. Zachary, *Frontier,* 30–32.

28. MIT, *Catalog, 1915–1916,* December 1915, vol. 51, 583.

29. Applied Mechanics (P); Elements of Electrical Engineering (F); Electrical Engineering Laboratory (P); Geology (P); Physics (P); Physical Laboratory (P); Railroad Drawing (P); Advanced Surveying (P); International Law (P); Lithology (the study of rocks) (C).

30. After graduating as valedictorian from Hamilton, Allen entered graduate school at Johns Hopkins to study the classics. He compiled an excellent record, but a worsening eye problem made it difficult, if not impossible, for him to read, and in effect ended a promising scholarly career. He returned to Seattle to live at home, taking economics and sociology courses at the University of Washington. A friend suggested he apply for a civil service position in the federal Department of Agriculture. Allen took the examination and was dispatched to Miami, Florida, in the fall of 1915. His work was excellent and he was selected to be sent to India. Just after the New Year he came to Washington and worked at the department's experimental farm near today's Pentagon to prepare for his new assignment. He contracted a quick-acting form of pneumonia and died a few days later. *Hamilton Literary Magazine,* February 1916, 262–263; AMG, *Appreciation.*

31. Allen's grave is in Section 2, grave 3754. Apparently, Chaplain Groves was able to have his son interred there as a dependent, in the expectation of using the grave site himself. While this never happened, the site became the place where the body of Lt. Gen. Leslie R. Groves and the ashes of Grace Wilson Groves are buried. In August 1933 Chaplain Groves had Gwen's remains (buried at Fort Lawton) disinterred for reburial in Pasadena, next to his second wife, Jane. When he died six years later he was buried alongside the two sisters. The First Congregational Church in downtown Washington was built in 1865 and torn down in 1957 to be replaced by a new building, which was finished in 1960.

32. LRG, *Grandchildren,* 103. Later Groves remembered that there were about 26 boys there, and about 250 active candidates nationwide competing for 12 appointments. Ermenc, *Scientists,* 208.

33. Representative questions are presented in *Official Register of the Officers and Cadets, United States Military Academy for 1916* (West Point, N.Y.: USMA, 1916), 69–73.

34. Though Dick never received a degree from MIT he considered himself a member of the class of 1917, and attended the 50th reunion at Chatham, Massachusetts, in the summer of 1967.

CHAPTER 4 West Point (June 1916–November 1918)

1. LRG, *Grandchildren,* 103.

2. The Regular Army is the permanent army, maintained in peacetime as well as in war. It is one of the components of the United States Army (USA) or Army of the United States (AUS), which includes the Regular Army, the National Guard, and the Army Reserve.

3. LRG, *Grandchildren,* 105.

4. Much of the description of the buildings of Groves's time is derived from the Historic American Buildings Survey project of the National Park Service, HABS No. NY-5708. See also Theodore J. Crackel, *The Illustrated History of West Point* (New York: Harry N. Abrams, 1991), 189–223, and Montgomery Schuyler, "The Works of Cram, Goodhue & Ferguson, The New West Point," *Architectural Record* 21, no. 1 (January 1911), 87–112.

5. Robie S. Lange, *An Overview of the History and Development of the United States Military Academy,* HABS No. NY-5708, 71.

6. The 1918 edition of *Bugle Notes* states, "The purpose of the Academy is to shape a man and a soldier — an honorable, loyal, courageous, self-reliant, disciplined, intelligent gentleman and officer." *Bugle Notes* 10 (1918–1919), 8–9.

7. "At West Point, graduates had all taken the same classes, undergone the same hazing, marched in the same formations. This common experience gave them a sense of community, of fraternity, that they could not bear to see destroyed. A change in the continuity at the Academy would have the effect of disturbing their bond. To tamper with West Point would be to tamper with the cement that held the army officer corps together." Ambrose, *Duty,* 207.

8. For the classes from 1910 through 1919 the average number of graduates was 140; from 1920 through 1929 it was 225; from 1930 through 1939 it was 456. For the classes from 1990 through 1999 the average number of graduates was 970.

9. Robert Charlwood Richardson Jr., *West Point: An Intimate Picture of the National Military Academy and the Life of the Cadet* (New York: G. P. Putnam's Sons, 1917), 103. Captain Richardson (class of 1904) was an assistant professor of English in Groves's plebe year. His book is an account of what it must have been like when Groves was there. See also Jeffrey Simpson, *Officers and Gentlemen: Historic West Point in Photographs* (Tarrytown, N.Y.: Sleepy Hollow Press, 1982).

10. *Plebe* is short for the Latin word *plebeian,* referring to a member of the lower or ordinary class — or, in the West Point dictionary, "less than nothing." *The Howitzer: The Yearbook of the Class of 1919,* 361.

11. Dick was one of 158 admitted on June 15, 1916. On July 10, 146 additional cadets (the "Juliettes") were admitted, and 18 more between July 11 and July 26. *Official Register of the Officers and Cadets, United States Military Academy for 1916,* 40–45. At the time of graduation and depending on class rank, each graduate is assigned a Cullum number, which provides a reference point for further biographical information. See annual editions of *Register of Graduates and Former Cadets of the United States Military Academy* (West Point, N.Y.: Association of Graduates, 1946–current). Indispensable are the last two of the earlier Cullum editions, often with detail left out of the successor editions. Bvt. Maj. Gen. George W. Cullum, *Biographical Register of the Officers and Graduates of the United States Military Academy at West Point, New York Since Its Establishment in 1802,* Supplement VIII, 1930–1940, edited by Lt. Col. E. E. Farman, retired, and Supplement IX, 1940–1950, edited by Col. Charles N. Branhan, retired. Beginning with the class of 1978 the listings are alphabetical and not by General Order of Merit (that is, by class ranking). Leslie Richard Groves's Cullum number is 6032. See *Biographical Register,* Supplement VI-B, 2010; VII, 1338; VIII, 382; IX, 371.

12. "Graduation Day Exercises at the Military Academy," *News of the Highlands,* June 17, 1916. The weekly newspaper of the garrison town, Highland Falls, that adjoins West Point has much interesting detail. As we shall see, 26 years later Styer would be instrumental in choosing Dick to build the atomic bomb.

13. Theoretically it ended with his class. *The Howitzer: The Yearbook of the Class of 1920* says that they "were the last plebe class to go thru the mill under the old regime of intensified hazing," 14.

14. Brig. Gen. William W. Ford, "Plebe Life 1918," *Assembly,* June 1982, 37.

15. From October 1942 until the end of the war Conrad held various positions with G-2, headquarters, European theater of operations, up to and including

assistant chief of staff. When Groves sent Horace K. Calvert to London to establish a liaison office in London in early 1944 he first met with Colonel Conrad. Calvert had a desk in Conrad's office where he could examine the raw intelligence that came through. LRG, *Now,* 194–195. Conrad also assisted the Alsos mission in its movements about northern Europe from the fall of 1944 until V-E Day.

16. The Young Men's Christian Association beginning in 1908 originally published *Bugle Notes* annually. In 1924 it was edited and distributed by the upper-classmen. The plebe was refunded his money if he knew the contents.

17. *Regulations for the United States Military Academy,* April 1, 1918, Paragraph 109; *Bugle Notes* 10 (1918–1919), 96–97.

18. *Regulations for the United States Military Academy,* April 1, 1918, Paragraphs 178–195; *Annual Report of the Superintendent, United States Military Academy, 1918,* 39; Frederick P. Todd, *Cadet Gray: A Pictorial History of Life at West Point as Seen Through Its Uniforms* (New York: Sterling Publications, 1955).

19. *The Howitzer: The Yearbook of the Class of 1917,* 362. Whereas the class entered with 322 cadets, by the end of fourth class year 257 were left.

20. "The six general principles of the code are: (1) lying, quibbling, evasion, or a resort to technicalities in order to shield guilt or defeat the spirit of justice are not tolerated; (2) a cadet who intentionally violates the honor system should resign at once and offenders are never granted immunity; (3) anything to which a cadet signs his name means irrevocably what is said, both as to letter and spirit; (4) no intentional dishonesty is condoned; (5) every man is honor bound to report any breach which comes to his attention; and (6) the Corps, individually and collectively, is the guardian of its honor system." Ambrose, *Duty,* 280.

21. Nye, *USMA.*

22. A compilation of data of the academic standings of all West Point graduates who held the rank of major general or higher in the Civil War, World War I, and World War II (a total of 618), shows a correlation between academic profi-ciency and later career success. Of the 275 cadets who later became major gen-erals or higher in World War II, 119 graduated in the top third of their class (43 percent), 98 in the middle third (36 percent), and 58 in the bottom third (21 percent). The percentages are comparable for the Civil War and World War I. Col. Charles P. Nicholas, "Six Hundred and Eighteen Major Generals," *Assembly,* January 1952, 10–11.

23. The courses he took in his plebe year and the textbooks he used are described in *Official Register of the Officers and Cadets, United States Military Academy for 1916* (West Point, N.Y.: USMA, 1916), 75–79.

24. *Official Register of the Officers and Cadets, United States Military Academy for 1917* (West Point, N.Y.: USMA, 1917), 34, 52–53. A distinguished cadet is one who exceeds 92 percent of the possible total. The Hon. John C. Calhoun, secretary of war, instituted the practice of reporting distinguished cadets on February 10, 1818.

25. "Army Defeats Navy in Annual Football Game," *News of the Highlands,* December 2, 1916.

26. *The Howitzer: The Yearbook of the Class of 1920,* 15.

27. Dickson, *Chronicle,* 2; *News of the Highlands,* March 10, 1917.

28. But he made it to brigadier general and died, as the oldest living graduate, two

months shy of his 101st birthday in May 1996. *Assembly,* July–August 1996, 5.

29. The courses he took in his yearling year and the textbooks he used are described in *Official Register of the Officers and Cadets, United States Military Academy for 1917* (West Point, N.Y.: USMA, 1917).

30. *Official Register of the Officers and Cadets, United States Military Academy for 1918* (West Point, N.Y.: USMA, 1918), 36.

31. Nye, *USMA,* 115–124.

32. The class that entered in June and July of 1916 had nine companies, an increase from six two years earlier. Dickson, *Chronicle,* 1.

33. Keyes graduated in the class of 1913. In the small world of West Point, his son, Geoffrey Keyes Jr., was a classmate of General Groves's son, Richard H. Groves, in the class of 1945. To receive credit for playing a game the cadet must be in at least one full period of the game. *Bugle Notes* 10 (1918–1919), 67.

34. "Carlisle Easy for Army," *New York Times,* November 11, 1917. The eight games are described in some detail in *The Howitzer: The Yearbook of the Class of 1919,* 262–269.

35. And is pictured in *The 1917/18 Howitzer,* 413, and in *The 1919 Howitzer,* 288.

36. Letter, Pulsifer to LRG, October 19, 1945. Williamson was probably classmate George McKnight Williamson Jr., and G. B. may have been George Bryan Conrad. The phrase "my peoples" refers to goats, of which Pulsifer, not an engineer and near the bottom of his class, was one.

37. "Spanish Influenza — What It Is and How It Should Be Treated," *News of the Highlands,* November 2, 1918. See also *Annual Report of the Superintendent, United States Military Academy, 1919,* 39.

38. *Official Register of the Officers and Cadets, United States Military Academy for 1919* (West Point, N.Y.: USMA, 1919), 22. See also *Annual Report of the Superintendent, United States Military Academy, 1918.*

39. The system provided a way for cadets in the middle or lower sections to improve their ranking: through special tests at the end of the term called Written General Reviews. A cadet who did well on WGRs could add points and thereby move upward.

40. *Register of Demerits,* Class of November 1918, Part 1 (A–L), USMA Archives; *Abstract of Delinquencies,* 11 vols., June 1, 1916–November 30, 1918, USMA Archives; *Regulations of the United States Military Academy,* April 1916.

41. Demerits were not recorded for one month after admission. For the remainder of the period to May 31, one-third of the number of demerits received each month was deducted. *Regulations for the United States Military Academy,* April 1, 1916, 81.

42. For well over a century since its founding graduates of West Point received only a commission in the army. In 1933 all graduates also received a bachelor of science degree, retroactively conferred to those who had graduated after the academy had been accredited by the American Association of American Universities in 1925.

43. *Forms D, Graduated Nov. 1, 1918 (Class of 1920),* USMA Archives; *Official Register of the Officers and Cadets, United States Military Academy for 1919* (West Point, N.Y.: USMA, 1919), 17. The cadets exceeding 92 percent were regarded as "distinguished," a group of the top 15.

44. A list published in 1949 ranked 1,262 USMA graduates from the classes 1821–1949 who attained 90 percent or above in order of merit. In this ranking

of 135 classes Groves ranked 401st overall. Given the differences of how cadets were graded in the 19th and 20th centuries, a more accurate accounting might be to limit the graduates to the classes from 1900 to 1945. In this subset Groves ranks 159th. Many in this group are familiar as Groves's mentors, cohorts, and protégés, officers he worked for or those who worked for him. Of the 20th-century graduates, at number one is Frederick Smith Strong (class of 1910), with a percentage of 99.4998. As we shall see, Colonel Strong replaced Groves as chief, Operations Branch, OCE, in March 1942, after Groves was appointed deputy chief of construction. *Honors Register: Graduates of the United States Military Academy Who Were Credited with 90%, or Above, of Their Respective Course Maxima — in Order of Merit,* September 1, 1949.

45. In a letter to Chaplain Groves, Owen reported Dick saying, "He claims that if they stay there the full four years, he will emerge at the top of his class. If they are cut short he expects to be third or fourth." Letter, OGG to Chaplain Groves, June 1917, LC. In June 1917 Owen visited his brother at West Point.

46. Fourteen of the class of 1920 who chose infantry served with the 27th and 31st Infantry Regiments, which constituted the American Expeditionary Force Siberia, sent to try to crush the Bolshevik Revolution. The move proved unsuccessful, and the last of the regiments sailed from Vladivostok to the Philippines in April 1920. Dickson, *Chronicle,* 6.

47. Vidal was also Gore Vidal's father.

48. The authorized strength of the Corps of Cadets for 1915 was 705. For 1916 this nearly doubled, to 1,332.

49. *New York Times,* May 12, 1917; *New York Times,* May 7, 1918: "'Blue Devils' of France Pay a Visit to West Point," *News of the Highlands,* May 11, 1918.

50. James A. Blackwell, *On Brave Old Army Team* (Novato, Calif.: Presidio Press, 1996), 85–86.

51. The class of 1920 (Groves's class) graduated on November 1, 1918, 19 months early, 10 days before the armistice, after having spent 28 months at West Point. The class was originally called the class of 1920. Later when there was another class that actually graduated in 1920, it became the first class of 1920. Finally it became the class of November 1918. Dickson, *Chronicle,* 3.

52. When World War I broke out there were 4,900 officers of the Regular Army who had at least one year of commissioned service. When the U.S. entered the war over 193,000 emergency officers were commissioned. Thomas J. Fleming, *West Point: The Men and Times of the Military Academy* (New York: William Morrow & Company, Inc., 1969), 303.

53. *Historical Statistics of the United States, Colonial Times to 1970,* Bicentennial Edition (Washington, D.C.: U.S. Department of Commerce, 1975), pt. 2, 1141.

54. Ambrose, *Duty,* 253–255. As the class history noted, "The Army Reorganization Act of 1920 buried our class behind thousands of temporary officers seeking regular army commissions by a new ruling that service as Cadet, USMA, no longer counted as military service. Our prospect was that, without another war, we would die as majors or, at best, lieutenant colonels and no such war was dreamed of in the 1920's. As a result the class spent 16 years in the grade of first lieutenant." Dickson, *Chronicle,* 8.

55. Brig. Gen. William W. Ford, "Plebe Life, 1918," *Assembly,* June 1982, 38.

56. *The Howitzer: The Yearbook of the Class of 1920,* 16.

57. Proving once again that it is not always easy to predict when a war might end.

"West Point Breaks Graduation Record," *New York Times,* November 2, 1918, 13.

58. *Sixty-eighth Annual Report of the Association of Graduates of the United States Military Academy at West Point, New York,* June 11, 1937, 89.

CHAPTER 5 Caught Behind the "Hump" (December 1918–June 1921)

1. Dickson, *Chronicle,* 8. By 1926, 68 graduates of Dick's class had left to pursue other careers. In general, "Pay was stagnant, the duty was often tedious, and promotion was glacially slow. Thirteen years was the normal interval between attaining the rank of first lieutenant and promotion to captain in the interwar army, and some captains spent seventeen years at the latter rank. By the time the lucky survivors of the system reached the rank of general, they were normally at least 59 years of age. General officers could serve at most two or three years before reaching mandatory retirement age. Even a brilliant and well-connected officer like George C. Marshall served ten years at the grade of major and did not reach brigadier-general until he was 56 years old." Ronald Spector, "The Military Effectiveness of the US Armed Forces, 1919–39," in Allan R. Millett and Williamson Murray, eds., *Military Effectiveness,* vol. 2, *The Interwar Period* (Boston: Allen & Unwin, 1988), 77.

2. William M. Griffith (1858–194?) was the third child and second son of Owen and Jane Griffith, seven years older than Gwen.

3. Maj. James A. Dorst, Corps of Engineers, "Ft. Humphreys and Historical Belvoir," *Military Engineer* XV, no. 82 (July–August 1923), 332–337.

4. General Black was first in the class of 1877, Richard Wilson's class. He was chief of engineers from March 7, 1916, to October 31, 1919.

5. The Engineer School predates the Military Academy, tracing its roots to the American Revolution. In 1778 a school of engineering was started at West Point. With the end of the war the school closed; it reopened in 1794 then closed again in 1798 after a fire destroyed many facilities. In 1801 the War Department revived the school, and less than a year later Congress authorized the Corps of Engineers and constituted it at West Point as a Military Academy. For the next 64 years the corps supervised the Military Academy, with a predominantly engineering curriculum. Following the Civil War, supervision of the academy passed to the War Department. In 1868 engineers at Willets Point, New York, started an informal School of Application. In 1885 the school received formal recognition by the War Department, and in 1890 the name was changed to the U.S. Engineer School. In 1901 the school moved from Willets Point to Washington Barracks in Washington, D.C. In 1920 it moved to Camp Humphreys/Fort Belvoir, and in 1988 to Fort Leonard Wood, Missouri.

6. Maj. Gen. W. M. Black, chief of engineers, to Col. S. E. Tillman, superintendent, USMA, January 9, 1919. Quoted in RHG, *DNO,* 205–206.

7. *The Howitzer: The Yearbook of the Class of 1920,* 165.

8. LRG, *Grandchildren,* 110.

9. LRG *Official Records,* 64. Major General Peterson (class of 1908) served as inspector general of the army from February 27, 1940, to June 4, 1945.

10. USACE, *History.*

11. Military and civilian needs were entwined. In the aftermath of the War of 1812 the Corps of Engineers "reported that national defense should rest upon four pillars: a strong Navy at sea; a highly mobile regular Army supported by reserves and National Guard; invincible defenses on the seacoast; and improved

rivers, harbors and transportation systems that would permit rapid armed concentration against an invading enemy and swifter, more economical logistical lines." USACE, *History,* 37.

12. To name a few: Sylvanus Thayer (1785–1872), Joseph Gilbert Totten (1788–1864), William Gibbs McNeil (1801–1853), Dennis Hart Mahan (1802–1871), John Gross Barbard (1815–1882), Quincy Adams Gillmore (1825–1888), and Henry Larcom Abbot (1831–1927).

13. The corps supervised and built an enormous variety of buildings, bridges, and monuments as well as parks, streets, and water projects, most of which continue to make up modern Washington. The buildings include: the Old Executive Office Building, Capitol wings and dome, Library of Congress, Pension Building, Natural History Museum, Army War College (now National Defense University), Department of Agriculture Building, and Government Printing Office. The Corps of Engineers completed the Washington Monument and built the Lincoln Memorial. Bridges include the Chain, Key, Memorial, and 14th Street Railroad Bridges over the Potomac, the Navy Yard Bridge over the Anacostia, and the Taft and Massachusetts Avenue Bridges over Rock Creek Park. The Corps of Engineers laid out Rock Creek, Potomac, and Glover Parks, and supervised the planting of 3,800 cherry trees around the Tidal Basin. Albert E. Cowdrey, *A City for the Nation: The Army Engineers and the Building of Washington, D.C., 1790–1967* (Washington, D.C.: Historical Division, Office of Administrative Services, Office of the Chief of Engineers, 1979); Pamela Scott and Antoinette J. Lee, *Buildings of the District of Columbia* (New York: Oxford University Press, 1993). Engineers, especially those who live and work in Washington, carry with them a strong sense of shaping their environment.

14. Meigs's public works projects are perhaps more lasting contributions to the nation. He built Washington's water supply system, supervised the extension of the Capitol and installation of the iron dome, constructed the monumental Pension Building (after he retired from active service in 1882), and chose Arlington as the site of a national cemetery, perhaps as a continuing reminder to civilian and military leaders alike of the human costs of war.

15. The account comes from LRG, *Grandchildren,* 112–120; *The Howitzer: The Yearbook of the Class of 1920,* 165–167 (published in 1928); a series of 18 letters from LRG to his father and stepmother (dated June 30 to September 7, 1919); LRG, Diary Notes.

16. LRG, Diary Notes, 1.

17. LRG, Diary Notes, 11.

18. LRG, Diary Notes, 11–12.

19. LRG, Diary Notes, 16.

20. Letter, LRG to Dada (his stepmother), Verneuil, July 16, 1919. For the Americans' encounters with French culture, see Meigs, *Optimism,* 69–106.

21. Letter, LRG to Dada, Verneuil, July 16, 1919.

22. LRG, Diary Notes, 20.

23. Groves and his classmates were not alone touring European military and cultural sites. American soldiers in very large numbers remained in France for the six months following the war and with very little to do became tourists. Meigs, *Optimism,* 69–106.

24. LRG to Dada, September 4, 1919, from the USS *Kroonland.*

25. LRG to Chaplain Groves, August 25, 1919, from Paris.

26. Nancy K. Bristow, *Making Men Moral: Social Engineering During the Great War* (New York: New York University Press, 1996).

27. Meigs treats the sexual attitudes in *Optimism,* 107–142. In an effort to deter marriage and keep women in general away from the soldiers: "The army and the American popular press did the best they could to push the French woman into the role of harlot. Magazines, developing a new version of an old stereotype of the French woman, wise and cynical in the ways of love, thrilled readers with articles about the possibilities of Americans debauched by French women but remained silent on the subject of American men and French women marrying." Ibid., 125.

28. Edward P. Craypol, *America for Americans: Economic Nationalism and Anglophobia in the Late Nineteenth Century* (Westport, Conn.: Greenwood Press, 1973); John E. Moser, *Twisting the Lion's Tail: American Anglophobia Between the World Wars* (New York: New York University Press, 1998).

29. Headquarters, the Engineer School, Final Standing in Basic Course, June 6, 1921, Folder, LRG, Commendations and Qualifications, Box 3, Entry 7530C, Papers of LRG, RG 200, NARA II. Groves's average mark was 87.91, 2.23 points from being first. Ranking third was Capt. James C. Marshall. Marshall, as we shall shortly learn, was the first head of the Manhattan Engineer District in June 1942, before Groves succeeded him in September 1942.

30. *LRG Official Records,* 76–80.

31. *LRG Official Records,* 82–83.

CHAPTER 6 Married with Children (July 1921 – July 1931)

1. LRG, *Grandchildren,* 130–131. From December 1913 to May 1916 Col. Richard Wilson commanded the 14th Infantry Regiment and Fort Lawton.

2. LRG, *Grandchildren,* 130. The words are hers and reflect an upbringing in a military family moving from post to post. Grace never went to a regular school or did many things that young girls do.

3. LRG, *Grandchildren,* 130.

4. According to Grace, "I told Rocky maybe I wasn't sure about marrying him: we had some trying scenes. He returned to his destroyer and I returned his ring and presents and I had a sad, sad letter telling how the sailors had to restrain him from jumping overboard." LRG, *Grandchildren,* 130.

5. LRG, *Grandchildren,* 131. After staying in the Hotel St. Francis on Union Square the night of March 3, according to Grace they planned to move into Dick's three-room apartment at the Presidio on March 4. Letter, GWG to OGG, March 4, 1922, LC.

6. LRG, *Grandchildren,* 133 (emphasis in original).

7. The captain of the new steamer, H. F. Alexander, was determined to beat the time it took a train to travel to Seattle. On its maiden voyage, in a heavy fog, the ship hit a rock, but the boat did not sink. Grace and the other passengers were transferred to a Coast Guard ship and proceeded on to Seattle. LRG, *Grandchildren,* 132–133.

8. Schulz was also division engineer of the Hawaiian Division. His daughter Caroline married John Stewart Service in 1932. Service was one of the State Department's "Old China Hands" fired as a security risk during the McCarthy era. She died in 1997, and Service died at age 89 in 1999.

9. "Engineers Win National Team Matches," *Military Engineer* XVI, no. 90

(November–December 1924), 472–473. The captain of the team was Maj. C. L. Sturdevant, whom we shall meet again soon.

10. *Annual Report of the Association of Graduates of the USMA,* 1930, 345–347.

11. RHG, *War Stories.* These two volumes are an effort to record events in his life that occurred mainly within the family.

12. AG 330.13, Commendation of Construction Work on Kahuku-Pupukea Trail, April 12, 1924, *LRG Official Records,* 104. The commendation was signed by Maj. Gen. Charles Pelot Summerall, commanding general of the Hawaiian Division. Groves's Efficiency Report during this period was mostly above average except a below average for Tact — "The faculty of being considerate and sensible in dealing with others." Ibid., 105. Summerall was born in 1867 and graduated from West Point in the class of 1892. He participated in the Philippine Insurrection in 1899–1900 and in the China Relief Expedition in 1900–1901, with the forces that attacked Peking, and knew Chaplain Groves. From November 1926 to November 1930 he was chief of staff of the army.

13. To avoid confusion I will refer to Richard Hulbert Groves as "Richard."

14. According to family legend, during the trip there was a near catastrophe. Since babies were not allowed in the dining room, Mrs. Groves left Richard in the stateroom while she hurriedly ate her meals. One day as she returned from lunch she found her young son partway through an open porthole.

15. LRG, *Grandchildren,* 133.

16. For the period November 4, 1925, to June 30, 1926, Major Schley wrote, "A very solid, thoroughly dependable man, with unusually sound judgment and excellent mental qualities. He has the highest standard of moral character." And for the period July 1, 1926, to July 1, 1927, Lieutenant Colonel Schley wrote, "An exceptionally able and industrious officer. He has an excellent mind and mature judgment. He is a good organizer, executive and administrator, and shows ability as an engineer. He is moral, cultured and genteel." *LRG Official Records,* 122, 124. Schley was a bachelor and often had Mrs. Groves, sometimes with Dick, chaperone his dates with his lady friend. Some socializing took place on a former Vanderbilt yacht that had been left to the army years before.

17. RHG, *War Stories,* 15–17, 271–274. Aunt Bossie, née Florence Mary Chaffin, was Grace Chaffin Wilson's youngest sister. Duncan Hines was born in 1880 and spent the early part of his career as a traveling salesman, getting to know well which eating places were to be avoided and which were worth a return visit. Later the Hineses traveled throughout North America, discovering restaurants that served delicious yet reasonable meals. They compiled a personal list of 167 recommended eating places in 30 states and the District of Columbia, and in 1935 mailed it (instead of the usual Christmas card) to friends and family. They then received requests for it from hundreds of people they had never heard of and realized there was a demand for a guide to good eating. In 1936 *Adventures in Good Eating* was published, describing 2,000 of their finds. Florence died in 1939, and Duncan remarried in 1946. He authored several other books *(Lodging for a Night* and *Duncan Hines' Vacation Guide)* aimed at a new and growing phenomenon: the vacationing American, exploring the country by automobile. A food company was established in 1949 producing food under the Duncan Hines label. In 1956 Proctor & Gamble purchased all of Hines's interest. Though Duncan Hines died in 1959, he remains a household name to this day.

18. Lynn M. Alperin, *Custodians of the Coast: History of the United States Army Engineers at Galveston* (Galveston, Tex.: USACE, 1977), 112. The footnote says, "Telephone interview with Jack Beck, October 1974."

19. "The floodwaters took eighty-four lives and destroyed thirty-million dollars in property in Vermont alone, fifty-five deaths in the Winooski Valley." William Edward Leuchtenburg, *Flood Control Politics: the Connecticut River Valley Problem, 1927–1950.* (Cambridge, Mass.: Harvard University Press, 1953), 29.

20. "Raging Winooski River Carries Away Pontoon Bridge," *Burlington Free Press and Times,* December 2, 1927, 8. It should be noted that the placement of the bridge was not in a preferred position and was not chosen by Groves. Furthermore, the bridge itself was not in very good condition.

21. *LRG Official Records,* 132. The Red Cross, on the other hand, commended him.

22. Groves provided a vivid description of what happened in a two-page type-written account, n.d., ca. February 1928, LC.

23. "Winooski, City Regrets Death of Sergeant Littlefield," *Burlington Free Press and Times,* February 4, 1928, 9; "Winooski, Flags Displayed at Half Mast in Honor of Memory of Sergeant Littlefield," *Burlington Free Press and Times,* February 5, 1928, 9. A wife and two children survived the 35-year-old World War I veteran.

24. In a letter to his brother he describes his injuries in great detail. Letter, LRG to OGG, February 11, 1928, LC.

25. *LRG Official Records,* 155–156.

26. RHG, *DNO,* 364.

27. How the scientists learned about this obscure event in Groves's career is unclear.

28. RHG, *War Stories,* 41.

29. RHG, *War Stories,* 42.

30. *LRG Official Records,* 160–175.

31. Lieutenant Colonel Sultan's final report, *U.S. Army Interoceanic Canal Board Report,* was written in part by Lieutenant Groves. The provisional battalion was composed of three survey companies: A Company (Company A, 1st Engineers), B Company (Company A, 29th Engineers), and C Company (Company F, 11th Engineers), a headquarters and service platoon, and added detachments from the Medical and Finance Departments and the quartermaster and signal corps. The total strength was 25 officers and 295 enlisted men. Several of the engineer officers would work with Groves in the future, most notably 2nd Lt. Kenneth D. Nichols. First Lt. Kenner F. Hertford (fifth in the class of 1923) later served as the deputy commander of the Armed Forces Special Weapons Project, Sandia Base, Albuquerque, New Mexico, from 1948 to 1952. Second Lt. William E. Potter — later a major general and governor of the Panama Canal Zone — served under Groves in A Company and called him a "hard taskmaster." *Engineer Memoirs: Major General William E. Potter* (Washington, D.C.: Office of History, U.S. Army Corps of Engineers, July 1983), 7.

32. "When we were staying with my parents in Seattle during DNO's absence in Nicaragua, Father taught you to read in just a few days. . . . He enjoyed having a little boy to teach and he took you to see all the things of interest to a little boy — the ships in the harbor etc. — etc. You were the son he never had!" Letter, GWG to RHG, ca. July 1981.

33. LRG, *Grandchildren,* 140.

34. LRG, *Grandchildren,* 141.

35. Upon his return he would write an article about the battalion's adventures for the *National Geographic,* complete with photographs. Lt. Col. Dan I. Sultan, "An Army Engineer Explores Nicaragua," *National Geographic* LXI, no. 5 (May 1932), 593–627. "We lost one member of our party [to a shark], who was attacked before he could get ashore from a capsized boat off the bar at Greytown." Ibid., 597.

36. Stimson, Bundy, *Service,* 110–116; Hodgson, *Colonel,* 98–116.

37. Sandino and his guerrilla forces fought the American marines, the Guardia, and the Nicaraguan government for the next six years until, in 1934, he was betrayed and murdered on the orders of Anastasio Somoza Garcia. Hodgson, *Colonel,* 117. The incident would not be forgotten.

38. The American-built *Victoria* had been sailing the waters of Lake Nicaragua since 1884. Its hull was pockmarked with bullet holes from past revolutions. During the weeklong voyage from Granada to San Carlos and back again, its cargo included every manner of item, from coconuts and chickens to cacao and cattle. A photograph of this craft and a description of the voyage are in Lt. Col. Dan I. Sultan, "An Army Engineer Explores Nicaragua," *National Geographic* LXI, no. 5 (May 1932), 613–615. The arrival of the boat at each port was the occasion for a fiesta. "Not only do the natives come to buy and to sell but they come for their one day of 'whoopee' per week, and for a couple of shots of *aguardiente,* or *'aguaro',* as the local firewater is called. *Aguaro* boasts a revolution in every gallon, and after a couple of drinks a native is ready for anything: either to clasp you around the neck as a friend or to swing at you with a *machete,* depending upon his inclination." B. B. Talley, "From Corinto to San Juan del Norte," *Military Engineer* XXII, no. 124 (July–August 1930), 349.

39. With further assistance from Grandfather Wilson. Groves's efforts to advance his son a grade were unsuccessful. The normal school year ran from May until mid-February, with a ten-week vacation during the dry season. Richard's abbreviated school year started in August 1930 and ended in February 1931, followed by sessions with the tutor during the "vacation" period. Upon resumption of classes in May 1931 Richard remained in the first grade. Nevertheless, he left in July with a certificate stating that his grade was "Muy Bueno." RHG, *War Stories,* 90–92.

40. Lt. Col. Dan I. Sultan provided a graphic account, "The Managua Earthquake," *Military Engineer* XXIII, no. 130 (July–August 1931), 354–361.

41. A clause in the U.S. Constitution (Article I, Section 9, Clause 8) prohibits U.S. Army officers from accepting any foreign decorations or medals without the consent of Congress. In Groves's case it took until May 19, 1936, for Congress to approve his award, and before it was delivered to him. Letter, Adjutant General to LRG, July 24, 1936, Folder, LRG, Commendations and Qualifications, Box 3, Entry 7530C, Papers of LRG, RG 200, NARA. Colonel Sultan and Maj. Paul R. Hawley, chief surgeon, also received the Presidential Medal of Merit, as did more than a dozen navy and marine corps officers. During World War II as chief surgeon, European theater of operations, Major General Hawley, at Groves's behest, issued cover orders in May 1944 alerting headquarters if the Germans resorted to radiological warfare during the Normandy invasion. See LRG, *Now,* 202–203.

42. Folder, LRG, Commendations and Qualifications, Box 3, Entry 7530C, Papers of LRG, RG 200, NARA.

CHAPTER 7 Learning the Ropes (July 1931–June 1935)

1. William Addleman Ganoe, *The History of the United States Army* (New York: D. Appleton–Century Company, 1943), 463–529; Russell F. Weigley, *History of the United States Army* (Bloomington: Indiana University Press, 1984), 395–420.

2. Patrick J. Hurley, secretary of war, *Annual Reports of the War Department — 30 June 1932.*

3. Given the size of the armed forces during World War II and for the 50 years of the Cold War, it is sometimes difficult to appreciate just how small the U.S. Army was between the wars. The peak number of army commissioned officers in mid-1945 was 778,316, with a total strength of 8,266,373. Marvin A. Kreidberg and Merton G. Henry, *History of Military Mobilization in the United States Army, 1775–1945* (Department of the Army Pamphlet No. 20-212, November 1955), 379.

4. *Army List and Directory, October 1, 1931* (Washington: U.S. War Department, Adjutant General's Office, 1931), 96–100. By comparison, there were 3,711 officers in the infantry, 1,510 in field artillery, 1,384 in the air corps, 1,002 in coastal artillery, and 979 in the cavalry. For the entire army there were only 10 major generals, 11 brigadier generals, and 500 colonels. Twelve years later in the middle of World War II, the *Army Directory of April 20, 1943,* listed the number of officers in the Corps of Engineers as 945, a surprisingly small figure.

5. Brown's *Annual Report for the Fiscal Year Ending June 30, 1933,* lists 18 officers in the Office, Chief of Engineers.

6. Maj. Gen. Lytle Brown, "Engineers, Builders and Executives," *Military Engineer* XXIII, no. 128 (March–April 1931), 187.

7. There were also Cathedral Mansions, North, and Cathedral Mansions, South, located on either side.

8. LRG, *Grandchildren,* 153.

9. At the time Colonel Graves (USA, retired) was chairman of the Interoceanic Canal Board, after having retired in 1921 for a physical disability (deafness). The board was established under the same legislation that authorized the survey in 1929. President Hoover appointed Lt. Gen. Edgar Jadwin (USA, retired) as chairman, Graves, and three others. Jadwin was chief of engineers until he retired for age in 1929. On March 2, 1931, Jadwin died of a cerebral hemorrhage and Graves became chairman. Lieutenant Colonel Sultan's report as well as the report of the Interoceanic Canal Board (November 30, 1931) are printed in *United States Army Interoceanic Canal Board,* Washington, D.C., 1931.

10. Graves had played fullback for four years at the University of North Carolina before coming to the academy. At West Point he played fullback during the 1901 season and then tackle, and was captain in 1904. *The Howitzer* for the class of 1905 says: "In football he stands high, plays low, slugs hard, and never gets caught. He made an annual habit of eating young Navies until they begged to have him muzzled." There are differences of opinion about how good a coach he was. Some felt that he destroyed his teams in practice, leaving them too battered to play well in the real games. He would tell his players in scrimmage, "I want to see blood." Obituary, *Assembly,* July 1954, 72–74. The game was brutal in those days. In 1905, 18 men were killed playing college football. The Army-Navy game of that year resulted in 11 serious injuries. In 1909, 30 were killed across the country. James A. Blackwell, *On Brave Old Army Team: The Cheating Scandal That Rocked the Nation, West Point, 1951* (Novato, Calif.: Presidio Press, 1996), 107.

As head coach for the 1906 and 1912 seasons, Graves's combined record was seven victories, eight defeats, and one tie. For the 1913–1916 seasons Charles D. Daly, another legend and one of Pot's classmates, was brought back as head coach, with Graves coaching the line. They posted a record of 31 wins, 9 losses, and 1 tie, beating Navy all four years and becoming National Campions in 1914. Gene Schoor, *100 Years of Army-Navy Football* (New York: Henry Holt and Co., 1989), 241. Pot also played baseball for four years and tug-of-war for two.

11. One famous story retold by his son is about building warehouses in France. Graves told the men to cut back on the amount of lumber and nails. They did. He told them to cut back again, and then a third time, causing some to fear that the warehouses would fall down. Graves said, "That's great. Add a pound of nails and two boards and build the rest of them that way." The lesson being that in war you should use the minimum amount of material needed to accomplish the mission, because everything is at a premium. *Engineer Memoirs: Lieutenant General Ernest Graves, U.S. Army* (Alexandria, Va.: Office of History, U.S. Army Corps of Engineers, August 1997), 5–6. See also Ernest Graves, "Military Road Building Not a Science — Only a Job," *Military Engineer* XII, no. 62 (March–April 1920), 161–168, 177; Maj. Ernest Graves, *Construction in War: Lessons Taught by the World War, 1917–1919,* no. 64, Occasional Papers, Engineer School, U.S. Army (Washington, 1921).

12. Jadwin graduated first in the class of 1890. He was selected by General Goethals as an assistant in the construction of the Panama Canal. He served in France during World War I, overseeing major construction. Graves considered him an exemplary engineer. Jadwin was chief from June 27, 1926, to August 7, 1929, succeeded by Maj. Gen. Lytle Brown from October 1929 to October 1933, and by Maj. Gen. Edward M. Markham from October 1933 to October 1937.

13. LRG, *Grandchildren,* 153. In the four generations of the family that attended West Point they ranked either first or second in their class. Pot's father-in-law was Rogers Birnie, first in the class of 1872. Pot's son Ernie Jr. was second in the class of 1944, and his grandson Ralph Henry Graves was first in the class of 1974.

14. After the lease was up at Cathedral Mansions in the summer of 1932, the family moved to a first-floor apartment at 2601 Calvert Street, Northwest, the westernmost of three buildings situated midway between the Wardman Park and Shoreham Hotels, which together were known as the Wardman Annex. The family lived there until they left Washington in June 1935.

15. I would like to thank Lt. Gen. Ernest Graves Jr., USA, retired, for providing me with a copy. The passages quoted here are from pages 16–17, 73, and 166–167.

16. Interview with Lt. Gen. Ernest Graves Jr. (USA, retired) by Robert S. Norris, December 29, 1997.

17. There is much evidence to suggest that Groves kept the entire Manhattan Project in his head.

18. Letter, Lt. Gen. Ernest Graves Jr., USA, retired, to Robert S. Norris, January 6, 1998.

19. New Dealer and Secretary of the Interior Harold L. Ickes would later say that the Army Corps of Engineers "is the most powerful and most pervasive lobby in Washington, . . . the political elite of the army, . . . the perfect flower of bureaucracy." Foreword to Arthur Maass, *Muddy Waters: The Army Engineers and*

the Nation's Rivers (Cambridge, Mass.: Harvard University Press, 1951), ix.

20. One of the more famous examples of a senior officer taking a direct and personal role as mentor in the education of a junior officer is that of Maj. Gen. Fox Conner (1874–1951, class of 1898) and Maj. Dwight D. Eisenhower, though almost every general who assumes high rank has the experience.

21. John Clifford Hodges Lee was born in 1887 and graduated 12th in the class of 1909. He was in Washington from September 1931 to 1938 and got to know Groves. Some used his initials as the nickname Jesus Christ Himself, sometimes shortened to just Himself. Thomas Bernard Larkin was born in 1890 and graduated 21st in the class of 1915. He was with Pershing in Mexico in 1916 and probably first met Graves there. He served in France during World War I and may have had further contact with Graves. In October 1933 he undertook, as district engineer, the construction of the Fort Peck Dam. When Groves was in Kansas City as assistant to the division engineer he no doubt came to know Larkin. Philip Bracken Fleming was born in 1887 and graduated first in the class of 1911. In 1932 he was made deputy administrator of the Public Works Administration. Raymond Albert Wheeler was born in 1885 and graduated fifth in the class of 1911. He was chief of engineers from October 4, 1945, to February 28, 1949. Glen Edgar Edgerton was born in 1887 and graduated first in the class of 1908. From 1933 to 1936 he was in charge of the Rivers and Harbors Division in Washington. Max Clayton Tyler was fourth in the class of 1903, the class in which Douglas MacArthur was first. John Jennings Kingman was the son of a former chief of engineers and graduated fourth in the class of 1904. From August 1930 to September 1933 Lieutenant Colonel Kingman was chief of the Rivers and Harbors Division, OCE. He later served as assistant chief of engineers until his retirement on November 30, 1941.

22. Charles E. Kirkpatrick cited many examples of World War II generals being sponsored in the 1920s and 1930s by older officers, with George Marshall's name appearing with great regularity. Kirkpatrick, "Gaps," 15–19.

23. Two of Groves's contemporaries had similar experiences. Thomas Dodson Stamps was perhaps Groves's best friend. Three years older than Dick, Stamps graduated seventh in the class of August 1917. They got to know one another during overlapping service with the 1st Engineers at Fort DuPont, Delaware, from May 1928 to October 1929. Later they were together in Washington in the chief's office from August 1933 until June 1935. From 1938 until 1956 Stamps served as professor and head of the Department of Civil and Military Engineering at West Point (in 1942 the name was changed to Military Art and Engineering) and then dean of the academic board. Upon retirement in 1957 he settled in the Washington area, where he remained until his death in April 1964. Lucius Du Bignon Clay was one year older than Groves and graduated 27th in the class of June 1918. At certain junctures they served together and learned under some of the same mentors, overlapping assignments in the chief's office from mid-1933 to mid-1935.

24. The button, dating from the War of 1812, shows the masonry of the bastion of a marine battery, embrasured, casemated, and crenellated, surrounded by water, a rising sun with rays, all surmounted by a soaring eagle bearing in its beak a streamer displaying the motto ESSAYONS. The French word is easily translated as "let us try." According to an official history the self-confident corps stretched its meaning to "let us succeed." Blanche D. Coll, Jean E. Keith, and Herbert H.

Rosenthal, *The Corps of Engineers: Troops and Equipment* (Washington, D.C.: Office of the Chief of Military History, Department of the Army, 1958), 4. The buttons of the other officers are inscribed with E PLURIBUS UNUM, "out of many, one," the motto of the United States. Two other insignia identify the corps. The Turreted Castle was adopted in the 1840s. The device was worn as a collar adornment beginning in 1892. The official Seal of the Corps, sometimes referred to as the Coat of Arms, was adopted shortly after the Civil War to commemorate the consolidation of the Corps of Topographical Engineers with the regular Corps of Engineers established in 1802. The larger shield is divided into three horizontal sections; the top is solid blue while the bottom is vertical red and white stripes. The center sections show the original shields of the two historic corps. An eagle stands atop and a garland surrounds the larger shield.

25. He also had the additional responsibility of being quartermaster for transportation for the OCE until he was relieved on October 25, 1934. Basically this meant taking care of the OCE's travel arrangements.

26. Wilby, third in the class of 1905 (just behind Pot Graves), later served as superintendent of West Point from January 1942 to September 1945, when Groves's son Richard was a cadet. Obituary, *Assembly,* summer 1966, 106.

27. *Annual Report of the Chief of Engineers,* June 30, 1932.

28. Lt. Col. James B. Cress, Engineers, USA, *The Development of Anti-Aircraft Searchlights,* Occasional Paper of the Engineer School, U.S. Army, (Washington, 1940).

29. Groves provided some interesting detail in a late interview. "The Engineers at that time were responsible for antiaircraft searchlights, and I was not a bit satisfied with the sound locators. As the altitude of planes grew, the sound locator became rather hopeless. So I started a project on heat detection and it was one of the first large, secret contracts that was ever issued by the War Department as far as I know. It was supposed to be completely secret except for the general staff section that approved it and got the money. Even the contract papers were sealed and put in a safe in one section of the Chief of Engineers' office. It was written on the envelope that it could not be opened except by order of the Chief of Engineers or by me, which was most unusual in those days." Interview with LRG by Al Christman, May 1967, Naval Historical Center, Washington, D.C., 1.

30. Groueff, *Manhattan,* 278–280.

31. Like all other government employees, he did not work on Wednesday afternoons but did on Saturday mornings.

32. RHG, *War Stories,* 126. The Giants won the series in five games.

33. His travel orders, as compiled by his son, show a grueling schedule. In 1933 there were 7 trips, in 1934 there were 34, and there were 10 more in the first half of 1935. *LRG Official Records,* 218–281.

34. D. W. Niven to Chief of Engineers, June 26, 1935. Folder, LRG, Commendations and Qualifications, Box 3, Entry 7530C, Papers of LRG, RG 200, NARA.

35. On September 11, 1934, Groves appeared before the Washington Promotion Examining Board to determine his physical fitness for active duty and promotion. His medical records were submitted. They showed: Vincents angina, November 7–21, 1919; measles, February 8–24, 1923; a description of the wounds resulting from the accident in Vermont on February 2, 1928; and a short bout with dengue fever January 24–30, 1931, while in Nicaragua. On

September 18 the board recommended his promotion, and so, after 14 years as a first lieutenant, Groves was promoted to captain on October 31, 1934, with date of rank October 20, 1934. For years the War Department appealed to Congress to provide for more flexibility in the officers' promotion system, which was based exclusively upon seniority in length of commissioned service for grades up to and including colonel. Finally, in 1934 Congress enacted into law a system that accelerated promotion for every officer below the grade of colonel. For a discussion of the long-standing problem and its resolution see the *Annual Reports of the Chief of Staff* (Gen. Douglas MacArthur) for 1933, 1934, and 1935.

CHAPTER 8 **Finishing Schools (June 1935–July 1939)**

1. Tyler, *Leavenworth;* Boyd L. Dastrup, *The US Army Command and General Staff College: A Centennial History* (Manhattan, Kans.: Sunflower University Press, 1982); Arthur J. Stanley Jr., "Fort Leavenworth: Dowager Queen of Frontier Posts," *Kansas Historical Quarterly* 42, no. 1 (spring 1976), 1–23. When Groves attended it was called the Command and General Staff School. The name was changed in 1946.

2. RHG, *War Stories,* 173–207.

3. Groves family legend has it that Mr. Roberts baptized Sacagawea, the Indian girl who accompanied Lewis and Clark. The family historian, Richard H. Groves, after researching the topic in some detail, believes that the elderly woman who turned up on the reservation in the late 19th century was not the Lewis and Clark Sacagawea.

4. It was started as Hoover Dam. For a period in the 1930s it was known as Boulder Dam.

5. One reason for accelerating Richard's education at every point may have been to save money by having him enter Deerfield, or another prep school, as a sophomore.

6. For the 1928–1929 academic year the course of instruction was lengthened to two years; seven classes, from 1930 to 1936, graduated from the two-year course. To help meet the goal for more trained officers, the C&GSS reverted to a one-year course in 1935 and doubled the annual output of staff officers. The classes from 1936 to 1940 were one year. Timothy K. Nenninger, "Creating Officers: The Leavenworth Experience, 1920–1940," *Military Review,* November 1989, 58–68; *Annual Report of the Command and General Staff School, Fort Leavenworth Kansas, 1935–1936* (Fort Leavenworth, Kans.: C&GSS Press, June 20, 1936). Two students in the second-year class who would go on to hold high positions were Maj. Lucian K. Truscott Jr. and Capt. Albert C. Wedemeyer.

7. Of the eight two were lieutenant colonels, four were captains, and two were majors. One major apparently did not go to West Point, as he is not listed in the *Register of Graduates.* Of the six others, besides Groves, who did go to West Point, four were in classes prior to his, one was in a later class, and one was a classmate. The seven ranked in their classes as follows: 1st, 4th (Groves), 5th, two at 7th, one at 17th, and one at 25th. Over their careers three made brigadier general, and one major general; Groves retired as a lieutenant general. One died in the war and another retired disabled as a colonel. In short, the engineers that year were well represented, and most of them went on to have distinguished careers.

8. Infantry sent 42 students; field artillery, 18; air corps, 17; and cavalry and coastal artillery, 10 each.

9. Nenninger, "Leavenworth and Critics," 201–202.

10. Helmer W. Lystad (class of 1920) was a tennis-playing friend of Dick's, taking the second year of the two-year course. He shot himself on February 1, 1936, while being treated for a nervous breakdown. The obituary in the *Annual Report of the Association of Graduates of the USMA, 1936,* 268–270, does not mention how he died. His widow, Marcia, was cadet hostess when Richard was at West Point in the 1940s, and many years later at Distaff Hall was one of Mrs. Groves's best friends.

11. Letter, Jane Griffith Groves to MNG, July 23, 1923, LC. Chaplain Groves chain letter, November 3, 1925, LC.

12. Nenninger, "Leavenworth and Critics," 225; *Schedule for 1935–1936 One-Year Class, The Command and General Staff School, Fort Leavenworth, Kansas, 1935.* Among the instructors that year were Lt. Col. Lewis H. Brereton and Maj. Lewis A. Pick. Lieutenant General Brereton was chairman of the Military Liaison Committee from August 1946 to April 1948, when Groves served for a year as one of the army members. Pick was the chief of engineers from March 1, 1949, to January 26, 1953, a job that Groves had his eye on.

13. The Germans developed this system, known as the Kriegspiel, a war game played out on a map to teach military tactics to officers. It is the basis for the so-called case study method that the Harvard Business School claims to have discovered, which it did — at Leavenworth.

14. Tyler, *Leavenworth,* 4. "The entire curriculum emphasized the command process involving interaction between commanders and general staff officers, and tactical decision making." Nenninger, "Leavenworth and Critics," 203.

15. Nenninger, "Leavenworth and Critics," 226. Groves's son Richard has described the core of the training. "The primary vehicle of instruction was the map exercise. The student was required to play a role (e.g., Division, G-3) and, employing the problem-solving techniques he had been taught through the estimating process, to decide what he should do and then produce the necessary written orders for doing it. In the course of the year, he would have gone through this routine hundreds of times — enough so that it remained a part of him throughout the remainder of his service." Interview with RHG by Robert S. Norris, December 15, 1998.

16. The only American battle that qualifies as a Cannae is the battle of Nashville, where Gen. George H. Thomas's Union forces annihilated John Bell Hood's Rebel troops on December 15–16, 1864. In a press conference during the Gulf War of 1990–1991, Gen. Norman Schwartzkopf referred to the battle of Cannae to help him describe his strategy to defeat the Iraqi army.

17. Alfred von Schlieffen (1833–1913) was German chief of staff from 1891 to 1906. His plan — based on the Cannae principle of envelopment — for invading Belgium and France was not understood by his successor, Helmuth von Moltke the Younger (1848–1916), whose alterations ultimately destroyed it in the German attack of August 1914, setting the stage for the four years of trench warfare that followed. Gunther E. Rothenberg, "Moltke, Schlieffen, and the Doctrine of Strategic Envelopment," in Peter Paret, ed., *Makers of Modern Strategy: From Machiavelli to the Nuclear Age* (Princeton, N.J.: Princeton University Press, 1989), 296–325. Shortly after Schlieffen's death, his published

articles and public speeches were collected under the title *Gesammelte Schriften,* 2 vols. (Berlin 1913). An abbreviated edition of these collected writings appeared in 1925 under the title *Cannae.* The Command and General Staff School Press at Fort Leavenworth, Kansas, published an abbreviated English translation of the Cannae articles in 1931 along with a box of 101 maps.

18. In Groves's time, upon graduation students were rated superior, excellent, very satisfactory, and satisfactory. A recommendation was also made as to eligibility for further military education. Groves's marks could not have been better at the corps level, where he was rated superior in all five staff positions. At the division level he received three excellent and two very satisfactory ratings. He was also graded superior in academic rating and was recommended for further training in high command and General Staff duty. *LRG Official Records,* 301. Army Chief of Staff Gen. Malin Craig spoke at graduation.

19. Kirkpatrick, "Gaps," 13. Kirkpatrick argued that the C&GSS "was not the place that educated the judgment of the great leaders of World War II." "[T]he curriculum and methods of instruction did not lend themselves to that purpose.... Education in this sense ... was never the intent of Leavenworth, and it would be unfair to condemn the institution for failing to do something that was never its task." He concluded that if men such as Patton, Eisenhower, and Groves came away from Leavenworth with greater knowledge and abilities than their peers, it is because they probably entered with greater talents and worked harder.

20. There has been an ongoing debate, then and now, about the course of instruction at Leavenworth in the 1920s and 1930s. While many believe that Leavenworth (and the War College) kept the army alive in the interwar years, there are critics who think that the army schools were provincial, discouraged innovative thinking, rewarded conformity, and did not do as much as they could have to identify future combat commanders. "[T]he instruction rarely challenged the superior officer, and some even considered it frankly rote. The course was certainly difficult, as many contemporary accounts demonstrate, but the difficulty lay rather in the assimilation of a vast amount of diverse material, rather than in mastering subjects that were themselves complex. Students mastered the multifarious details of military organization and operations planning, learning approved methods of staff procedure and a common approach to problem solving. This methodology was far more important than the actual solution to any problem. The faculty produced a consensus 'school solution' to each tactical problem, and graded students on how closely their answers approximated the school solution." Kirkpatrick, "Gaps," 5–6. See also Nenninger, "Leavenworth and Critics." The combat generals who would gain fame in World War II were well read in military history and biography. Kirkpatrick provided a survey of their reading habits. Ibid., 20–32. There is ample documentation that the ongoing study of the profession of arms, supplemented by the formal schools, was a key contributor to their success. "Diligent and thorough independent study of the profession of arms, singly and in groups and occasionally aided by mentors, accounts for the qualities the very best American generals displayed." Ibid., 33.

21. Kirkpatrick, "Gaps," 32; Robert H. Berlin, "United States Army World War II Corps Commanders: A Composite Biography," *Journal of Military History* 53, no. 2 (April 1989), 147–167.

22. *LRG Official Records,* 294–295.

23. It was designed by Henry Tanner, a Kansas City architect who had gone broke in the depression and ended up working for the Army Corps of Engineers. He also designed the planned community of Fort Peck City. Montgomery, "Dust," 95.

24. Robert L. Branyan, *Taming the Mighty Missouri: A History of the Kansas City District Corps of Engineers, 1907–1971* (Kansas City, Mo.: USACE, 1974), iii–iv; *Army List and Directory,* U.S. War Department, Adjutant General's Office, October 20, 1936, 238.

25. Richard Curtis Moore was born in 1880 and graduated 12th in the West Point class of 1903. He had a varied career with the Corps of Engineers, and assumed high staff position. From July 22, 1940, to March 8, 1942, he was deputy chief of staff to Gen. George C. Marshall, a crucial period of Groves's life. He also served as the army member on the NDRC and OSRD from January 17, 1941, to March 31, 1942. Obituary, *Assembly,* winter 1967, 100. Groves felt that at an early stage Moore did not act quickly enough toward mobilization.

26. Fort Peck was a Works Progress Administration (in 1939 renamed Work Projects Administration, WPA) and Public Works Administration (PWA) project. Executive Order 7034 established the WPA on May 6, 1935.

27. The dam was virtually finished two years earlier, when on September 22, 1938, five million cubic yards of the earthen dam slid into the Missouri River. Eight men lost their lives. Six of them remain buried somewhere in the dam. A board was convened to assess the causes and work resumed, though the design was completely changed. In all 59 men lost their lives at Fort Peck. On October 11, 1940, the last load of material was dumped, topping it out at 250 feet. "Fort Peck — A Half-Century and Holding," *District News* (Omaha District), Special Edition 11, no. 2 (summer 1987). Montgomery, "Dust," 94–111.

28. "We have forgotten, in this service-industry, computer-driven age, that engineers once carried authority, that great civil works were not regarded with suspicion, that changing the landscape was the patriotic thing to do." Montgomery, "Dust," 100. See also Todd Shallat, "Engineering Policy: The U.S. Army Corps of Engineers and the Historical Foundation of Power," *Public Historian* 11, no. 3 (summer 1989), 7–27.

29. "10,000 Montana Relief Workers Make Whoopee on Saturday Night," *Life,* November 23, 1936, 9–16. Almost all of the photographs depict what the editors called "a human document of American frontier life" in the midst of the depression. Ernie Pyle, later a Pulitzer Prize–winning war correspondent, visited, and his account was published in the *Washington Daily News,* September 1, 1936. In Wheeler, Pyle found, "The taverns open at 8 in the evening and run 'til 6 in the morning. At night the streets are a melee of drunken men and painted women." The cover photo was mislabeled. Actually, it is not the dam but the concrete piers for the highway bridge and release gates at the spillway.

30. Clark Kittrell was born on March 18, 1893, and graduated 10th in the class of April 1917. He was an instructor in mathematics at West Point from 1918 to 1922. After graduating with a bachelor of science degree in civil engineering from MIT in 1923, he was at Schofield Barracks commanding Company D, 3rd Engineers, from 1923 to 1926. From 1933 to 1940 he was construction engineer and later district engineer, Fort Peck Dam. During the Big Slide Kittrell was almost killed when the dam crumbled.

31. Wesley G. Jones, "A Conversation with an Engineer Legend," *Assembly*, September 1992, 20 (emphasis added). I will return to this perceptive sentence later. Among Chorpening's many duties at Fort Peck was serving as public relations officer. He escorted Margaret Bourke-White while she took her photographs for *Life* magazine. He was somewhat dismayed that there were numerous pictures of town life and only one (the cover) of the dam. Chorpening died in 1997. See also *Engineer Memoirs, Major General Claude H. Chorpening*, Oral History Interview, October 12–14, 1981, Office of History, Headquarters, U.S. Army Corps of Engineers, Fort Belvoir, Va. Other November 1918 classmates at Fort Peck were Arthur William Pence, who graduated just ahead of Groves and was in charge of the tunnels. David A. D. Ogden (13th in the class and went with Groves to the Engineer School and Europe) was chief of construction, under Larkin, until 1937. Two other classmates were Richard Lee (number 20), and John Bell Hughes (number 70). Ewart G. Plank (class of 1920) served with Groves in Nicaragua on the canal survey. Plank worked at Fort Peck for five years, where he unofficially became known as "Mayor" of Fort Peck, Montana.

32. His great-grandfather Amiel Weeks Whipple (class of 1841) has his name on Battle Monument, West Point's most prestigious adornment. His grandfather Charles W. (class of 1868) and uncle Sherburne (class of 1904) also graduated, though his father did not.

33. *Autobiography, William Whipple, Jr.*, 80. A copy is at the Office of History, Headquarters, U.S. Army Corps of Engineers, Fort Belvoir, Va.

34. Letter, William Whipple Jr. to Robert S. Norris, March 3, 1998.

35. Ball, *Command*; Phyllis I. McClellan, *Silent Sentinel on the Potomac: Fort McNair, 1791–1991* (Bowie, Md.: Heritage Books, 1993), 131–146.

36. Ball, *Command*, 68–69.

37. War Department, Adjutant General's Office, Detail as Students, 1938–1939, Course, Army War College, November 17, 1937. Folder, LRG, Commendations and Qualifications, Box 3, Entry 7530C, Papers of LRG, RG 200, NARA.

38. Groves was the youngest of the engineers. Lt. Col. Joseph C. Mehaffey was 3rd in the class of 1911, Lt. Col. Douglas L. Weart was 10th in the class of 1915, Capt. Clarence L. Adcock was 18th in the class of June 1918. Maj. Lewis A. Pick, an eventual chief of engineers after the war, was born in 1890 and did not attend West Point.

39. Among his classmates were Dr. Paul R. Hawley, who had been to Nicaragua in 1929–1931 with Groves. They soon would go there again for a short trip, and have a very interesting contact later in the war working on Project Peppermint. Another classmate was Oliver P. Echols from the air corps. Beginning in 1943 Groves would work closely with Echols to modify a number of B-29 bombers for atomic bomb delivery. Classmate Hoyt S. Vandenberg graduated 240th in the class of 1923. After rising fast through the ranks during the war, he served as assistant chief of staff, G-2 (Intelligence Division), in 1946. He was then appointed director of the Central Intelligence Group (CIG) for its short life before it was absorbed into the newly established Central Intelligence Agency in September 1947. In September 1947 the air force was created as a coequal service; Vandenberg became the first vice chief of staff (under Spaatz) and then chief of staff (1948–1953). As we shall see, Groves's MED foreign intelligence files were transferred to the CIG in 1947.

40.Ball, *Command,* 248.

41.Ball, *Command,* 249.

42.*Douglas Southall Freeman on Leadership,* edited with commentary by Stuart W. Smith (Shippensburg, Penn.: White Mane Publishing Co., 1993), 164–174. In 1942–1944 Freeman's three-volume work, *Lee's Lieutenants,* was published, a work perhaps better known than the Lee biography and still in print. Freeman then turned to George Washington. In 1958 his seven-volume biography (completed by his assistants) was awarded the Pulitzer Prize, though he had died in 1953.

43.Letter, LRG to Douglass Southall Freeman, August 18, 1948. Papers of LRG, Folder Shi-Str, Box 9, Entry 7530B, RG 200, NRA.

44.Roger H. Stuewer, "Bringing the News of Fission to America," *Physics Today,* October 1985, 49–56. For a contemporary account see "The Fifth Washington Conference on Theoretical Physics," *Science,* February 24, 1939, 180–182.

45.*LRG Official Records,* 321–322. Groves and DeWitt would soon meet again on the issue of Japanese-American internment after Pearl Harbor.

CHAPTER 9 Final Rehearsal (July 1939–Summer 1942)

1. *The Army Directory* of April 20, 1940, lists 110 officers. Groves was one of only five from the Corps of Engineers. Of the 14 captains serving, 4, including Groves, were from the class of November 1918. He would soon be involved with several others: Lt. Col. Thomas T. Handy, Maj. Clayton L. Bissell, and Capt. John H. Hinds. His football coach, Lt. Col. Geoffrey Keyes, and Lt. Col. Omar Bradley also served at the same time. *The Army Directory,* April 20, 1940 (Washington, D.C.: War Department, Adjutant General's Office).

2.LRG, *Grandchildren,* p. 156. While he was away his wife and son moved the family possessions from 3132 Oliver Street, Northwest, after the lease expired, to 3508 Thirty-sixth Street, Northwest.

3.The courtship was a lengthy one. In a letter to her brother Owen in June 1928 she informed him that they were engaged. Letter, Gwen Groves to Owen Groves, June 1928, LC. Gwen was not happy with a wedding gift sent by Grace's sister, Mary Wilson. "A gift arrived from Mary with 17 cents postage and an engraved card. Inside was the tackiest china wedding couple imaginable. Cost 15 cents at the most. It was not even good enough to put on the wedding cake. Just after I had written a note of acknowledgement Mary's note came telling me not to take the trouble to thank her 'for the sweet little bridal couple.' We have decided she is nuts or were they trying to be insulting?" Letter, Gwen Groves Brown to MNG, July 12, 1940, LC.

4.Groves would spend five months (December–April) with the Operations and Training Division serving in the Mobilization Branch and the last two months (May and June) in the Air Branch. Interview with General Groves, June 19, 1956, USACE, Office of History, Remington and Fine, Military Construction, Interviews and Comments, VII, Box 32. The assistant chief of staff (G-3) from August 4, 1939, to November 22, 1940, was Brig. Gen. Frank Maxwell Andrews, the first airman to have such an assignment. General Andrews was killed in an aircraft accident in Iceland on May 3, 1943. On February 7, 1945, Camp Springs Army Air Field, outside Washington, D.C., was renamed Andrews Field. After the establishment of the air force in 1947, it became Andrews Air Force Base.

5.Gerhard L. Weinberg, *A World at Arms: A Global History of World War II* (New

York: Cambridge University Press, 1994), 84. For an extensive review of current trends in the historiography of the period, with particular attention to U.S. policy, see Justus D. Doenecke, "Historiography: U.S. Policy and the European War, 1939–1941," *Diplomatic History* 19, no. 4 (fall 1995), 669–698.

6. The nation's air force was known as the Army Air Corps from 1926 until June 20, 1941, when it became the U.S. Army Air Forces. On September 18, 1947, the U.S. Air Force became a separate military service.

7. Harold I. Gullan, "Expectations of Infamy: Roosevelt and Marshall Prepare for War, 1938–41," *Presidential Studies Quarterly* 28, no. 3 (summer 1998), 510–522.

8. "If war came, [engineer] troops were to clear the way and build; to survey and map; to supply water and electricity; to develop materials and techniques for camouflage; to operate railroads." Blanche D. Coll, Jean E. Keith, and Herbert H. Rosenthal, *The Corps of Engineers: Troops and Equipment* (Washington, D.C.: Office of the Chief of Military History, Department of the Army, 1958), 2–3. In November 1942 railroad operations were transferred to the newly created transportation corps.

9. Fine, Remington, *Corps,* 244–272.

10. Fine, Remington, *Corps,* 84. In September 1939 the QMC consisted of less than 12,000 military personnel and approximately 37,000 civilians. By war's end it would number 500,000 military and 75,000 civilians. Erna Risch, *The Quartermaster Corps: Organization, Supply, and Services,* vol. 1 (Washington, D.C.: Office of the Chief of Military History, 1953), 7–8.

11. Fine, Remington, *Corps,* 111. In the fall of 1939 army personnel numbered approximately 200,000. By November 1944 the army had housed and trained six million troops.

12. On September 16, 1940, President Roosevelt signed the Selective Service and Training Act. A month later 16,400,000 Americans had registered for the draft. Under the draft law, no men could be conscripted until "shelter, sanitary facilities, water supplies, heating and lighting arrangements, medical care, and hospital accommodations" had been provided for them. Fine, Remington, *Corps,* 150.

13. Fine, Remington, *Corps,* 519. It is worth mentioning that there were differences of opinion within the General Staff about how fast mobilization should occur and how much money should be spent.

14. Fine, Remington, *Corps,* 151.

15. Keith E. Eiler, *Mobilizing America: Robert P. Patterson and the War Effort, 1940–1945* (Ithaca, N.Y.: Cornell University Press, 1997).

16. The assistant chief of staff, G-3 (Operations and Training), was Brig. Gen. Frank M. Andrews. Under him was Lt. Col. Harry L. Twaddle, Groves's superior. Brig. Gen. Richard C. Moore, former Missouri River Division engineer, was the G-4 (Supply) until he was replaced by Col. Eugene Reybold, also an engineer. Moore became deputy chief of staff on July 22, 1940, and served until March 8, 1942. He was promoted to major general on February 2, 1941. Reybold would replace Major General Schley as chief of engineers on October 1, 1941, and serve until October 4, 1945.

17. Fine, Remington, *Corps,* 159.

18. Memo to File, Assessment of General Bryan [*sic*] B. Somervell, His Strengths and His Weaknesses, n.d., Folder Somervell, Box 5, Entry 7530J, Papers of LRG, RG 200, NARA.

19. *LRG Official Records,* 326.

20. Schley (class of 1903) was district engineer at Galveston beginning in 1924. After eight years in the Panama Canal Zone he was appointed chief of engineers in October 1937 and served until his retirement on September 30, 1941. Obituary, *Assembly,* fall 1965, 81.

21. Memo to the Adjutant General, July 11, 1940, Folder, Biographies and Records Concerning the Life of LRG, Box 1, Entry 7530C, Papers of LRG, RG 200, NARA.

22. "By direction of the President, Major Leslie R. Groves, Jr., (Corps of Engineers), General Staff Corps, is relieved from detail as member of the General Staff Corps, from assignment to the War Department General Staff, and from duty in the Office of the Chief of Staff, Washington, DC, effective 22 July 1940, and is assigned to duty in the Office of the Quartermaster General, Washington, DC, and will report to The Quartermaster General for duty." WDSO No. 171, July 22, 1940, Folder, Biographies and Records Concerning the Life of LRG, Box 1, Entry 7530C, Papers of LRG, RG 200, NARA. After hearings and deliberations throughout the summer of 1940, Congress authorized the National Guard and reserves to federal duty for 12 months on August 27, and on September 16 passed the Selective Service and Training Act, requiring all men between the ages of 21 and 35 to register.

23. Fine, Remington, *Corps,* 159. Interview with General Groves, June 19, 1956, USACE, Office of History, Remington and Fine, Military Construction, Interviews and Comments, VII, Box 32.

24. LRG Appointment Book, November 13, 1940; interview with General Groves, June 19, 1956, USACE, Office of History, Remington and Fine, Military Construction, Interviews and Comments, VII, Box 32. On September 12, 1940, Groves requested that his name be changed in all official records from Leslie R. Groves Jr. to Leslie R. Groves, in view of the recent death of his father. *LRG Official Records,* 343.

25. Roscoe C. Crawford, number three in the class of 1912 and the relevant engineer in this calculation, was appointed colonel on October 16, 1940, less than a month before Groves (November 13, 1940), who was eight classes behind. Francis K. Newcomer and William H. Holcombe, first in the classes of 1913 and 1914, respectively, were both appointed colonel on June 26, 1941.

26. Groves was not unique in this regard. Notable at the time was the advance of Laurence Sherman Kuter (44th in the class of 1927), who went from captain to brigadier general in less than two years.

27. Michael J. (Jack) Madigan was born in approximately 1895 in Danbury, Connecticut, but grew up in western Pennsylvania. He received little formal education and from an early age worked for coal-mining companies. He moved to New York and came to the attention of Robert Moses. Madigan's firm, Madigan-Hyland, played a leading role in engineering the Henry Hudson and Belt Parkways, the Marine and Cross Bay Parkway Bridges, the Bronx-Whitestone Bridge, and the Tappan Zee Bridge, helping make Madigan a millionaire. Robert Patterson was assistant secretary of war from July 31, 1940, to December 15, 1940, and on December 16 assumed the new position of undersecretary of war. In September 1940 he appointed Madigan his special assistant.

28. Fine, Remington, *Corps,* 257–258. The number of people employed under the Construction Division went from 5,380 in July to 396,255 in December. Ibid., 222.

29. Groves later recalled saying at this point "that Somervell would be a very fine man for the job if he could be obtained but that he might be a little difficult to handle as he had never displayed too much of the trait of loyalty toward his superiors." LRG, Memo to File, Somervell, March 30, 1967, Folder Somervell, Box 5, Entry 7530J, Papers of LRG, RG 200, NARA.

30. Groves had no choice, but Somervell apparently took it as an affront. As Groves later recounted, he did not want to have his orders changed. "He [Somervell] appeared to be quite upset by this decision which I made instantly and it was quite apparent to me that he felt that anyone who was offered the opportunity to be one of his assistants should jump at it no matter what other prospects he might have." Memo to File, Assessment of General Bryan *[sic]* B. Somervell, His Strengths and His Weaknesses, n.d., Folder Somervell, Box 5, Entry 7530J, Papers of LRG, RG 200, NARA.

31. Ohl, *Supplying;* Millett, *Organization,* 1–8; Fine, Remington, *Corps,* 256; USACE, Office of History, Biographies; Charles J. V. Murphy, "Somervell of the S.O.S." *Life,* March 8, 1943, 82–94; Obituary, *Assembly,* October 1955, 63.

32. Somervell was one of only nine officers who had been awarded both, making him someone to watch. His immediate superior in France was Lt. Col. Ernest Graves. In his fitness report of Somervell, Graves noted, "this is the best officer I ever saw, or hope to see." Ohl, *Supplying,* 15, 13. The circumstances of his receiving the DSC were as follows. "Leaving his construction assignment on October 24, [1918] Somervell decided first to visit some friends at the headquarters of the 89th Division, which was taking part in the Meuse-Argonne Offensive. . . . Somervell volunteered his services to the division's chief of staff, Colonel John C. H. Lee. . . . The final push to the Meuse River began on November 1, and by November 5, the 89th was at the Meuse in the proximity of Pouilly. A reconnaissance early in the morning of November 5 indicated that all bridges in this area had been demolished by the retreating Germans, but another later that day indicated the bridge at Pouilly could be crossed. Lee and Somervell went forward to assess the situation, and, with the help of a sergeant, determined that men could apparently make their way across the bridge if they went single file. Lee then asked Somervell to conduct his own reconnaissance. Just before midnight, Somervell and two scouts wormed their way across the bridge. Once on the far side of the river they encountered a detachment of Germans and in a brief flurry of shooting, drove them off. In all, Somervell and the two scouts had advanced more than 500 yards beyond the American outposts, a feat of 'extraordinary heroism in action with the enemy.'" Ohl, *Supplying,* 13–14.

33. The enormous project called for a 195-mile canal (four times the length of the Panama Canal) across northern Florida, with 90 miles of it cut through earth and rock. Existing waterways would need to be widened, deepened, and straightened as well. In 1936 Congress rejected any further funding. Ohl, *Supplying,* 23–24.

34. The Works Progress Administration was established by Executive Order 7034 on May 6, 1935, and headed by Hopkins. There were 49 programs, one for each state and one for New York City. It was renamed the Work Projects Administration in 1939.

35. As administrator he oversaw thousands of projects, kept more than 200,000 employed, and spent money at a rate of $10 million per month. His most

ambitious and expensive project was building North Beach Airport, renamed
La Guardia in 1940. Ohl, *Supplying,* 32.

36. Ohl, *Supplying,* 38.

37. Fine, Remington, *Corps,* 257; Interview with Michael J. Madigan, June 18, 1956,
USACE, Office of History, Remington and Fine, Military Construction,
Interviews and Comments, VII, Box 32, 5.

38. "[H]e had won a reputation as the kind of tough, uncompromising, ruthless
expediter that Marshall needed for a nasty job. 'He was efficient; he shook the
cobwebs out of their pants,' General Marshall declared after the war." Forest C.
Pogue, *George C. Marshall, Ordeal and Hope: 1939–1942* (New York: Viking Press,
1966), 297. At the end of 1940 Groves outranked Somervell, but on January 29,
1941, Somervell skipped colonel and was promoted to brigadier general.

39. See chart 6 in Fine, Remington, *Corps,* 261.

40. Fine, Remington, *Corps,* 260. This would be precisely the style that Groves
would soon adopt in running the Manhattan Project. In Somervell he was
learning from a master.

41. Fine, Remington, *Corps,* 363. In another encroachment upon the quartermasters' turf, on November 18 General Marshall ordered that construction at all
Army Air Corps stations (except those in Panama) be transferred to the engineers. See Ibid., 255, 259–267. "Actually, Somervell did not personally get rid of
many individuals. It always appeared to me that it distressed Somervell when he
had to personally handle matters of this kind, although he heartily approved of
such actions. For example, while most if not all of the twenty-odd
Constructing Quartermasters released were relieved by my direction, Somervell
never objected to such action." Comments on Draft of the History of Military
Construction, LRG to Maj. Gen. A. C. Smith, Office of the Chief of Military
History, July 22, 1955, Chapter VI, 2. Office of History, Chief of Engineers.

42. Groves's Operations Branch incorporated the former Fixed Fee, Lump Sum,
Procurement and Expediting, and Repairs and Utilities Branches.

43. Ohl, *Supplying,* 11. The speed of Somervell's rise through the ranks was breath-
taking. He went from lieutenant colonel to brigadier general on January 29,
1941, and a little more than 13 months later to lieutenant general, on March 9,
1942. Groves later said, "Somervell, like most aggressive and brilliant leaders,
resented control. He wanted to be independent and he was constantly working
in that direction." LRG Comments on Fine, Remington, ca. July 1955, Chapter
6, 3. Office of History, Chief of Engineers.

44. Fine, Remington, *Corps,* 377–378.

45. Many years later Groves said of Somervell, "He was by all odds the best
prospect in 1940 of the officers of his age group to be Commander in Europe
. . . [h]e had a brilliant record in World War I . . . [h]e had had wide experience
in the handling of important matters. He had always shown decisiveness and
personal courage in handling difficult situations. If he had been in command in
Europe, I believe that he would have been much more successful than General
Eisenhower was, and his abilities would have enabled us to win with fewer
casualties. I doubt if he would ever have been surprised in the Battle of the
Bulge. He was a most dynamic individual and extremely hard working.
Everyone who served under him respected these qualities." Memo to File,
Somervell, March 30, 1967, Folder Somervell, Box 5, Entry 7530J, Papers of
LRG, RG 200, NARA. In an earlier interview Groves felt there was, "no compar-

ison between Somervell and Eisenhower as to intelligence." Groves said he didn't think that Somervell liked him very much. He offered a reason that says more about Groves that it does about Somervell. Groves attributed this to the fact that he (Groves) was too smart; Somervell didn't want anyone so smart so close. Interview with General Groves, June 19, 1956, USACE, Office of History, Remington and Fine, Military Construction, Interviews and Comments, VII, Box 32, 5.

46. Fine, Remington, *Corps*, 373–378.

47. Fine, Remington, *Corps*, 280. There would be approximately 500 in the Operations Branch by mid-1941. Ibid., 441.

48. Fine, Remington, *Corps*, 293.

49. Fine, Remington, *Corps*, 294.

50. The GOCO model of ownership and management had been adopted earlier. The approach had been used to a limited degree by the United States, and by Great Britain, during World War I. Great Britain had called its approach the national factory system. The modern example in the United States was largely created to develop the ammunition base during the mobilization in World War II. Unlike other U.S. plants (such as those for tank production), these were to be government owned and contractor operated because no commercial use could be projected for them in peacetime. The War and Navy Departments built and owned the military-industrial facility. The services retained established corporations to operate the facilities, assuming responsibility for personnel, production schedules, quality control, and other tasks. See Robert H. Bouilly, "The Development of the Ammunition Industrial Base, 1940–1942," in Judith L. Bellafaire, general ed., *The U.S. Army and World War II: Selected Papers from the Army's Commemorative Conferences* (Washington, D.C.: U.S. Army, Center of Military History, 1998), 103–117; Fine, Remington, *Corps*, 309–341. Groves would use this model for the Manhattan Project; it would last into the 21st century.

51. The War Department divided its ammunition production facilities into ordnance works, which produced ordnance, and ordnance plants, which loaded ordnance. The War Department constructed 25 load, assemble, and pack (LAP) plants, 21 high-explosive/smokeless powder works, 12 chemical works, and 13 small-arms-ammunition manufacturing plants, at a total cost of $4.3 billion, twice as much as the Manhattan Project. Most were located in the interior of the country to minimize the dangers from enemy air raids.

52. Robert H. Bouilly, "The Development of the Ammunition Industrial Base, 1940–1942," in Judith L. Bellafaire, general ed., *The U.S. Army and World War II: Selected Papers from the Army's Commemorative Conferences* (Washington, D.C.: U.S. Army, Center of Military History, 1998), 105. Du Pont had contracts totaling more than $600 million for eight ordnance works: Alabama (Sylacauga), Chickasaw (Millington, Tennessee), Gopher (Rosemont, Minnesota), Indiana (Charlestown), Kankakee (Joliet, Illinois), Morgantown (West Virginia), Oklahoma (Pryor), and Wabash River (Newport, Indiana). See William Voigt Jr., Ordnance War Administration, Study No. 11, Monograph No. 1, GOCO Facilities Directory (April 5, 1945).

53. The War Production Board was established on January 16, 1942. In the summer of 1942 new priority ratings, ones that Groves would use for the Manhattan Project, came into effect and caused some confusion. In early September 1942 the War Production Board stripped the ANMB of its power to assign priorities.

54. From November 1940 to April 1, 1941, the quartermaster corps transferred approximately 80 air corps projects with a cost of $200 million to the Corps of Engineers. The projects included airfield construction, aircraft assembly plants, pilot schools, and the like. Fine, Remington, *Corps,* 267–272.

55. Groves used these telephone conversations as a method of control. "All telephone conversations between myself and members in the field were normally recorded and then transcribed. The purpose of this was to enable other members of my organization to learn of the situation at various points and particularly to enable them to fulfill promises which I made to the field. . . . Because of the volume, the transcriptions were hastily made. After they were made, these records were checked by my chief secretary (various persons until early 1941, from which time on, Mrs. O'Leary) and all copies marked. A copy was then sent to each officer responsible for the fulfillment of a promise. Upon fulfillment, the officer would report back to Mrs. O'Leary. If the report back was not received within the time limit given in the promise, Mrs. O'Leary would then call and continue to check until the promise had been fulfilled." Comments on Draft of the History of Military Construction, LRG to Maj. Gen. A. C. Smith, Office of the Chief of Military History, July 22, 1955, Chapter 5, 7. Office of History, Chief of Engineers.

56. Fred Sherill to LRG, with attachment, October 28, 1947, Folder Fred Sherill, Box 9, Entry 7530B, Papers of LRG, RG 200, NARA.

57. Numerous examples could be cited to show the constant pressure from Capitol Hill. On July 10, 1941, House Majority Leader McCormick called Groves about an appointment with representatives of a construction company. LRG Appointment Book. Groves has numerous examples of pressure by senators and congressman to build facilities in their states in his comments on the Fine and Remington manuscript.

58. Four years later at a tense moment when President Truman met Groves he said, "Oh we have known each other for a long time." Groves did not have a very high regard for the Truman Committee. As he later said, "It was not an investigating committee, but a prosecuting committee hunting for headlines, and that fact should be recognized." Comments on Draft of the History of Military Construction, LRG to Maj. Gen. A. C. Smith, Office of the Chief of Military History, July 22, 1955, Chapter 6, 10. Office of History, Chief of Engineers. Groves first knew Truman from his time in Kansas City. Some years later General Somervell wrote to Groves saying that a professor who wanted to write a book about the Truman Committee had approached him. Somervell said he was considering telling the professor the committee's real origins. "[T]hey threatened to form the committee to make our lives miserable if we did not give the contract for the St. Louis Small Arms Plant to a contractor named Deal, of St. Louis." Somervell said he did not want to embarrass Groves (or Styer or Gregory) about all of this, but felt "that perhaps the best thing is for the country to know the truth about the committee's birth and the motivation which activated the investigation." Groves replied with further details of a very involved episode. He did not have any firsthand knowledge of the threat, but remembered being told about it by Somervell at the time and urged him to tell what he knew. "Don't hesitate because of any possible embarrassment to me — I don't embarrass that easily." Somervell to LRG, July 24, 1952, and LRG to Somervell, September 8, 1952, Folder, Shi–Str, Box 9, Entry 7530B, Papers of LRG, RG 200, NARA.

59. Doyle-Russell & Wise would soon be chosen to build the Pentagon. In 1943 Groves selected J. A. Jones to build the steam power plant and the K-25 gaseous diffusion plant at Oak Ridge. The card notebooks and card catalog are in Box 2, Entry 7530E, Papers of LRG, RG 200, NARA.

60. Fine, Remington, *Corps,* 513.

61. For the six-month period from April through September the total was $3.3 billion. Fine, Remington, *Corps,* 606. In 2001 dollars this would be approximately $66 billion.

62. To name a few, the Iowa Ordnance Plant and the Pantex Plant, both built by Groves, would eventually become the two major nuclear weapon final-assembly plants. The Holston Ordnance Plant provided the high explosive for U.S. nuclear weapons for decades.

63. The Pentagon Data Book I and II (compiled by Gavin Hadden), and Folder, Blueprints and Miscellaneous Data, The Pentagon Project, Control Division, Army Service Forces, June 25, 1944, Box 1, Entry 7530E, Papers of LRG, RG 200, NARA; Goldberg, *Pentagon;* Fine, Remington, *Corps,* 431–439; Janet A. McDonnell, "Constructing the Pentagon," in Barry W. Fowle, ed., *Builders and Fighters: U.S. Army Engineers in World War II* (Fort Belvoir, Va.: Office of History, U.S. Army Corps of Engineers, 1992), 107–116.

64. *The WPA Guide to Washington, DC, The Federal Writers' Project Guide to 1930s Washington,* with a new introduction by Roger G. Kennedy (New York: Pantheon Books, 1983, originally published 1942), 224–225.

65. On June 11, in testimony before a House subcommittee, Chief of Staff General Marshall had proposed the Arlington site. In 1941 Arlington Cemetery did not extend as far eastward as it does today. The eastern boundary was Arlington Ridge Road (no longer there) running north–south from the Ord & Weitzel Gate to the Memorial Gate, to the McClellen Gate, to today's Service Gate. The Boundary Channel creating Columbia Island stretching from the Theodore Roosevelt Island to the lagoon prevented the eastward approach to the Potomac. At a later date it was partially filled in. Groves's brother Allen was working at the Experimental Farm when he died in 1916.

66. Groves and Casey had known each other since West Point. Hugh J. "Pat" Casey was one year ahead of Groves, graduating third in the class of June 1918 (originally the class of 1919). He also attended Engineer School and went on the European battlefield tour with Groves, and was in OCE in the Rivers and Harbors Division when Groves was also in Washington in the early 1930s. In September 1941, when Douglas MacArthur was recalled to active duty, he asked for Casey to be his engineer; Casey would eventually rise to be chief engineer officer for the Southwest Pacific Area. Somervell tried to keep him, saying, "Now look, Pat . . . if you do that, you're going to be going to somebody who's reached the top and won't go any further, whereas you stick with me, and I'm on the way to the top and I'd advise you to just carry on with me." *Engineer Memoirs: Major General Hugh J. Casey, US Army* (Washington, D.C.: Office of History, U.S. Army Corps of Engineers, December 1993), 141.

67. Bergstrom had designed the Hollywood Bowl, the Los Angeles Museum of History, Science, and Art, and large commercial structures. Goldberg, *Pentagon,* 16–17.

68. The new 750-acre Washington National Airport, just to the south at Gravelly Point, much of it on land dredged from the Potomac, was put into service in

June 1941. President Roosevelt had dedicated the terminal building in September 1940.

69. As assistant secretary of the navy in 1917, Roosevelt had urged President Wilson to build the "temporary" Munitions and Navy Buildings on the Mall, a mistake he did not want to make again. At a press conference on August 19, Roosevelt said of his role, "It was a crime . . . for which I should be kept out of Heaven, for having desecrated the loveliest city in the world — the Capital of the United States." Quoted in Goldberg, *Pentagon,* 27. The depot location kept the War Department building somewhat out of sight from the Washington side. See the political cartoon, Fine, Remington, *Corps,* 436. Work had just begun on the quartermaster depot. It was discontinued and transferred to a new site in Alexandria, at Cameron Station.

70. Ohl, *Supplying,* 50; Goldberg, *Pentagon,* 28. Three parcels made up the building site and adjoining grounds, all of it already owned by the government. The 305 acres included approximately 57 acres from the Experimental Farm, 146.5 acres from the airport, and 102 acres acquired for the quartermaster depot. Col. John J. O'Brien, Chief of Engineers, to Commanding General, Services of Supply, Acquisition of Land for New War Building, August 5, 1942, Folder, Pentagon Data Book I, Box 1, Entry 7530E, Papers of LRG, RG 200, NARA.

71. Fine, Remington, *Corps,* 437.

72. Somervell's original plan on July 21 was for five million square feet of gross interior floor space for up to 40,000 workers. At the end of August Roosevelt ordered this cut in half. Disregarding this, Somervell's revised goal as he broke ground was four million square feet. With additions along the way, the building eventually ended up being 6.24 million square feet.

73. McShain is the subject of a biography by Carl M. Brauer, *The Man Who Built Washington: A Life of John McShain* (Wilmington, Del.: Hagley Museum and Library, 1996). McShain's papers are at the Hagley Museum and Library. After the war McShain built numerous other buildings in the Washington area, including the White House renovation of 1950–1951, the National Shrine of the Immaculate Conception (1959), the British Embassy addition (1958), and the Kennedy Center for the Performing Arts (1971). McShain died on September 9, 1989, at the age of 90.

74. *LRG Appointment Book.*

75. When the building was proposed for the Arlington Farms site, it was bounded by five roads, which shaped the building into a pentagon. See an early sketch in Goldberg, *Pentagon,* 15. Though eventually the building was placed at the depot site, the pentagonal shape was retained. In a May 9, 1942, memo to all concerned the adjutant general stated that "the building in Arlington, now under construction for the War Department, is the 'Pentagon Building.'" Other shapes were considered, but many old forts were pentagonal and may have influenced Bergstrom and the others. Goldberg, *Pentagon,* 19.

76. Renshaw was born in 1907 and graduated from the USMA in the class of 1929. He served with the quartermaster corps on various construction projects during the 1930s. In August 1941 he was appointed to oversee the design and construction of the Pentagon. After the war until his retirement in 1960 Renshaw was in charge of numerous other large construction projects. Obituary, *Assembly,* March 1981, 128–129.

77. After Bergstrom resigned, Witmer became chief architect, on April 11, 1942. Goldberg, *Pentagon,* 36.

78. The December 1942 issue of *Airlanes* magazine was devoted to the Pentagon. Included were profiles of a few of the key people, including Groves, who by this time was four months into the Manhattan Project. There were also short pieces about concrete, electrical power, air-conditioning, roads and bridges, and advertisements taken out by the subcontractors, who described their contribution in the building of the Pentagon. "On this mammoth project we have completed the following, 6,000,000 cubic yards of excavation, 200,000 lineal feet of trench work grading, landscaping," said Potts & Callahan contractors, of Baltimore.

79. Pictures of Stimson's new offices were published in *Life,* December 21, 1942, 83–84, showing the private dining room for 26, kitchen, bathroom, personal wardrobe room, and private elevator, the only one in the Pentagon. Today 3E880 is the office of the secretary of defense, and 3E994 is the office of the deputy secretary of defense.

80. The addition of the three rings of the fifth floor extended the schedule by two months. The navy, though invited in late 1942, did not move into the Pentagon until August 1948. During the war it took over the vacated Munitions Building on the Mall. The Joint Chiefs of Staff remained in their offices in the four-story, white marble Public Health Service Building between 19th and 20th on Constitution Avenue, across the street from the Munitions Building. Today it is the Interior South Building. The Joint Chiefs moved to the Pentagon in April 1947.

81. Memo to the Commanding General, Army Service Forces, from LRG, Basic Data on the Pentagon, Arlington, Va., April 3, 1943, Folder, Pentagon Data Book I, Box 1, Entry 7530E, Papers of LRG, RG 200, NARA; Anthony F. Leketa, "The Pentagon Renovation Program," *Military Engineer,* July 1991, 31–35; Goldberg, *Pentagon,* 183–184.

82. "The location of the Pentagon in Virginia raised the question of racial segregation early in the construction. In March 1942 Groves inquired of Renshaw whether separate toilet facilities were being provided for whites and blacks as required by Virginia law." There was also the question of separate eating facilities. The engineers opted for separate facilities. "Moreover, Groves was either unaware of or ignored the president's Executive Order No. 8802 of 25 June 1941 which forbade discrimination because of 'race, creed, color, or national origin' in the federal government and by federal contractors." Roosevelt resolved the issue after a tour of the nearly completed building. Goldberg, *Pentagon,* 62.

83. McShain himself loved to see concrete poured and claimed that he supervised the record pour of 2,875 yards in one day. Carl M. Brauer, *The Man Who Built Washington: A Life of John McShain* (Wilmington, Del.: Hagley Museum and Library, 1996), 84.

84. They include, in ranking order, Roosevelt, Stimson, Patterson, Marshall, Somervell, Reybold, Robins (chief of construction), Groves, Renshaw, Bergstrom, Witmer, the contracting companies, and Hauck.

85. Louis Stockstill, "A Statistical Grandeur," *Journal of the Armed Forces,* May 18, 1968, 13.

86. Albert Joseph Engel was a Republican representative from Michigan who served from 1935 to 1951.

87. Fine, Remington, *Corps,* 460–476.
88. The War Plans Division of the General Staff, which set new goals of 4,150,000 men by the end of 1942 and 10,380,000 by June 30, 1944, revised this. Ohl, *Supplying,* 56.
89. Fine, Remington, *Corps,* 499. "Between December 1941 and August 1945, the Corps of Engineers called upon private architect-engineers and constructors to undertake emergency contracts totaling $8.5 billion [approximately $170 billion in 2001 dollars] — one third of all new construction performed in the United States during that period." Ibid., 562. This does not include $2 billion (approximately $40 billion in 2001 dollars) for the Manhattan Project and $550 million in civil projects.
90. Hewes, *Root,* 57–128; Frederick S. Haydon, "War Department Reorganization, August 1941–March 1942," *Military Affairs,* pt. 1 spring 1952, 12–29, pt. 2 fall 1952, 97–114.
91. We recall that in early January 1941 Somervell was a lieutenant colonel. For his role in creating his new job and lobbying to get it see Ohl, *Supplying,* 60–62. The ASF was called Services of Supply (SOS) for its first year. On March 12, 1943, the name of the command was changed to Army Service Forces. The army official history uses ASF from March 9, 1942, onward. Millett, *Organization,* 1.
92. See Millett, *Organization.* These included the quartermaster corps, transportation corps (created out of the QMC), Ordnance Department, Medical Department, signal corps, and Chemical Warfare Service. The ASF was also assigned responsibility for supervising the army's four administrative services: the Adjutant General's Office, the Office of the Judge Advocate General, the Finance Department, and the Office of the Provost Marshal General.
93. Fine, Remington, *Corps,* 506. From April to November 1943 Strong was chief engineer, SOS, China-India-Burma theater, until he was replaced by Col. Thomas F. Farrell. See also USACE, Office of History, Headquarters, Biographies. As noted in chapter 4, of the 20th-century graduates, at number one is Frederick Smith Strong (class of 1910), with a percentage of 99.4998. Strong died in 1986 at the age of 98. Obituary, *Assembly,* July 1988, 167.
94. *LRG Appointment Book,* April 7, 1942.
95. "The Army Corps of Engineers placed nearly $12 billion worth of construction inside the United States. The corps was responsible for 3,000 command installation projects undertaken in this program, including 230 cantonments: 140 ports of embarkation and staging areas, 210 general depots representing either new installations or expansion work at existing bases. The Corps of Engineers assumed maintenance responsibility for a vast military area, which included over a billion square feet of buildings, 62,640 miles of roads, 22,000 miles of electric lines, 12,550 miles of sewer lines, 4,260 miles of railroads, 12,860 miles of water mains, 2,860 miles of gas lines, and 1,590 miles of steam lines. In addition, the corps supervised construction of nearly 300 major industrial projects." American Public Works Association, Ellis L. Armstrong, ed., *History of Public Works in the United States, 1776–1976* (Chicago: American Public Works Association, 1976), 598.
96. Fine, Remington, *Corps,* 519–521. Construction in the United States declined as the overseas effort increased. See also Blanche D. Coll, Jean E. Keith, and Herbert H. Rosenthal, *The Corps of Engineers: Troops and Equipment* (Washington, D.C.: Office of the Chief of Military History, Department of the Army, 1958), 132–145.

97. As General Reybold said, "to back the fighting front, to speed the hour of triumph, to reduce the awful toll of war. . . ." Fine, Remington, *Corps,* 605.

98. As the official historians delicately put it, "Colonel Groves . . . was noted more for forcefulness than for diplomacy. Critical and demanding, he was as unsparing of himself as he was of others." Fine, Remington, *Corps,* 500. Of those he passed by, one had been a temporary brigadier general when Groves was a cadet. Another had been on the board to consider Groves's promotion from second to first lieutenant. All were field officers when he was a first lieutenant. Now they took orders from him. Comments on Draft of the History of Military Construction, LRG to Maj. Gen. A. C. Smith, Office of the Chief of Military History, July 22, 1955, Chapter 9, 14–15. Office of History, Chief of Engineers.

99. "Absolutely cold-blooded in removing officers who failed to deliver and capable of running down those who got in his way, Somervell pushed reorganization plans to rapid fruition." Forest C. Pogue, *George C. Marshall, Ordeal and Hope: 1939–1942* (New York: Viking Press, 1966), 297.

100. Kai Bird, *The Chairman: John J. McCloy, The Making of the American Establishment* (New York: Simon & Schuster, 1992), 147–174.

101. Telephone conversation, LRG and Col. Stanley L. Scott, August 15, 1942, quoted in Fine, Remington, *Corps,* 516. For confirmation of Groves's prediction see House Committee on Interior and Insular Affairs, *Personal Justice Denied, Report of the Commission on Wartime Relocation and Internment of Civilians* (March 1992).

102. J. L. DeWitt, *Final Report: Japanese Evacuation from the West Coast, 1942* (Washington, D.C.: U.S. Government Printing Office, 1943); Stetson Conn, Rose C. Engelman, Byron Fairchild, *The Western Hemisphere: Guarding the United States and its Outposts* (Washington, D.C.: Office of the Chief of Military History, Department of the Army, 1964), 115–149. General DeWitt had been quartermaster general from February 3, 1930, to February 2, 1934, and was commandant of the Army War College from 1937 to 1939, when Groves was in attendance. It was said that he was under consideration to be chief of staff in 1940, along with General Drum and General Marshall. DeWitt initially recommended that German and Italian "enemy" aliens be interned simultaneously with the Japanese. Stephen C. Fox, "General John DeWitt and the Proposed Internment of German and Italian Aliens During World War II," *Pacific Historical Review* 57, no. 4 (November 1988), 407–438.

103. House Committee on Interior and Insular Affairs, *Personal Justice Denied, Report of the Commission on Wartime Relocation and Internment of Civilians* (March 1992), 149.

104. "Standards and Details — Construction of Japanese Evacuee Reception Centers (June 8, 1942)," DeWitt, *Final Report: Japanese Evacuation from the West Coast, 1942* (Washington, D.C.: U.S. Government Printing Office, 1943), appendix 4, 584–591.

105. Reybold succeeded Maj. Gen. Julian L. Schley on October 1, 1941. General Reybold remained chief until October 4, 1945.

106. Groves never received the DSM for his work with the Construction Division. As we shall see, the issue caused enormous controversy and bitterness in the months after the end of the war. Somervell wrote two of Groves's Efficiency Reports, which are worth quoting given his role in choosing him to head the Manhattan Project. For the one covering the period November 14, 1940, to June 30, 1941, Somervell said of Groves that he was "A superior officer of fine

professional attainments and character, who prosecutes his work with ability and courage and who is not afraid to assume responsibility. He has handled a difficult position exceptionally well. . . . I rate him as among the most able officers I know." For the report July 1, 1941, to November 26, 1941, Somervell wrote, "A superior officer, forceful, intelligent, resourceful, loyal, not afraid to assume responsibility and possessing both initiative and a large capacity for work under trying conditions." *LRG Official Records,* 351–354.

CHAPTER 10 **Fateful Decisions**

1. This argument is developed at greater length in Goldberg, "Inventing," 429–452.
2. Zachary, *Frontier.*
3. The other members were: James Bryant Conant, president of Harvard University; Karl Taylor Compton, president of MIT; Frank Baldwin Jewett, president of the National Academy of Sciences; Conway Peyton Coe, commissioner of Patents; Richard Chace Tolman, professor of physical chemistry and mathematical physics, California Institute of Technology; Brig. Gen. George V. Strong, representing the army; and Rear Adm. Harold G. Bowen, representing the navy. The composition of the committee remained the same for the year and a day of its existence with one exception. Maj. Gen. Richard C. Moore, deputy chief of staff — and Groves's boss in Kansas City — succeeded Strong on January 17, 1941. The central offices of the NDRC, and its successor the OSRD, were established at 1530 P Street, Northwest, in the Carnegie Institution Building.
4. Rhodes, *Making,* 303–424; Hewlett, Anderson, *New World,* 9–83; Jones, *Manhattan,* 12–46.
5. Conant, "History." This 47-page history was classified until 1979.
6. Conant, "History," 3. The committee was chaired by Compton and included Ernest O. Lawrence, J. C. Slater, J. H. Van Vleck, W. D. Coolidge, and Bancraft Gheradi.
7. Hershberg, *Conant.*
8. At its peak the OSRD included a force of 25,000 or more scientists and technologists. Two-thirds of the physicists in the United States worked for Bush during this period. The OSRD was a catalyst in the transformation of the federal sponsorship of science. The wartime focus was the development of new weapons for the military. Larry Owens, "The Counterproductive Management of Science in the Second World War: Vannevar Bush and the Office of Scientific Research and Development," *Business History Review* 68 (winter 1994), 515–576.
9. Briggs remained chairman, and George Braxton Pegram was vice chairman. The other members were Gregory Breit, Harold Clayton Urey, Samuel King Allison, Henry DeWolf Smyth, Edward Uhler Condon, and Ross Gunn, the liaison officer from the Naval Research Laboratory. The title changed to Section on Uranium and finally, with the word *uranium* too sensitive, to S-1.
10. Conant, "History," 9. The MAUD Committee was established in April 1940. The reports are reprinted in Gowing, *Britain,* 394–436.
11. Conant, "History," 16.
12. Conant, "History," 24, 18. Compton was chairman. Warren Kendall Lewis, R. S. Milliken, and George B. Kistiakowsky were added.

13. Groves, *Now*, 10; Jones, *Manhattan,* 34–35; Zachary, *Frontier,* 199.
14. Quoted in Zachary, *Frontier,* 199.
15. Quoted in Zachary, *Frontier,* 200, 201.
16. Quoted in Zachary, *Frontier,* 201.
17. Also under consideration was William E. R. Covell, first in the famous class of 1915. Marshall told Bundy he wanted him to remain where he was, "filling a very important position in New York controlling the construction in all of the leased bases." Marshall to Bundy, February 12, 1942, Folder, Biographies and Records Concerning the Life of LRG, Box 1, Entry 7530C, Papers of LRG, RG 200, NARA.
18. Styer was promoted to brigadier general on March 10, and at that time was designated by General Marshall and the secretary of war as the officer to become familiar with the program, looking forward to the time when the army would take it over. Conant, "History," Part 2, 9. Bush's account has Stimson selecting Styer. Bush then goes on to describe Styer's introduction to what the secret project was about. Vannevar Bush, Oral History, 1964, Institute Archives, MIT Library, 461. "Once the military purpose of the project became governing Somervell and Styer entered the picture." LRG, *Now,* xv.
19. James Creel Marshall, born in 1897, was in the June 1918 class at the USMA, five months ahead of Groves; he went on the battlefield tour of Europe in the summer of 1919 with him. Like Groves, he was caught behind the "Hump" and gradually rose through the ranks to be Syracuse District Engineer before being tapped to head the Manhattan Project in June 1942. He retired in 1947, worked for several construction and engineering companies, and died in 1977. Obituary, *Assembly,* December 1981, 126–128; James Creel Marshall, USACE, Office of History, Biographical Files. Interview with Brig. Gen. James C. Marshall, April 19, 1968, USACE, Office of History, Remington and Fine; Colonel Marshall wrote a detailed diary, Chronolgy of District "X" (17 June 1942–28 October 1942), hereafter refered to as *Marshall Diary.* Lt. Col. Kenneth D. Nichols, Capt. Robert C. Blair (assistant executive officer), and Capt. Allan C. Johnson also made entries in the diary, to keep each other informed. The diary actually goes to June 16, 1943. In an interesting memo of May 5, 1942, Styer lists 40 "Engineer officers of varying degrees of ability, all considered by me to be above average, who might be considered for important assignments." In what is apparently ranked order Groves heads the list, but James Marshall and Kenneth Nichols do not appear on it at all. General Styer's Correspondence, Box 44, Commanding General's file, Headquarters ASF, RG 160, NARA. The selection was probably made by the chief of engineers, Maj. Gen. Eugene Reybold, his assistant Maj. Gen. David McCoach, and Styer. LRG to Styer, August 16, 1961, Folder Stu–Sz, Box 10, Entry 7530B, Papers of LRG, RG 200, NARA.
20. Groves also noted the June 18 meeting, but said he "did not reveal many of the details of his work." LRG, *Now,* 11. See also Interview with Brig. Gen. James C. Marshall, April 19, 1968, USACE, Office of History, Remington and Fine. David McCoach (class of 1910) served as executive officer to the chief of engineers and as assistant chief of engineers from May 1941 to October 1943.
21. Jones, *Manhattan,* 40. Marshall's written directive for the project came primarily in the form of a series of recent letters by Bush, the enclosures all endorsed by Styer to Marshall with the words "referred to you for your information and appropriate action." *Marshall Diary,* June 19, 1942.
22. Several other people whom Marshall recruited would remain and work with

Groves, including Charles Vanden Bulck and John Harman. Marshall's secretary, Virginia Olsson, would later work for Nichols.

23. Later the MED offices were moved to 261 Fifth Avenue, at Twenty-ninth Street.

24. Jones, *Manhattan,* 43–44, citing War Department, OCE, GO 33, August 13, 1942. Unlike most Corps of Engineers districts, which are limited to a section of the country, the MED's geographical scope included all of the United States with extensions to Europe and the Pacific. *DSM* continued to be used.

25. Nichols had first met with the president of Stone & Webster on the morning of June 27 and selected the company. Further decisions about Stone & Webster were made on June 29 at an afternoon meeting in Groves's office. *Marshall Diary,* June 27, 1942, 17; June 29, 1942, 19–21; July 9, 1942, 42. See also K. D. Nichols, Narrative Reports — DSM Project, June 18, 1942–August 22, 1942, Folder 22, Narrative Reports: DSM Project 1942–1946, Box 1, Nichols Papers, Office of History, USACE.

26. Isotopes are different forms of the same element in which the atoms have the same number of protons but different numbers of neutrons.

27. Thomas B. Cochran, William M. Arkin, Milton M. Hoenig, *Nuclear Weapons Databook.* Vol. 1 of *U.S. Forces and Capabilities* (Cambridge, Mass.: Ballinger Publishing Company, 1984), 22–23.

28. Four enrichment methods to concentrate the U-235 were being considered. Prof. Ernest O. Lawrence of the University of California, Berkeley, was working on the electromagnetic method. Dr. Harold Urey of Columbia University was exploring the gaseous diffusion method. Eger V. Murphree, the director of research for Standard Oil of New Jersey, was using the centrifuge method, and the Naval Research Laboratory was exploring enrichment through thermal diffusion. At the University of Chicago, under the direction of Arthur H. Compton, studies and experiments were being carried out on how to design reactors to produce plutonium. Rhodes, *Making,* 388–389.

29. Later Groves had this to say about Marshall: "Colonel Marshall should not have been chosen for the position. He was just too nice a person and was lacking in brashness and self-confidence necessary to fight and win his way in Washington against the opposition which such an enormous project would naturally encounter. He would present his case well but would accept adverse decisions from his seniors in government. He would also defer to the scientists on all scientific matters even as was usually the case they were merely guessing without adequate data to back up their views. He would also accept their advice to delay decisions until further data could be collected." LRG, Comments on *The New World,* 9, Box 3, Entry 7530J, Papers of LRG, RG 200, NARA.

30. *Marshall Diary,* July 31, 1942, 84; August 22, 1942, 113.

31. *Marshall Diary,* September 2, 1942, 131; September 10, 1942, 144.

32. According to Styer, Bush complained to the president about the progress of the project. He cited four areas in which he was displeased with Marshall. Marshall had set up the main headquarters in New York rather than Washington. He had not solved the priority situation, or acquired the Tennessee site. Finally, Bush wanted a higher-ranking officer who knew his way around wartime Washington to get the job done. Nichols, *Trinity,* 49.

33. Jones, *Manhattan,* 74.

34. Zachary, *Frontier,* 202.

35. LRG, "Atom General," 16.

36. LRG, *Now,* 3. He also said that he "had agreed by noon that day [September 17] to telephone [his] acceptance of a proposed assignment to duty overseas." LRG, "Atom General," 15. While there is no reason to doubt Groves's word, there is no evidence of him being offered such an assignment in his service record, or in the records of the Corps of Engineers or the adjutant general. Preparations at the time were under way for Operation Torch, the invasion of northwestern Africa, which began on November 8 when joint Anglo-American forces landed in French Morocco and Algiers. Several of Groves's classmates, colleagues, and mentors were already in England staging for Torch. In May 1942 Maj. Gen. John C. H. Lee was made commanding general, service forces, for the European theater of operations. His chief of staff was Brig. Gen. Thomas B. Larkin, who was appointed commanding general, Mediterranean base sector, North African theater of operations, in October. Groves may have heard through the grapevine that he was being considered to replace Larkin, or to serve in some other position under Lee, but it never reached the formal stage.

37. Much later Groves asked Styer to tell him the background of his appointment. Groves wrote up Styer's reply. "General Marshall told me [Styer] that I was to take the job. I reported this to Somervell, who said he was opposed as he did not want to lose me. Somervell then discussed the matters with General Marshall. Marshall said that it would either have to be I, or Somervell would have to come up with someone who was entirely suitable. Somervell and I discussed the matter and decided that you [Groves] would be the victim. This was approved by Marshall, Stimson and the President." Memo to File, Folder Stu–Sz, Box 10, Entry 7530B, RG 200, Papers of LRG, NARA. An additional reason that Styer did not want the job was that it appeared to have a slim chance of success; he would be giving up a good job for an uncertain one.

38. Memo, W. D. Styer to the Chief of Military History, Department of the Army, August 15, 1961, Folder Stu–Sz, Box 10, Entry 7530B, Papers of LRG, RG 200, NARA.

39. Fine, Remington, *Corps,* 661.

40. Fine, Remington, *Corps,* 662. General Robins and General Reybold were not enthusiastic about having the Corps of Engineers involved. Interview with Brig. Gen. James C. Marshall, April 19, 1968, USACE, Office of History, Remington and Fine, 8.

41. *LRG Appointment Book,* September 17, 1942. It appears that a transcript of the short hearing was not printed.

42. Groves, *Now,* 3–4; Fine, Remington, *Corps,* 660. Later Groves speculated on why he was chosen. "I have always assumed, however, that it was because of my record in handling large affairs. For about two years I had been directly responsible for the construction operations of the Army in the United States. This had entailed the employment of over a million men and the actual construction of completed work amounting to about eight billions of dollars. It was probably realized too that I was thoroughly familiar with governmental operations in Washington not only in the War Department but in many other executive branches and Congress as well. Another consideration may have been that I was accustomed to operating with as little supervision as possible. In other words I did not lean on my superiors for decisions." Letter, LRG to Helen R. Shelley, January 25, 1969, Folder S–Sheri, Box 8, Entry 7530B, Papers of LRG, RG 200, NARA.

43. Nichols, *Trinity*, 49.

44. "It never occurred to me, at the time, that I was also about to become the impresario of a two-billion-dollar grand opera with thousands of temperamental stars in all walks of life." LRG, "Atom General," 16.

45. "As a result of both my experience with Groves in Nicaragua and having gotten to know him better in the early months of the project, I did not relish being under his direct control by being in the same office." Nichols, *Trinity*, 56.

46. Nichols, *Trinity*, 50.

47. LRG, "Atom General," 16.

48. Groves depicts himself as passive, saying "that General Marshall had directed that I be made a brigadier general." LRG, *Now*, 5. Nichols's account of the meeting has Groves initiating the promotion. Nichols, *Trinity*, 51. Major Groves had used a similar argument in November 1940 to General Gregory when he assumed new responsibilities for army construction. Then the argument had been that those under his command and the civilian contractors would only respect him if he were a full colonel. In twenty-six months he went from captain to brigadier general.

49. LRG, *Now*, 4.

50. Groves's orders are reproduced in LRG, *Now*, 417–418. Besides *DSM*, the Manhattan Project was also called the *Uranium Committee* (under the administrative umbrella of the National Bureau of Standards) from October 1939 to June 1940 and *S-1* under the National Defense Research Committee — precursor to the OSRD — from June 1940 to June 1942. Sometimes it was referred to as *Tube Alloys,* a name used by the British counterpart to the OSRD. Less used was *SAM* (Substitute Alloy Materials), after the name given to the S-1 laboratory at Columbia University.

51. LRG, *Now*, 20; Nichols, *Trinity*, 55–56; Zachary, *Frontier,* 201–202. Styer had told Groves and Nichols to return a letter Bush had sent Styer about priorities and explain to Bush a plan to get a new rating.

52. Bush's comments in his oral history more than 20 years later confuse the chronology and make it seem as though he had some say in the decision. "At the time he [Groves] was appointed, when Styer first recommended his appointment and I'd looked him up, I said to Styer that it wouldn't do at all; I was sure that Groves would get into trouble with the scientists. That held up Groves' appointment as the Head of the Manhattan District and occasioned his coming around to pay his respects to me when he hadn't been appointed at all. After some time, Styer sat down with me again and he said, 'Look, there isn't another man in the U.S. Army that can do this job with the skill that Groves would use. We'll *have* to appoint him, and then you and I'll have to sit on the lid to be sure that he doesn't get into trouble with the scientists.' We did just that, to the best of our ability." Vannevar Bush, Oral History, 1964, Institute Archives, MIT Library, 420.

53. Later Groves wrote that he was sure that Bush and Conant were dissatisfied with the choice. "[W]hen I was brought into the picture they realized that I had too much [drive] and that many toes would be stepped on. And I am sure they were [dissatisfied], despite my invariably sweet disposition." Groves to Styer, August 16, 1961, Folder Stu–Sz, Box 10, Entry 7530B, Papers of LRG, RG 200, NARA.

54. Bush to Bundy, September 17, 1942, Records of the Manhattan Engineer

District, Entry 2, Harrison-Bundy Papers, Folder 7, RG 77, NARA; Bush to Conant, S-1 Files, Bush-Conant Papers, Folder 15, RG 227, NARA. Bush also wrote to Conant on September 17 and 21. Hershberg, *Conant* 161, footnote 18. Harvey Hollister Bundy was born in 1888. He attended Yale, was tapped for Skull and Bones, and went to Harvard Law School.

55. The September 18 memo from Somervell to Marshall read, "It is planned to relieve Colonel Groves from his present duties and assign him to special duty to take complete charge of some work of the greatest importance and urgency at this time. Because of the importance of this work, and the many complicated relationships with other agencies of our Government and with Allied Governments, it is considered essential that he be given appropriate rank to effectively carry out this work." *LRG Official Records,* 363.

56. The president's June 1942 statement said, "As soon as the preparation of detailed plans and designs has progressed sufficiently, the highest priority should be given to the plant or plants which at that time show the most promise of success and which in demands for critical materials will have the least serious effect on other urgent programs." *Manhattan District History,* bk. 1, vol. 9: *Priorities Program,* 2.1.

57. Jones, *Manhattan,* 57–58, 81–82; *Manhattan District History,* bk. 1, vol. 9: *Priorities Program,* appendix A, documents 3 and 4. Clay believed that the atomic project could not possibly succeed during the war, and this attitude colored his decisions on the priority question as it affected Manhattan. At the 1945 Army-Navy football game Clay came up to Groves to congratulate him on the success of the bomb program and admitted his judgment had been wrong.

58. *Marshall Diary,* July 31, 1942.

59. Colonel Nichols wrote, "I prepared a draft of a letter for Colonel Groves which is to be addressed to Donald Nelson and signed by General Marshall. This procedure was subsequently modified to be a letter from Donald M. Nelson to Colonel Groves stating that this project should be given authority to issue AAA priorities." *Marshall Diary,* September 17, 1942, 151.

60. LRG, Comments on *The New World,* 30, Box 3, Entry 7530J, Papers of LRG, RG 200, NARA.

61. LRG, *Now,* 22; Jones, *Manhattan,* 81.

62. "This time our discussion was cordial and uninhibited." LRG, *Now,* 21–22.

63. Sapolsky attributes the navy's exclusion from the Manhattan Project primarily to friction between Vice Adm. Harold G. Bowen, director of the Naval Research Laboratory at the start of World War II, and civilian scientists, especially Vannevar Bush. Bowen was a difficult man and had numerous enemies and rivals within the navy as well. Bowen's goal was to develop nuclear propulsion for naval vessels. Harvey M. Sapolsky, *Science and the Navy: The History of the Office of Naval Research* (Princeton, N.J.: Princeton University Press, 1990), 10–19. Bush, in a letter to Vincent Davis dated June 9, 1960, wrote, "The decision to take the [atomic bomb] program up with the Army rather than the Navy was my own, and it was based on the general attitude of the Services in regard to relations with the civilian research carried on in my own organization." In a later letter Bush elaborated on the navy's aloof attitude. Prior to Pearl Harbor the navy believed it had matters well in hand and felt it did not need any help in scientific and technical research matters from civilians. Vincent Davis, *The Admirals Lobby* (Chapel Hill: University of North Carolina Press,

1967), 175, footnote 45. Davis also quoted Bush as saying, "Navy officers, especially some in NRL and the Bureaus, lack sufficient respect for or ability to work cooperatively with civilian scientists." Vincent Davis, *The Politics of Innovation: Patterns in Navy Cases,* vol. 4, monograph 3 (Denver, Col.: University of Denver, 1966–1967), 25n.

64. Purnell was born on September 6, 1886, attended the Naval Academy, and graduated in the class of 1908. He retired on October 1, 1946, and died on March 5, 1955.

65. LRG, *Now,* 23; Stimson, *Diary,* September 23, 1942. "In the afternoon we held a very important meeting on S-1 and arranged for the organization and immediate prosecution of the work in regard to it."

66. As we shall see, Groves used the Real Estate Branch of the Construction Division under the direction of Colonel O'Brien to purchase most of the land at Oak Ridge, Los Alamos, and Hanford. His liaison officers contacted the other technical services so as not to duplicate what was already there to be used. The signal corps could put in the communications system; the transportation corps could assist with obtaining priority railroad reservations or the air transport command with airplanes.

67. Vannevar Bush, *Pieces of the Action* (New York: William Morrow and Co., 1970), 61–62.

68. In the words of the memo signed that day, "Groves will sit with the Committee and act as Executive Officer to carry out the policies that may be determined." Memorandum A, September 23, 1942, War Department.

69. Vannevar Bush, *Pieces of the Action* (New York: William Morrow and Co., 1970), 62.

70. Later Groves said, "No officer I ever dreamed of had the free hand I had in this project; no theater commander ever had it and I know of no one in history who has had such a free hand." Quoted in Lt. Col. John F. Moynahan, *Atomic Diary* (Newark, N.J.: Barton Publishing Co., 1946), 75.

71. "A scientist could choose to help or not to help build nuclear weapons. That was his only choice." Rhodes, *Making,* 378.

72. Zachary, *Frontier,* 202.

73. Nichols, *Trinity,* 60.

CHAPTER 11 His Own Construction: Atomic Factories and American Industry, Oak Ridge and Hanford

1. Consodine to Horace K. Corbin, n.d., but ca. 1945–1946, Groves, L. R., Folder 201, Box 2, Entry 7530C, Papers of LRG, RG 200, NARA.

2. One of the better books about the Manhattan Project does this: Groueff, *Manhattan.* A more recent work by art historian Peter Bacon Hales describes how the military aspects of the project influenced daily life at the three major sites: Hales, *Spaces.*

3. Top Secret, File 25, Folder B, Tab 8. The document is also included in a booklet that Groves compiled titled *The Atomic Bomb, 1942–1945, Reproductions of Some of the Critical Documents Pertaining to the Development and Use of the Atomic Bomb, with Explanatory Notes by Leslie R. Groves, Commanding General Manhattan Engineer District During World War II.*

4. Hodgson, *Colonel,* 5. See also David F. Schmitz, *Henry L. Stimson: The First Wise Man* (Wilmington, Del.: SR Books, 2001).

5. Stimson, Bundy, *Service,* 40.

6. LRG interview with Fred Freed, n.d., Box 4, Entry 7530B, Papers of LRG, RG 200, NARA.

7. Never before or since has there been a commanding general of an engineer district.

8. Groves later wrote, "Marshall was relieved as District Engineer in 1943 by me after I had been approached by McCoach. McCoach said there was a position for Marshall overseas and that he would be made a brigadier if I was willing to relieve him from the MED. McCoach also stated, of course, that he did not see any reason why I should have both Marshall and Nichols. . . . I acceded quite readily, even happily, to McCoach's suggestion because I thought Nichols would do a better job, and he did." LRG to Styer, August 16, 1961, Folder Stu–Sz, Box 10, Entry 7530B, Papers of LRG, RG 200, NARA.

9. LRG, *Now,* 28.

10. Jones, *Manhattan,* 517.

11. The earlier 10-step (A-1 to A-10) system that Groves was familiar with in the 1940–1942 period (see chapter 9) broke down under increased wartime mobilization. One tendency was "rating inflation," whereby numerous projects kept being assigned higher and higher ratings, thus negating the essential purpose. For a discussion see R. Elberton Smith, *The Army and Economic Mobilization* (Washington, D.C:. Office of the Chief of Military History, 1959), 505–596.

12. "As soon as the preparation of detailed plans and designs has progressed sufficiently, the highest priority should be given to the plant or plants which at that time show the most promise of success and which in demands for critical materials will have the least serious effect on other urgent programs." *Manhattan District History,* bk. 1, vol. 9: *Priorities Program,* appendix B.

13. "The importance of the highly secret work being carried on under the Manhattan District, Corps of Engineers, dictates that it continue to have the preferred position with respect to all other war programs. This position was established at the initiation of the work, in accordance with directives of the President of the United States." Robert P. Patterson to Donald M. Nelson, January 31, 1944, Folder, Second Red Folder, Box 5, Entry, Manhattan Project, RG 319, NARA. Apparently Groves drafted the letter for Patterson's signature. "I understand that there has been some speculation within the War Production Board as to what effect, if any, the progress of our army in Europe will have upon the Manhattan District project. The completion in as short a time as possible of this highly secret work is still of the utmost importance and the project must continue to have the preferred position with respect to all other war programs." Robert P. Patterson to J. A. Krug, Acting Chairman, War Production Board, September 25, 1944, Folder, First Red Folder, Box 5, Entry, Manhattan Project, RG 319, NARA. For Patterson's role during the war see Keith E. Eiler, *Mobilizing America: Robert P. Patterson and the War Effort, 1940–1945* (Ithaca, N.Y.: Cornell University Press, 1997). "He authorized (or signed) many of the large, unusual contracts that set the project in motion, diverted startling amounts of money from the general 'Expediting Production' funds at his disposal, wangled special priorities and allocations for unheard-of quantities of scarce material, and arranged assistance for the recruitment, movement, and nonunionization of armies of workers." Ibid., 438–439.

14. Ruhoff (1908–1973) was educated at the University of Wisconsin and received a Ph.D. from Johns Hopkins University in 1932. He joined the U.S. Rubber

Company in New Jersey as a research chemist from 1935 to 1937, then took a position with the Mallinckrodt Chemical Works in St. Louis, where he worked until he retired in 1972. He was inducted into the army and served in the Corps of Engineers from 1942 to 1946.

15. The comptroller general testified before the Senate Special Committee on Atomic Energy on April 4, 1946, and said: "We have audited, or are auditing, every single penny expended on this project. . . . We audited on the spot and kept it current, and I might say it has been a remarkably clean expenditure . . . from the very beginning he [General Groves] has insisted upon a full audit and a full accountability to the General Accounting Office." *Manhattan District History,* bk. 1, vol. 4, 2.4.

16. In Groves's vast correspondence to his subordinates he always used the rank of the person in the salutation — for instance, "Dear Colonel" or "Major" — or sometimes just the person's last name, "Dear Nichols." The trend toward familiarity that he observed, especially in the postwar period, was unwelcome. He considered it a "vicious tendency" to use "first names in official correspondence, directives or conversations. It is a modern habit not consistent with sound military procedure." LRG, Comments on Fine and Remington, ca. July 1955, Chapter 8, 16, Office of History, Corps of Engineers.

17. "One basic principle established in the early days was that of non–empire building. We never did anything ourselves that we could have anyone else do for us. We took full advantage of the capacities of private industry, of universities, and of other government resources. We reasoned that a going organization could do a better job than a newly created one. We did not duplicate any competent existing organization." LRG, "The A-Bomb Program," in Fremont E. Kast and James E. Rosenzweig, eds., *Science, Technology and Management* (New York: McGraw-Hill, 1963), 39–40.

18. LRG, Comments on Fine and Remington, July 22, 1955, Chapter 9, 5, Office of History, Corps of Engineers. On O'Brien see Fine, Remington, *Corps,* 363, 395, 493.

19. Army Regulation 100-61 (Corps of Engineers — Real Estate Acquisition, September 15, 1942).

20. Interview, Robert S. Norris with Coni O'Leary Watson, April 4, 1998.

21. The two rooms would grow to five, and then — at the very end, in July 1945 — five or so more rooms were added to take care of the reporters when the news of the bombing broke. LRG to Richard F. Newcomb, July 24, 1964, Box 3, Entry 7530J, Papers of LRG, RG 200, NARA.

22. Interview with Jean O'Leary by Jones and Falk, July 12, 1960, Folder, Manhattan Interviews, Box 7, Entry, Manhattan Project, RG 319, NARA.

23. On a trip to Hanford she checked on living conditions in the women's camp. Thayer, *Management,* 183.

24. "'General Groves was his father's favorite son,' Mrs. Groves told me. 'He gave his son a strict Presbyterian upbringing. He gave him a relish for scholarship and his own exceptional emotional stability. Of all those gifts, General Groves was to need the latter most. His ability to fall asleep the minute he lay down was, for example, indispensable.'" Robert De Vore, "The Man Who Made Manhattan," *Collier's,* October 13, 1945, 67.

25. As we shall see, Groves did not take a break until 1947, after the dissolution of the MED.

26. See, for instance, *LRG Appointment Book,* July 12, 1943.

27. LRG, Memo to File, Secret Papers, n.d., West Point Collection.

28. *Building the Holston Ordnance Works,* booklet by Fraser-Brace Engineering Co., New York, n.d. A copy is in Folder, Contractors C-4, Box 7, Entry 7530A, Papers of LRG, RG 200, NARA.

29. The chemical name for RDX is cyclotrimethylenetrinitramine, or cyclonite. The chemical name for TNT is trinitrotoluene.

30. This management technique later became known by the term *concurrency.* Such overlapping and compression of the development, testing, and production phases, with the goal of early operation, is evident in many other large Pentagon projects undertaken after the war. For an account of Gen. Bernard Schriever's management of the Atlas ICBM program, see John Clayton Lonnquest, "The Face of Atlas: General Bernard Schriever and the Development of the Atlas Intercontinental Ballistic Missile, 1953–1960," Ph.D. dissertation, Department of History, Duke University, 1996. The risky practice does not always have happy results: Weapons systems sometimes are fielded before enough development and testing have taken place, and the measures thus required to correct the problems inflate budgets.

31. The prime contractor was Fraser-Brace Engineering Co., which Groves had just used to complete the Weldon Springs Ordnance Works, in Missouri, probably the largest TNT plant in the world. The Tennessee Eastman Corporation, which Groves would soon use again to operate Oak Ridge, designed and operated the plant under a contract with the Army Ordnance Department. The Holston District was under the general supervision of Col. C. L. Hall, division engineer, Ohio River division. The initial district engineer in direct charge of construction was Lt. Col. E. R. Gates.

32. Fine, Remington, *Corp,* 595.

33. *Building the Holston Ordnance Works,* booklet by Fraser-Brace Engineering Co., New York, n.d., 16. A copy is in Folder, Contractors C-4, Box 7, Entry 7530A, Papers of LRG, RG 200, NARA. 16.

34. The plant was shut down at the end of the war. It reopened and for decades produced the various kinds of high explosive that would be used in tens of thousands of U.S. nuclear weapons.

35. As we shall see in chapter 18, a third method for enriching uranium was undertaken in 1944.

36. Jones, *Manhattan,* 117–148; Hewlett, Anderson, *New World,* 141–167.

37. L. Compare and W. L. Griffith, *The U.S. Calutron Program for Uranium Enrichment: History, Technology, Operations, and Production,* ORNL-5928 (Oak Ridge, Tenn.: Oak Ridge National Laboratory, October 1991), 7.

38. Hewlett, Anderson, *New World,* 151.

39. Nichols, *Trinity,* 42.

40. *Manhattan District History,* bk. 5, vol. 4: *Silver Program.* "Free" silver was a silver reserve not used for coinage or to back silver certificates. See Stimson to Secretary of the Treasury, August 29, 1942, appendix B.

41. Jones, *Manhattan,* 133.

42. LRG, *Now,* 98. "Always we were driven by the need to make haste." Ibid, 96.

43. LRG, *Now,* 103.

44. LRG to Conant, June 5, 1944, quoted in Jones, *Manhattan,* 146. At the end of 1944, on Groves's orders Kelley was reassigned, switching jobs with Lt. Col.

John R. Ruhoff, head of the Madison Square Area Office. Ruhoff had a Ph.D. in chemical engineering from Johns Hopkins and worked for Mallinckrodt Chemical Works in St. Louis before being inducted into the army.

45. Herbert Childs, *An American Genius: The Life of Ernest Orlando Lawrence* (New York: E. P. Dutton and Co., 1968), 345–346. Childs has John Hecker, TEC production manager, recounting the interchange. The story pops up again with a different source in a much less reliable book by Stanley A. Blumberg and Gwinn Owens, *Energy and Conflict: The Life and Times of Edward Teller* (New York: G. P. Putnam's Sons, 1976), 120. According to Teller's account it happened during Groves's first trip to Berkeley in October 1942. Teller makes it seem as though he were a witness, though he was not in California at the time.

46. Jones, *Manhattan,* 149–171; Hewlett, Anderson, *New World,* 120–141.

47. P. C. Keith, "The Role of the Process Engineer in the Atomic Bomb Project," *Chemical Engineering* 53 (1946), 112–120.

48. Beth Laney Smith and Karen Trogdon Kluever, *Jones Construction: Looking Back, Moving Forward* (Charlotte, N.C.: Laney-Smith, 1989), 26, 39–40.

49. Groueff, *Manhattan,* 214.

50. Groueff, *Manhattan,* 90–93, 124–127, 179–181, 262–275.

51. Quoted in Groueff, *Manhattan,* 272.

52. Ibid.,, 272.

53. "Atom City," *Architectural Forum,* October 1945, 102–116.

54. Nichols, *Trinity,* 108. Nichols later recounted his opinion to Gene Vidal. Vidal was then living in Connecticut, as was Groves, and had the opportunity to pass on what Nichols had said about him to his classmate. Groves expressed shock at each sentence, repeating, "Did Nichols say that?" When Vidal got to the punch line Groves "broke out into a big grin and beamed with pleasure." *Ibid.*

55. Folder 22, Narrative Reports DSM Projects 1942–1946, Box 1, Nichols Papers, Office of History, USACE.

56. For accounts of daily life at Oak Ridge see Charles W. Johnson and Charles O. Jackson, *City Behind a Fence: Oak Ridge, Tennessee 1942–1946* (Knoxville: University of Tennessee Press, 1981); James Overholt, ed., *These Are Our Voices: The Story of Oak Ridge 1942–1970* (Oak Ridge, Tenn.: Children's Museum of Oak Ridge, 1987); Hales, *Spaces;* George O. Robinson Jr., *The Oak Ridge Story: The Saga of a People Who Share in History* (Kingsport, Tenn.: Southern Publishers, 1950); Rachel Fermi and Esther Samra, *Picturing the Bomb: Photographs from the Secret World of the Manhattan Project* (New York: Harry N. Abrams, 1995).

57. *Hanford Site Historic District* (Richland, Wash.: USDOE, 2000); Thayer, *Management;* Goldberg, "Hanford," 39–89.

58. In a minor way Du Pont was already involved, because James Conant had arranged for Charles M. Cooper to report to the Met Lab to assist Arthur Compton on large-scale plutonium extraction. Cooper arrived on August 3 and remained there throughout the war. Cooper had been director of MIT's School of Chemical Engineering before joining Du Pont in 1935. Thayer, *Management,* 30; Compton, *Quest,* 132. Conant had been a consultant to Du Pont from mid-1929 to 1933, when he became president of Harvard. Hershberg, *Conant,* 56. See also Hounshell, "Du Pont," 245.

59. David A. Hounshell and John Kenly Smith Jr., *Science and Corporate Strategy: Du Pont R&D, 1902–1980* (Cambridge: Cambridge University Press, 1988), 338–346; Thayer, *Management,* 30–74.

60. Besides Clinton and Hanford Du Pont built 53 other plants during the war, including $500 million worth of military explosives plants. By comparison Du Pont spent about $350 million on Hanford. Hounshell, "Du Pont," 253. Earlier that year Groves had worked with Ackart building munitions plants.

61. Interview with Matthias. Thayer, *Management,* 169.

62. Thayer, *Management,* 66–74, 197–198. Related methods are program evaluation and review technique (PERT) and CAD-CAM. These techniques are routinely taught in university engineering courses.

63. Charles W. Cheape, *Strictly Business: Walter Carpenter at Du Pont and General Motors* (Baltimore: Johns Hopkins University Press, 1995), 190–191.

64. Groves took into account the possibility of a reactor accident that might threaten Knoxville, only 25 miles away.

65. Warren Kendall Lewis (1882–1975) has been called the father of modern chemical engineering and was the first head of the Department of Chemical Engineering at MIT. For his role in the 20th century, see Robert L. Pigford, "Chemical Technology: The Past 100 Years," *Chemical & Engineering News,* April 6, 1976, 190–203. Lewis was a chemistry professor at MIT when Groves attended there.

66. The other members of the committee were three Du Pont men, Crawford H. Greenewalt, Tom Crumley Gary, and Roger Williams, along with Eger Murphree of Standard Oil Development Corporation. In 1982 Greenewalt confirmed that the Lewis Committee was established solely for the benefit of Du Pont. Hounshell, "Du Pont," 246, footnote 17.

67. Carlisle, *Supplying,* 23.

68. LRG, *Now,* 70–71. "Although not an original siting requirement, Hanford's arid environment and soil features allowed large amounts of liquid waste to be released to the ground without immediately descending to groundwater." R. H. Gray and C. D. Becker, "Environmental Cleanup: The Challenge at the Hanford Site, Washington, USA," *Environmental Management* 17, no. 4 (1993), 462.

69. F. T. Matthias to Jesse A. Remington, April 28, 1964, Folder, Matthias Comments and Observations, Box 12, Entry, Manhattan Project, RG 319, NARA. Interview with Franklin T. Matthias, et al., by Stanley Goldberg, January 14, 1987, for the Smithsonian Video History Program. Matthias left Hanford in February 1946 and worked for Kaiser Engineers building dams and tunnels in Brazil, Canada, and the United States. He retired in 1973 and died in 1993. Groves originally had intended Matthias to be his executive officer in Washington. LRG, *Now,* 74.

70. Sanger, *Working,* 19–20.

71. The sites in California were Pit River, Needles, and Blythe. Carlisle, *Supplying,* 24.

72. Interview with Matthias. Thayer, *Management,* 164–165.

73. "The soil appeared to be mostly sand and gravel, which is almost ideal for heavy construction." LRG, *Now,* 75.

74. "Groves himself represented a larger tradition of military bureaucracy devoted to the manipulation and transformation of American spaces to the goal of greater efficiency and the control of nature, imbedded within the Army Corps of Engineers." Hales, *Spaces,* 3.

75. LRG, *Now,* 69.

76. John M. Whiteley, "The Hanford Nuclear Reservation: The Old Realities and

the New," in Russell J. Dalton, et al., *Critical Masses: Citizens, Nuclear Weapons Production, and Environmental Destruction in the United States and Russia* (Cambridge, Mass.: MIT Press, 1999), 29–57.

77. Jones, *Manhattan,* 331–342; LRG, *Now,* 75–77; *Manhattan District History,* bk. 4, vol. 4.

78. Matthias's diary of almost 600 pages — covering the period from February 1943 to December 1945 — is an important, but underutilized, source of information about developments at Hanford. There is a copy at the Office of History, USACE. Groves's policy was that no one was allowed to keep a diary. Apparently an exception was made for Matthias.

79. Interview with Matthias, in Thayer, *Management,* 167. "At Hanford the pace was literally killing; early in July [1944] . . . Kadlec died of a heart attack, an apparent victim of strain and pressure." Fine, Remington, *Corps,* 321, 514, 690.

80. Jones, *Manhattan,* 332. The town of Hanford was named for Judge Cornelius Holgate Hanford. The judge and several Seattle and Tacoma industrialists formed the Hanford Irrigation and Power Company. They constructed a power plant at the foot of Priest Rapids that generated 20,000 horsepower of electricity, which in turn pumped water for irrigation. Eight miles downriver from the old White Bluffs ferry slip, the company located a power substation and the town of Hanford. Lisa Scattaregia, "The Disappearance: How Did Two Entire Towns Vanish?" *Courier,* February 1991, 3; Lisa Scattaregia, "The Disappearance: How Did Two Entire Towns Vanish?" *Courier,* April 1991, 5–7.

81. LRG, *Now,* 76.

82. Stimson Diary, June 17, 1943. In an entry for March 13, 1944, Stimson wrote, "Truman is a nuisance and an untrustworthy man. He talks smoothly, but he acts meanly."

83. Stimson Diary, June 17, 1943.

84. Jones, *Manhattan,* 340.

85. Jones, *Manhattan,* 341.

86. There were 2,500 regular houses and 1,804 prefabs built. Unit costs for the best units, four-bedroom houses, were $6,025 to $6,580. Monthly rent, including utilities, was $62.50 to $67.50 (unfurnished). Hanford Engineer Works, *History of Operations (1 January 1944 to 20 March 1945),* OUT-1462 (Richland, Wash., 1945), 104.

87. Thayer, *Management,* 16–17; Col. F. T. Matthias, "Building the Hanford Plutonium Plant," *Engineering News Record,* December 13, 1945, 118–124. A long-lost film about Hanford was discovered in 1999. An 89-minute videotape, *War Construction in the Desert,* has been produced by the Columbia River Exhibition of History, Science and Technology of Richland and is available for purchase from the facility.

88. Scientist John Marshall has recounted a story about his colleague Leo Szilard. "At one point, I'm told, [Szilard] went to Arthur Compton and said the [Du Pont] engineers were getting intolerable. Either they had to be fired or he was going to quit. Compton said, 'You have just resigned.' After half an hour Compton cooled off and came back and said, 'I didn't mean that, Leo.'" Sanger, *Working,* 49.

89. Groueff, *Manhattan,* 139. Later Read would be assistant chief engineer and chief engineer. Another Du Pont engineer who had worked with Groves building munitions plants was Roger W. Fulling.

90. Thayer, *Management,* 89–93.

91. Groves has some harsh things to say about labor unions before, during, and after the war. In his dealings with them during the mobilization period: "The War Department did surrender to the unions and it was absolutely true that the labor unions put extra compensation above responsibility, loyalty and patriotism to the United States. It is also true that the situation was one that could be termed properly 'creeping socialism,' or worse, in this country." Comments on Fine and Remington, ca. July 1955, chapter 7, 6, Office of History, Corps of Engineers.

92. Sanger, *Working,* 68.

93. Name bands such as Kay Kayser and Louis Armstrong were brought into the camp for dances and concerts. The Harlem Globe Trotters put on a basketball exhibition, and the Portland and Seattle professional baseball teams played a game at Hanford. F. T. Matthias to Jesse A. Remington, April 28, 1964, 16g Folder, Matthias Comments and Observations, Box 12, Entry, Manhattan Project, RG 319, NARA.

94. One summary of incidents dealt with by the Hanford police between March 15, 1943, and August 28, 1944, shows 3,156 incidents of intoxication and 1,124 incidents of burglary, the two largest. Sanger, *Working,* 96.

95. Atomic Energy Commission, *Operational Accidents and Radiation Exposure Experience Within the AEC, 1943–1975* (Washington, D.C.: USAEC, fall 1975), 41–43. There were six other fatalities at five other MED sites.

96. This happened in two instances at Hanford. On January 13, 1944, four died when two on-site locomotives, operating in a fog, collided. On June 14, 1944, five welder chippers were crushed to death when a 49-ton steel tank fell on them.

97. Thayer, *Management,* 45–46, 142–157.

98. Du Pont's refusal to operate the semiworks was a calculated step in its withdrawal from nuclear physics research. After the war, Du Pont informed Groves that it intended not to continue to operate Hanford after its contract terminated in May 1946. Groves vigorously protested but this time did not get his way. The best he could do was to extend the contract to October 31, 1946, after which General Electric operated Hanford.

99. Sanger, *Working,* 39.

100. How or why they were so designated remains a mystery. See Carlisle, *Supplying,* 235, footnote 45. The preliminary plan called for six reactors, A–F. When cut back to three, perhaps B, D, and F remained. They started excavation of B first because the topography at A was not good.

101. The reactor was housed in a large reinforced-concrete and concrete-block structure, designated a 105 building, roughly 250 feet long by 230 feet wide by 95 feet high. Around the reactor were walls three to five feet thick to provide radiation shielding. *Hanford Site Historic District* (Richland, Wash.: USDOE, 2000), chapter 2, section 3, Reactor Operations.

102. The consensus view is that it was Graves who proposed the safety factor of extra tubes. See the references in Thayer, *Management,* 53, and Sanger, *Working,* 57–58. As the first operations manager, Walter O. Simon, recounted, "Du Pont was a conservative organization. . . . if someone asked for a two-story building Du Pont design would put enough steel in it for four stories, being convinced that sooner or later someone would add an extra floor or two." Sanger, *Working,* 152.

103. LRG, Comments on *The New World,* 97, Box 3, Entry 7530 J, Papers of LRG, RG 200, NARA. The scientists never acknowledged their mistake.

104. Production operations started at 10:48 P.M. on September 26, with 901 of the 2,004 tubes charged with uranium slugs. Pile power was increased to nine megawatts; after 18 hours, operations were halted. After six hours they were restarted, but it was apparent that the neutrons were being absorbed at a greater rate than they were being created, and power diminished. Hanford Engineer Works, *History of Operations (1 January 1944 to 20 March 1945),* OUT-1462 (Richland, Wash., 1945), 76–78.

105. As one company history put it, "Had Du Pont designed the reactor as desired by Wigner and his recalcitrant colleagues with no space for additional uranium slugs, it would have proven to be an expensive, sophisticated heap of junk." David A. Hounshell and John Kenly Smith Jr., *Science and Corporate Strategy: Du Pont R&D, 1902–1980* (Cambridge: Cambridge University Press, 1988), 341. "The Chicago physicists had never operated the Clinton pile at full power long enough to observe this phenomenon." Hounshell, "Du Pont," 251.

106. The water table is approximately 240 feet below the surface of the central plateau, compared with 50 to 75 feet below the 100 Area. *Hanford Site Historic District* (Richland, Wash.: USDOE, 2000), chapter 2, section 1.

107. The Du Pont design project manager for the chemical separation buildings was Raymond P. Genereaux. Sanger, *Working,* 58–65.

108. The television camera was mounted on the crane for the operators to see what they were doing. It was possibly the first industrial use of television.

109. Thomas B. Cochran, William M. Arkin, Robert S. Norris, and Milton Hoenig, *U.S. Nuclear Warhead Production.* Vol. 2 in *Nuclear Weapons Databook* (Cambridge, Mass.: Ballinger, 1987); Thomas B. Cochran, William M. Arkin, Robert S. Norris, and Milton Hoenig, *U.S. Nuclear Warhead Production Facility Profiles.* Vol. 3 in *Nuclear Weapons Databook* (Cambridge, Mass.: Ballinger, 1987).

110. Warren Bennis and Patricia Ward Biederman, *Organizing Genius: The Secrets of Creative Collaboration* (Reading, Mass.: Addison-Wesley Publishing Co., 1997). Among the great groups analyzed were Xerox's Palo Alto Research Center; Apple Computer, whose work led to the Macintosh; Lockheed's Skunk Works; and the Walt Disney Studio animators.

111. Groves provided a concise digest of his management philosophy and practice at a 1962 conference on national advanced-technology management. LRG, "The A-Bomb," 31–40.

112. Ibid., 39.

113. Ibid., 39.

114. "As was always my custom I set completion dates which I was sure could not be met. It was only in this way that I could be certain that every effort would be made and no one could think of easing up if he had too easy a schedule." LRG, Comments on *The New World,* 54, Box 3, Entry 7530J, Papers of LRG, RG 200, NARA.

115. LRG, "The A-Bomb," 31.

116. LRG, "Blueprint for a Crash Program," *This Week,* June 26, 1958.

CHAPTER 12 His Own Science: Oppenheimer, Los Alamos, and the Scientists

1. The account of the Chicago meeting is found in LRG, *Now,* 39–41; Groueff,

Manhattan, 33–34; Wyden, *Day One,* 59–60; Kurzman, *Day of Bomb,* 114–117; Lanouette, *Genius,* 237–238; Compton, *Quest,* 114–115.

2. According to Seaborg, who was visiting California that week, the others in the room were Samuel K. Allison, Thomas V. Moore, Eugene P. Wigner, Martin D. Whitaker, John A. Wheeler, Edward S. Steinbach, and Leo Szilard. *The Plutonium Story: The Journals of Professor Glenn T. Seaborg, 1939–1946,* edited and annotated by Ronald L. Kathren, Jerry B. Gough, and Gary T. Benefiel (Columbus, Ohio: Battelle Memorial Institute, 1994), 195–198.

3. *The Plutonium Story: The Journals of Professor Glenn T. Seaborg, 1939–1946,* edited and annotated by Ronald L. Kathren, Jerry B. Gough, and Gary T. Benefiel (Columbus, Ohio: Battelle Memorial Institute, 1994), 196; Hewlett, Anderson, *The New World,* 181.

4. The clash has been described many times. Rhodes, *Making,* 431 ff; Groueff, *Manhattan,* 35 ff; Lanouette, *Genius,* passim; Compton, *Quest,* 114 ff; Lawren, *General,* 76 ff; Carol S. Gruber, "Manhattan Project Maverick: The Case of Leo Szilard," *Prologue,* summer 1983, 73–87.

5. As he told Lawrence in one of his many trips to Berkeley, "You know, I'm not the least bit interested in the scientific knowledge of the world, except insofar as it gets this job done." Lawrence responded, "No, I thoroughly agree. As a matter of fact, I'm about as pure as anybody. I haven't any interest in doing any work for the sake of scientific knowledge. I'm deadly serious on this thing. I'm not in it for fun right now. It's deadly serious business to me." Minutes of Coordination Meeting, April 16, 1943, Folder, Minutes, First 2 Ring Folder, Box 19, Entry MED, Lawrence Berkeley Laboratory Archives.

6. LRG, Comments on *The New World,* 18, Box 3, Entry 7530 J, Papers of LRG, RG 200, NARA.

7. LRG, Comments on *The Virus House* by David Irving, January 6, 1970, Folder David Irving, Box 2, Entry 7530J, Papers of LRG, RG 200, NARA.

8. LRG, *Now,* 39.

9. LRG, Comments on Mrs. Gowing's Book, Box 2, Entry 7530J, Papers of LRG, RG 200, NARA; LRG, Memo, June 1, 1945, Top Secret (M1109), File 12F; LRG, Memo to File, Szilard Obituary, August 20, 1968, Folder Feld, Bernard, Box 2, Entry 7530J, Papers of LRG, RG 200, NARA.

10. LRG, Memo to File, Szilard, October 17, 1963, Folder Stu–Sz, Box 10, Entry 7530B, Papers of LRG, RG 200, NARA. "[H]e wouldn't have him at Harvard no matter if I endowed the Chair for life." LRG, Undated Memo to File, Szilard, Folder, Biographies of LRG, Box 1, Entry 7530C, Papers of LRG, RG 200, NARA.

11. Lanouette, *Genius,* 240.

12. LRG, Undated Memo to File, Szilard, Folder, Biographies of LRG, Box 1, Entry 7530C, Papers of LRG, RG 200, NARA. He added, "Whether I would have done it or not I don't know as there was the danger that it might stir up trouble among our scientists in Chicago."

13. LRG, Memo to Secretary of War, October 29, 1945, Leo Szilard, Top Secret, Folder 12H.

14. Groueff, *Manhattan,* 34. The quote gets repeated; see, for instance, Goodchild, *Oppenheimer,* 60; Wyden, *Day One,* 60. Groves does not indicate that he said anything like this in his memoir; nor is it in the official history. Hewlett, Anderson, *New World,* 181.

15. Groves, Hailey Interview, December 1957, D/1, Folder Hailey Interview, Box 2, Entry 7530J, Papers of LRG, RG 200, NARA.

16. Groueff, *Manhattan,* 34; Goodchild, *Oppenheimer,* 61.

17. "His small mustache and portly bearing, his too-tight collar straining at the button, made him look to the scientists like an Oliver Hardy in uniform, the very caricature of an Army officer. But he knew his business and was confident of his destiny. From the beginning, he distrusted the scientists, particularly the foreigners with their strange accents and their tendency to break into incomprehensible languages when they talked to one another in his presence." Bernard J. O'Keefe, *Nuclear Hostages* (Boston: Houghton Mifflin Co., 1983), 65.

18. Lansdale, Manuscript, 83. Lansdale was replaced by Lt. Col. Charles Harvey Banks (USMA, class of 1940).

19. Ibid., 84.

20. "Groves's authoritarian, anti-Semitic views cast Szilard as a pushy Jewish busybody." Lanouette, *Genius,* 237. "To the anti-Semitic Groves, who had come to despise Szilard because he considered him a pushy Jewish busybody." Barton J. Bernstein, in Helen S. Hawkins, G. Allen Greb, and Gertrude Weiss Szilard, eds., *Toward a Livable World: Leo Szilard and the Crusade for Nuclear Arms Control* (Cambridge, Mass.: MIT Press, 1987), xxxii. "Groves seems to have attributed Szilard's brashness to the fact that he was a Jew." Rhodes, *Making,* 502.

21. No difficulties, that is, beyond his view that most scientists had high opinions of themselves and thought they were experts on everything. On Teller: "He never got over the fact that Bethe was selected instead of him to head the Theoretical Division of Physics at Los Alamos." On Bethe: " . . . like so many experts he considered himself an expert on everything. I think if given a chance he could have expounded at great length on the failings of Ted Williams as a batter and why his average should have been at least two or three hundred points higher. Like almost all of the other scientists he had very little knowledge of military history and therefore was incompetent to judge this matter [Japanese surrender discussions]." LRG, Memo to File, August 1, 1965, Folder, Comments on *The New World,* Box 3, Entry 7530J, Papers of LRG, RG 200, NARA.

22. For an account of the European-born scientists see Laura Fermi, *Illustrious Immigrants: The Intellectual Migration from Europe, 1930–41,* 2nd ed. (Chicago: University of Chicago Press, 1971), 174–214.

23. Emilio Segrè, *A Mind Always in Motion* (Berkeley: University of California Press, 1992), 182. Groves clearly knew that Segrè and Fermi were Italian and recognized the language when he heard it. He was fluent in Spanish, and spoke passable French and German. If the incident did happen, he may have been pulling their leg(s). Segrè's account might be another in a long line of apocryphal stories by the scientists about the buffoonery of General Groves.

24. The December 11 (4:20 P.M.) entry reads, "Due to race and the fact that he can't become a citizen difficulties would be encountered after the war. Oppenheimer advised that he is an expert on some parts of the job. Have been unsuccessful in persuading him before and don't know what luck we will have now. Will check into this and let you know how necessary, etc." LRG Appointment Book. Chandrasekhar, who died on August 21, 1995, is regarded as one of the great physicists of the 20th century, cowinner of the Nobel Prize in 1983. Born in Lahore, India, in 1910, he obtained a Ph.D. in physics from Trinity College, Cambridge University, and moved to the University of

Chicago in 1937, where he remained until his death. He became an American citizen in 1958.

25. "Prejudice against Jews and other minorities used to be taken as a matter of course until after the end of World War II." Leonard Dinnerstein, *Antisemitism in America* (New York: Oxford University Press, 1994), 247.

26. Quoted in Montgomery C. Meigs, "Managing Uncertainty: Vannevar Bush, James B. Conant and the Development of the Atomic Bomb," Unpublished Ph.D. dissertation, University of Wisconsin — Madison, 1982, 15.

27. Zachary, *Frontier,* 218–239.

28. Vannevar Bush, Oral History, Institute Archives, MIT Library, Reel 7A, 422.

29. LRG to Conant, December 9, 1942, Folder 86, Historical Special, Bush-Conant Papers, RG 227/S1, NARA.

30. The eight who had already won prizes were Niels Bohr (1922), James Franck (1925), Arthur H. Compton (1927), James Chadwick (1935), Enrico Fermi (1938), Ernest O. Lawrence (1939), and Isidor I. Rabi (1944), all in physics, as well as Harold C. Urey in chemistry (1934). After the war more than a dozen Manhattan Project scientists would win Nobel Prizes: Sir John Cockcroft (1951), Emilio Segrè (1959), Owen Chamberlain (1959), Eugene P. Wigner (1963), Maria Goeppert Mayer (1963), Richard P. Feynman (1965), Hans A. Bethe (1967), Luis W. Alvarez (1968), Val F. Fitch (1980), and Norman F. Ramsey (1989), all won in physics. In chemistry there was Glenn T. Seaborg (1951), Edwin M. McMillan (1951), Willard F. Libby (1960), and Robert S. Mulliken (1966). Joseph Rotblat shared the Nobel Peace Prize in 1995. John H. Van Vleck (1977), as a member of the Harvard physics department, helped recruit promising graduate students to send to Los Alamos, including Soviet spy Theodore Hall.

31. "Throughout the life of the project, vital decisions were reached only after the most careful consideration and discussion with the men I thought were able to offer the soundest advice. Generally, for this operation, they were Oppenheimer, von Neumann, Penney, Parsons and Ramsey." LRG, *Now,* 343.

32. LRG, Comments Made from Notes Made in 1960, September 20, 1963, Folder, Oppenheimer, J. Robert, Box 4, Entry 7530J, Papers of LRG, RG 200, NARA.

33. "About three days later back in Berkeley, Oppie and I were working in his office when Groves came in, followed by Col. Kenneth D. Nichols. The first thing Groves did was take off his jacket, hand it to the colonel and say 'Find a tailor or dry cleaner and get this pressed,' and Nichols took it and walked out of the door. I was quite impressed by the way a general could treat a colonel. But that was Groves's way: I think it was a matter of policy to be as nasty as possible to his subordinates." Serber, *Peace,* 72. In an earlier version Serber said, ". . . Groves came in with a colonel in tow, probably Ken Nichols. Groves walked in, unbuttoned his tunic, took it off, handed it to Nichols, and said, 'Take this and find a dry cleaner and get it cleaned.' Treating a colonel like an errand boy. That was Groves's way." Robert Serber, *The Los Alamos Primer* (Berkeley: University of California Press, 1992), xxxii. Nichols does not, as might be expected, describe the incident in his memoir, *The Road to Trinity,* nor did he remember it when asked about it in an interview with the author. Interview with Maj. Gen. Kenneth D. Nichols by Robert S. Norris, May 16, 1998.

34. Van Vleck to Lawrence, October 25, 1941, enclosure Minutes of October 21, 1941, Meeting, Lawrence Papers, Bancroft Library. The meeting was held in Dr. Coolidge's office at the General Electric Research Laboratory.

35. Smith, Weiner, *Oppenheimer,* 224.

36. John H. Manley, "A New Laboratory Is Born," in Badash, et al., *Los Alamos,* 21–40; J. H. Manley, "Assembling the Wartime Labs," *Bulletin of the Atomic Scientists,* May 1974, 42–48.

37. The attendees included Oppenheimer, Hans Bethe, Edward Teller, Emil Konopinski, Felix Bloch, Stanley S. Frankel, Eldred C. Nelson, Robert Serber, and John Van Vleck. Hawkins, *Trinity,* 3–4. See also Serber, *Peace,* 71–72. It was at this conference that Teller first raised the idea of a fusion/hydrogen weapon to be set off by an atomic bomb. Among the other topics discussed was whether a fission bomb might ignite the earth's atmosphere. It was decided that this would not happen, but the question lingered.

38. LRG to Oppenheimer, September 27, 1960, Oppenheimer Papers, Manuscript Division, LOC; Nichols, *Trinity,* 73; LRG, Comments Made from Notes Made in 1960, September 20, 1963, Folder, Oppenheimer, J. Robert, Box 4, Entry 7530J, Papers of LRG, RG 200, NARA.

39. "He was as lacking in confidence as anyone I have ever seen. He was never sure of what was going to happen and always feared the worst." "He was not a doer himself and could never make decisions. At heart he was a coward." LRG, Folder, Comments on *The New World,* 20, Box 3, Entry 7530J, Papers of LRG, RG 200, NARA.

40. Groves's general rule when it came to knowledgeable personnel with leftist backgrounds was to keep them on the project rather than let them go. They were easier to keep track of.

41. Vannevar Bush, Oral History, Reel 7A, 419, Bush Papers, Institute Archives, MIT Library.

42. A sample would include speeches by Robert Serber, Victor Weisskopf, Abraham Pais, and Glenn Seaborg in *Physics Today* 20, no. 10 (October 1967), 34–57. These tributes were later expanded into a book, I. I. Rabi, et al., *Oppenheimer* (New York: Charles Scribner's Sons, 1969). See also Hans Bethe, "J. Robert Oppenheimer, 1904–1967," *Biographical Memoirs of Fellows of the Royal Society,* Vol. 14 (1968), 391–416; Bethe, "Oppenheimer," *Science,* March 3, 1967, 1080–1084; Rudolf Peierls, "J. Robert Oppenheimer," *Dictionary of Scientific Biography,* Vol. 10 (New York: Charles Scribner's Sons, 1974), 213–218.

43. Philip M. Stern, *The Oppenheimer Case: Security on Trial* (New York: Harper & Row, 1969), 41.

44. Goodchild, *Oppenheimer;* Nuel Pharr Davis, *Lawrence & Oppenheimer* (New York: Simon & Schuster, 1968); Smith, Weiner, *Oppenheimer;* James W. Kunetka, *Oppenheimer: The Years of Risk* (Englewood Cliffs, N.J.: Prentice Hall, 1983). There are several forthcoming biographies of Oppenheimer by Gregg Herken, Priscilla Johnson McMillan, Kai Bird and Martin Sherwin, and Patrick Marham.

45. Lansdale saw this as well. "Moreover, it was perfectly evident that both he [Oppenheimer] and his wife regarded this project as his outstanding career opportunity. I became convinced that not only was he loyal, but that he would let nothing interfere with the successful accomplishment of his task and thus his place in scientific history." Lansdale, Manuscript, 35.

46. *Manhattan Project History,* bk. 8; *Manhattan District History;* Edith C. Truslow, and edited by Kasha V. Thayer, *Nonscientific Aspects of Los Alamos Project Y, 1942 Through 1946* (Los Alamos, N.M.: Los Alamos Historical Society, 1991).

47. Dudley's preferred site was Oak City, Utah. John H. Dudley, "Ranch School to Secret City," in Badash, et al., *Los Alamos,* 1–11.

48. The date of November 16 was established by McMillan based on Dudley's travel records. Smith, Weiner, *Oppenheimer,* 346, footnote 19. See also LRG to McMillan, March 9, 1970. Groves flew back from Albuquerque the evening of the 16th and returned to his office at noon on the 17th. According to McMillan, he and Oppenheimer met Dudley at Jemez Springs, whereupon the three rented horses from a nearby ranch and went on horseback to inspect the area. They were later joined by Groves, who immediately rejected the site, and the four drove on to Los Alamos. See Folder, Manhattan Interviews, Box 7, Entry, Manhattan Project, RG 319, NARA.

49. Oppenheimer first came to New Mexico in 1922 and knew the area around the Pecos Valley well. Later his father leased a small ranch for Robert and Frank in Cowles. They called it Perro Caliente and eventually bought it in the mid-1930s, spending part of several summers there. Smith, Weiner, *Oppenheimer,* 8, 142, 165, 214. Ashley Pond, a Detroit businessman whose purpose was to provide an outdoor-oriented education for city-bred boys, mostly from wealthy Midwest families, founded the Ranch School in 1917. In its 25 years of existence about 600 boys went there, with a peak enrollment of 47 students in the fall of 1940. Other well-known alumni include William S. Burroughs (1925–1931), William S. Farish Jr. (1926–1930), Percival Keith (Dobie's son, 1941–1942), William L. Veeck Jr (1931–1932), and Gore Vidal (1939–1940). For a history of the Ranch School see Fermor S. Church and Peggy Pond Church, *When Los Alamos Was a Ranch School,* 2nd ed. (Los Alamos, N.M.: Los Alamos Historical Society, 1998).

50. Edwin M. McMillan, "Early Days at Los Alamos," in Badash, et al., *Los Alamos,* 13–19; Smith, Weiner, *Oppenheimer,* 238. Oppenheimer, McMillan, and Lawrence, with Colonel Dudley, visited Los Alamos again on November 20 to reassure themselves and to begin laying out the site. Oppenheimer to LRG, November 23, 1942, Folder, Large Red Folder II, Box 6, Entry, Manhattan Project, RG 319, NARA. The spectacular location of Los Alamos probably had an influence on morale and work, as it has been remarked upon by several scientists. "With all the procurement and transportation difficulties that were associated with such a remote spot, I am sure that the landscape let us, perhaps even made us, work more intensely than we otherwise could have and saved time and money." Cyril Stanley Smith, "Some Recollections of Metallurgy at Los Alamos, 1943–45," *Journal of Nuclear Materials* 100 (1981), 3.

51. That evening a letter from Secretary of War Stimson to Director Albert J. Connell was read to the students and faculty assembled at a special meeting. It read, "You are advised that it has been determined necessary to the interests of the United States in the prosecution of the War that the property of Los Alamos Ranch School be acquired for military purposes." Fermor S. Church and Peggy Pond Church, *When Los Alamos Was a Ranch School,* 2nd ed. (Los Alamos, N.M.: Los Alamos Historical Society, 1998), 24. On January 21, 1943, diplomas were awarded to four graduates who had completed the regular requirements, among them Stirling A. Colgate, who became a theoretical physicist and later worked at Los Alamos.

52. On March 15, 1944, these responsibilities were transferred to the Manhattan Engineer District. The design contracts totaled almost $908,000 through 1946

and construction $39.7 million, a minuscule amount compared to Oak Ridge or Hanford. *Manhattan Project History,* bk. 7, vol. 1, sections 4 and 5.

53. The choice went to J. E. Morgan and Sons and to Robert E. McKee, both of El Paso. Kruger continued as architect-engineer.

54. Oppenheimer to Hans and Rose Bethe, December 28, 1942, in Smith, Weiner, *Oppenheimer,* 245.

55. USAEC, *Oppenheimer,* 171–172. According to Colonel Tyler's notes of a visit by the general, "General Groves reminded Col. Tyler that we are not in the chain of command under Oak Ridge, that we should absolutely take no orders from Oak Ridge, but should only look to them for whatever assistance that they can give us. At any time, when they give us orders which are in conflict with our policy, or which will interfere with us in anyway, we are to let the General know about it. We should send almost nothing through the District." Notes Taken During Visit of LRG, February 7–9, 1945, Folder 96, Manhattan Project Data, 1939–1948, Box 6, Nichols Papers, Office of History, USACE.

56. LRG to Colonel T. D. Stamps, August 21, 1946, Folder 201, 1946, G-O-B, Box 2, Entry 7530C, Papers of LRG, RG 200, NARA.

57. Interview with Colonel Ashbridge, March 3, 1964, USACE, Office of History, Remington and Fine, Folder Ashbridge, Box 3, Remington-Fine 1958–1970, Manhattan Project.

58. Ashbridge attended in 1918–1919, one year after it opened.

59. Interview with Colonel Ashbridge, March 3, 1964, USACE, Office of History, Remington and Fine, Folder Ashbridge, Box 3, Remington-Fine 1958–1970, Manhattan Project.

60. The two other commanding officers that Groves chose were Col. Lyle E. Seeman, who served from November 1945 to September 1946, followed by Col. Herbert C. Gee.

61. *Manhattan Project History,* bk. 8, vol. 1, S19. The ratio of military to civilian changed from 5 percent military and 95 percent civilian in August 1943 to 50:50 two years later. War Department, *Complications of the Los Alamos Project,* November 12, 1946, RG 319, NARA. The Los Angeles Area Office under Maj. Stanley L. Stewart was responsible for procurement.

62. LRG to Crane, February 6, 1964, Folder Coo–Cz, Box 2, Entry 7530B, Papers of LRG, RG 200, NARA.

63. For example, on March 11, 1943, Groves called Father O'Donnell, president of Notre Dame, to request the release of Dr. Bernard Waldman. He was successful. Waldman worked on airborne diagnostic measurements at Trinity with Alvarez. The two were scheduled to be flown over ground zero at the Trinity test to drop gauges by parachute to measure blast, but bad weather prevented the planes from moving into position. At Tinian Waldman flew on the observer plane *The Great Artiste* during the Hiroshima mission. He carried a Fastax camera to film the explosion but was unsuccessful. Hoddeson, et al., *Critical,* 393–394.

64. There were also SEDs at Oak Ridge, more than 1,250 of them working at K-25, Y-12, and the Clinton Laboratory. June Adamson, "Special Engineer Detachment," in James Overholt, ed., *These Are Our Voices: The Story of Oak Ridge 1942–1970* (Oak Ridge, Tenn.: Children's Museum of Oak Ridge, 1987), 118–125; June Adamson, "The SED in Oak Ridge, 1943–1946," *Tennessee Historical Quarterly* 57, no. 3 (fall 1997), 196–210.

65. John S. Rigden, *Rabi: Scientist and Citizen* (New York: Basic Books, 1987), 149–151.

66. At the outset Oppenheimer had agreed to a military lab and was willing to join the army, at a rank of lieutenant colonel. Oppenheimer had been measured for a uniform and had taken his physical at the Presidio. Groves, worried that Oppenheimer might not pass it, ordered Fidler to see to it that the army doctor not flunk him. I would like to thank Gregg Herken for generously sharing with me notes of his interview with Harold Fidler, December 16, 1992.

67. Hawkins, *Trinity,* 495–497; Hershberg, *Conant,* 168–169. In January 1944 Groves wrote to Oppenheimer quoting from the latest report to the president on his mission. "The mission is to carry on the research and experiment necessary for the final purification of the production material, its fabrication into suitable active components, the combination of these components into a fully developed usable weapon, and to complete the above in time to make effective use of the weapon as soon as the necessary amount of the basic material has been manufactured." LRG to Oppenheimer, January 26, 1944, Folder Groves, L. R., Box 36, Oppenheimer Papers, LOC.

68. Interview by Verne Stadtman with Robert M. Underhill, July 25, 1966, Bancroft Library, 236.

69. Transcript of Robert Underhill Memories, 1975, Bancroft Library, 2.

70. J. L. Heilbron and Robert W. Seidel, *Lawrence and His Laboratory: A History of the Lawrence Berkeley Laboratory* (Berkeley: University of California Press, 1989).

71. LRG, Folder, Comments on *The New World,* Box 3, Entry 7530J, Papers of LRG, RG 200, NARA.

72. Lawrence, and his counterpart in Chicago, Compton, represented a developing tendency in 20th-century science, the scientist as entrepreneur. The interaction of the Manhattan Project with science accelerated what would be called Big Science after the war.

73. Lawrence to LRG, March 23, 1944, Lawrence Papers, 72/117C, Carton 28, Folder 37, Bancroft Library. See also Lawrence to LRG, November 1, 1946, and Lawrence to Harrison, July 18, 1945, in *ibid.*

74. Compton, *Quest,* 112–113.

75. Compton, *Quest,* 112.

76. LRG, Memo to File, Niels Bohr, December 13, 1962, Folder Bohr, Niels, Box 1, Entry 7530J, Papers of LRG, RG 200, NARA.

CHAPTER 13 His Own Intelligence: Domestic Concerns

1. For example, Peer de Silva, Lyle Edward Seeman, and Henry S. Lowenhaupt.

2. LRG to RHG, June 27, 1958, Folder Groves, Richard, Box 4, Entry 7530B, Papers of LRG, RG 200, NARA.

3. "This compartmentalization policy became a far more pervasive influence in the project after the Army assumed full responsibility for its administration. Where the OSRD had applied compartmentalization primarily to research and development organizations, the Army incorporated it into virtually every type of activity undertaken by the project." Jones, *Manhattan,* 268–269.

4. *Manhattan District History,* bk. 1, vol. 14: *Intelligence & Security,* December 31, 1945. "It was apparent that the successful completion of the American Project prior to the German might decide the course of the war and be a dominant factor in the post-war peace." Ibid., 1.1. Of the eight problems that faced Groves upon his appointment, the "Establishment and maintenance of a security system that would brook no violation of secrecy" was listed as number

one. LRG, "History of the Manhattan Project," Speech, Fort Belvoir, Va., September 23, 1946.

5. Lawrence Freedman and Saki Dockrill, "Hiroshima: A Strategy of Shock," in Saki Dockrill, ed., *From Pearl Harbor to Hiroshima: The Second World War in Asia and the Pacific, 1941–45* (New York: St. Martin's Press, 1994), 191–212; Sadao Asada, "The Shock of the Atomic Bomb and Japan's Decision to Surrender — A Reconsideration," *Pacific Historical Review,* 1999, 477–512.

6. LRG, *Now,* 141; Goldberg, "Compartmentalization," 38–43. "It was something that I insisted upon to the limit of my capacity." USAEC, *Oppenheimer,* 164.

7. Letter, W. A. Akers to M. W. Perrin, Attitude of Military Authorities, November 16, 1942, PRO, London, AB1/128. See also Letter, W. A. Akers to J. B. Conant, S1 Project, December 15, 1942, PRO, London, AB1/128. ". . . although we appreciate the desire to avoid as much as possible leakage of information on the S1 project, there is a strong feeling among the British group that the division of work into watertight compartments can be carried to the point at which inefficiency may be considered to outweigh the gain in secrecy." As it turned out, it was not Akers's call to make. Primarily as a result of the acrimonious relations with Groves, Akers was fired as scientific director in October 1943 and replaced by the pragmatic and conciliatory James Chadwick, with whom Groves got on well. Groves was suspicious of Akers from the start, as was Bush. He was research director at Imperial Chemical Industries, and for Groves that was evidence enough that the British were in it for their own commercial interests.

8. LRG, Memo to File, David Greenglass and His Treasonable Conduct at Los Alamos, Box 2, Entry 7530J, Papers of LRG, RG 200, NARA.

9. Ibid. In early 1943 Condon spent six weeks as associate director, more than enough time to gain Groves's animosity. Groves reproduced Condon's April 1943 letter of resignation to Oppenheimer in his memoir, *Now It Can Be Told,* 429–432. Condon found the "extreme concern with security . . . morbidly depressing." In an effort to overcome compartmentalization he had begun a weekly lecture-seminar, or colloquium, to provide a forum for the exchange of ideas. Groves unwillingly permitted these colloquia. Oppenheimer had promised the scientists that they would be held. As Groves commented, "to abandon the idea would have been a breach of faith and would have shown that Oppenheimer was to be a mere puppet in my hands," which could not be permitted.

10. See the diagrams in LRG, *Now,* 2, and Nichols, *Trinity,* 15–17. "No one person, other than myself, had access to all the facts and access did not mean knowledge — the field was too great, time was too short, and I was seeking accomplishment not knowledge." LRG, Speech to the Institute of Radio Engineers and the American Institute of Electrical Engineers, January 23, 1946, Box 4, Entry 7530I, Papers of LRG, RG 200, NARA. "General Groves personally established and supervised the security policies and so compartmentalized was the construction, operational and production phases of the venture that only a few persons ever knew its full and complete implications and objectives." Press Release, "Major General Leslie R. Groves Directs Vast Project," August 6, 1945.

11. Quoted in Jones, *Manhattan,* 272.

12. Groves estimated that a "thousand or so" were aware of the "major outlines" of

the project without defining what that meant. LRG, Comments on *The New World,* 2, Box 3, Entry 7530J, Papers of LRG, RG 200, NARA.

13. Marcus L. E. Oliphant's observations of Groves during a meeting on September 13, 1943, are worth quoting. "General Groves gave an account of his security measures. . . . The system of compartmentalization, which has been adopted, was designed to prevent any individual, apart from himself and Tolman, from having a complete picture of the project. . . . He explained that he, as executive officer, made available to his seniors in the army itself only such restricted and general information as was necessary in order to further the general scheme. In fact only 5 or 6 senior officers knew what he was doing. . . . In fact, it became rapidly evident as discussion proceeded that he is an absolute dictator, in who his superiors place great faith. He is obviously a very able man." Oliphant, Notes on Conversations with Americans in Washington, at the New War Department, on Monday, September 13th, 1943, Fast Neutron Physics and the Bomb, PRO, London, AB1/376.

14. While Groves knew everything about the Manhattan Project, he did not know about some other secret programs that had a bearing on his own. Other secret programs had their own levels of compartmentalization, and Groves, without a need to know, did not receive the information. Groves did not know about the Ultra, Magic, or Venona intercepts. *Venona* was an arbitrary code word for a highly secret program that intercepted approximately 3,000 encrypted telegraphic cables sent and received by the Soviet Union to its spies in the United States and elsewhere. The Army Signal Security Agency began to intercept Russian diplomatic traffic on February 1, 1943. It was only in December 1946 that American code breakers first translated a KGB message. Over the next few years the deciphering of certain messages led to the arrest of Klaus Fuchs and Julius and Ethel Rosenberg for passing atomic secrets to the Soviet Union. It took decades for many of the cables to be deciphered in part or in full, and only beginning in 1995 were the first ones made public. Many of them still have not been decrypted. Robert Louis Benson and Michael Warner, eds., *Venona: Soviet Espionage and the American Response, 1939–1957* (Washington, D.C.: National Security Agency and Central Intelligence Agency, 1996); John Earl Haynes and Harvey Klehr, *Venona: Decoding Soviet Espionage in America* (New Haven, Conn.: Yale University Press, 1999).

15. Stewart, *Organizing,* 27–31. Stewart was the secretary of the NDRC and deputy director of the OSRD.

16. "Top Secret" covered certain documents, information, and materials, "the security aspect of which is paramount, and whose unauthorized disclosure would cause exceptionally grave damage to the nation." Covered under this was information such as production figures of end product and materials, the military use, successful methods, plant locations.

"Secret" classification referred to "information or features, the disclosure of which might endanger national security, cause serious injury to the interest or prestige of the nation or any government activity, or would be of great advantage to a foreign power." Covered under this was information involving technical design, processes, construction methods, scope and size of the project, maps, and photographs.

"Confidential" classification referred to "information or features the disclosure of which might be prejudicial to the interest or prestige of the United States, a

governmental activity, or an individual, or be of advantage to a foreign nation."
Covered under this was information involving statistics on the materials and
personnel required to construct the facility as well as the construction and production progress.

"Restricted" classification referred to information "for official use only, or
when disclosure should be limited for reasons of administrative privacy, or
denied the general public." Covered under this was administrative and organizational information, such as that in security bulletins. *Manhattan District
History*, bk. 1, vol. 14, appendix B-7, 2–3; Stewart, *Organizing*, 250–252.

17. "In the spring of 1942, the Army took over from the Navy the responsibility
for personnel security investigation with a few limited exceptions." Stewart,
Organizing, 248. Stewart says that the number of investigations conducted for
OSRD for all divisions and sections was in the neighborhood of 45,000 to
50,000. Ibid., 250. Though Stewart does not provide the number for just S-1 it
could not have been more than a few hundred, perhaps a thousand. The
number of investigations conducted under the Manhattan Project was approximately 400,000.

18. Accounts of these crucial days in 20th-century physics have been told in great
detail. Rhodes, *Making*, 247–258; Ruth Lewin Sime, *Lise Meitner: A Life in
Physics* (Berkeley: University of California Press, 1996), 231–258; Abraham Pais,
Niels Bohr's Times, In Physics, Philosophy, and Polity (Oxford: Clarendon Press,
1991), 452–464.

19. Enrico Fermi, "Physics at Columbia University," *Physics Today* 8 (November
1955), 13.

20. Smyth Report, 45–46.

21. Ironically, all or most of these articles constituted a major espionage windfall for
the Soviet Union. Sometime before July 1943, someone who had access to them
passed the titles and summaries of 286 articles to a Soviet agent or agents. That
person has not been identified. Knowledge of this comes from the archives of
Russian foreign intelligence. V. P. Visgin, ed., "Roots of the Soviet Atomic
Project: Role of Intelligence, 1941–1946," *Problems in the History of Natural Science
and Technology*, 1992, 97–134. The treasure trove was provided to Igor V.
Kurchatov, head of the recently established Soviet atomic bomb program. In a
24-page memo to M. G. Pervukhin, dated July 3, 1943, Kurchatov analyzed
summaries of 237 papers covering the various aspects of the "uranium problem,"
as they were written about by American and émigré scientists. Almost one-
quarter of those he surveyed were devoted to the various methods of isotope
separation: gaseous diffusion (29), centrifuge (18), electromagnetic (4), thermal
(6), and the general problems involved in separation (5). Thirty-two were
devoted to the problems of a heavy-water pile and another 29 to the graphite
pile. Fifty-five were devoted to the chemistry of uranium, 10 to constructing a
bomb out of U-235, 14 to the transuranic elements (neptunium and plutonium),
30 to the neutron physics of nuclear fission, and 3 each to the isotopes U-232
and U-233, and the physiological effects of uranium. All in all a handy com-
pendium of the state of knowledge about the American bomb program.

22. Lansdale, Manuscript; Hershberg, *Conant*, 157–158; USAEC, *Oppenheimer*,
258–281. Lee was military attaché in London and then served as assistant chief
of staff, G-2, from February 1, 1941, to May 4, 1942.

23. Lansdale, Manuscript, 22. Maj. Gen. George Veazey Strong had had a long and

interesting career before becoming G-2 on May 5, 1942. Born in 1880, Strong graduated from West Point in the class of 1904 and joined the cavalry. His first assignment was to Fort Meade, South Dakota, where he saw service against the Ute Indians. In 1916 he transferred to the judge advocate general's department. After service in France during World War I he became a professor of law at West Point. He had several assignments on the General Staff War Plans. Marshall sent him to England to observe and report upon the battle of Britain. He served as assistant chief of staff, G-2, until February 6, 1944, when he retired for age and died in January 1946 at age 65. Obituary, *Assembly*, January 1947, 6. In the next chapter we will meet his successor, Maj. Gen. Clayton L. Bissell, who served until January 25, 1946.

24. Otto L. Nelson Jr., *National Security and the General Staff* (Washington, D.C.: Infantry Journal Press, 1946), 525.

25. Lansdale, Manuscript, 82; Jones, *Manhattan,* 255.

26. The eleven branch offices were in Boston, New York, Baltimore, Cleveland, Chicago, Grand Junction (Colorado), Santa Fe, Berkeley, Richland (Washington), Washington, D.C., and Pasadena. Normally they worked out of the area engineer's office.

27. Lansdale, Manuscript, 43.

28. "The inspector general's report led to the immediate unraveling of the Counter Intelligence Corps. The counter subversive program was terminated, and most CIC agents in the continental United States were merged with the criminal investigators of the Provost Marshal's Office to form a new Security Intelligence Corps that operated under the control of the service commands. Although CIC detachments continued to serve with the Army Air Forces, the Manhattan Project, and tactical units, the presence of the Counter Intelligence Corps on the home front was effectively eliminated." John Patrick Finnegan, *Military Intelligence,* (Washington, D.C.: United States Army Center of Military History, 1998), 75. See also Jones, *Manhattan,* 257.

29. Letter, Strong to Commanding General, ASF, December 18, 1943, Organization of Counter Intelligence Corps Detachment for Manhattan Engineer District; U.S. Army Intelligence Center, *History of the Counter Intelligence Corps*, vol. 8: *The Counter Intelligence Corps with Special Projects* (Fort Holabird, Md.: U.S. Army Intelligence Center, December 1959), appendix 2. In a letter dated December 27, 1943, General Strong relieved the service commands of counterintelligence responsibilities "with respect to the DSM project." Ibid., appendix 3.

30. *Manhattan District History,* bk. 1, vol. 14: *Intelligence & Security,* December 31, 1945; *History of the Counter Intelligence Corps*, vol. 8: *The Counter Intelligence Corps with Special Projects* (Fort Holabird, Md.: U.S. Army Intelligence Center, December 1959); Jones, *Manhattan,* 253–279. Had enemy agents known of the bomb program and the location of key facilities, they might have tried to destroy or disrupt it through physical means. A well-placed bomb at the gaseous diffusion plant or the electromagnetic plant at Oak Ridge, or one at the B pile at Hanford, would have delayed, if not prevented, delivery of the scarce fissile material to Los Alamos.

31. U.S. Army Intelligence Center, *History of the Counter Intelligence Corps,* Vol. 8: *The Counter Intelligence Corps with Special Projects* (Fort Holabird, Md.: U.S. Army Intelligence Center, December 1959), 85. Arrest records turned up by these checks revealed some applicants to have a range of charges, from speeding

to arson, rape, and murder. One individual indicated he had been arrested 116 times for bootlegging. Several escaped convicts, parole violators, and deserters were apprehended. Ibid.

32. LRG, Memo to the Commanding General, Army Service Forces, Prisoners-of-War Camps, May 11, 1945, Folder, Manhattan District, Box 10, Washington Liaison Office, Official Correspondence 1944–1948, RG 326, AEC, Atlanta Record Center. See also Matthias Diary, USACE, Office of History, March 29, April 3, April 4, 1945.

33. On May 21, 1940, President Roosevelt authorized the FBI to conduct warrantless electronic surveillance of people suspected of espionage. It is likely that many or most of the phones were also tapped at Columbia's SAM lab and at the University of Chicago's Met Lab.

34. Interview with Boris Pash by Stanley Goldberg, April 18, 1990.

35. An additional wire was added that bypassed the disconnect, thereby turning the telephone into a microphone. Lansdale, Manuscript, 29. "At one point Pash with the assistance of the local telephone company had a switchboard installed in the basement of a rented house in Oakland manned 24 hours a day where a number of these clandestine microphone installations were monitored and recorded." Ibid.

36. Memo, Welch to Ladd, CINRAD, July 23, 1943, FBI Bureau File 100-190625-22A.

37. "If he [Groves] could coop these people up in one place, it would be a lot easier to control their talking." John H. Manley, "A New Laboratory Is Born," in Badash, et al., Los Alamos, 26.

38. Another reason Groves allowed Oppenheimer to have his way on this matter was that otherwise they would have a difficult time recruiting scientists from other projects.

39. Bernice Brode provides a vivid description. "Tales of Los Alamos," in Badash, et al., Los Alamos, 133–159.

40. For example, Norman Ramsey's Delivery Group did not give regular reports at the colloquia. The operational details were kept from most of the scientists. Norman F. Ramsey, Oral History Interview, July 19–August 4, 1960, Columbia University Oral History Collection, 94.

41. Letter, Richard C. Tolman to LRG, June 11, 1943, Folder Tolman, R. C., Box 17, Bush-Conant Papers, RG 227/S1, NARA. After the item, "Time schedules on the production of usual amounts of material," Groves wrote, "This information must be held absolutely secret among Oppenheimer, Parsons and Thomas."

42. Norman F. Ramsey, Oral History Interview, July 19–August 3, 1960, Columbia University Oral History Collection, 68.

43. Hearings Before the Senate Committee on Atomic Energy, 79th Cong., 1st Sess., part 2, December 5, 6, 10, and 12, 1945, 290–291.

44. LRG, Working Paper for Now It Can Be Told, Folder, Working Paper, Box 4, Entry 7530M, Papers of LRG, RG 200, NARA.

45. Robert Chadwell Williams, Klaus Fuchs, Atom Spy (Cambridge, Mass.: Harvard University Press, 1987); Michael Dobbs, "Code Name 'Mlad,' Atomic Bomb Spy," Washington Post, February 25, 1996, A1; Michael Dobbs, "New Documents Name American as Soviet Spy," Washington Post, March 6, 1996, A1; Joseph Albright and Marcia Kunstel, Bombshell: The Secret Story of America's Unknown Atomic Spy Conspiracy (New York: Times Books, 1997); Richard Rhodes, Dark Sun: The Making of the Hydrogen Bomb (New York: Simon & Schuster, 1995).

46. USMA, *Biographical Register,* Supplement, Vol. 9, 1940–1950, 1144; Obituary, *Assembly,* June 1983, 140–141.

47. De Silva to Pash, September 2, 1943, Folder, AEC, Oppenheimer, Box 76, Strauss Papers, Herbert Hoover Presidential Library.

48. Philip M. Stern, *The Oppenheimer Case: Security on Trial* (New York: Harper & Row, 1969), 49. Pash's June 29, 1943, memo is in USAEC, *Oppenheimer,* 821–823.

49. On August 26, 1942, Pash and Lt. Lyall Johnson had interviewed Oppenheimer in Johnson's office in the New Class Room Building on the Berkeley campus.

50. Lansdale, Manuscript, 33.

51. Letter to Director, FBI, from N. J. L. Pieper, June 30, 1943, FBI, Bureau File 100-190625-206. This had been one of three recommendations that Pash offered to Lansdale in his memo of the day before: "That every effort be made to find a suitable replacement for subject and that as soon as such replacement is trained that subject be removed completely from the project and dismissed from employment by the United States Government." USAEC, *Oppenheimer,* 822.

52. USAEC, *Oppenheimer,* 170.

53. USAEC, *Oppenheimer,* 273–275.

54. De Silva had a long career in U.S. intelligence. After serving in the Manhattan Project, he joined the Strategic Services Unit in 1946, which had inherited the vestiges of the wartime Office of Strategic Services. When SSU and its successor the Central Intelligence Group were absorbed into the newly formed Central Intelligence Agency in September 1947, de Silva began a long association with the CIA, including serving as chief of station in Vienna, Seoul, Hong Kong, Saigon, and Australia. He retired in January 1973 and died in August 1978. De Silva replaced 2nd Lt. Curtis L. Clark as Los Alamos security officer and served from October 1943 to April 1945. Lt. Thomas O. Jones, who had been chief of the Santa Fe Intelligence Branch, replaced de Silva.

55. Quoted in Philip M. Stern, *The Oppenheimer Case: Security on Trial* (New York: Harper & Row, 1969), 78–79.

56. "Though Parsons officially reported to Nichols Parsons kept Groves appraised of all developments." Jones, *Manhattan,* 258.

57. These issues came up in an interesting interchange between Hoover and Groves. In a detailed five-page letter of April 1, 1946, Hoover recounted the past history of problems. Groves drafted a detailed response dated April 8 but did not send it, believing that it was inadvisable to enter into a controversy, and also that Hoover was forwarding the views of one of his subordinates and not personally concerned about the matter. See Letter, Hoover to LRG, April 1, 1946, and Draft Letter, LRG to Hoover, April 8, 1946, Folder Hoover 1, Box 4, Entry 7530B, Papers of LRG, RG 200, NARA. The letters were also discussed in an FBI memorandum; see Memo, D. M. Ladd to the Director, The Manhattan Engineer District (MED), July 15, 1947, File 100-190625 (obtained through the FOIA).

58. LRG, Memo to File, MED Operations, Security, October 4, 1965, Folder Hoover 1, Box 4, Entry 7530B, Papers of LRG, RG 200, NARA.

59. Lansdale, Manuscript, 34.

60. Memo, Surveillance of Mr. STARR, October 2, 1943, Folder 201, Groves, General Correspondence 1941–47, Box 86, Entry 5, MED Decimal Files 1942–1948, RG 77, NARA.

61. During World War II the United States observed year-round Daylight Saving

Time (from February 2, 1942, to September 30, 1945) to save energy. To inspire the American public and to constantly remind them that they were at war the time was expressed as Eastern, Central, Mountain, or Pacific War Time.

62. Memo, Physical surveillance of LRG, October 2, 1943, October 5, 1943, Folder 201, Groves General Correspondence 1941–47, Box 86, Entry 5, MED Decimal Files 1942–1948, RG 77, NARA.

63. LRG, Undated Memo to File, Secret Papers, Box 1, Entry 7530C, Papers of LRG, RG 200, NARA. On November 2, 1943, Groves was issued a belt and two holsters, one for a .45-caliber and the other for a smaller-caliber pistol or revolver.

64. LRG to David Kahn, August 16, 1961, Folder K, Box 5, Entry 7530B, Papers of LRG, RG 200, NARA; David Kahn, *The Codebreakers: The Story of Secret Writing* (New York: Scribner, 1996, first published 1967), 546–547.

65. Compton, *Quest,* 183–184.

66. Daniel Lang, *Early Tales of the Atomic Age* (Garden City, N.Y.: Doubleday and Co., 1946), 21.

67. LRG to Lawrence, July 29, 1943, Lawrence Papers, 72/117C, Carton 28, Folder 37, Bancroft Library.

68. Washburn, "Attempt"; Michael S. Sweeney, *Secrets of Victory: The Office of Censorship and the American Press and Radio in World War II* (Chapel Hill: University of North Carolina Press, 2001). 197–207.

69. Soon after the war he wrote of his experience in the chapter "Best Kept Secret of the War." Theodore F. Koop, *Weapon of Silence* (Chicago: University of Chicago Press, 1946).

70. The note was sent to 2,000 daily newspapers, 11,000 weeklies, and all radio stations in the United States. Washburn, "Attempt," 7. At the suggestion of General Strong several harmless elements were included to camouflage the imoportance of the sensitive ones, especially uranium. Michael S. Sweeney, *Secrets of Victory: The Office of Censorship and the American Press and Radio in World War II* (Chapel Hill: University of North Carolina Press, 2001), 200.

71. John W. Raper, "Forbidden City, *The Cleveland Press,* March 15, 1944. Raper was on vacation in New Mexico.

72. Washburn, "Attempt," 11.

73. Lawren, *General,* 188.

74. LRG to Marshall, August 16, 1944, Top Secret (M1109), File 12A.

75. LRG, Memo to File, Breach of Security by Arthur Hale, August 16, 1944, Top Secret (M1109), File 12A.

76. LRG, Memo to File, October 10, 1963, Folder Koop, Theodore F., Box 2, Entry 7530J, Papers of LRG, RG 200, NARA.

77. Washburn, "Attempt," 16.

78. Ibid.

79. Paul Filipkowski, "Postal Censorship at Los Alamos, 1943–1945," *American Philatelist* (April 1987), 345–350. See also Bernice Brode, "Tales of Los Alamos," in Badash, et al., *Los Alamos,* 140–141.

80. Other box numbers were used. The military technical staff used Box 180, the local MED staff and nontechnical staff used 1539, and the Provisional Military Police Detachment No. 1 used Box 527. Groves and the Washington office used P.O. Box 2610, Washington D.C. The original project address was P.O. Box 1722, Santa Fe.

81. USAEC, *Oppenheimer,* 262. Testimony of John Lansdale.

82. The lists cover only the period August–October 1943, and are part of the CINRAD (Communist Infiltration in the Radiation Laboratory, University of California, Berkeley, California) files. FBI Bureau File 100-190625 (obtained through the FOIA).

83. CINRAD files, FBI Bureau File 100-190625 (obtained through the FOIA).

84. The five were Chairman Clarence Cannon (D-MO), Albert J. Engel (R-MI), John Taber (R-NY), George H. Mahon (D-TX), and J. Buell Snyder (D-PA). Jones, *Manhattan,* 274; LRG, *Now,* 363–365. A Senate delegation visited Hanford after V-J Day. On the earlier visits to the Hill see Memo, Bush to Bundy, February 24, 1944, Top Secret (M1109), File 12B, and Bush Memo, June 10, 1944, Top Secret (M1109), File 12A.

85. USAEC, *Oppenheimer,* 170.

86. "Scoffing at the dangers involved in conducting an espionage ring in the United States during the early 1930s, at a time when no effective American counterintelligence agency existed, Nadya [Ulanovskaya, a former Soviet spy] said: "If you wore a sign saying 'I am a spy,' you might still not get arrested in America when we were there." Allen Weinstein, *Perjury: The Hiss-Chambers Case* (New York: Random House, 1997), 108. "The US government response to the extended Soviet espionage effort that reached its peak in the war was limited and slow. With a few exceptions, until 1939 both executive-branch officials and their Congressional counterparts displayed little interest in the data that Soviet technicians were collecting in American plants. Contrary to the long prevalent view of J. Edgar Hoover, the FBI Director was not always obsessed with Communists." Katherine A. S. Sibley, "Soviet Industrial Espionage Against American Military Technology and the US Response, 1930–1945," *Intelligence and National Security* 14, no. 2 (summer 1999), 105.

CHAPTER 14 His Own Intelligence: Foreign Concerns

1. According to Boris Pash, Groves once told him that even if a tiny country such as Uruguay possessed the weapon, "it could dictate its terms to the rest of the world." Pash, *Alsos,* 10.

2. "The advent of the atomic bomb and nuclear diplomacy thrust science into all subsequent deliberations about foreign policy and foreign intelligence." Ronald E. Doel and Allan A. Needell, "Science, Scientists, and the CIA: Balancing International Ideals, National Needs, and Professional Opportunities," in Rhodri Jeffreys-Jones and Christopher Andrew, eds., *Eternal Vigilance? 50 Years of the CIA* (London: Frank Cass, 1997), 59–81. The quote is on page 74.

3. Per F. Dahl, *Heavy Water and the Wartime Race for Nuclear Energy* (Philadelphia: IOP, 1999); Kurzman, *Blood;* Powers, *Heisenberg,* 196–202; Arnold Kramish, *The Griffin* (Boston: Houghton Mifflin Co., 1986), 159–178; Irving, *German,* 130–170, 192–211.

4. The four-man crew of the second Halifax bomber made it safely back to Scotland.

5. "In fact about a ton of heavy water, in concentrations ranging from 10.5 to 99.3 per cent, had been lost from the damaged cells, the equivalent of about 350 kilograms of pure heavy water." Irving, *German,* 170.

6. Groves's account of his role in the February Norsk-Hydro attack is a bit confused. He mistakenly takes credit for instigating Gunnerside when in fact he

did not know about it until two weeks after it was over. He says, "At my instigation [General George V.] Strong [G-2], with the approval of General H. H.
Arnold and Major General T. T. Handy, had brought this matter to the personal
attention of General Eisenhower, and suggested that the Rjukan plants be
either bombed or sabotaged. The first attempt to put these works out of commission involved the use of guerrilla forces. Some five months after my request,
three Norwegians, especially trained in sabotage techniques, and wearing
British uniforms, parachuted into Norway, where local guerrillas met them.
After nearly a week of hard cross-country skiing, they arrived at Rjukan and
attacked the factories there on February 27, 1943." LRG, *Now,* 188. Groves is
no doubt referring to a September 8, 1942, memo from Strong to Eisenhower
in which these actions are "called to [Eisenhower's] attention" and "should seriously be considered." Top Secret, File 7E, no. 17. September 8, of course, is nine
days before Groves was selected to head the Manhattan Project. While he may
have learned of the Strong memo early on, there is no evidence that he insti-
gated Gunnerside, or had any knowledge at all of Freshman until the spring of
1943. He apparently transposed some of the more assertive actions he did take
in the spring and summer of 1943 to the earlier attack.

7. According to his aide on intelligence matters, Robert R. Furman, Groves was
   ignorant of Freshman until then. Kurzman, *Blood,* 185.
8. Memo, Marshall to Dill, April 3, 1943, Top Secret, File 7E, no. 13.
9. A translation of the article in *Svenska Dagbladet* is in Top Secret, File 7E, no. 16.
10. LRG, *Now,* 188.
11. "Discoverer Doubts Use of 'Heavy Water' in War," *New York Herald Tribune,*
    April 5, 1943. Elsewhere Urey responded, "The story in *The Times* was certainly
    weird enough to be a real hair raiser. . . I am quite sure that it cannot be used
    as a deadly explosive of any kind. I think the best thing that can be done with
    such stories is to ignore them." Top Secret, File 7E, no. 6.
12. To JSM, From War Cabinet Offices, April 6, 1943, Top Secret, File 7E, no. 10; To
    Dill from Air Ministry Special Signals Office, April 7, 1943, Top Secret, File 7E,
    no. 14.
13. Paraphrase of Telegram Just Received from a Reliable Source, August 13, 1943,
    Top Secret, File 7E, no. 4.
14. Memo, Marshall to Dill, Interference with German Operations, n.d., Top
    Secret, File 7E, no. 13.
15. The Norwegian government-in-exile in London was not informed of the
    attack and was shocked by the raid.
16. Strong to Marshall, Memorandum for the Chief of Staff, August 13, 1943, Top
    Secret, File 7E, no. 3.
17. Powers, *Heisenberg,* 213.
18. "I formulated a plan for gathering scientific information at the battle front and
    in General Strong's name 'lobbied' through the War Department authority to
    put it into effect. This became the Alsos mission." Lansdale, Manuscript, 38.
19. Goudsmit, *Alsos;* Pash, *Alsos;* Leo J. Mahoney, *A History of the War Department
    Scientific Intelligence Mission (Alsos), 1943–1945* (Unpublished Ph.D. dissertation,
    Kent State University, 1981).
20. "Alsos was named by someone in G-2. I was very much annoyed when I
    learned of it but did not want to change it and thus attract attention. The mis-
    sion's main objective was atomic information. For security reasons, it was nom-

inally under G-2. Actually, it was operated in accordance with my wishes at all times." LRG, Comments on *The Virus House* by David Irving, January 6, 1970, Folder David Irving, Box 2, Entry 7530J, Papers of LRG, RG 200, NARA.

21. Oppenheimer to Furman, September 22, 1943, Folder Furman, Major Robert R., Box 34, Papers of JRO, Manuscript Division, LOC. A year later an Alsos member collected a few bottles of water from the River Rhine. Furman shipped the bottles back to Washington and included a bottle of French wine, intended as a joke, for testing as well. A few days later the message came back, "Water negative, wine positive, send more." Suspicious of everything, Washington thought perhaps that there might be a secret German laboratory in the French wine country and wanted more samples, which were supplied. Pash, *Alsos*, 133–134; Goudsmit, *Alsos*, 22–24.

22. Pash, *Alsos*, 11.

23. Pash tells the tale well in his *Alsos*.

24. Pash made a trip to Italy June 4-10, 1944, arriving in Rome shortly after the Americans liberated it. Three Italian scientists were apprehended and interrogated.

25. LRG to Assistant Chief of Staff, G-2, March 10, 1944, Top Secret, File 26N.

26. LRG, *Now*, 185. Elsewhere Groves dates it a bit later, either December 1943 or early January 1944. See Letters, LRG to Irving, November 29, 1965, and LRG to Irving, January 14, 1966, and LRG, Comments on *The Mare's Nest*, all in Folder Irving, David, Box 5, Entry 7530B, Papers of LRG, RG 200, NARA. At approximately this time Stimson too was troubled by the duplication and competition among the three separate intelligence agencies: G-2, Office of Naval Intelligence, and Office of Strategic Services. On November 9, 1943, he called in General Marshall to discuss the problem. Stimson Diary, November 10, 1943.

27. Anthony Cave Brown, *The Last Hero: Wild Bill Donovan* (New York: Vintage Books, 1984), 304–310; Stanley P. Lovell, *Of Spies & Stratagems* (Englewood Cliffs, N.J.: Prentice Hall, 1963), 163–169.

28. Groves's diary shows a visit to the office by Furman (and Lansdale and Calvert) on August 24.

29. LRG, Memo to File, General Marshall, October 16, 1963, Folder M, Box 6, Entry 7530B, Papers of LRG, RG 200, NARA.

30. Victor Weisskopf, *The Joy of Insight: Passions of a Physicist* (New York: Basic Books, 1991).

31. Heisenberg traveled to several German-occupied countries and neutral Switzerland during the war years. He functioned as "a kind of roving ambassador for German science, visiting German Cultural Institutes and speaking at universities. These cultural institutes were set up in the occupied countries for the purpose of exposing the natives to the merits of German science and culture. Ernst von Weizsacker, Carl-Frederich's father, played an active role in establishing these centers." Jeremy Bernstein, Letter, *New York Review of Books*, June 22, 1991, 64; Mark Walker, "Physics and Propaganda: Werner Heisenberg's Foreign Lectures Under National Socialism," *Historical Studies in the Physical Sciences* 22 (1992), 339–389. As it turned out, Heisenberg arrived in Zurich on November 17, also went to Basel and Bern, lectured almost daily, and returned to Germany on November 25. His trip to Copenhagen in September 1941, accompanied by von Weizsäcker, had been to attend an astrophysics conference sponsored by the German Cultural Institute, and not solely to see Bohr.

32. As Bethe said, "I have to confess that Victor Weisskopf and I thought of kidnapping Heisenberg at the latest in 1942." Dawidoff, *Catcher,* 191.

33. Weisskopf to Oppenheimer, October 28, 1942, Folder Weisskopf, V. F., Box 77, Oppenheimer Papers, Manuscript Division, LOC.

34. Oppenheimer to Weisskopf, October 29, 1942, Folder Weisskopf, V. F., Box 77, Oppenheimer Papers, Manuscript Division, LOC.

35. Oppenheimer to Bush, October 29, 1942, Folder Bush, Box 23, Oppenheimer Papers, Manuscript Division, LOC.

36. Dawidoff, *Catcher,* 161.

37. This information was supplied in part in a series of letters from Lt. Col. H. W. Dix, Technical Section, OSS, to Major Furman or to his deputy Maj. Francis J. Smith.

38. Dawidoff, *Catcher,* 161.

39. In early 1944 Philip Morrison "expressed the thought that it would be wise to kidnap a man such as Von Weizsacker." Report, January 12, 1944, Folder 32.60.-1, Summary Reports, 1944, Box 170, Entry 22, RG 77, NARA.

40. Stanley P. Lovell, *Of Spies & Stratagems* (Englewood Cliffs, N.J.: Prentice Hall, 1963), 108.

41. Dawidoff, *Catcher,* 200.

42. LRG, *Now,* 216. For evidence that others besides Berg were involved see Dawidoff, *Catcher,* 194–196.

43. LRG, Comments on *The Virus House* by David Irving, January 6, 1970, Folder David Irving, Box 2, Entry 7530J, Papers of LRG, RG 200, NARA.

44. LRG to Compton, May 30, 1943; Cable, Compton to LRG, June 1, 1943; Memo, LRG to Strong, June 2, 1943, DSM Project, Folder, Liaison Reports to British, Box 168, Entry 22, RG 77, NARA.

45. Strong to Marshall, Memo for the Chief of Staff, August 13, 1943, Top Secret, File 7E, no. 3. See also Powers, *Heisenberg,* 210–211.

46. On December 3, 1943, the British had bombed Leipzig and the Institute for Theoretical Physics. Most of Heisenberg's papers and his house were destroyed. He was spending most of his time in Berlin, in Hechingen, and with his family in Urfeld in the Bavarian Alps. Powers, *Heisenberg,* 336.

47. Powers, *Heisenberg,* 337–339.

48. LRG, Memo to File, October 7, 1963, Folder R, Box 8, Entry 7530B, Papers of LRG, RG 200, NARA.

49. LRG, Comments on *The Virus House* by David Irving, January 6, 1970, 13, Folder David Irving, Box 2, Entry 7530J, Papers of LRG, RG 200, NARA.

50. Dix to Furman, May 11, 1944, Folder 32.7003.2 July 1942–June 1944, Box 171, Entry 22, RG 77, NARA.

51. Goudsmit, *Alsos,* 75.

52. One of the most important agents was Paul Rosbaud, a scientific editor at Springer whose position kept him knowledgeable regarding German research developments. Arnold Kramish, *The Griffin* (Boston: Houghton Mifflin Co., 1986).

53. On March 21, 1944, Sir John Anderson informed Churchill that "fortunately all the evidence we have goes to show that the Germans are not working seriously on the project." Quoted in Irving, *German,* 224–225.

54. LRG, *Now,* 187. "I could not help but believe that the Germans, with their extremely competent group of first-class scientists, would have progressed at a rapid rate and could be expected to be well ahead of us."

55.LRG, *Now,* 207–223, 230–249; Pash, *Alsos;* Goudsmit, *Alsos; Manhattan District History,* bk. 1, vol. 14: *Foreign Intelligence,* Supplement 1.

56.Peppermint documents are in Top Secret, File 7D. See also *Manhattan District History,* bk. 1, vol. 14: *Foreign Intelligence,* Supplement 2; Memo, LRG to Bradley, February 28, 1948, with attached paper, Defensive Measures Against Possible Uses by Germans of Radioactive Warfare — Operation Peppermint, Folder 201, LRG, Box 86, Entry 5, RG 77, NARA; Maj. A.V. Peterson, "Peppermint," in Brown, MacDonald, *Secret History,* 234–238; Ferenc M. Szasz, "Peppermint and Alsos," *Military History Quarterly,* spring 1994, 42–47. On the general topic of U.S. interest in radiological weapons during World War II see Barton J. Bernstein, "Radiological Warfare: The Path Not Taken," *Bulletin of the Atomic Scientists,* August 1985, 44–49.

57.Memo, LRG to Marshall, March 22, 1944, Folder F, Box 3, Entry 7530B, Papers of LRG, RG 200, NARA.

58.Eisenhower to Marshall, Top Secret, Folder 18, Tab A.

59.About a dozen soldiers and scientists carried Geiger counters onto the Normandy beaches. Other sensing equipment had been shipped to the European theater to be used if there was an emergency, and more was ready to be shipped from the United States. Barton J. Bernstein, "Radiological Warfare: The Path Not Taken," *Bulletin of the Atomic Scientists,* August 1985, 47–48.

60.Pash and his comrades would often be the first Americans to reach liberated territory in the coming months. Occasionally enemy troops surrendered to them; they were often given a hero's welcome in liberated French villages.

61.The London Alsos office was consolidated with the Paris office on September 20 and soon moved, on September 25, to the sixth floor of 9 Rue Presbourg, where the Navy Mission in France was located. A final move took place on April 10, 1945, to 124 Boulevard Bineau, Neuilly-sur-Seine. When in Paris, Pash and most of the others lived at the Royal Monceau Hotel on Avenue Hoch.

62.Groves was deeply suspicions of Joliot, believing that he collaborated fully until it became clear the Germans could not win and only then joined the Resistance. LRG, Comments on *The Virus House* by David Irving, January 6, 1970, Folder David Irving, Box 2, Entry 7530J, Papers of LRG, RG 200, NARA.

63.Pash, in his travels far and wide, provided Conrad with intelligence about such things as the status of the ports of Antwerp and Bordeaux, and partisan activities in southern France. Pash, *Alsos.*

64.Pash has said, "Much later I was told that the Oolen shipment and the subsequent shipment found in France were to be dropped on Hiroshima — in somewhat altered form, to be sure." *Alsos,* 99.

65.Goudsmit, *Alsos,* 70–71.

66.Pash, *Alsos,* 158.

67.See Shurcliff's reports in Folder Tolman, Box 9, Entry OSRD S-1, RG 227, NARA.

68.For Oranienburg documents see Top Secret, File 7C; LRG, *Now,* 230–231.

69.LRG to Marshall, April 23, 1945, Folder, Germany, Box 8, Entry Manhattan Project, RG 319, NARA. "The capture of this material, which was the bulk of uranium supplies available in Europe, would seem to remove definitely any possibility of the Germans making use of an atomic bomb in this war." A second motive for removal is clear from a memo from Calvert to Groves in early February. "Belgian material if still in Stassfurt presents very difficult problem as there are equivalent to approximately 1,000 tons uranium oxide. Russian known reserves 200 to 300 tons so that capture of Belgian material may be of very great importance to them." Calvert to LRG, February 2, 1945, CM-IN-1229, Folder 55, Harrison-Bundy Files.

70. Throughout these operations Alsos was about 12 hours ahead of the French troops. While no difficulties were experienced, Pash, for one, did not think very much of their fighting ability. "Having leaned on the Americans throughout the war, now that the enemy was on the run they were feeling their oats and had become pompous in their attitude if not their results." Pash, *Alsos,* 202.

71. Late in July Goudsmit and a small Alsos group went to Berlin, to the Kaiser Wilhelm Institute for Physics, and found some remnants of the uranium research that had gone on there prior to the evacuation to Hechingen. In the sub-basement was the pit in which the pile had been constructed. Goudsmit, *Alsos,* 123–127.

72. Goudsmit was responsible for determining which of the scientists should be held in military custody and which could remain at large. He selected 10: Hahn, von Weizsäcker, Wirtz, Bagge, Korsching, von Laue, Heisenberg, Harteck, Gerlach, and Diebner.

73. See the picture opposite page 108 in Goudsmit, *Alsos.*

74. The cable is reprinted in Pash, *Alsos,* facing the title page.

75. Goudsmit, *Alsos,* 108, 160–186.

76. Notes on Discussion with Dr. Chadwick, Friday — June 1, 1945, Folder 001, Meetings, Box 33, Entry 5, RG 77, NARA. The memo to the file is unsigned, but Groves's office diary shows a 6:00 P.M. meeting with Chadwick.

77. Goudsmit, *Alsos,* xiv. R. V. Jones tells the same story in his earlier *Wizard War,* 481–482.

78. R. V. Jones says they were moved to Farm Hall from Versailles at his suggestion, because "an American general had said the easiest way of dealing with the postwar developments in nuclear physics in Germany would be to shoot the physicists." Goudsmit, *Alsos,* xiv; Jones, *Wizard War,* 481.

79. Goudsmit, *Alsos,* xiv; Jones, *Wizard War,* 481–482.

80. The Washington set was used for *Operation Epsilon: The Farm Hall Transcripts,* introduced by Sir Charles Frank (Berkeley, Calif.: University of California Press, 1993). Among the British listed in the PRO set to receive copies were Michael W. Perrin, Lt. Cmdr. Eric Welsh, and Sir James Chadwick. See the archival note, 10-13. Jeremy Bernstein provided extensive annotations and other material in his *Hitler's Uranium Club: The Secret Recordings at Farm Hall* (Woodbury, N.Y.: American Institute of Physics, 1996).

81. Cassidy introduction, xvii, in Jeremy Bernstein, *Hitler's Uranium Club: The Secret Recordings at Farm Hall* (Woodbury, N.Y.: American Institute of Physics, 1996).

82. There is correspondence from the chief archivist of the World War II Records Division of the National Archives to Groves providing him with extracts from the Farm Hall conversations that he had requested. Sherrod East to LRG, December 5, 1961, Folder E, Box 3, Entry 7530B, Papers of LRG, RG 200, NARA.

83. Early in their internment on July 6, Heisenberg dismissed the possibility. Diebner wondered whether microphones were installed. Heisenberg replied, "Microphones installed? (laughing) Oh no, they're not as cute as all that. I don't think they know the real Gestapo methods; they're a bit old fashioned in that

respect." By November, partly as a result of some wires found in the back of a cupboard, there were suspicions that they might be "bugged." *Operation Epsilon: The Farm Hall Transcripts,* introduced by Sir Charles Frank (Berkeley, Calif.: University of California Press, 1993), 33, 2, 3.

84. Jones, *Wizard War.*

85. The major works in the current debate are Paul Lawrence Rose, *Heisenberg and the Nazi Atomic Bomb Project: A Study in German Culture* (Berkeley: University of California Press, 1998); Mark Walker, *German National Socialism and the Quest for Nuclear Power, 1939–1949* (Cambridge: Cambridge University Press, 1989); Mark Walker, *Nazi Science: Myth, Truth, and the German Bomb* (New York: Plenum Press, 1995); David Cassidy, *Uncertainty: The Life and Science of Werner Heisenberg* (New York: W. H. Freeman, 1992); Powers, *Heisenberg;* and a steady stream of articles and letters in *Nature, American Scientist, Science, Physics Today,* and the *New York Review of Books,* among others. See the bibliography in Rose, 325–345.

86. LRG, Comments on *The Virus House* by David Irving, January 6, 1970, 1, 2, 7, 17, Folder David Irving, Box 2, Entry 7530J, Papers of LRG, RG 200, NARA.

87. LRG, *Now,* 335–336. In fact, in the Farm Hall transcripts someone else uses the figure *two tons.* But it does not matter, since kilograms and not tons were needed.

88. Rose provided the argument and evidence in extensive detail. Paul Lawrence Rose, *Heisenberg and the Nazi Atomic Bomb Project: A Study in German Culture* (Berkeley: University of California Press, 1998).

89. See Heisenberg's article in *Nature,* "Research in Germany on the Technical Applications of Atomic Energy," August 16, 1947, 211–215. Goudsmit replied in "Heisenberg on the German Uranium Project," *Bulletin of the Atomic Scientists* 3 (1947), 343, and in "Nazis' Atomic Secrets," *Life* (October 20, 1947), 123–134, a preview of the book.

90. Goudsmit, *Alsos,* 76.

CHAPTER 15 **His Own Air Force**

1. Interview with LRG by Al Christman, May 1967, 1, Naval Historical Center, Washington, D.C.

2. LRG, *Now,* 160. Earlier in the day Commander Parsons, who had been looking forward to a combat assignment, had been ordered to the chief of naval operations' office. Much like Groves's brief meeting with Somervell, Parsons's few minutes with Admiral King would change his life. The admiral informed Parsons that the services of an ordnance officer were needed "to supervise the actual production of an atomic bomb," and that he was the choice for this "important and urgent wartime duty." Christman, *Target,* 108.

3. Interview with LRG by Al Christman, May 1967, 54, Naval Historical Center, Washington, D.C.

4. Christman, *Target;* VADM Jerry Miller, USN, retired, "The Warrior Ethos," *Shipmate* (January–February 2001), 16–19.

5. It is there that he received the nickname that would last his entire life. Playing on the name *Parsons,* the midshipmen gave him the name *Deacon,* which soon shortened to *Deak.* Christman, *Target,* 15.

6. Jerry Miller, *Nuclear Weapons and Aircraft Carriers: How the Bomb Saved Naval Aviation* (Washington, D.C.: Smithsonian Institution Press, 2001).

7. Captain Frederick L. Ashworth, USN, "Dropping the Atomic Bomb on Nagasaki," *USNI Proceedings,* January 1958, 15.

8. Parsons to Leis, February 5, 1948, Folder William S. Parsons, Box 1, William S. Parsons Papers, LOC.

9. Roscoe Wilson to LRG, November 3, 1959, Folder W, Box 10, Entry 7530B, Papers of LRG, RG 200, NARA.

10. Norman F. Ramsey, Oral History Interview, July 19–August 4, 1960, Columbia University Oral History Collection; Norman F. Ramsey, Oral History Interview, June 20, 1991, *Rad Lab: Oral Histories Documenting World War II Activities at the MIT Radiation Laboratory* (IEEE, 1995); Norman F. Ramsey, Oral History Interview, May 9, 1995, IEEE.

11. Norman F. Ramsey, Oral History Interview, July 19–August 4, 1960, Columbia University Oral History Collection, 130.

12. Norman F. Ramsey Jr. to Roy Chadwick, October 23, 1943, Folder Dr. Norman Ramsey, Box 6, Tolman Files, RG 227/S1, NARA.

13. Memo, N. F. Ramsey to Capt. W. S. Parsons, October 14, 1943, Lancaster Aircraft, Folder Dr. Norman Ramsey, Box 6, Tolman Files, RG 227/S1, NARA.

14. Originally the uranium gun-type bomb was designated Thin Man. Some explanation was needed for those who were brought into contact with the project but could not be told its real purpose. According to one account the word *pullman* was used as part of the cover story to suggest that a specially modified Pullman car would be used by President Roosevelt (Thin Man) and Prime Minister Churchill (Fat Man) in a secret tour of the United States to visit defense plants. Bowen, *Silverplate,* 96.

15. Col. Donald L. Putt was responsible for the actual modification of the B-29, in collaboration with his deputy Capt. R. L. Roark. Bowen, *Silverplate,* 96; Amy C. Fenwick, *History of Silverplate Project* (Wright-Patterson AFB, Historical Office, Air Material Command, June 1952).

16. Bowen, *Silverplate,* 100–101.

17. According to Werrell, Col. Roscoe C. Wilson and Brig. Gen. Frank Armstrong were in contention. Kenneth P. Werrell, *Blankets of Fire: U.S. Bombers Over Japan During World War II* (Washington, D.C.: Smithsonian Institution Press, 1996), 213–214. Groves wanted Wilson, but Arnold vetoed it because Wilson lacked combat experience. R. C. Wilson to LRG, February 9, 1953, and LRG, Memo to File, September 25, 1963, Folder Hadden, Gavin, Box 3, Entry 7530B, Papers of LRG, RG 200, NARA. Tibbets recounted that "the field had been narrowed to three people: a brigadier general, a full colonel, and me, a lieutenant colonel." The names were given to General Arnold, who chose Tibbets "and apparently never offered any explanation for his decision." Brig. Gen. Paul W. Tibbets, USAF, retired, "Training the 509th for Hiroshima," *Air Force Magazine* (August 1973), 51. Much later Groves indicated that Brig. Gen. Mervin E. Gross (Operations, Commitments & Requirements) recommended Tibbets to Arnold. LRG, Oral History Transcript, April 3, 1970, "Hap" Arnold, Murray Green Collection, Box 66, Folder 3, USAF Academy, 10. In an August 29, 1944, memo to Groves, Captain Derry stated that General Ent had recommended a Colonel Montgomery (deputy A3 for the 20th Bomber Command), but that General Giles decided against him and nominated Tibbets. Derry to LRG, August 29, 1944, Top Secret, File 5.C.A.

18. LRG, Oral History Transcript, April 3, 1970, "Hap" Arnold, Murray Green Collection, Box 66, Folder 3, USAF Academy, 47.

19. Coster-Mullen, *Atom Bomb,* 11.
20. Bowen, *Silverplate,* 104.
21. By May 24 it stood at 205 officers and 1,624 enlisted men, exceeding its authorized strength. Bowen, *Silverplate,* 106. In June the total increased with the addition of the 1st Technical Detachment, War Department Miscellaneous Group, a team of scientists and technicians, some military, some civilian.
22. Charles G. Hibbard, "Training the Atomic Bomb Group," *Air Power History,* fall 1995, 24–33; Leonard J. Arrington and Thomas G. Alexander, "World's Largest Military Reserve: Wendover Air Force Base, 1941–63, *Utah Historical Quarterly* 31 (fall 1963), 324–335. Ent was the leader of the first Ploesti oil field raid.
23. James Les Rowe, *Project W-47* (Livermore, Calif.: JaARo Publishing, 1978).
24. Groves later told a possibly apocryphal story about Admiral Nimitz being flown — in an army plane — to an island in the Pacific. As they were about to land they realized it was still held by the Japanese. After hearing this story, whether it was true or not, Groves and Parsons agreed that a large part of the 509th's training would be over water. LRG, Oral History Transcript, April 3, 1970, "Hap" Arnold, Murray Green Collection, Box 66, Folder 3, USAF Academy, 13.
25. Gerrard-Gough, Christman, *Experiment.*
26. Ibid., 210.
27. Ibid.
28. John T. Hayward and C. W. Borklund, *Bluejacket Admiral: The Navy Career of Chick Hayward* (Annapolis, Md.: Naval Institute Press, 2000), 121–122. As a teenager in 1923 Hayward was an assistant batboy for the New York Yankees, one of his proudest accomplishments.
29. King to Nimitz, January 27, 1945, Top Secret, File 5.B.C.
30. Ashworth to LRG, February 24, 1945, Top Secret, File 5.C.K.
31. Kimble graduated last in the USMA class of June 1918 and was a long-standing friend of General Groves. His chief of staff was Frank L. Beadle (USMA, class of June 14, 1922). Neither was aware that Kirkpatrick worked for Groves.
32. Memo, Lt. Col. Peer de Silva to Capt. W. S. Parsons, USN, Security Considerations Pertinent to Combat Personnel of the 509th Composite Group, July 11, 1945, LANL Archives, A-84-019, 9-7.
33. For LeMay's knowledge of the atomic bomb see General Curtis E. LeMay with MacKinlay Kantor, *Mission with LeMay: My Story* (Garden City, N.Y.: Doubleday and Co., 1965), 379–390.
34. General Curtis E. LeMay with MacKinlay Kantor, *Mission with LeMay: My Story* (Garden City, N.Y.: Doubleday and Co., 1965), 381, 387.

CHAPTER 16 His Own State and Treasury Departments

1. Edward R. Landa, "The First Nuclear Industry," *Scientific American* (November 1983), 180–193.
2. On April 1, 1946, Groves nominated Sengier for the Presidential Medal of Merit. Groves noted that "Mr Sengier's sound judgment, initiative, resourcefulness and unfailing cooperation contributed greatly to the success of the atomic bomb program." Top Secret, Folder 20, Tab V. The medal was presented in a private ceremony.
3. Another 3,500 tons of uranium ore remained in Belgium and was captured by the Germans in June 1940 before it could be shipped to safety in Britain or the United States.

4. Negotiations to obtain Canadian ore were begun in 1942 with Eldorado Gold Mines, Ltd. (later Eldorado Mining and Refining, Ltd.). To more easily accomodate American and British orders and maintain secrecy Eldorado was purchased by the Canadian government. Eldorado mined uranium at its Great Bear Lake mine in the Northwest Territories and refined the ore at its facility at Port Hope, Ontario. By 1946 more than 4,000 tons of ore concentrate containing approximately 1,100 tons of U_3O_8 in the form of black oxide had been delivered to the MED. The uranium story and the personalities involved are extensively treated in Vilma R. Hunt, *Uranium Merchants* (McGraw-Hill, forthcoming). See also James Eayrs, *In Defence of Canada: Peacemaking and Deterrence* (Toronto: University of Toronto, 1972), 258–281.

5. Most of the uranium in the United States was in carnotite ores on the Colorado Plateau, but the high-grade deposits had already been mined earlier, primarily for their radium content. The heavy demand for vanadium during the war created a source of uranium oxide as a by-product of the vanadium processing. The tailings were of such low uranium content that it was necessary to concentrate them at or near the mine before shipment. Plants in Uravan and Durango processed the tailings, producing a sludge containing 15 to 20 percent black uranium oxide. This then went to Grand Junction, Colorado, for processing to yellow cake, containing 10 to 15 percent U_3O_8, which in turn went to the Linde refinery in Tonawanda, New York.

6. Brown, *Neutron,* 234–235.

7. Septimus H. Paul, *Nuclear Rivals: Anglo-American Atomic Relations, 1941–1952* (Columbus: Ohio State University Press, 2000), 41.

8. Hershberg, *Conant,* 187.

9. "With unrestricted exchange, the Americans would inevitably pass a great deal more information to the British than they could expect to receive in return." Brown, *Neutron,* 230.

10. Much else was decided at Quebec as well, including Roosevelt's commitment from Churchill for a May 1, 1944, cross-channel invasion of Europe, termed Overlord, and that the supreme Allied commander should be American.

11. U.S. Department of State. *Foreign Relations of the United States: The Conferences at Washington and Quebec, 1943.* (Washington, D.C.: U.S. Government Printing Office, 1970), 1117–1119. For the connection between Roosevelt's agreement for an atomic alliance and Churchill's agreement for the Normandy invasion, see Brian Villa, "The Atom Bomb and the Normandy Invasion" in *Perspectives in American History,* Vol. 2, 1977–1978 (Cambridge: Harvard University Press, 1978), 463–502.

12. Septimus H. Paul, *Nuclear Rivals: Anglo-American Atomic Relations, 1941–1952* (Columbus: Ohio State University Press, 2000), 52; see also James Eayrs, *In Defence of Canada: Peacemaking and Deterrence* (Toronto: University of Toronto, 1972), 271.

13. Hewlett, Anderson, *New World,* p. 282.

14. Szasz, *British.* Szasz focused primarily on the prominent group at Los Alamos. A list titled "British Mission," apparently compiled in 1953, has 93 names (including a few Frenchmen, Canadians, and other nationalities) with arrival and departure dates. Many worked at more than one place, such as Berkeley and Y-12, or New York and K-25. Some worked a few months, others a few years. Folder 7, Box 1, Entry 7530T, Papers of LRG, RG 200, NARA.

15. LRG to Gowing, March 26, 1965, Box 2, Entry 7530J, Comments, Interviews and Reviews 1949–70, Papers of LRG, RG 200, NARA. Groves read her book, *Britain and Atomic Energy, 1939–1945,* very carefully, thought it "splendid," wrote detailed comments, and annotated his copy with notes in the margins. Groves's longer comments are in *ibid.* His copy of the book was given to West Point. Fuchs had been naturalized in the summer of 1942.

16. Brown, *Neutron,* 254.

17. Quoted in Powers, *Heisenberg,* 473. The incident is also recounted in Joseph Rotblat, "Leaving the Bomb Project," *Bulletin of the Atomic Scientists,* August 1985, 16–19, but he does not quote Groves's exact words. At the end of 1944, when it became evident that Germany was not working seriously on the bomb, Rotblat requested permission to leave Los Alamos; he returned to Britain.

18. Donald H. Avery, *The Science of War: Canadian Scientists and Allied Military Technology During the Second World War* (Toronto: University of Toronto Press, 1998), 176–202.

19. Chadwick tried to mollify Groves and the Americans by choosing as director John Cockcroft, a renowned British scientist, to replace Hans von Halban. He also got Groves to agree to having a group of Canadian scientists come to Chicago's Met Lab. The meeting — held in January 1944 and chaired by Groves and Chadwick — did not go well for the Canadians. Much to the disappointment of Chadwick and his Canadian colleagues, Groves reaffirmed that there would be no collaboration on plutonium chemistry or plutonium separation.

20. Donald H. Avery, *The Science of War: Canadian Scientists and Allied Military Technology During the Second World War* (Toronto: University of Toronto Press, 1998), 190.

21. LRG, *Now,* 224–229; Harrison-Bundy Files, Folder 36; Top Secret, Folders 12 and 26; Hewlett, Anderson, *New World,* 331–335; Bertrand Goldschmidt, *Atomic Rivals,* translated by Georges M. Temmer (New Brunswick and London: Rutgers University Press, 1990).

22. In this instance as in others Groves often had to compromise for the good of the project. When Oppenheimer informed the general that the services of George Placzek and his group, who had also been working in Canada, were absolutely essential at Los Alamos, Groves acceded to his request.

23. Sir John's reply was sent in the form of an aide-memoire. Harrison-Bundy Files, Folder 18, Tab J. Groves sent a memo to Winant dated October 31, 1944, also in Folder 18, Tab J.

24. LRG to Secretary of War, December 14, 1944, Harrison-Bundy Files, Folder 36, Tab L.

25. "In Dec. 1944, FDR, discouraged by reason of the Battle of the Bulge, did tell me he thought the war might be prolonged and that we should be prepared to use the bomb against Germany." LRG, Memo to File, August 1, 1965, Folder, Comments on *The New World,* Box 3, Entry 7530J, Papers of LRG, RG 200, NARA. Elsewhere he said, "Mr. Roosevelt told me to be ready to do it." Ermenc, *Scientists,* 252. Vannevar Bush felt the same way. "If he [FDR] had any doubts, he never expressed them to me." Quoted in Zachary, *Frontier,* 214.

26. LRG, *Now,* 229.

27. Various drafts of the trust agreement and related papers are in Harrison-Bundy Files, Folder 48.

28. General Styer headed the important Technical Subcommittee of the CPC. Its reports furnished much of the data upon which the CPC made its decisions.

29. Minutes of 18 CDT meetings from June 30, 1944, to July 3, 1946, are in Top Secret, Folder 9B. "[T]he Trust was inevitably dominated by its Chairman, General Groves, and though the British admired his extraordinarily powerful personality they realised that he had every intention of making it his own show as much as he could." Gowing, *Britain,* 301.

30. Stimson Diary, October 17, 1944.

31. Stimson Diary, October 21, 1944.

32. In 2001 dollars this would be approximately $450 to $500 million.

33. Groves had conversations, conferences, or correspondence with the following Treasury Department officials between October 1944 and December 1947: Secretary of the Treasury Henry Morganthau Jr.; Secretary Fred M. Vinson (1945–1946); Secretary John W. Snyder (1946–1949); Undersecretary D. W. Bell; Assistant to the Undersecretary W. Heffelfinger; Fiscal Assistant Secretary E. F. Bartelt. Hadden to LRG, December 2, 1953, Folder Hadden, Gavin, Box 3, Entry 7530B, Papers of LRG, RG 200, NARA.

34. Interview with Gerry Elliot by Stanley Goldberg, August 30, 1991.

35. Interview with Anne Wilson Marks by Robert S. Norris, October 26, 2000.

36. Consodine to Horace K. Corbin, n.d., but ca. 1945–1946, Folder 201, Groves, L. R., Box 2, Entry 7530C, Papers of LRG, RG 200, NARA.

37. Lansdale, Manuscript, 48.

38. Stettinius recorded in his diary that Roosevelt first informed him of the bomb on January 19. On January 22 he went to Stimson's office, where he spent two hours with the secretary, Forrestal, McCloy, and Bundy learning more about it. Stettinius appointed Assistant Secretary of State James C. Dunn to be liaison with the military on all atomic matters. Edward R. Stettinius Jr., *Roosevelt and the Russians: The Yalta Conference,* Walter Johnson, ed. (Garden City, N.Y.: Doubleday and Co., 1949), 33–35.

39. "General Marshall will have with him an army officer named Considine [*sic*] who will have all the basic papers. He has been instructed to give me these papers if I ask him and he will look me up at M [Malta]." Thomas M. Campbell and George C. Herring, eds., *Diaries of Edward R. Stettinius, Jr., 1943–1946* (New York: New Viewpoints, 1975), 218.

40. Memo, LRG to Bundy, February 6, 1945, Harrison-Bundy Files, Folder 27; also quoted in Helmreich, *Ores,* 50–51.

41. Allen Weinstein and Alexander Vassiliev, *The Haunted Wood: Soviet Espionage in America — The Stalin Era* (New York: Random House, 1999), 157–168; John Earl Haynes and Harvey Klehr, *Venona: Decoding Soviet Espionage in America* (New Haven: Yale University Press, 1999), 138–145.

42. Hadden to LRG, December 2, 1953, Folder Hadden, Gavin, Box 3, Entry 7530B, Papers of LRG, RG 200, NARA.

43. LRG to Bartelt, November 27, 1953, Folder Bac–Bri, Box 1, Entry 7530B, Papers of LRG, RG 200, NARA.

44. Helmreich, *Ores,* 110.

45. Lilienthal, *Journals,* 10.

46. Verne W. Newton, *The Cambridge Spies: The Untold Story of Maclean, Philby, and Burgess in America* (Lanham, Md.: Madison Books, 1991). Former director of

Central Intelligence Richard Helms believed that Maclean was the most valuable known Soviet agent ever to operate in the West.

47. Christopher Andrew and Vasili Mitrokhin, *The Sword and the Shield: The
    Mitrokhin Archive and the Secret History of the KGB* (New York: Basic Books,
    1999), 115.

48. Quoted in Christopher Andrew and Vasili Mitrokhin, *The Sword and the Shield:
    The Mitrokhin Archive and the Secret History of the KGB* (New York: Basic Books,
    1999), 127.

49. Groves Collection, West Point.

CHAPTER 17 The Groves Family During the War

1. Photo, "General's Wife Christens Ship," *Atlanta Constitution,* February 1,
   1945, 11.

2. Interview with John L. Hadden by Robert S. Norris, January 9, 1999.

3. Memo, SAC [Special Agent in Charge] to Director, FBI, Owen Groves, April
   23, 1947. Given the context of this memo and others it is obvious that they are
   referring to Marion Groves.

4. FBI Memo, M. A. Jones to Mr. Nichols, September 14, 1953. The two deleted
   passages are likely "Marion, his" and "or his wife."

5. LRG to Gwen Groves Brown, September 5, 1944.

6. LRG to RHG, September 19, 1944.

7. Graduate John McPhee wrote of Boyden's influence on young minds and
   bodies during his 64 years as headmaster. *Frank L. Boyden of Deerfield* (Noonday
   Press, 1985). The tuition at the time was $1,600 a year, but Boyden set $600 for
   Richard. On an army officer's salary even this amount was a sacrifice, but one
   that was worth it.

8. RHG, *War Stories,* 215–216.

9. RHG, *War Stories,* 274–275.

10. LRG to RHG, December 11, 1944. The position was actually "in home."

11. RHG, *War Stories,* 265–266.

12. The circumstances were that Rep. John C. W. Hinshaw and Sen. Hiram W.
    Johnson, both from California, offered Richard an opportunity to compete for
    an appointment through a civil service exam. Though Richard studied hard
    and won the competition by his high scores, the congressman and the senator
    each left it to the other to make an appointment, leaving the Deerfield senior
    empty handed. His father knew nothing of this, because it took place while he
    was in Nicaragua, from August to November 1939. Upon his return and
    learning of the unfortunate outcome he tried to have the decision reversed, but
    to no avail. RHG, *War Stories,* 288–289.

13. LRG to RHG, April 25, 1941, Folder Groves, Richard H., Box 4, Entry 7530B,
    Papers of LRG, RG 200, NARA.

14. LRG to Stamps, April 26, 1941, Folder Groves, Richard H., Box 4, Entry
    7530B, Correspondence 1941–70, Papers of LRG, RG 200, NARA.

15. Part of the strategy was to possibly major in physics. Groves wrote to his son's
    physics professor, Henry DeWolf Smyth, in early January 1942, and had breakfast
    with him in Washington on January 16 to discuss what courses his son should
    take. Groves would soon meet Professor Smyth again in another capacity.

16. RHG, *War Stories,* 357–358.

17. Richard entered the academy as a member of the class of 1946. In the fall of

1942, due to the war, the course was accelerated for all of the classes, and Richard graduated in three years as a member of the class of 1945.

18. RHG, *War Stories,* 374.

19. A series of letters was generously provided to the author by RHG.

20. Author's collection.

21. LRG to RHG, September 9, 1944. On June 5, 1945, Richard graduated 86th in a class of 852.

22. LRG to RHG, January 27, 1944, author's collection.

23. LRG to RHG, March 13, 1944, author's collection.

24. LRG to RHG, April 20, 1944, author's collection.

25. LRG to RHG, June 19, 1944, author's collection. To his sister he related another sad Christmas that he just experienced. "As usual, I got very little for Christmas so that your gift [of dates] was particularly appreciated. Also as usual, the rest of the family had large amounts of presents showered on them, particularly Patsy [his daughter-in-law]." LRG to Gwen Groves Brown, January 2, 1947.

26. LRG to RHG, November 15, 1944, author's collection.

27. LRG to RHG November 1, 1944, author's collection. "I enjoyed our weekend together in New York very much, and I hope you found me not too inferior to your usual blond companions." LRG to RHG, November 15, 1944, author's collection.

28. On October 14, 1945, in the Chapel of the Naval Hospital in Chelsea, Massachusetts.

29. LRG to GGR, October 7, 1946. He continued, "I know you will appreciate knowing that our house is now overrun with women. It was bad enough before when the ratio was 2 to 1, but now it is 3 to 1. Fortunately one of them does not yet wear nylons and hang them where they will annoy me."

30. LRG to GGR, November 6, 1946.

31. LRG to GGR, January 13, 1947 (emphasis in original). There were further offenses. Cigarette butts were found in the ashtray, and candy wrappers all over the floor. The latter were potentially incriminating evidence against the general if Mrs. Groves had found them first.

32. Helen O. Mankin, "Tennis with Daughter Helped Relax Chief of Atom Bomb Project," *Philadelphia Inquirer,* August 24, 1945.

33. Letter, GWG to RHG, ca. July 1981.

34. Letters, GWG to RHG, probably 1980.

35. In April 1944 Grace received a check for $7,523.94 as her share of the estate. In July 1944 the house was sold for $6,500 ($2,000 down and the balance at the rate of $50 per month). Letters, GWG to R. A. Sawyer, April 20, 1944, and July 15, 1944, Folder Groves, L. R., Personal and Business Matters, Box 4, Entry 7530C, Papers of LRG, RG 200, NARA.

36. "It was just as much a surprise to me as to everyone else," he told the press. "Atom Bomb Chief Mum to Belvoir Son," *Belvoir Castle,* August 10, 1945.

37. GWG to MNG, August 13, 1945, LC.

CHAPTER 18 **Supplying Atomic Fuels: Enough and in Time**

1. Hoddeson, et al., *Critical;* Hoddeson, "Change."

2. Cyril Stanley Smith, "Some Recollections of Metallurgy at Los Alamos, 1943–45," *Journal of Nuclear Materials* 100 (1981), 4. The first unambiguous plutonium metal was made in Chicago on November 6, 1943.

3. Hoddeson, "Change," 281.

4. Hewlett, Anderson, *The New World,* 251.

5. Thomas was responsible for coordinating the metallurgical work on plutonium and directing the final purification process.

6. Groves flew to Santa Fe the evening of July 26 and was back in his office the afternoon of July 28, spending July 27 at the lab. Four days later he repeated the trip, spending August 2 at Los Alamos and returning the next day.

7. Report, Dr. J. B. Conant to LRG, Visit to Los Alamos on August 17, 1944, Folder, Design and Testing of Bomb, Box 8, Entry, Manhattan Project, RG 319, NARA.

8. Report, Dr. J. B. Conant to LRG, Visit to Los Alamos on August 17, 1944, Folder, Design and Testing of Bomb, Box 8, Entry, Manhattan Project, RG 319, NARA. Conant described two other designs. One, designated the Mark II, was an implosion bomb without explosive lenses, with the center filled with hydrogen and either the metal or the hydride suspended in the very center. This is a very inefficient bomb, requiring either approximately seven kilos of U-235 or nine kilos of plutonium to yield only 100 to 500 tons TNT equivalent. It was a backup design to be used only if all other implosion methods failed. The Mark III bomb would be a "no lens" implosion bomb using no hydrogen, but the chances it would have worked were slim.

9. Christy was born in 1916 in Vancouver, British Columbia. He received his Ph.D. in theoretical physics from Berkeley in 1941. After spending most of the war years at Los Alamos, he taught at Cal Tech from 1946 to 1986.

10. Hoddeson, "Change," 283.

11. Oppenheimer to LRG, April 28, 1944, Folder, Thermal Diffusion, Box 11, Entry, Manhattan Project, RG 319, NARA.

12. LRG, *Now,* 120.

13. Groueff, *Manhattan,* 316.

14. LRG, *Now,* 120–121.

15. LRG, Comments on *Reach to the Unknown,* February 25, 1967, 7, Barbara Storms, Trinity, Folder Carl Zimmerman, Box 6, Entry 7530J, Papers of LRG, RG 200, NARA.

16. Jane H. Hall to Brig. Gen. D. L. Crowson, USAF, director of military application, USAEC, November 30, 1964. For the last five months of 1945 the U-235 content rose to 93.6 percent. Sometime around 1950, 93.5 percent was decided upon as the enrichment level for weapon-grade uranium.

17. Thomas J. Ahrens, "Albert Francis Birch," *National Academy of Sciences, Biographical Memoirs,* vol. 74 (Washington, D.C.: National Academy Press, 1998), 12.

18. Coster-Mullen, *Atom Bombs,* 24.

19. Maj. J. A. Derry to Adm. W. S. DeLany, August 17, 1945, Top Secret, File 5.C.3.N.

20. Goldberg, "Hanford," 61–65.

21. H. Greenewalt to LRG, July 22, 1944, Folder, Security, Box 9, Entry, Manhattan Project, RG 319, NARA.

22. J. N. Tilley to Maj. W. L. Sapper, October 18, 1944, Top Secret, Folder 5, Tab B. Tilley was assistant manager of TNX, the explosives department responsible for plutonium production.

23. Daniel Grossman, "Hanford and Its Early Radioactive Atmospheric Releases," *Pacific Northwest Quarterly,* January 1994, 13. During the notorious Green Run

of 1949, the environment was badly polluted by treating the billets after 60, 30, and, in some cases, only 12 days.

24. Roger Williams to LRG, February 14, 1945, Top Secret, Folder 5, Tab I.

25. LRG to Roger Williams, February 15, 1945, Top Secret, Folder 5, Tab I.

26. Col. K. D. Nichols to LRG, March 9, 1945, Top Secret, Folder 5, Tab J. Groves penciled on this memo, "This schedule was not satisfactory and was revised at my meeting w/Evans of DuPont on March 22." See also W. O. Simon to J. N. Tilley, Production Inventory and Schedule, March 19, 1945, Top Secret, Folder 5, Tab K.

27. LRG to Roger Williams, March 22, 1945, Top Secret, Folder 5, Tab M. His second sentence is noteworthy, because it shows how closely Groves was following the delivery schedule. The Hanford records show that as of March 29 there would be 5,558.5 grams, though all of this would not be shipped until June 6.

28. Williams replied on April 9 with a proposed power increase program. Pile B would remain at 250 MW, while D and F would be raised in steps to 280 MW. Roger Williams to LRG, April 9, 1945, Top Secret, Folder 5, Tab M. Groves wrote back on April 12 saying that while he did not object he felt the program was "quite conservative." LRG to Roger Williams, April 12, 1945, Top Secret, Folder 5, Tab M. The B reactor reached 250 MW on February 4, 1945, and did not go higher. The D reactor reached 250 MW on February 15, 1945. For the speedup it went to 265 MW on May 12, and to 280 MW on June 12. The F reactor first reached 250 MW on March 9, 1945. For the speedup it went to 265 MW on June 13. The first material was delivered to 200 Area on August 4, 1945.

29. LRG to J. R. Oppenheimer, March 22, 1945, Top Secret, Folder 5, Tab L. Groves told Oppenheimer that 231.5 grams had already been shipped and that the schedule would be one kilo by April 20, two kilos by May 1, three kilos by May 10, four kilos by May 20, and five kilos by June 1. "If it should be found necessary to use more than 10 kilograms the next two kilograms would be at the rate of one per week." Groves's purpose is obvious. The first five kilos were needed for the test at Trinity (which would occur on July 16, 1945) and the second five for the first implosion bomb, to be dropped on Japan soon after.

30. Matthias Diary, March 24, 1945, 53.

31. Matthias Diary, May 3, 1945, 70.

32. Matthias Diary, May 9, 1945, 72.

33. Gerber said that "[t]he exact length of cooling time during this period is unknown or is classified, but it was less than thirty days." Col. Frederick J. Clarke, who would succeed Matthias, said it dropped as low as 15 days in mid-1945. Michele Stenehjem Gerber, *On the Home Front: The Cold War Legacy of the Hanford Nuclear Site* (Lincoln: University of Nebraska Press, 1992), 44, 236, footnote 56. Grossman found in the Hanford records that the first load of irradiated uranium, discharged in November, went to the T-Plant on December 26, having cooled 32 days, and that this became the normal practice. Grossman, "Hanford and Its Early Radioactive Atmospheric Releases," *Pacific Northwest Quarterly,* January 1994, 11.

34. Thayer, *Management,* 172–173.

35. J. N. Tilley, April 21, 1945, Product Inventory, Proposed Pushing Schedule from Piles, and Forecast for Delivery of Product from 231, Folder, Third Red Folder, Box 5, Manhattan Project, RG 319, NARA.

36. Sanger, *Working,* 195–198.
37. Interview with Lyall Johnson by Robert S. Norris, March 23, 1999.
38. *LRG Appointment Book,* July 4, 1945. The Trinity hemispheres were completed and delivered on July 2, 1945.
39. Matthias Diary, August 5, 1945, 98. Presumably a unit is 80 grams in a container.
40. A July 31, 1945, memorandum requested two planes be furnished on August 5, 7, and 9 to fly "extremely valuable" cargo from Naval Air Station, Pasco, Washington, to Santa Fe, New Mexico, in accordance with the verbal agreement between Lieutenant General George and Major General Groves. Memo, R. J. Weir to Assistant Chief of Staff, Priorities and Traffic, Air Transport Command, Request for Special Airplanes, July 31, 1945, Folder S1-8.1, Box 1, Entry, Washington Liaison Office, RG 326, Atlanta Records Center, NARA.
41. For an account of this important work see Edward Hammel, *Plutonium Metallurgy at Los Alamos, 1943–1945* (Los Alamos, N.M.: Los Alamos Historical Society, 1999); Richard D. Baker, Siegfried S. Hecker, and Delbert R. Harbur, "Plutonium: A Wartime Nightmare but a Metallurgist's Dream," *Los Alamos Science,* winter–spring 1983, 142–151; Cyril Stanley Smith, "Some Recollections of Metallurgy at Los Alamos, 1943–45," *Journal of Nuclear Materials* 100 (1981), 3–10; Cyril Stanley Smith, "Metallurgy at Los Alamos, 1943–1945," *Metal Progress* 65, no. 5 (May 1954), 81–89. The Smyth Report devotes one short paragraph to the work of the division.
42. The Los Alamos scientists were lucky in at least two respects. First, given the high assembly velocities of the implosion design, purity requirements could be somewhat relaxed from the extremely high standards that were first believed to be necessary. Second, the melting point of plutonium, first thought to be about 1,800 degrees Celsius, was actually 641 degrees, making it easier to work with. Interview with Morris Kolodney by Robert S. Norris, February 12, 1999.
43. Hoddeson, et al., *Critical,* 330–331.

CHAPTER 19 **Groves and the Use of the Bomb I: The Target and Interim Committees**

1. Giovannitti, Freed, *Decision,* 243 (emphasis in original).
2. "If one hopes to find a thoughtful review of the profound issues inherent in the decision to use an atomic bomb, one will not find it in the deliberations of the Interim Committee. Indeed the degree of muddle-headedness, confusion, and contradictory positions taken by key personalities is remarkable." Lawrence Lifschultz and Kai Bird, "The Legend of Hiroshima," in *Hiroshima's Shadow* (Stony Creek, Conn.: Pamphleteer's Press, 1998), liv.
3. Giovannitti, Freed, *Decision,* 245 (emphasis in original). In a 1949 interview Groves had this to say: "Now as to when was the decision reached to drop the bombs. That was not reached as it is now claimed by Mr. Truman. That decision was reached, in my opinion by Pres. Roosevelt when he approved the report around Christmas of 1942 that said that we should attack this problem in a major way and that it was going to cost hundreds of millions of dollars. From that time on it was impossible for anybody to back out. It was like having to lean on a toboggan and going down hill." Interview with LRG by George Carroll, June 15, 1949, Folder, Comments, Interviews and Reviews, Box 1, Entry 7530J, Papers of LRG, RG 200, NARA.
4. The term is from Sigal, *Fighting.* I have drawn heavily upon his interpretation.

5. "In 1945, American leaders were not seeking to avoid the use of the A-bomb. Its use did not create ethical or political problems for them. Thus, they easily rejected or never considered most of the so-called alternatives to the bomb." Bernstein, "Understanding," 235.

6. Harry S. Truman, *Year of Decisions* (Garden City, N.Y.: Doubleday and Co., 1955), 10, 11, 17. "Byrnes had already told me that the weapon might be so powerful as to be potentially capable of wiping out entire cities and killing people on an unprecedented scale. And he added that in his belief the bomb might well put us in a position to dictate our own terms at the end of the war." Ibid., 87.

7. LRG, Memo to File, Report of Meeting with the President, April 25, 1945, Folder F, Box 3, Entry 7530B, Papers of LRG, RG 200, NARA.

8. LRG to Truman, October 21, 1955, Folder F, Box 3, Entry 7530B, Papers of LRG, RG 200, NARA; LRG, Memo to File, February 25, 1966, Folder Truman, Harry S. (President), Box 5, Entry 7530J, Papers of LRG, RG 200, NARA.

9. As Groves said, "It is hard to say just when a war is won. To go to one that we all understand, our own civil war, we all know today and I think they knew then that after the capture of Vicksburg and the Battle of Gettysburg that the war was won but there were an awful lot of men killed afterwards."

10. Kenneth M. Glazier Jr., "The Decision to Use Atomic Weapons Against Hiroshima and Nagasaki," *Public Policy* 18 (1970), 463–516.

11. LRG, *Now*, 265.

12. Interview with LRG by George Carroll, 4, Folder, Comments, Interviews and Reviews, 1949–1970, Box 1, Entry 7530J, Papers of LRG, RG 200, NARA.

13. As Truman said, "From the time I first sat down in the President's chair I found myself part of an immense administrative operation. There had been a change of executives, but the machinery kept going on in its customary routine manner, and properly so. It would have been sheer nonsense to expect anything else." Harry S. Truman, *Year of Decisions* (Garden City, N.Y.: Doubleday and Co., 1955), p. 87. See also Sherwin, *Destroyed*, 147–148; Barton J. Bernstein, "Roosevelt, Truman, and the Atomic Bomb," *Political Science Quarterly*, spring 1975, 34–36. For evidence that Roosevelt would have dropped the bomb see Schaffer, *Wings*, 170–171.

14. Robert Jungk, *Brighter Than a Thousand Suns* (New York: Harcourt, Brace and Co., 1958), 208.

15. LRG to Styer, October 12, 1945, Folder Stu–Sz, Box 10, Entry 7530B, Papers of LRG, RG 200, NARA. He also told his old friend, "If you have seen any pictures of me, I hope you notice how thin I am getting." On Christmas Eve 1944 Groves told a dozen of his top aides, "If this weapons fizzles each of you can look forward to a lifetime of testifying before congressional investigating committees." Fletcher Knebel and Charles W. Bailey II, *No High Ground* (New York: Harper & Brothers, Publishers, 1960), 75.

16. Sigal, *Fighting*, 182.

17. John D. Chappel, *Before the Bomb: How America Approached the End of the Pacific War* (Lexington: University Press of Kentucky, 1997), 23–38 and passim; John W. Dower, *War Without Mercy: Race and Power in the Pacific War* (New York: Random House, 1986), 36, 52.

18. LRG, *Now*, 324. In a speech shortly after the war Groves told a large audience,

"As I look at the pictures of our men coming home from Japanese prisons and hear secondhand accounts and firsthand accounts of the experiences of the men who made the march from Bataan, I am not particularly worried about how hard this weapon hit the Japanese." LRG, Speech to IBM Luncheon, September 21, 1945, Folder, Mr. Watson's Luncheon, NY, Box 3, Entry 7530I, Papers of LRG, RG 200, NARA.

19. Gregg Herken, *The Winning Weapon: The Atomic Bomb in the Cold War, 1945–1950* (New York: Random House, 1981). See also Barton J. Bernstein, "Roosevelt, Truman, and the Atomic Bomb, 1941–1945: A Reinterpretation," *Political Science Quarterly* 90, no. 1 (spring 1975), 23–69; Bernstein, "The Atomic Bombings Reconsidered," *Foreign Affairs* 74, no. 1 (January–February 1995), 135–152; Alperovitz, *Decision*.

20. Giovannitti, Freed, *Decision,* 245 (emphasis in original).

21. Giovannitti, Freed, *Decision,* 63. Though it was originally inspired "so that Hitler would not get it first, the fact remains that the original decision to make the project an all-out effort was based upon using it to end the war." LRG, *Now,* 265.

22. Sigal, *Fighting,* 179–182.

23. As Sigal says about Truman, "it is not clear that he ever knew what the alternatives were, or exactly which option he had chosen, because the options were never clearly delineated, formally arrayed, and systematically compared." Sigal, *Fighting,* 181.

24. LRG, *Now,* 267. Arthur Compton confirmed Groves's role. "Already the strategy for the military use of the bomb has been carefully worked out. For shaping this strategy General Groves was primarily responsible." Compton, *Quest,* 221.

25. LRG, *Now,* 267. The following argument draws on Sigal, *Fighting,* 182–192.

26. Manhattan Engineer District, *The Atomic Bombings of Hiroshima and Nagasaki* (Washington, D.C.: GPO, 1946), 7. An Army Air Forces booklet of photographs of Hiroshima and Nagasaki described Hiroshima as "the most important administrative and military center in western Honshu and site of Army clothing and munitions depots, ordnance plants, and other industries." *The Atomic Bomb* (Washington, D.C.: Army Air Forces, n.d., but ca. October 1945).

27. "The American distaste for area bombing had evaporated in the face of a set of Japanese targets marked by the complex intermingling of industry and society and fire-prone wooden structures.... Nobody involved in the decision on the atomic bombs could have seen themselves as setting new precedents for mass destruction in scale — only in efficiency." Lawrence Freedman and Saki Dockrill, "Hiroshima: A Strategy of Shock," in Saki Dockril, ed., *From Pearl Harbor to Hiroshima: The Second World War in Asia and the Pacific, 1941–45* (New York: St. Martin's Press, 1994), 196. See also Schaffer, *Wings.*

28. LRG, *Now,* 267.

29. General Arnold said, "We had figured we would probably have to drop four atomic bombs or increase the destructiveness of our Super Fortress missions by adding the heavy bombers from Europe." H. H. Arnold, *Global Mission* (New York: Harper & Brothers, 1949), 260. Arnold was surprised at Japan's "abrupt surrender" after sustaining only two bombs. In an interview Groves said, ". . . it was not until December of 1944 that I came to the opinion that two bombs would end the war. Before that we had always considered more as being more

likely. Then I was convinced in a series of discussions I had with Admiral Purnell." Interview with LRG, May 1967, 15. Naval Historical Center, Washington, D.C.

30. LRG, *Now,* 162.

31. LRG, Memo to File, Brig. Gen. Thomas F. Farrell, dictated November 12, 1962, Folder, Comments on *The New World,* Box 3, Entry 7530J, Papers of LRG, RG 200, NARA.

32. Two others on the list were Col. T. Dodson Stamps, perhaps Groves's best friend, and Col. Clark Kittrell, a colleague and friend of more than 25 years whose relationship with Groves went back to West Point, Hawaii, and Fort Peck.

33. Farrell was five years older than Groves. A New Yorker, he had gone to Rensselaer Polytechnic Institute, received a civil engineering degree in 1912, and worked on the Panama Canal from 1913 to 1916. He entered the Corps of Engineers as a second lieutenant in November 1916 and served in France during World War I. After the war he was an instructor at the Engineer School at Fort Humphreys, Virginia, and at West Point. Farrell resigned from the Regular Army in 1926 and was appointed commissioner of canals and waterways of New York State. In January 1930 he was appointed chief engineer of the state of New York.

34. The Ledo Road linked the railhead of Ledo (Assam Province, India) with Kunming, China. Considered one of the toughest engineering jobs of the war, the road went through some of the most difficult terrain in the world. The engineers, under the command of Gen. Lewis A. Pick, had to contend with monsoons, mud, and malaria as they completed several hundred miles of road in January 1945. Early in February Pick led the first convoy into Kunming. Karl C. Dod, *The Corps of Engineers: The War Against Japan* (Washington, D.C.: Office of the Chief of Military History, U.S. Army, 1966), 432–439 and passim. Pick was one of Groves's instructors at Fort Leavenworth, was a classmate at the Army War College, and became chief of engineers in 1949. Farrell replaced Col. Frederick S. Strong. Strong, we remember, had replaced Groves as chief of the Operations Branch when Groves was promoted to deputy chief of the Construction Division in March 1942. It was a small world among the upper reaches of the engineer officer corps, and their paths crossed many times.

35. Notes on Initial Meeting of Target Committee, LANL Archives, A-84-019, 11-10. Groves is mistaken in his book, 268, saying the first meeting was on May 2. On May 2 Groves traveled from Detroit to St. Louis.

36. Test bombing raids using incendiaries were conducted against Nagoya on January 3, against Kobe on February 4, with a "conclusive test" against Tokyo on February 25, with 172 B-29s dropping more than 450 tons of bombs. Wesley Frank Craven and James Lea Cate, *The Army Air Forces in World War II,* vol. 5: *The Pacific: Matterhorn to Nagasaki, June 1944 to August 1945* (Chicago: University of Chicago Press, 1953), 572–573, 608–644; Kenneth P. Werrell, *Blankets of Fire: U.S. Bombers Over Japan During World War II* (Washington, D.C.: Smithsonian Institution Press, 1996), 150–168; Stanley L. Falk, "A Nation Reduced to Ashes," *Military History Quarterly* 7, no. 3 (spring 1995), 54–61.

37. Robert A. Pape, "Why Japan Surrendered," *International Security* 18, no. 2 (fall 1993), 165. See also United States Strategic Bombing Survey, *Summary Report, Pacific War* (Washington, D.C.: July 1, 1946); *Army Air Forces Statistical Digest, World War II* (December 1945), table 199.

38. Barton J. Bernstein, "Eclipsed by Hiroshima and Nagasaki: Early Thinking about Tactical Nuclear Weapons," *International Security* 15, no. 4 (spring 1991), 152.

39. Hiroshima, the 7th largest city in Japan, and Nagasaki, the 12th largest, were not among the 33 "Selected Urban Industrial Concentrations" on Hansell's or LeMay's target list up until June 15. Richard B. Frank, *Downfall: The End of the Imperial Japanese Empire* (New York: Random House, 1999), 261.

40. Notes on Initial Meeting of Target Committee, LANL Archives, A-84-019, 11-10, 4.

41. For Hiroshima this was left to bombardier Maj. Thomas W. Ferebee, who chose the T-shaped Aioi Bridge over the Ota River, easy to spot from 31,000 feet.

42. Maj. J. A. Derry and Dr. N. F. Ramsey to LRG, Summary of Target Committee Meetings on 10 and 11 May 1945, May 12, 1945, LANL Archives, A-84-019, 11-10. See also D. M. Dennison, Preliminary Report on Operational Procedures, May 9, 1945, LANL Archives, A-84-019, 29-14. A copy of Dennison's 12-page report went to Groves. The operational procedures Dennison outlined were quite close to what happened three months later.

43. LRG, *Now,* 269.

44. There were three types of fuses to ensure that the bombs exploded at the pre-scribed height. Barometric fuses detonate when the bomb reaches a certain height. Timing fuses detonate after the period of time that corresponds to the prescribed altitude. Radar fuses receive echoes from the ground to detect the prescribed altitude and detonate. For each type there were two fuses, for a total of six. If all of this redundancy failed, then an impact fuse would detonate when the bomb hit the ground.

45. Memo to LRG, Summary of Target Committee Meetings on 10 and 11 May 1945, May 12, 1945, 4, LANL Archives, A-84-019, 11-10. Kyoto was formerly the fourth largest city in Japan, having in 1940 a population of almost 1,100,000. By mid-1945 the number had declined to 800,000. USSBS, *Effects of Air Attack on Osaka — Kobe — Kyoto,* Report No. 58 (June 1947), 243.

46. Later the USSBS assessed the Kokura Arsenal complex as fourth in relative importance behind Tokyo, Osaka, and Sagami. USSBS, *Japanese Army Ordnance,* Report No. 45 (December 1946), 5.

47. Stimson, Bundy, *Service,* 617; Lawrence Freedman and Saki Dockrill, "Hiroshima: A Strategy of Shock," in Saki Dockril, ed., *From Pearl Harbor to Hiroshima: The Second World War in Asia and the Pacific, 1941–45* (New York: St. Martin's Press, 1994), 191–212. In a later interview Marshall spoke of the importance of shocking them into action. Forest C. Pogue, *George C. Marshall: Statesman* (New York: Viking, 1987), 19.

48. Douglas J. MacEachin, *The Final Months of the War with Japan: Signals Intelligence, U.S. Invasion Planning, and the A-Bomb Decision* (Washington, D.C.: CIA Center for the Study of Intelligence, 1998).

49. Richard B. Frank, *Downfall: The End of the Imperial Japanese Empire* (New York: Random House, 1999).

50. Minutes of Third Target Committee Meeting — Washington, 28 May 1945, LANL Archives, A-84-019, 11-10, 1.

51. With Marshall's informal directive, Groves's plan "would not be formally con-sidered by the Joint Chiefs of Staff or the Combined Chiefs" where opposition, especially from Admiral Leahy and the navy, could form. LRG, *Now,* 271.

52. LRG, *Now,* 273.

53. Hodgson suggests that Stimson had been reminded of the glories of Kyoto by the chance visit of a cousin, Henry Loomis, who dined at Woodley a few months before. Loomis had taken courses at Harvard on Japanese history and culture, and the evening's conversation had lodged in Stimson's mind. Hodgson, *Colonel,* 323–324. See also Otis Cary, "The Sparing of Kyoto — Mr. Stimson's 'Pet City,'" *Japan Quarterly* XXII, no. 4 (October–December 1975), 337–347; Schaffer, *Wings,* 143–146.

54. Groves's phrasing is less precise. "In the course of our conversation he gradually developed the view that the decision should be governed by the historical position that the United States would occupy after the war. He felt very strongly that anything that would tend in any way to damage this position would be unfortunate." LRG, *Now,* 274–275. On June 1 Stimson ordered Arnold to refrain from the conventional bombing of Kyoto without his permission.

55. LRG, *Now,* 275.

56. The estimated population of Nagasaki in the summer of 1945 was 230,000 in a built-up area of 3.3 square miles. USSBS, *Effects of the Atomic Bomb on Nagasaki, Japan,* Report No. 93, vol. 1, 28. Groves's report puts the figure lower, at 195,000. Manhattan Engineer District, *The Atomic Bombing of Hiroshima and Nagasaki* (Washington, D.C.: U.S. Government Printing Office, 1946), 18.

57. "I think that from that time on until he left for Potsdam for the Potsdam Conference, I must have seen him a half dozen times to a dozen times. On each occasion I asked him to reconsider on Kyoto and on each occasion he told me no." Interview with LRG by Fred Freed, n.d., 12–13, Box 4, Entry 7530B, Papers of LRG, RG 200, NARA.

58. *FRUS: Conference of Berlin,* II, 1372; Jones, *Manhattan,* 530. Arnold, LeMay, and the 20th Air Force kept pressing to firebomb the city with incendiaries.

59. *FRUS: Conference of Berlin,* II, 1372.

60. *FRUS: Conference of Berlin,* II, 1373, footnote 3.

61. H. H. Arnold, *Global Mission* (New York: Harper & Brothers, 1949), 585, 589.

62. *FRUS: Conference of Berlin,* II, 1373; LRG, *Now,* 27.

63. LRG, *Now,* 275.

64. Sigal, *Fighting,* 190.

65. Hewlett, Anderson, *New World,* 365.

66. Memo, Discussed with President Truman April 25, 1945, quoted in Henry L. Stimson, "The Decision to Use the Atomic Bomb," *Harper's,* February 1947.

67. Memo, Harrison to the Secretary of War, May 1, 1945, Harrison-Bundy Files, Folder 69, S-1 Interim Committee.

68. Leon V. Sigal, "Bureaucratic Politics & Tactical Use of Committees: The Interim Committee & the Decision to Drop the Atomic Bomb," *Polity* 10, no. 3 (spring 1978), 326–364; Compton, *Quest,* 233–234.

69. The invitations, sent on May 4, described the function of the committee: "to study and report on the whole problem of temporary war controls and later publicity, and to survey and make recommendations on post war research, development and controls, as well as legislation necessary to effectuate them." Bush-Conant Papers, Folder 19, RG 227, NARA.

70. Hewlett, Anderson, *New World,* 344–345. Stimson cleared the proposed members with the president on May 2. On May 3 Stimson suggested Byrnes; the president agreed and personally telephoned him in South Carolina to invite him to serve.

71. General Groves's diary for May 9 has him at 9:30 at the secretary of war's office for the first meeting of the Interim Committee. The notes for that meeting, as recorded by 2nd Lt. R. Gordon Arneson — Mr. Bundy's assistant and secretary of the committee — and dated May 17, do not show him as present.

72. LRG, Comments on Article by Secretary of War Henry L. Stimson, *Harper's,* February 1947. General Marshall attended only two meetings, by invitation. Of course, at the time the American people had no knowledge of the atomic bomb or which of its leaders was considering its use. It would be many years before the notes of the committee were declassified.

73. Leon V. Sigal, "Bureaucratic Politics & Tactical Use of Committees: The Interim Committee & the Decision to Drop the Atomic Bomb," *Polity* 10, no. 3 (spring 1978), 331, 345.

74. "He was deliberately left out because of my feeling that his advice would be unstable. This opinion did not meet with any objection from Bush and Conant due to all the troubles we had during the project because of Urey's frequent displays of indecision and lack of confidence." LRG, Comments on Article by Secretary of War Henry L. Stimson, *Harper's,* February 1947. Groves has a tendency to take credit for many things. With regard to choosing the four scientists Bush, and especially Conant, probably had a greater role.

75. Hewlett, Anderson, *New World,* 356–360.

76. Robertson, *Sly and Able,* 390–413.

77. "Domestic political concerns were primary. Congress would 'expect results' for the two billion dollars invested for the secret weapon of the Manhattan Project, and both the public and their representatives would be outraged if the Truman administration later were shown to have displayed any reluctance to win the war with Japan as quickly as possible by forgoing the use of this weapon." Robertson, *Sly and Able,* 405.

78. For a discussion of how the incompatible ideas of American superiority and international control were expressed by Truman, see S. David Broscious, "Longing for International Control, Banking on American Superiority: Harry S. Truman's Approach to Nuclear Weapons," in John L. Gaddis, ed., *Cold War Statesmen Confront the Cold War: Nuclear Diplomacy Since 1945* (New York: Oxford University Press, 1999), 15–38.

79. Compton, *Quest,* 238.

80. In a speech shortly after the war Groves said, "I have been asked if a demonstration of the power of the bomb on some barren island would not have been enough to cause Japan to capitulate and thus avoid the bombing of her cities. My answer has been and will continue to be 'Emphatically No!' And, I believe that any American soldier in the Pacific would give the same answer. A nation fighting for its very existence doesn't give up that easily. General Lee fought on for almost two years after Gettysburg and the Russians refused to surrender long after the Germans thought they should have at Stalingrad. The Germans were none too cooperative in the winter and spring of 1945. The reaction of the Japanese to that technique would have been to discount our determination to force them into submission as quickly as possible. They would have interpreted our failure to use the bomb as a sign of weakness." Speech to the Washington Club, November 5, 1946, Folder, Washington Club, Box 5, Entry 7530I, Papers of LRG, RG 200, NARA.

81. Groves knew of the visit, having had his agents follow the trio to South

Carolina. Robertson, *Sly and Able,* 409. Another scientist raising questions was James Franck. Franck's April 21 memo had been presented to Wallace and then passed on to Bush. Sigal, *Fighting,* 339.

82. Sadao Asada, "The Shock of the Atomic Bomb and Japan's Decision to Surrender — A Reconsideration," *Pacific Historical Review,* 1999, 477–512; Herbert P. Bix, *Hirohito and the Making of Modern Japan* (New York: HarperCollins, 2000), 487–530; Herbert P. Bix, "Japan's Delayed Surrender: A Reinterpretation," *Diplomatic History* 19, no. 2 (spring 1995), 197–225.

CHAPTER 20   **Groves and the Use of the Bomb II: Trinity, Hiroshima, and Nagasaki**

1. Oppenheimer, Memo on Test of Implosion Gadget, February 16, 1944, LANL, A-84-019, 22-7; Hoddeson, et al., *Critical,* 174–175.

2. Later Groves said that if Ickes had been brought into the picture, "this would have caused untold difficulties and would probably have resulted in complete disclosure to the whole world as well." LRG, Memo to File, May 13, 1963, Folder Ickes, Box 3, Entry 7530J, Papers of LRG, RG 200, NARA.

3. Carl Maag and Steve Rohrer, *Project Trinity, 1945–1946,* DNA 6028F (Washington, D.C.: Defense Nuclear Agency, Nuclear Test Personnel Review, 1986), 31.

4. LRG to Oppenheimer, October 17, 1962, and Oppenheimer to LRG, October 20, 1962, Folder Groves, L. R., Correspondence, Box 36, Oppenheimer Papers, Manuscript Division, LOC. Lamont has Bainbridge phoning Oppenheimer with news that the air force had turned over a section of the Alamogordo Bombing and Gunnery Range to the Manhattan Project (in September 1944). Oppenheimer is relaxing with a book of Donne's poems, having just read the "Batter my heart" sonnet, and when Bainbridge asks for a code name he responds, "We'll call it Trinity." Lansing Lamont, *Day of Trinity* (New York: Atheneum, 1965), 70. See also Ferenc Morton Szasz, *The Day the Sun Rose Twice* (Albuquerque: University of New Mexico Press, 1984), 40–41.

5. Goldberg, "Words," *Invention & Technology,* fall 1991, 48–54; Oppenheimer to LRG, March 10, 1944, LANL, A-84-019, 22-7.

6. Helen Weigel Brown, "The Army's Biggest One-Man Office," *Liberty,* March 24, 1945, 10–11, 62.

7. The idea of constructing railroad track from Pope to the installation point was investigated, but at a cost of $520,000 it was rejected. The estimated cost for the trailer method was $150,000. Memo, G. R. Tyler to LRG, Transportation Contract, Trinity Project, February 7, 1945, Folder, 400, Equipment, Box 66, Entry 5, RG 77, NARA.

8. Rogers Brothers Corporation of Albion, Pennsylvania, built the trailer in 30 days upon specifications furnished by the Eichleay Corporation, experts in moving heavy machinery.

9. See the B. F. Goodrich advertisement, "$12,000 worth of tires for one 30-mile trip," in the January 19, 1946, issue of *Saturday Evening Post.*

10. The Cowpuncher Committee was established in early March and was composed of Samuel K. Allison, Robert F. Bacher, George B. Kistiakowsky, Charles C. Lauritsen, William S. Parsons, and Hartley Rowe.

11. Memo, Oppenheimer to All Group Leaders, June 14, 1945, LANL Archives, A-84-019, 10-07.

12. Memo, Oppenheimer to All Group Leaders, June 14, 1945, LANL Archives, A-84-019, 11-14.
13. *LRG Appointment Book*, July 2, 1945.
14. Stafford L. Warren, "The Role of Radiology in the Development of the Atomic Bomb," in Kenneth D. A. Allen, ed., *Radiology in World War II, Medical Department in World War II: Clinical Series* (Washington, D.C.: Office of the Surgeon General, Department of the Army, 1966), 848. Groves's appointment book shows a call to Dr. Chapman about Warren on March 10 and being in Rochester on March 19.
15. Hacker, *Dragon*, 84.
16. At North 10,000 the pairs were Robert R. Wilson and Dr. Henry Barnett, at West 10,000 John Manley and an unknown doctor, and at South 10,000 Frank Oppenheimer and Dr. Louis Hemplemann.
17. Joseph O. Hirschfelder, "The Scientific and Technological Miracle at Los Alamos," in Badish, et al., *Los Alamos*, 72–78; Hacker, *Dragon*, 90–91.
18. Hacker, *Dragon*, 91.
19. Brig. Gen. B. G. Holzman to LRG, April 17, 1963, Folder Haw–Hol, Box 4, Entry 7530B, Papers of LRG, RG 200, NARA. The following year Holzman wrote to Groves with a newspaper clipping from the *San Francisco Chronicle* of September 11, 1964, about Hubbard being dismissed from San Francisco State College for unprofessional conduct and dishonesty. Brig. Gen. B. G. Holzman to LRG, September 16, 1964, Folder Haw–Hol, Box 4, Entry 7530B, Papers of LRG, RG 200, NARA.
20. The literature describing the test is extensive: Ferenc Morton Szasz, *The Day the Sun Rose Twice* (Albuquerque: University of New Mexico Press, 1984); Groves, "Some Recollections of July 16, 1945," *Bulletin of the Atomic Scientists*, June 1970, 21–27; Kenneth T. Bainbridge, "A Foul and Awesome Display," *Bulletin of the Atomic Scientists*, May 1975, 40–46; Lansing Lamont, *Day of Trinity* (New York: Atheneum, 1965); Wyden, *Day One*, 203–234. The War Department's press release remains a vivid account, reprinted in Henry DeWolf Smyth, *Atomic Energy for Military Purposes* (Stanford, Calif.: Stanford University Press, 1989), appendix 6, 247–254.
21. The visits were February 4–7, February 27–March 2, April 17–20, June 7–10, and June 26–28.
22. Interview with Admiral John "Chick" Hayward by Gregg Herken, March 6, 1996. I would like to thank Gregg Herken for sharing the interview with me.
23. Conant has them arriving at around 8:00 P.M. His account of his experience at Alamogordo, "Notes on the 'Trinity' Test," July 17, 1945, 4:30 P.M., is reprinted in Hershberg, *Conant*, 758–761. Szasz has Groves and Farrell arriving at noon, and Wyden at 7:00 P.M. The distance from Albuquerque to the base camp was approximately 150 miles, in those days probably a four-hour trip.
24. LRG, Comments on *Reach to the Unknown*, February 25, 1967, 4–5, Barbara Storms, Trinity, Folder Carl Zimmerman, Box 6, Entry 7530J, Papers of LRG, RG 200, NARA.
25. LRG, Comments on *Reach to the Unknown*, February 25, 1967, Barbara Storms, Trinity, Folder Carl Zimmerman, Box 6, Entry 7530J, Papers of LRG, RG 200, NARA.
26. "The tent had been put up by men who either weren't experienced in putting up tents or else they thought we'll take care of that General. At any rate

the ropes were loose. There was a very high wind and it slapped pretty badly, slapped enough so that I thought at one time maybe I'd better go out and tighten up those tent ropes. But I didn't and I went to sleep." Interview with LRG by Fred Freed, n.d., 33, Box 4, Entry 7530B, Papers of LRG, RG 200, NARA.

27. In response to a sentence that claimed that one dissenting vote — among Bainbridge, Hubbard, Farrell, and Oppenheimer — could have called off the test, Groves stated, "There was only one dissenting vote that could have called off the test and that was my own. This operation was not run like a faculty meeting. Advice was sought and carefully considered but then decisions were made by those responsible. There was no one but myself to carry this responsibility." LRG, Comments on *Reach to the Unknown,* February 25, 1967, 6, Barbara Storms, Trinity, Folder Carl Zimmerman, Box 6, Entry 7530J, Papers of LRG, RG 200, NARA.

28. General Farrell noted that one of the reasons Groves left the control point was "because of our rule that he and I must not be together in situations where there is an element of danger, which existed at both points." LRG, *Now,* 436. The base camp was 17,000 yards from ground zero, the control point 10,000 yards.

29. LRG, *Now,* 296.

30. There is an apocryphal story about this moment. As Vannevar Bush witnessed the bright light he exclaimed, "It looks like the sun to me." Conant said, "No, it looks like the moon to me." Groves then said, "No, you are both wrong, it looks like a star to me." Wyden's book presented it as fact, but did not document it. He had Bush saying that the blast seemed brighter than a star, whereas Groves pointed to his two stars and said, "Brighter than two stars." Wyden, *Day One,* 213.

31. LRG, *Now,* 296. Conant described the moment: "As soon as I had lowered my dark glass and before rising I shook Gen. Groves hand who said, 'Well, I guess there is something in nucleonics after all.'" Conant, "Notes on the 'Trinity' Test," July 17, 1945, 4:30 P.M., in Hershberg, *Conant,* 760.

32. LRG, *Now,* 438–39. One of the most famous of the Niagara Falls daredevils was Jean François Gravelet (1824–1873), better known as "the Great Blondin, the greatest wonder of the age." On June 30, 1859, Blondin walked across the Niagara River Gorge on a tightrope, completing the 1,000-foot trip in 20 minutes. During that summer he completed eight more crossings, one time carrying his manager on his back. He returned the following year for more crossings, the last of which was completed on September 8, 1860. Though his career was short, his reputation lasted well into the 20th century.

33. "About 40 seconds after the explosion the air blast reached me. I tried to estimate its strength by dropping from about six feet small pieces of paper before, during and after the passage of the blast wave. Since at the time, there was no wind I could observe very distinctly and actually measure the displacement of the pieces of paper that were in the process of falling while the blast was passing. The shift was about 2½ meters, which, at the time, I estimated to correspond to the blast that would be produced by ten thousand tons of T.N.T." E. Fermi, "My Observations During the Explosion at Trinity on July 16, 1945," LANL Archives, A-84-019, 22-6.

34. LRG, *Now,* 298. The *New York Times* science reporter William L. Laurence had Groves saying, "Yes, it is over as soon as we drop one or two on Japan." *Dawn Over Zero: The Story of the Atomic Bomb* (New York: Alfred A. Knopf, 1946), 187.

35. Interview with LRG by Al Christman, May 1967, Naval Historical Center, Washington, D.C. 15–16.

36. Maxwell D. Taylor, *The Uncertain Trumpet* (New York: Harper & Brothers, Publishers, 1960), 3.

37. Almost two years earlier the outlines of this idea were developing, though more than two seemed necessary at the time. In a memo to the vice president, the secretary of war, and the chief of staff, Groves wrote, on behalf of the Military Policy Committee, "What is needed is one decidedly powerful bomb, plus the ability to follow it up with others. If the enemy is wavering, this might readily end the war." LRG to Vice President, Secretary of War, Chief of Staff, August 21, 1943, Top Secret, Folder 23.

38. LRG Appointment Book, July 16, 1945.

39. *FRUS: Conference of Berlin,* II, 1360. According to Groves's diary, Harrison called Groves at 9:50 A.M. Washington time. The time of the outgoing cable was 1524Z, or 10:24 A.M. in Washington. Groves had arranged to send messages directly to Stimson, using Harrison, without going through the normal staff channels. To have done that would have aroused suspicion about why a major general was communicating with the secretary of war.

40. *FRUS: Conference of Berlin,* II, 1360.

41. LRG, *Now,* 303.

42. Waiting in the outer office as Groves came in was James Edward Westcott, a photographer who took thousands of photographs of Oak Ridge, and some of the better-known ones of Groves. Westcott has remembered that "a rough looking guy" came in the door and went to the rear office. Mrs. O'Leary then said to Westcott that he could go in. Groves said to Westcott, "Ed we can't do it right now, I have to prepare a report for the Secretary of War, can we do it later?" After quickly cleaning himself up and changing his uniform Groves was out the door again. Westcott spent 25 years at Oak Ridge and then transferred to AEC headquarters in 1966, and retired in 1977. Interview with James E. Westcott by Robert S. Norris, March 2, 1998.

43. *FRUS: Conference of Berlin,* II, 1360–1361.

44. LRG, *Now,* appendix 8, 433–440. This version does not include the sentence, "It resulted from the atomic fission of about 13½ pounds of plutonium which was compressed by the detonation of a surrounding sphere of some 5000 pounds of high explosive." Sherwin, *Destroyed,* 308. Groves says that it was finished about 2:00 or 2:30 in the morning, typed and retyped by two exhausted secretaries. Interview with LRG by Fred Freed, n.d., 40–41, Box 4, Entry 7530B, Papers of LRG, RG 200, NARA.

45. LRG, Comments on *Japan Subdued* by Herbert Feis, 16, Folder Farrell, Thomas, Box 2, Entry 7530J, Papers of LRG, RG 200, NARA.

46. "Bronx" shipments were those of "irreplaceable material — loss of which would have a serious effect on the war effort." "Bowery" shipments were those of "replaceable material — loss of which would result in several weeks delay." Draft of a Proposed Letter from Assistant CNO for Material (or COMINCH) to Commander Western Sea Frontier, May 23, 1945, Top Secret, Folder 5C, Tab 1, P. The Bowery is a section of Lower Manhattan and the Bronx is a borough of New York City, north of Manhattan.

47. Details about the configuration of the "male" target and the "female" projectile were first published in Coster-Mullen, *Atom Bombs,* 24.

48. The journey is told in some detail in Richard F. Newcomb, *Abandon Ship! Death of the U.S.S.* Indianapolis (Bloomington: Indiana University Press, 1976, first published 1958), 22–45.

49. A later book says that long after the war Capt. Charles B. McVay III would claim that he was told. He was given a sealed envelope and read it when the ship left San Francisco. Dan Kurzman, *Fatal Voyage: The Sinking of the USS* Indianapolis (New York: Atheneum, 1990), 13.

50. Maj. J. A. Derry to Adm. W. S. DeLany, August 17, 1945, Top Secret, File 5.C.3.N.

51. De Silva to LRG, July 21, 1945, Box 19, Entry 3, RG 77, NARA. Pierce was executive officer to the chief of intelligence and security branch.

52. Harlow W. Russ, *Project Alberta: The Preparation of Atomic Bombs for Use in World War II* (Los Alamos, N.M.: Exceptional Books, 1984).

53. Ramsey to Farrell and Parsons, Summary Status Report as of 1 August 1945, August 1, 1945, Box 17, Entry 3, RG 77, NARA. For details of the arming and fusing systems see Coster-Mullen, *Atom Bombs,* 17–19.

54. Unit F33 was dropped near Tinian on August 8, as the last test unit. Unit F32 was the casing for the third bomb. It was returned to Los Alamos in October and disassembled.

55. Maj. J. A. Derry to Adm. W. S. DeLany, August 17, 1945, Top Secret, File 5.C.3.N.

56. Marshall to Handy, July 22, 1945, CM-IN-22566, Top Secret, Folder 5D, Tab 4,E,C4.

57. Marshall to Handy, July 25, 1945, CM-IN-24908, Top Secret, Folder 5D, Tab 4,E,C.1.

58. Handy to Spaatz, July 25, 1945, Folder 5B, Directives, Memos, Etc., Top Secret, RG 77, NARA.

59. The order is Tab B of a July 24 memo Groves wrote to General Marshall describing the plan of operations. Memo, LRG to Chief of Staff, Plan of Operations — Atomic Fission Bomb, July 24, 1945, Folder, Dropping the Bomb, Box 8, Entry, Manhattan Project, RG 319, NARA. Groves went to Handy's office on July 21.

60. Wesley Frank Craven and James Lea Cate, *The Army Air Forces in World War II,* Vol. 5: *The Pacific: Matterhorn to Nagasaki, June 1944 to August 1945* (Chicago: University of Chicago Press, 1953), 36–40.

61. Bowen, *Silverplate,* 113–114 (emphasis added).

62. LRG to RHG, June 16, 1959, Folder Groves, Richard H., Box 4, Entry 7530B, Papers of LRG, RG 200, NARA. Two other official historians note the ambiguity and in passing mention its cause: "The command channels, indeed, were highly irregular . . . the JCS as a body was not involved, and the two important officials above Arnold were Groves and his civilian chief Secretary of War Henry L. Stimson . . . The chain of command was ill understood at Tinian and apparently at Guam; even after the surrender an official report from USASTAF declared that 'due to the fact that the atomic bomb circumvented established command channels for the most part, because of its necessary secrecy, little is known at this level of the authority, which must have originated at a level of approximately the Big Three.'" Wesley Frank Craven and James Lea Cate, *The Army Air Forces in World War II,* vol. 5: *The Pacific: Matterhorn to Nagasaki, June 1944 to August 1945* (Chicago: University of Chicago Press, 1953), 707. As noted,

the command channels excluded the Joint Chiefs of Staff and of course the Combined Chiefs of Staff. The chiefs were informed individually at various times about the bomb but did not deal with it as a body. "During the war years the Joint Chiefs of Staff had been generally aware of the development of atomic research, although this very secret project was not under their authority and JCS discussion of it, if there was any, was never recorded." Grace Person Hayes, *The History of the Joint Chiefs of Staff in World War II: The War Against Japan* (Annapolis, Md.: Naval Institute Press, 1982), 723.

63. In a 1949 letter to Nichols, Groves comments on a misstatement that Robert Bacher had made recently in testimony to Congress. "That was his reference to the fact that the improved bomb used at Eniwetok had not been more than a gleam in someone's eye. He apparently has forgotten entirely Oppenheimer's urgent recommendation to me soon after July 16th, 1945 that we go directly to this model, instead of the previously planned unit scheduled for the second delivery." LRG to Nichols, July 13, 1949, Folder N, Box 6, Entry 7530B, Papers of LRG, RG 200, NARA. Groves was referring to Operation Sandstone in 1948, when composite core designs were first tested.

64. LRG to Oppenheimer, July 19, 1945, Top Secret, File 5.B.I.

65. In Top Secret at 5D.I, Series I, Part I, Folder 5I, Tab A. There are six pages of longhand notes on yellow paper and four pages of original typed copy.

66. Stimson to Harrison, July 23, 1945, CM-IN-23487, Top Secret, Folder 5D, Tab 4, E, C2.

67. Harrison to Stimson, July 24, 1945, CM-OUT-37350, Top Secret, Folder 5D, Tab 4, E, A11.

68. Memo, LRG to Chief of Staff, July 30, 1945, Top Secret, Folder 5B, Tab O.

69. Memo, LRG to Chief of Staff, Plan of Operations — Atomic Fission Bomb, July 24, 1945, Folder, Dropping the Bomb, Box 8, Entry, Manhattan Project, RG 319, NARA.

70. Among the accounts of the mission to Hiroshima are Gordon Thomas and Max Morgan Witts, *Ruin from the Air: The* Enola Gay's *Atomic Mission to Hiroshima* (Chelsea, Mich.: Scarborough House, 1977); Gordon Thomas and Max Morgan Witts, Enola Gay — *Mission to Hiroshima* (Belton, U.K.: White Owl Press, 1995); Joseph L. Marx, *Seven Hours to Zero* (New York: G. P. Putnam's Sons, 1967); Paul W. Tibbets, *Flight of the* Enola Gay (Reynoldsburg, Ohio: Buckeye Aviation Book Company, 1989). Besides Tibbets, other crew members have written accounts, including Jacob Beser, *Hiroshima & Nagasaki Revisited* (Memphis, Tenn.: Global Press, 1988), and George R. Caron and Charlotte E. Meares, *Fire of a Thousand Suns* (Westminster, Col.: Web Publishing Company, 1995).

71. Interview with LRG by Al Christman, May 1967, Naval Historical Center, Washington, D.C., 30–31.

72. LRG, *Now,* 321.

73. LRG, *Now,* 322. Z refers to Greenwich time, which was 6:15 P.M., Sunday evening, August 5, in Washington, while Groves was at the Army-Navy Club having dinner, or 8:15 A.M. Monday morning, August 6, in Hiroshima.

74. John T. Hayward and C. W. Borklund, *Bluejacket Admiral: The Navy Career of Chick Hayward* (Annapolis, Md.: Naval Institute Press, 2000), 139.

75. Lt. Col. J. F. Moynahan to LRG, May 23, 1946, Memo, History, Psychological Warfare, Folder, Dropping the Bomb, Box 8, Entry Manhattan Project, RG 319,

NARA. Lt. Col. John F. Moynahan, *Atomic Diary* (Newark, N.J.: Barton Publishing Co., 1946), 41–43.

76. Crocker Snow, *Log Book: A Pilot's Life* (Privately published, 1997).

77. Capt. Frederick L. Ashworth, USN, "Dropping the Atomic Bomb on Nagasaki," *USNI Proceedings,* January 1958, 12–17.

78. Joseph Laurance Marx, *Nagasaki: The Necessary Bomb?* (New York: The Macmillan Co., 1971); Frank W. Chinnock, *Nagasaki: The Forgotten Bomb* (New York: World Publishing Co., 1969).

79. Tibbets's choice of Sweeney to command the second mission, rather than a more experienced combat pilot, was controversial from the start and has remained so ever since. The latest round came in response to the publication of Sweeney's book, a work with many errors of fact and interpretation, and one that should be used with caution. Maj. Gen. Charles W. Sweeney, with James A. Antonucci and Marion K. Antonucci, *War's End: An Eyewitness Account of America's Last Atomic Mission* (New York: Avon Books, 1997). For a discussion of the errors and the controversy see Coster-Mullen, *Atom Bombs,* 11, 121–122.

80. Coster-Mullen, *Atom Bombs,* 121.

81. The question of who was in charge, Major Sweeney or Commander Ashworth, is another contentious issue of the Nagasaki mission. While Sweeney was aircraft commander, Ashworth, as weaponeer, was responsible for where, when, and how the bomb was dropped. The possibility for confusion of command between Tibbets and Parsons never arose on the error-free Hiroshima mission.

82. A second source obscuring the city may have been the use of a pipe that diverted steam from a power plant into the river, creating condensation clouds. Coster-Mullen, *Atom Bombs,* 66.

83. The issue of whether they should have aborted the mission after the first unsuccessful run on Kokura is also a contentious one among members of the 509th. For details see Coster-Mullen, *Atom Bombs,* 66–67.

84. They had already passed over the original aim point, which was 1.3 miles away.

85. According to Groves, "the aiming point was in the city, east of the harbor," about a mile and a half south of where it was detonated. LRG, *Now,* 343, 345. Groves noted that "the damage was not nearly so heavy as it would have been if the correct aiming point had been used," 346.

86. Farrell to LRG, August 9, 1945, Cable APCOM 5479, Envelope G, Box 20, Entry 3, RG 77, NARA.

87. Groves's copy to Marshall, August 10, 1945, is in Top Secret, File 5.B.R. Marshall's copy with his handwritten sentence is at the George C. Marshall Library in Lexington, Virginia.

88. Stanley Weintraub, *The Last Great Victory: The End of World War II, July/August 1945* (New York: Truman Talley Books/Dutton, 1995), 504–528.

89. John Morton Blum, ed., *The Price of Vision: The Diary of Henry A. Wallace, 1942–1946* (Boston: Houghton Mifflin Co., 1973), 474.

90. Groves later contended that had there been a third mission, Tibbets was to fly it. Groves, Oral History Transcript, April 3, 1970, 16, Folder 3, Box 66, "Hap" Arnold, Murray Green Collection, U.S. Air Force Academy. The attack style and sequence would have been different from the first two missions. Following Nagasaki, the Japanese were aware that the B-29s had come from Tinian and that they flew in a small formation. In response they were taking aggressive defensive measures for a possible third attack. For these reasons Parsons,

Ashworth, and Ramsey recommended that, based upon discussions with "qualified and responsible air force tactical officers . . . the only thing that has a chance of getting through defenses of an alerted suspicious enemy is a single independent aircraft . . . without accompanying observational aircraft." Cable NR-544, August 27, 1945, LANL Archives, A-84-019, 29-2. Russ reported that while they were assembling Fat Man during midmorning of August 8, an unidentified ship appeared off the coast. As it approached men were seen diving from the deck, heading for shore. "We could see clearly that they were not Americans." Security guards arrived and fired on the swimmers and the ship driving them away. Russ and his colleagues suspected that this was a Japanese commando operation. Harlow W. Russ, *Project Alberta: The Preparation of Atomic Bombs for Use in World War II* (Los Alamos, N.M.: Exceptional Books, 1984), 67.

91. LRG to Arnold, August 10, 1945, Top Secret, File 5.B.Q.
92. LRG to Farrell, August 11, 1945, Top Secret, File 5.C.2.N.
93. Message, TA-2324 from Oppenheimer to Major Derry, August 9, 1945, LANL Archives A-87-019,29-13; Hoddeson, et al., *Critical,* 397.
94. LRG, Comments on *The New World,* 145, Box 3, Entry 7530J, Papers of LRG, RG 200, NARA.
95. "Bomb Boss:'Don't Share,'" *Boston Daily Record,* October 15, 1945, 28.
96. Bert J. Cillessen, "Embracing the Bomb: Ethics, Morality, and Nuclear Deterrence in the US Air Force, 1945–1955," *Journal of Strategic Studies* 21, no. 1 (March 1998), 96–134.
97. Quoted in Schaffer, *Wings,* 185.
98. Stimson, Bundy, *Service,* 633.

CHAPTER 21 War Hero for a Day (Mid–August–December 1945)

1. One item in a planning document concerns alternate plans. "It is recommended that the Interim Committee be asked to prepare two alternate plans for release of the story. These two plans should cover: 1. Surrender of Japan before use. 2. Unpredictable result in the test at Trinity." Memo, Consodine to LRG, MED Public Relations Program, June 27, 1945, 3, Folder 000.71, Releasing Information, Box 31, Entry 5, RG 77, NARA. I did not find any documents that concerned the preparation of these alternate plans.
2. Moynahan wrote an account of his experiences soon after the war. Lt. Col. John F. Moynahan, *Atomic Diary* (Newark, N.J.: Barton Publishing Co., 1946). He was assigned as chief of the press branch of Operation Crossroads.
3. Memo, Consodine to LRG, MED Public Relations Program, June 27, 1945, 3, Folder 0000.71, Releasing Information, Box 31, Entry 5, RG 77, NARA.
4. Stimson Diary, July 30, 1945.
5. *LRG, Appointment Book,* April 13, 1945; William L. Laurence, *Dawn Over Zero: The Story of the Atomic Bomb* (New York: Alfred A. Knopf, 1947), xi.
6. "Vast Power Source in Atomic Energy Opened by Science," *New York Times,* May 5, 1940, 1; William Laurence, "The Atom Gives Up," *Saturday Evening Post,* September 7, 1940. William Leonard Laurence (1888–1977) was born in Lithuania and came to the United States in 1905. He served overseas in the Army Signal Corps during World War I. He joined the *New York Times* in 1930. He first won a Pulitzer Prize for reporting in 1937, sharing it with four others for coverage of the Harvard tercentenary conference. He succeeded Waldemar Kaempffert as science editor in 1956, retired in 1964, and died at the age of 89

in Majorca, Spain. Laurence wrote Groves much later that his early article "was written for the express purpose of arousing our leaders in Washington to the great danger to us and to the free world in general." Laurence to LRG, July 21, 1968, Folder Lawrence *(sic)*, William, Box 2, Entry 7530B, Papers of LRG, RG 200, NARA. Groves's first choice had been Lockhart from the Office of Censorship. Lockhart felt he could not be spared and recommended Laurence. Groves, *Now,* 325.

7. Meyer Berger, *The Story of the New York Times: The First 100 Years, 1851–1951* (New York: Simon & Schuster, 1951), 510–524.

8. In his eyewitness account of the mission to Nagasaki he mistakenly thought that the aircraft carrying the bomb was *The Great Artiste,* when in fact it was the *Bockscar.* He was actually aboard *The Great Artiste.* In early August Groves alerted Turner Catledge of the *New York Times* that the paper should be prepared to soon cover a very big story and that its War Department correspondent Sidney Shalett should be ready. On August 7 the *Times* devoted 10 of the newspaper's 38 pages to articles about the atom bomb. Another *Times* correspondent, William H. Lawrence, reported on the bombings from Guam. Ever after Laurence would be referred to as Atomic Bill and Lawrence as Non-Atomic Bill.

9. The War Department's Bureau of Public Affairs released the Nagasaki story. Later Laurence sent the 11 stories to Groves in a form of a booklet titled *The Story of the Atomic Bomb.* Laurence inscribed on the cover, "To General Leslie R. Groves, With profound gratitude for the opportunity he gave me to write these stories about one of the greatest achievements of man — of which he was the major architect."

10. Robert DeVore, "The Man Who Made Manhattan," *Collier's,* October 13, 1945, 12–13ff.

11. Caroline Moseley, "Princeton and the Bomb: Five Professors Recall Their Days at Los Alamos," *Princeton Alumni Weekly,* June 7, 1995, 12–17.

12. Groves reminded them in the foreword that "Persons disclosing or securing additional information by any means whatsoever without authorization are subject to severe penalties under the Espionage Act."

13. H. D. Smyth, "The 'Smyth Report,'" *Princeton University Library Chronicle* 37, no. 1 (spring 1976), 181.

14. "Neither Stimson, Harrison, Bundy, nor Groves wanted to issue the report for its own sake. On the other hand, they recognized that news of the bomb was bound to generate a tremendous amount of excitement and reckless statements by independent scientists. Groves, whom Stimson considered a very conservative man, argued that the report was the lesser evil. Carefully contrived not to reveal anything vital, it would permit the War Department to seize the center of the stage from irresponsible speakers." Hewlett, Anderson Jr., *New World,* 400.

15. Henry DeWolf Smyth, *Atomic Energy for Military Purposes* (Stanford, Calif.: Stanford University Press, 1989).

16. "The Russian translation was set in type by the middle of November, and published early in 1946 in an edition of 30,000. The Smyth Report was distributed widely to scientists and engineers in the Soviet project. It provided them with an overall picture of what the United States had done. Along with the information from intelligence it exercised an important influence on the technical choices made in the Soviet project." David Holloway, *Stalin and the Bomb: The*

Soviet Union and Atomic Energy, 1939–1956 (New Haven, Conn.: Yale University Press, 1994), 173.

17. LRG, Memo to File, The Smyth Report: What It Is and Why It Was Published, Folder, Smyth Report, Box 14, Entry 7530A, Papers of LRG, RG 200, NARA.

18. LRG, Memo to File, The Smyth Report: What It Is and Why It Was Published, 11, 34, Folder, Smyth Report, Box 14, Entry 7530A, Papers of LRG, RG 200, NARA.

19. Marshall to MacArthur, August 12, 1945, Harrison-Bundy Files, Folder 64.

20. Included in the party were Capt. James Nolan, Col. Peer de Silva, Col. Ashley W. Oughterson, Dr. Junod of the Swiss Red Cross, and two Japanese medical officers who were to serve as interpreters. Also present were members of the International Red Cross, army medical corps, MacArthur's staff, and two representatives of the Japanese government. James Newman graduated from the USMA in June 1918, the class ahead of Groves's, and served as an engineer. One of his early assignments was to direct construction of the Washington National Airport.

21. Jane Warren Larson, "Mission to Japan," in James Overholt, ed., *These Are Our Voices: The Story of Oak Ridge 1942–1970* (Oak Ridge, Tenn.: Children's Museum of Oak Ridge, 1987), 236–245.

22. Stafford L. Warren, "The Role of Radiology in the Development of the Atomic Bomb," in Kenneth D. A. Allen, ed., *Radiology in World War II, Medical Department in World War II: Clinical Series* (Washington, D.C.: Office of the Surgeon General, Department of the Army, 1966), 888. Warren first went to Hiroshima arriving on September 10.

23. LRG to Farrell, August 20, 1945, Folder 201F, Box 85, Entry 5, RG 77, NARA.

24. At Farrell's request Colonel Warren came from Okinawa on September 7.

25. Cable, CM-IN-7928, Farrell to LRG, September 10, 1945, Folder, Envelope B General Farrell, Box 17, Entry 3, Tinian Files, RG 77, NARA. The information was sent on to Oppenheimer and Parsons at Los Alamos. Cable, Washington Liaison Office to Commanding Officer, Clear Creek, September 18, 1945, LANL Archives, A-84-019, 29-16.

26. Richard Tanter, "Voice and Silence in the First Nuclear War: Wilfred Burchett and Hiroshima," in Ben Kiernan, ed., *Burchett Reporting the Other Side of the World, 1939–1983* (London: Quartet Books, 1986), 13–40.

27. Moynahan accompanied the journalists and described the day. Col. John F. Moynahan, *Atomic Diary* (Newark, N.J.: Barton Publishing Co., 1946), 62–67.

28. In May 1942 the army inaugurated its A Award program for war work in government and private plants. The navy had had its own E Award program since 1906. In July 1942 both programs, along with the Army-Navy Munitions Board star-award program, were combined into the Army-Navy Production Award, generally known as the Army-Navy E Award. Stars were added for second and further achievements of the award.

29. LRG to Bethe, September 27, 1945, Folder 201B, Box 84, Entry 5, RG 77, NARA.

30. "Woman Aide to Atomic Chief, Four Officers Are Decorated," *Washington Post,* October 31, 1945, 1. At the ceremony in Groves's offices, Brig. Gen. Thomas Farrell received the Distinguished Service Medal, and Lt. Col. Peer de Silva, Lt. Col. John Derry, and Lt. Walter A. Parish received the Legion of Merit.

31. War Department Press Release, "54 Officers Receive Awards for Work on Atomic Bomb Project," for morning papers, November 26, 1945.

32. *LRG Official Records.*

33. Robins to Adjutant General, October 23, 1945, Folder, Biography & Records Concerning the Life of LRG, Box 1, Entry 7530C, Papers of LRG, RG 200, NARA.

34. Lutes to Adjutant General, Award for LRG, March 1, 1946, Folder, Biographies and Records of the Life of LRG, Box 1, Entry 7530C, Papers of LRG, RG 200, NARA.

35. Marshall to the President, September 1945, Folder, Biographies and Records Concerning the Life of LRG, Box 1, Entry 7530C, Papers of LRG, RG 200, NARA.

36. LRG, Speech, September 21, 1945, Folder, Mr. Watson's Luncheon, Box 3, Entry 7530I, Papers of LRG, RG 200, NARA.

37. Over the next decade Groves received six additional honorary degrees: Williams College (June 16, 1946); Hobart College (November 10, 1948); Pennsylvania Military College (June 5, 1951); Saint Ambrose College (June 1, 1952); Hamilton College (June 5, 1955); and Ripon College (November 6, 1955).

38. Memo, LRG to Chief of Staff, August 23, 1945, Folder, Third Red Folder, Box 5, Entry, Manhattan Project, RG 319, NARA.

39. LRG, Memo to File, April 1, 1968, Folder Seaborg, Glenn T., Box 5, Entry 7530J, Papers of LRG, RG 200, NARA. James Conant had offered Oppenheimer a job in the physics department at Harvard but he declined, saying he wanted to live in California. That was not to be: He accepted the position of director of the Institute for Advanced Study at Princeton, New Jersey, a post he held until his death in February 1967.

40. "I believed then, as I do now, that most big advances are made by very young men who have not accomplished anything particularly outstanding and who are therefore not afraid of impairing their reputations." LRG, Memo to File, April 1, 1968, Folder Seaborg, Glenn T., Box 5, Entry 7530J, Papers of LRG, RG 200, NARA.

41. LRG, *Now,* 376. Reflecting on the years since the war Groves said, "Unfortunately, the policy of super-selection has not endured. Its demise is one of the principal reasons why the Army is no longer supreme within the military establishment in the area of nuclear warfare." *Ibid.*

42. *Manhattan District History,* bk. 1, vol. 4, chapter 5, 5.5.

43. The account draws upon LRG, *Now,* 187, 367–372; Harrison-Bundy Files, Folders 7 and 70; Yoshio Nishina, "A Japanese Scientist Describes the Destruction of His Cyclotrons," *Bulletin of the Atomic Scientists,* June 1947, 145, 167; Oral History Interview with R. Gordon Arneson, June 21, 1989, Harry S. Truman Library.

44. John W. Dower, "'NI' and 'F': Japan's Wartime Atomic Bomb Research," in John W. Dower, ed., *Japan in War and Peace* (New York: New Press, 1993), 64. There are a few conspiracy theorists who think that the Japanese program progressed much farther, and that there were cover-ups about it. A recent article laying the myth to rest is Walter E. Grunden, "Hungnam and the Japanese Atomic Bomb: Recent Historiography of a Postwar Myth," *Intelligence and National Security* 13, no. 2 (summer 1998), 32–60.

45. Charles Weiner, "Retroactive Saber Rattling?" *Bulletin of the Atomic Scientists,* April 1978, 11.

46. Quoted in Charles Weiner, "Retroactive Saber Rattling?" *Bulletin of the Atomic Scientists,* April 1978, 11.

47. Arneson, who was working in Harrison's office at the time, recounts that he read the message that Major Britt had brought to Secretary of War Robert Patterson. It had Groves's approval for the immediate destruction of the cyclotrons. Arneson took the message to Harrison, realizing there was a mistake. They tried to contact Van Bush to see what he thought, but he had left for the day. They then called Oppenheimer, who said, according to Arneson, "Ah hell, if this is what Groves wants, let him have it." The message went out with Patterson's signature. Oral History Interview with R. Gordon Arneson, June 21, 1989, Harry S. Truman Library.

CHAPTER 22 **Caught in the Middle: Fights Over Domestic Policy (1946)**

1. The literature on the scientists' movement is vast. For a sampling see Hewlett, Anderson, *New World;* Smith, *Peril;* Lawrence Badash, *Scientists and the Development of Nuclear Weapons* (Atlantic Highlands, N.J.: Humanities Press, 1995); Lawrence S. Wittner, *The Struggle Against the Bomb,* Vol. 1: *One World or None: A History of the World Nuclear Disarmament Movement Through 1953* (Stanford, Calif.: Stanford University Press, 1993): Daniel J. Kevles, *The Physicists: The History of a Scientific Community in Modern America* (Cambridge, Mass.: Harvard University Press, 1995), esp. 287–366; Paul Boyer, *By the Bomb's Early Light: American Thought and Culture at the Dawn of the Atomic Age* (New York: Pantheon Books, 1985), 49–106. For Niels Bohr's role as "a precursor and model for physicists' engagement in the postwar political arena" see Finn Asserud, "The Scientists and the Statesman: Niels Bohr's Political Crusade During World War II," *Historical Studies in the Physical and Biological Sciences* 30, no. 1 (1999), 1–47.
2. Hearings, House Military Affairs Committee, October 9, 1945.
3. Groves appeared on November 28.
4. Memo, LRG to Secretary of War, June 2, 1945, Harrison-Bundy Files, Folder 62.
5. LRG to O'Connell, June 2, 1945, Folder, Letters to Friends, Box 4, Entry 7530C, Papers of LRG, RG 200, NARA; LRG, Memo, December 17, 1962, Box 3, Entry 7530J, Papers of LRG, RG 200, NARA; Memo, LRG to Bundy, June 15, 1945, Harrison-Bundy Files, Folder 62. Several other Manhattan scientists had been invited as well, including Franck, Compton, and Lawrence, among others. They were asked not to attend and complied. Groves said he would have objected to anyone visiting Russia who possessed the type of knowledge that Condon possessed at the time. Unlike the other scientists, "Condon vociferously insisted on making the trip to Moscow." Groves knew of Condon's past affiliations with "a great number of Communist front organizations" and his "insistence on going to Russia led me to lose confidence in the trust we could place in his loyalty to the U.S." LRG, Statement to the FBI, April 22, 1948, Folder, Condon Affair, Box 2, Entry 7530B, Papers of LRG, RG 200, NARA.
6. LRG to Patterson, November 23, 1945, Folder, Post War Legislation, Box 4, Entry, Manhattan Project, RG 319, NARA.
7. Senate Report No. 1211, Special Committee on Atomic Energy, "Atomic Energy Act of 1946," 79th Cong., 2nd Sess., April 19, 1946.
8. Senate Report No. 1211, Special Committee on Atomic Energy, "Atomic Energy Act of 1946," 79th Cong., 2nd Sess., 7–8.

9. Hewlett, Anderson, *New World,* 439–443, 450–451.

10. Other important questions were only briefly discussed or not seriously addressed at all. From day one, when it came to atomic weapons and the Constitution, accountability and democracy normally took second place. Enormous grants of power were given to new bodies under the imprimatur of national security with little oversight and public participation. For early recognition of this see Herbert S. Marks, "The Atomic Energy Act: Public Administration Without Public Debate," *University of Chicago Law Review* 15, no. 4 (summer 1948), 839–854. As the Cold War heated up the civilian-military divide soon became a difference without a distinction.

11. See March 18, 1946, radio address, printed in *Congressional Record* 92, pt. 10 (app.), March 19, 1946, 1467–1468.

12. LRG, Comments on *The New World,* 146, Box 3, Entry 7530J, Papers of LRG, RG 200, NARA.

13. LRG, *Now,* 305.

14. *Washington Post,* January 4, 5, 8, 1946. In March Albert Friendly wrote a lengthy series of articles in the *Washington Post* that was basically antimilitary and pro-scientist, claiming that the army security system hampered development.

15. The following section draws on Little, *Foundations,* 48–76; Rearden, *Formative,* 111–115.

16. LRG, *Now,* 395.

17. Ralph E. Lapp, *Atoms and People* (New York: Harper & Brothers, 1956), 79–80.

18. Lilienthal, *Journals,* 105–110.

19. Lilienthal, *Journals,* 116.

20. LRG, Memo to AEC, November 21, 1946, quoted in *FRUS, Intelligence,* 458–460.

21. Memo, Director of Central Intelligence (Vandenberg) to the National Intelligence Authority, August 13, 1946, quoted in *FRUS, Intelligence,* 394–395. The National Intelligence Authority (CIG's oversight body) was composed of the secretaries of state, war, and navy and the president's representative, Admiral Leahy. The DCI was responsible to the NIA and was a nonvoting member.

22. Minutes of the Sixth Meeting of the National Intelligence Authority, August 21, 1946, quoted in *FRUS, Intelligence,* 395–400; Telegram from the President's Chief of Staff (Leahy) to President Truman, August 21, 1946, Ibid., 401–402.

23. The commission requested additional names of the secretary. Lauris Norstad, chief of the Plans and Operations Division of the General Staff, was responsible for assembling a list of candidates to be presented to the commission. The commission selected Army Air Forces Col. James McCormack, a West Point graduate and Rhodes scholar. Others on the list included Lt. Gen. Wilhelm D. Styer and Lt. Col. Andrew J. Goodpaster. Hewlett, Duncan, *Shield,* 33.

24. Arthur B. Darling, *The Central Intelligence Agency: An Instrument of Government, to 1950* (University Park: Pennsylvania State University Press, 1990), 165. The NIA directive authorized the DCI to coordinate "all intelligence information related to foreign atomic energy developments and potentialities affecting the national security, and to accomplish the correlation, evaluation, and appropriate dissemination within the Government of the resulting intelligence." *FRUS, Intelligence,* 510–511. On July 25, 1947, the NIA agreed to a proposal from Chairman Lilienthal that the AEC become a permanent member of the Intelligence Advisory Board, and on August 5 he designated Rear Adm. John E. Gingrich the AEC director of intelligence, to serve as well on the IAB.

25. Bowen, *Silverplate.*
26. Bowen, *Silverplate,* 161.
27. Also serving were Brig. Gen. W. A. Borden and Col. C. H. Bonesteel of the army, along with Capt. G. W. Anderson Jr., Capt. V. I. Pottle, and Cmdr. W. S. Parsons of the navy.
28. Bowen, *Silverplate,* 193.
29. "Operation Crossroads arose out of a bitter rivalry between the Navy and the Army Air Force over the impact of the atomic bomb on their future budgets, missions and prestige." Jonathan Weisgall, *Operation Crossroads: The Atomic Tests at Bikini Atoll* (Annapolis, Md.: Naval Institute Press, 1994), 8. Lloyd J. Graybar agreed, arguing against those who suggest that Crossroads was an instance of atomic diplomacy in the developing Cold War. "The 1946 Atomic Bomb Tests: Atomic Diplomacy or Bureaucratic Infighting," *Journal of American History* 72, no. 4 (March 1986), 888–907.
30. Little, *Foundations,* 174–209.
31. A question facing the air force was whether a crew that had already dropped a bomb should do so again, or whether a new crew should gain practical experience. Col. Paul W. Tibbets and Maj. Thomas W. Ferebee competed, had the best scores, but were not chosen. Bowen, *Silverplate,* 258–277. For Tibbets's account of the military politics and intrigue behind the selection, see, Paul W. Tibbets, *Flight of the* Enola Gay (Reynoldsburg, Ohio: Buckeye Aviation Book Company, 1989), 245–253.
32. Paul Tibbets recounts speaking with the pilot, Maj. Woodrow P. Swancutt, just before takeoff and telling him the ballistic data they were going to use were wrong. Paul W. Tibbets, *Flight of the* Enola Gay (Reynoldsburg, Ohio: Buckeye Aviation Book Company, 1989), 250–251. An official air force history lays some of the blame on Groves. "Manhattan's reluctance to release complete ballistic information on the FM bomb had contributed to a serious error at CROSSROADS which placed the AAF in an embarrassing light." Little, *Foundations,* 194.
33. Bowen, *Silverplate,* 264.
34. Neal O. Hines, *Proving Ground: An Account of the Radiobiological Studies in the Pacific, 1946–1961* (Seattle: University of Washington Press, 1962), 20–49.
35. LRG, Memo to the Joint Chiefs of Staff, Test "C" Operation Crossroads, August 7, 1946, Decimal File 471.6, Box 5, RG 165, NARA.
36. Memo, Col. K. D. Nichols to LRG, Operations and Production, March 1, 1946, Folder, Fiscal, Box 8, Entry, Manhattan Project, RG 319. NARA.

CHAPTER 23 **"The Best, the Biggest, and the Most": Fights Over International Control**

1. During this period Groves spoke to the Army Navy Staff College (September 20, 1945; February 6, 1946), the Army Industrial College (October 31, 1945; December 11, 1945; January 10, 1946), at West Point (November 12, 1945), Fort Leavenworth (April 18, 1946; September 19, 1946), Fort Belvoir (Atomic Energy Conference, September 8, 1947), Camp Lee (QM School, November 24, 1947), and Fort Benning (Infantry Conference, June 24, 1946), among others. Many of these speaking engagements were classified as secret and not open to the public or press.
2. "Our Army of the Future — As Influenced by Atomic Weapons," U.S. Department of State, *Foreign Relations of the United States,* 1946, vol. 1, (Washington, D.C.: U.S. Government Printing Office, 1972), 1197–1203. In a

memo six days later he repeated many of the themes, LRG to Brig. Gen. C. C. Alexander, On the Effect of Atomic Energy on the Army of the Future, January 8, 1946, Folder, First Red Folder, Box 5, Entry, Manhattan Project, RG 319, NARA.

3. Russell D. Buhite and Wm. Christopher Hamel, "War for Peace: The Question of an American Preventive War Against the Soviet Union, 1945–1955," *Diplomatic History* 14, no. 3 (summer 1990), 367–384. Many others held similar views. Not surprising were Gens. Henry H. Arnold, Ely Culbertson, Carl Spaatz, and Curtis E. LeMay of the air force, but civilians such as W. Stuart Symington and even Bertrand Russell joined their ranks. The idea retained currency. An attack on China to prevent it from becoming a nuclear power was considered by Presidents Kennedy and Johnson and their advisers in the early 1960s. William Burr and Jeffrey T. Richelson, "Whether to 'Strangle the Baby in the Cradle': The United States and the Chinese Nuclear Program, 1960–64," *International Security* 25, no. 3 (winter 2000–2001), 54–99.

4. Talk by LRG before War Department Protective Construction Board, May 24, 1946, Classified Secret, Folder, Protective Construction Board, War Department, Box 4, Entry 7530I, Papers of LRG, RG 200, NARA.

5. Remarks by LRG before the Command and General Staff School, Fort Leavenworth, Kans., September 19, 1946, 26, Folder, Fort Leavenworth, Box 4, Entry 7530I, Papers of LRG, RG 200, NARA.

6. Robert S. Norris, William M. Arkin, and William Burr, "Where They Were," *Bulletin of the Atomic Scientists,* November–December 1999, 26–35.

7. Remarks by LRG before the Command and General Staff School, Fort Leavenworth, Kans., September 19, 1946, 28, Folder, Fort Leavenworth, Box 4, Entry 7530I, Papers of LRG, RG 200, NARA.

8. Robert Jungk, *Brighter Than a Thousand Suns* (New York: Harcourt, Brace and Co., 1958), 231. His estimates were never hard and fast. In October 1947 he predicted 15 to 20 years if they did it in secrecy without aid from the United States, Britain, or Switzerland. United Press, "Russians Far from A-Bomb?" October 3, 1947. In the fall of 1948 he estimated not before 1955. Joseph L. Myler (UP), "Reds Got Atom Secrets, Couldn't Unravel Them!" October 27, 1948. In the spring of 1949 Groves told the local chamber of commerce that it would be 11 more years. "Russian Atom Bomb? — Not for 11 Years, Says General Groves," *Utica Daily Press,* April 20, 1949.

9. Charles A. Ziegler, "Intelligence Assessments of Soviet Atomic Capability, 1945–1949: Myths, Monopolies and *Maskirovka,*" *Intelligence and National Security* 12, no. 4 (October 1997), 1–24.

10. James Byrnes told Leo Szilard, "General Groves tells me there is no uranium in Russia." Donald Fleming and Bernard Bailyn, eds., *The Intellectual Migration: Europe and America, 1930–1960* (Cambridge, Mass.: Harvard University Press, 1969), 126.

11. Between 1945 and 1950 there were three major discoveries in the Soviet Union. Just one of them — the Vismut (or Wismut) mines, as they were known — operated from 1946 until 1990, yielding approximately 220,000 tons of low-grade yellowcake. Groves was unaware of these early activities. David Holloway, *Stalin and the Bomb: The Soviet Union and Atomic Energy, 1939–1956* (New Haven, Conn.: Yale University Press, 1994), 111–112.

12. Merle Miller, *Plain Speaking: An Oral Biography of Harry S. Truman* (New York:

Berkeley Publishing, 1973), 228. Groves gave speeches at the Waldorf on February 26, 1946, and on March 14, 1946. In his published speech given on February 26 there is no mention of Russia. *Chemical & Metallurgical Engineering*, March 1946, 101, 104–105.

13. LRG to Nat C. Finney, July 8, 1949, Folder, 1948–49 Speeches, Box 6, Entry 7530I, Papers of LRG, RG 200, NARA.

14. LRG, "Atom General," 101.

15. Conant to LRG, February 9, 1949, Folder Cas–Col, Box 2, Entry 7530B, Papers of LRG, RG 200, NARA.

16. Charles A. Ziegler and David Jacobson, *Spying Without Spies: Origins of America's Secret Nuclear Surveillance System* (Westport, Conn.: Praeger, 1995), 4.

17. *The Roswell Report: Fact Versus Fiction in the New Mexico Desert* (Washington, D.C.: U.S. Government Printing Office, 1995).

18. Benson Saler, Charles A. Ziegler, and Charles B. Moore, *UFO Crash at Roswell: The Genesis of a Modern Myth* (Washington, D.C.: Smithsonian Institution Press, 1997).

19. LRG to Considine, July 7, 1947, Reel 2947, RG 18, NARA. I would like to thank Capt. James McAndrew for bringing this to my attention. The first part of his quote was published. AP, "Game of Spotting 'Flying Saucers' Sweeps Country as Mystery Holds," July 8, 1947.

20. *Public Papers of the Presidents, Harry S. Truman, 1945* (Washington, D.C.: General Services Administration; National Archives and Records Service, Office of the Federal Register, 1961), 212–213.

21. Paragraph 1 states: "Every treaty and every international agreement entered into by any Member of the United Nations after the present Charter comes into force shall as soon as possible be registered with the Secretariat and published by it."

22. LRG, *Now*, 411.

23. Steven M. Neuse, *David E. Lilienthal: The Journey of an American Liberal* (Knoxville: University of Tennessee Press, 1996), 168–177. This came at the recommendation of Acheson's assistant, Herbert Marks.

24. There was no love lost in Lilienthal's feelings toward Groves either. Groves wondered if it might have been because he was not consulted about Oak Ridge.

25. The final report is *A Report on the International Control of Atomic Energy*. Prepared for the Secretary of State's Committee on Atomic Energy (Washington, D.C.: U.S. Government Printing Office, March 7, 1946, 3.

26. The evolution of the Acheson-Lilienthal Report is treated extensively in Hewlett, Anderson, *New World*, 531–554; Joseph I. Lieberman, *The Scorpion and the Tarantula: The Struggle to Control Atomic Weapons 1945–1949* (Boston: Houghton Mifflin Co., 1970), 235–259.

27. Lilienthal, *Journals*, 30 (March 19, 1946). Lilienthal went on, "We need a man who is young, vigorous, not vain, and whom the Russians would feel isn't out simply to put them in a hole, not really caring about international cooperation. Baruch has none of these qualifications."

CHAPTER 24 Chief of Special Weapons (1947–1948)

1. At about this time he was asked what was his favorite hymn. He responded it was a Protestant hymn of the 17th century by John Bunyan, "He Who Would Valiant Be 'Gainst All Disaster." The thoughts expressed in the verses are about an individual carrying a heavy burden:

> *Who so beset him round*
> *With dismal stories,*

> *Do but themselves confound,*
> *His strength the more is.*
> *No foes shall stay his might,*
> *Though he with giants fight;*
> *He will make good his right*
> *To be a pilgrim.*

Letter, Lambert to Strauss, October 16, 1946, Folder, Thru December '46, Box 2, Entry 7530C, Papers of LRG, RG 200, NARA.

2. The commanding officer of Pratt, Col. Clyde M. Beck, wrote to Groves on January 10, trying to clarify whether he was coming as a patient or for a vacation. Beck had obviously been ordered by the Office of the Chief of Staff to provide accommodations for Groves (and Mrs. Groves), and from the tone of his letter was not too happy about it. Beck to LRG, January 10, 1947, Folder Groves, L. R. — Letters to Friends, Box 4, Entry 7530C, Papers of LRG, RG 200, NARA.

3. LRG to Baruch, February 19, 1947, Folder Bac–Bri, Box 1, Entry 7530B, Papers of LRG, RG 200, NARA. "As I told you, I will not make any definite decision as to my future until my return from Florida some time next month, but I can now see no alternative except to retire from the Army."

4. Lampert would later be superintendent of the USMA from June 1963 to January 1966. Groves was president of the Association of Graduates from 1961 to 1965.

5. I would like to thank Maj. Gen. Robert P. Young, USA, retired, for sharing with me his correspondence with Groves for the period December 1946 through October 1962. General Holbrook was Groves's classmate and a good friend.

6. Interview with Lt. Gen. Kenneth Cooper, USA, retired, by Robert S. Norris, November 24, 1997. Groves's appointment book shows that he attended an MLC meeting the morning of July 22, and spent the afternoon at the Army-Navy Country Club.

7. "I considered it entirely impractical and dangerous to national security for the AEC rather than the military to have custody of weapons." Nichols, *Trinity,* 245.

8. Necah Stewart Furman, *Sandia National Laboratories: The Postwar Decade* (Albuquerque: University of New Mexico Press, 1990), 119–145.

9. Z Division was named for Jerrold R. Zacharias. Zacharias had worked with the radar group at MIT's Radiation Laboratory for five years. Oppenheimer had chosen him to replace Parsons and head the division that would now carry out the functions once performed by O Division. Jack S. Goldstein, *A Different Sort of Time: The Life of Jerrold R. Zacharias, Scientist, Engineer, Educator* (Cambridge, Mass.: MIT Press, 1992), 66–68.

10. Little, *Foundations,* 77–80.

11. *First History of AFSWP 1947–1954,* Vol. 1: *1947–1948,* chapter 4, Headquarters, section 1. After October 6–7, 1947, AFWSP moved to the second floor, corridor 2, which had better security with a single guarded doorway for entrance and exit.

12. Little, *Foundations,* 83.

13. LRG to Brig. Gen. C. C. Alexander, On the Effect of Atomic Energy on the Army of the Future, January 8, 1946, 6–7, Folder, First Red Folder, Box 5, Entry, Manhattan Project, RG 319, NARA.

14. U.S. Army Corps of Engineers Unit History of the 2761st Engineer Battalion (Special), 19 August 1946–31 December 1947.

15. *Engineer Memoirs, Colonel Gilbert M. Dorland, U.S.A. Retired,* Oral History interview, April 1–3, 1987, Office of History, Headquarters, U.S. Army Corps of Engineers. On May 1, 1947, the 2761st Engineer Battalion (Special) was redesignated the 38th Engineer Battalion (Special), and in July Brig. Gen. Robert M. Montague was assigned as commanding general, Sandia Base.

16. The history of the bases under the Manhattan District is *Manhattan District History,* bk. 1, vol. 4, Top Secret Supplement: Storage Project. The history of bases under AFSWP for the years 1947 and 1948, *First History of AFSWP 1947–1954,* vol. 1: *1947–1948,* chapter 4, Headquarters, 4.4.10–4.4.24; chapter 5, Sandia Base, 5.43–5.44; chapter 6, Storage Bases.

17. Little, *Foundations,* 101, 505–536.

18. The Mark IV would not enter the stockpile until mid-1949.

19. Hawkins, *Trinity,* 240.

20. Office of the Assistant to the Secretary of Defense (Atomic Energy), *History of the Custody and Deployment of Nuclear Weapons, July 1945 through September 1977* (Washington, D.C.: Department of Defense, February 1978); Little, *Foundations,* 66–73; Rearden, *Formative,* 425–432.

21. Joint Committee on Atomic Energy, *Atomic Energy Legislation Through 93d Congress, 1st Session,* July 1974, 282.

22. Little, *Foundations,* 524.

23. "During this period the physical possession of the slowly increasing atomic stockpile was held by the ASWP, [deleted]. Technical custody was transferred to the AEC in the fall of 1947, but no change was made in the location of the weapons, and the AFSWP remained responsible for their physical protection. Meanwhile the question of custody as a matter of policy remained a moot point between the military services and the AEC, to be finally clarified by a decision of the president in July 1948." Little, *Foundations,* 505–506.

24. *Engineer Memoirs, Colonel Gilbert M. Dorland, U.S.A. Retired,* Oral History interview, April 1–3, 1987, Office of History, Headquarters, U.S. Army Corps of Engineers, 150.

25. *Engineer Memoirs, Colonel Gilbert M. Dorland, U.S.A. Retired,* Oral History interview, April 1–3, 1987, Office of History, Headquarters, U.S. Army Corps of Engineers, 150. "Later on, when Paul Ellis got into it, I would still do the typing out, and put one copy in the safe, burn the carbon, seal the envelope, hand it to him or perhaps some other courier, who would take it to Washington and hand it to General Groves, not to anybody else. Hand it to him personally. There were two copies of what we had, and it varied from month to month because some of the components for the nuclear had short half-lives." Ibid., 151. Dorland is referring to polonium initiators; polonium has a 140-day half-life.

26. Rearden, *Formative,* 440; Gregg Herken, *The Winning Weapon: The Atomic Bomb in the Cold War, 1945–1950* (New York: Vintage Books, 1982), 197.

27. Lilienthal, *Journals,* 165–166. Because the briefing was delivered orally, there is no written record of the number he mentioned. David Alan Rosenberg estimates that there were probably seven complete bombs at the time. "U.S. Nuclear Stockpile, 1945–1950," *Bulletin of the Atomic Scientists,* 1982.

28. Steven T. Ross, *American War Plans, 1945–1950* (London: Frank Cass, 1996), 15. As Ross noted (p. 22), the documents were JIC 329, Strategic Vulnerability of

the U.S.S.R. to a Limited Air Attack, November 3, 1945, and JIC 329/1, December 3, 1945. The target list probably came from a memo dated September 15, 1945, from General Norstad to General Groves in which Norstad listed 66 targets. "They [the JIC] were not until 1947 cleared to receive stockpile data, and when the Chiefs required such information, they received oral briefings."

29. Estimated from official data, which shows that at the end of June 1946, on the eve of the Crossroads tests at Bikini, which would use up 2, there were 9 plutonium cores, and a year later only 13. By the middle of 1948 the number had risen to 50. Rearden, *Formative*, 439. The corresponding numbers of non-nuclear components were 9, 29, and 53, respectively.

CHAPTER 25 **Retirement and a New Career (1948–1961)**

1. Handwritten Memo, LRG to Brereton, July 13, 1947, and Draft Memo, (by LRG), Brereton to Lilienthal, July 13, 1947, both in Folder 201, LRG Correspondence 1941–47, Box 86, Entry 5, RG 77, NARA.

2. Martin Agronsky radio broadcast, July 24, 1947, Folder 201, LRG 1947 G-O-C, Box 2, Entry 7530C, Papers of LRG, RG 200, NARA.

3. Ibid. Arthur Sylvester was assistant secretary of defense for public affairs for six years, from 1961 to February 1967. Groves wrote a memo to the file after reading an article that Sylvester wrote for the *Saturday Evening Post* (November 18, 1967) titled "The Government Has the Right to Lie." Groves said that his experience with him 20 years ago convinced him "that he was probably the most untrustworthy newspaper reporter in Washington. He constantly endeavored to seek out secret information; the publication of which he knew would be detrimental to the best interests of the United States. . . . He was one of the few reporters that I cordially disliked, on the grounds of my size-up of his lack of integrity, as well as his generally disagreeable manner. . . . He was a typical McNamara selection for high office in the Defense Department." LRG, Comments on Sylvester Article, Folder Sylvester, Arthur, Box 5, Entry 7530J, Papers of LRG, RG 200, NARA.

4. LRG, Memo to File, Promotion to Lieutenant General, n.d., Folder Stu–Sz, Box 10, Entry 7530B, Papers of LRG, RG 200, NARA.

5. Ibid.

6. Lilienthal, *Journals,* 250.

7. Hershberg, *Conant,* 357.

8. Bush to Conant, November 4, 1946, Folder Conant, James B. 1939–46, Box 27, Bush Papers, LOC.

9. Conant to Oppenheimer, January 5, 1948, Folder Conant, James B., Correspondence, Box 27, Oppenheimer Papers, LOC. After leaving the MED as district engineer in February 1947, Nichols went to West Point as a professor of mechanics. He returned to Washington after Groves's retirement at the end of February 1948, replacing him as chief of the Armed Forces Special Weapons Project and as one of the two army members of the Military Liaison Committee.

10. Lilienthal, *Journals,* 288; Hewlett, Duncan, *Shield,* 151–152.

11. Efficiency Report, Leslie R. Groves, Rating Officer, J. Lawton Collins, March 8, 1948. *LRG Official Records.*

12. See Senate Bill S.2223 and Senate Report No. 1246, "Authorizing the Promotion of Lt. Gen. Leslie Richard Groves to the Permanent Grade of Major General, United States Army," May 6, 1948. Rep. W. Sterling Cole of New York introduced a similar bill on the House side. See H.R.5596 and House Report No. 2231, "Promotion of Leslie R. Groves to Major General," June 7, 1948.

13. Kenneth C. Royall to Sen. Chan Gurney, March 1, 1948; Royall to Rep. Walter G. Andrews, March 10, 1948. Folder, Biographies and Records Concerning the Life of LRG, Box 1, Entry 7530C, Papers of LRG, RG 200, NARA.

14. Private Law 394A, 80th Cong., June 24, 1948.

15. LRG to RHG, February 11, 1948. Author's collection.

16. "I am gradually getting settled in Connecticut having yesterday signed an agreement to purchase a house at an outlandish figure but it was the best I could do in this locality." LRG to Consodine, March 19, 1948, Folder, Condon Affair, Box 2, Entry 7530B, Papers of LRG, RG 200, NARA.

17. In addition to Groves, Gen. Douglas MacArthur would become chairman of the board of Remington Rand in 1952 after his political hopes faded. MacArthur was probably introduced to Rand in the 1930s by fellow West Pointer Robert E. Wood (class of 1900), head of Sears Roebuck. Rand traveled in conservative Republican circles. MacArthur's chief function as chairman was to announce dividends and speak at banquets criticizing the income tax, swollen federal budgets, the excessive power of labor unions, and the Socialistic tendencies in the government and society, all views firmly held by Rand.

18. James W. Cortada, *Historical Dictionary of Data Processing Organizations* (New York: Greenwood Press, 1987), "Remington Rand Corporation," 235–241.

19. Scott McCartney, ENIAC: *The Triumph and Tragedies of the World's First Computer* (New York: Walker and Co., 1999). ENIAC was an acronym for "electronic numerical integrator and computer." Eckert (1919–1995), Mauchly (1907–1980).

20. Since the Manhattan Project the designing of new nuclear weapons has been one of the most important driving forces in the development of ever-more-powerful computers. Donald Mackenzie, "The Influence of Los Alamos and Livermore National Laboratories on the Development of Supercomputing," *Annals of the History of Computing* 13, no. 2 (1991), 179–201; Francis H. Harlow and N. Metropolis, "Computing and Computers: Weapons Simulation Leads to the Computer Era," *Los Alamos Science,* winter–spring 1983, 132–141. The marriage of the two technologies is still very much intact. It has been estimated that a team of scientists using calculators of the 1940s would take five years to solve what takes a Cray computer one second. Department of Energy, *The Need for Supercomputers in Nuclear Weapons Design,* January 1986, 11.

21. George Gray, "Engineering Research Associates and the Atlas Computer," *Unisys History Newsletter* 3, no. 3 (June 1999).

22. The navy cryptology group was the Communication Supplementary Activity — Washington (CSAW), located on Nebraska Avenue in upper Northwest at Ward Circle. In 1947, with the creation of the National Security Agency, the CSAW became the naval component.

23. Sperry Rand later dropped the *Rand* part of its name and merged with the Burroughs Corporation in 1986.

24. From December 15, 1950, to January 25, 1951, they visited England, France, Germany, Belgium, Holland, Switzerland, and Italy. In March 1954 they went

to Jamaica, and the following March to Cuba. From May 16 to June 4, 1955, they toured Argentina, Chile, Brazil, Peru, Colombia, and Venezuela.

25. LRG, Draft Speech, The Atom Bomb, Folder, Public Relations Article Misc, Box 12, Entry 7530A, Papers of LRG, RG 200, NARA.

26. Wallace to LRG, March 7, 1951: Wallace to LRG, June 22, 1951; LRG to Wallace, July 3, 1951; Folder W, Box 10, Entry 7530B, Papers of LRG, RG 200, NARA. See also John C. Culver and John Hyde, *American Dreamer: The Life and Times of Henry A. Wallace* (New York: W. W. Norton, 2000), 513-516.

27. "General Groves Rips Foreign Policy Run by Crisis — It's Military, Economic Folly, He Says," *Chicago Daily Tribune,* April 18, 1952.

28. LRG, Radio Broadcast, We the People, February 10, 1950. Folder, Slater, Bill-Americans Speak Up, Box 5, Entry 7530 J, Papers of LRG, RG 200, NARA.

29. Paul Martin, "Leslie Groves May Run For U.S. Senate Next Year," November 4, 1949, GNS News Service. As it turned out, McMahon was reelected in 1950 but died on July 28, 1952. Prescott Sheldon Bush was elected as a Republican to fill the vacancy and served out the term. Bush was reelected in 1956 and served until January 1963.

30. LRG to Winthrop W. Aldrich, August 14, 1952, Folder Ac–Az, Box 1, Entry 7530B, Papers of LRG, RG 200, NARA.

31. Memorandum, LRG to the Secretary of War, March 24, 1947, Folder Oppenheimer, J. Robert, Box 4, Entry 7530J, Papers of LRG, RG 200, NARA.

32. LRG to Lilienthal, November 14, 1946, Folder 201, Lilienthal, David, F., Box 87, Entry 5, RG 77, NARA.

33. LRG to Oppenheimer, May 18, 1950, Folder Oppenheimer, J. Robert, Box 7, Entry 7530B, Papers of LRG, RG 200, NARA. Also LRG to Oppenheimer, May 18, 1950, Folder LRG, Box 36, Oppenheimer Papers, LOC.

34. Barton J. Bernstein, "The Oppenheimer Loyalty-Security Case Reconsidered," *Stanford Law Review* 42, no. 6 (July 1990), 1383–1484; Philip M. Stern, *The Oppenheimer Case: Security on Trial* (New York: Harper & Row, 1969); John Major, *The Oppenheimer Hearing* (New York: Stein and Say, 1971).

35. USAEC, *Oppenheimer,* 165, 167, 171.

36. LRG, Memo with Reference to Miss Robb's Article, Folder Oppenheimer, Box 4, Entry 7530J, Papers of LRG, RG 200, NARA.

37. Prior to 1946 it was called the *Biographical Register of the Officers and Graduates of the United States Military Academy,* or Cullum's Register, for Bvt. Maj. Gen. George W. Cullum, who initiated the work in the mid–19th century. The volumes are a great resource and were essential in tracing the many officers whose careers intersected with Groves's.

38. These include Willard A. Holbrook, November 1918; John R. Jannarone, 1938; Charles P. Gross, 1914; Kenneth D. Nichols, 1929; Hugh J. Casey, June 1918; Kenneth E. Fields, 1933. The 46th superintendent of the USMA — June 1963 to January 1966 — Maj. Gen. James B. Lampert (class of 1936), served as Groves's executive officer with the MED from March 19, 1946, to February 12, 1947, and then served as chief of the AFSWP's personnel and administration division from February 28, 1947, to June 15, 1948.

39. The first recipient of the Thayer Award was Ernest O. Lawrence, in 1958. Beginning in 1957 there was an annual award — named for Leslie R. Groves — given to the cadet with the highest rating in nuclear physics.

40. Minutes of the 93rd Annual Meeting, AOG, USMA, *Assembly,* summer 1962,

17. For an account of the day see 12–13; the speech is reprinted on 14–15. The Thayer Medals for 1963, 1964, and 1965 were awarded to John J. McCloy, Robert A. Lovett, and James B. Conant, respectively.

CHAPTER 26 **Last Years (1962–1970)**

1. The publisher, Harper & Brothers, insisted on the title even though it had published a book by the same title in 1920. The English author of that earlier volume, Philip Gibbs, was one of five journalists who had been selected by the government to become official World War I correspondents and given access to the front lines. Gibbs's account of the horrors of war made it a best-seller. One of the working titles that Groves proposed was *Calculated Risk.*

2. There is some confusion over the amount of records. According to an early army inventory the Manhattan District, records totaled approximately 12,300 feet in June 1946. The Adjutant General's Office, Inventory of the Records of the Manhattan Engineer District, 1942–1948, Inventory NR 303 (Part I), Washington, D.C., November 1956, 3. Presumably this meant cubic feet. A later NARA publication refers to 12,300 linear feet (2.3 miles!). NARA Microfilm Publication Pamphlet Describing M1109, Correspondence, Top Secret, of the Manhattan Engineer District, 1942–1946 (Washington, D.C., 1982), 3.

3. LRG to Hadden, September 20, 1948, Folder Hadden, Gavin, Box 3, Entry 7530B, Papers of LRG, RG 200, NARA.

4. National Archives Microfilm A1218, *Manhattan District History,* 14 reels.

5. Letter, LRG to Nichols, February 29, 1948, Folder N, Box 6, Entry 7530B, Papers of LRG, RG 200, NARA. Approximately 72 linear feet were retained by General Groves. The Adjutant General's Office, Inventory of the Records of the Manhattan Engineer District, 1942–1948, Inventory NR 303 (Part I), Washington, D.C., November 1956, 3.

6. Letter, LRG to Nichols, February 29, 1948, Folder N, Box 6, Entry 7530B, Papers of LRG, RG 200, NARA.

7. The 125 Top Projects are listed in *Engineering News-Record,* July 26, 1999. The criteria used were that the project must have been completed in the past 125 years; have used advanced construction methods or technology; have demonstrated outstanding design or overcome unusual design challenges; have been the first of its kind; have made a major positive impact on the quality of life; have overcome major construction challenges; have become larger than life over time; and have demonstrated geographic and market diversity. The 125 Top People are detailed in the August 30, 1999, issue of *ENR.*

8. Henry Guerlac, "The Atomic Age: Its Birth," *New York Herald Tribune,* March 25, 1962, 5.

9. David Hawkins, book review, *Saturday Review,* May 12, 1962. The book was also reviewed in the *New York Times Book Review,* April 1; *Chicago Tribune,* April 8; and *World Politics,* October 1962, 125. "Groves is a blunt, frank man and his book happily reflects these characteristics."

10. LRG, Comments on Lansing Lamont's *Day of Trinity,* n.d., Folder Savage and Storm, Box 5, Entry 7530J, Papers of LRG, RG 200, NARA.

11. As Groves pointed out, there are no senior individuals acknowledged in the credits. LRG, Comments on *The New World,* 2, Box 3, Entry 7530J, Papers of LRG, RG 200, NARA. From the references in his comments, it is evident that Groves wrote it in 1969.

12. LRG, Comments on *Japan Subdued* by Herbert Feis, Folder Farrell, Thomas, Box 2, Entry 7530J, Papers of LRG, RG 200, NARA.

13. LRG, Comments on *The New World,* Box 3, Entry 7530J, Papers of LRG, RG 200, NARA. "They relied almost entirely on the written record. Security requirements were such that written records were kept to a minimum and almost nothing was written that could be handled verbally. As a result of their attitude many errors are included in what could and should have been an extremely accurate account." LRG, Comments on *Reach to the Unknown,* February 25, 1967, Barbara Storms, Trinity, Folder Carl Zimmerman, Box 6, Entry 7530J, Papers of LRG, RG 200, NARA.

14. Not to be confused with Lauchlin B. Currie, a senior administrative assistant to President Roosevelt and a Soviet spy.

15. Michael Blow, *The History of the Atomic Bomb* (New York: American Heritage Publishing Co., 1968).

16. LRG to Thorndyke *(sic),* September 3, 1968, with Attachment, Folder Thorndyke, Joseph J., Box 5, Entry 7530J, Papers of LRG, RG 200, NARA.

17. LRG, Comments on Article by Secretary of War Henry L. Stimson, *Harpers* Magazine, February 1947, Explains Why We Used the Atomic Bomb, n.d., Papers of LRG, RG 200, NARA. For background to Stimson writing the article see Bernstein, "Seizing," 35–72; Kai Bird, *The Color of Truth: McGeorge Bundy and William Bundy, Brothers in Arms: A Biography* (New York: Simon & Schuster, 1998), 88–98.

18. LRG to Gottleib, ca. November 8, 1963, Folder G, Box 3, Entry 7530B, Papers of LRG, RG 200, NARA.

19. Gottleib to LRG, November 20, 1963, Folder G, Box 3, Entry 7530B, Papers of LRG, RG 200, NARA.

20. LRG to Kyle, December 1, 1963, Folder K, Box 3, Entry 7530B, Papers of LRG, RG 200, NARA.

21. LRG, Memo to File, November 16, 1963, Folder F, Box 3, Entry 7530B, Papers of LRG, NARA.

22. Kyle to LRG, December 13, 1963, Folder K, Box 3, Entry 7530B, Papers of LRG, RG 200, NARA.

23. Kyle to LRG, January 20, 1964, Folder K, Box 3, Entry 7530B, Papers of LRG, RG 200, NARA.

24. Richard Wilson Groves would follow his father and grandfather to West Point, where he graduated in the class of 1975.

25. LRG, Comments on an Editorial in the *New York Times* of February 15, 1969, Folder, General Views, Box 6, Entry 7530J, Papers of LRG, RG 200, NARA.

26. Fred Kaplan, *The Essential Gore Vidal* (New York: Random House, 1999), 713–714.

27. Letters and Memos, Folder Dzau, Linson, Box 1, Entry 7530J, Papers of LRG, RG 200, NARA.

28. Letter, To Whom It May Concern, Charles E. Rea, April 2, 1970. Folder, Biographies and Records Concerning the Life of LRG, Box 1, Entry 7530C, Papers of LRG, RG 200, NARA.

29. Walter Reed General Hospital, Crown Report 90, July 14, 1970.

30. USAEC Release, No. N-124, July 14, 1970.

31. *Washington Daily News,* July 16, 1970.

32. *Washington Post,* July 15, 1970.

33. *Chicago Daily News,* July 18–19, 1970.
34. Memorials to Groves are few in number. A remark made by one of his class-
    mates at their 30th reunion dinner in 1948 was prescient. "Groves has done a
    greater thing for his country than any other man in its history. He will never
    be truly appreciated." Note to File, 10-21-49, Box 4, Entry 7530M, Papers of
    LRG, RG 200, NARA. There is a park named for Groves on the banks of the
    Columbia River in Richland, Washington. There is a bust of the general in the
    West Point Library. It was sculpted by Walker Hancock and unveiled in a cere-
    mony on April 19, 1976. The donor was Mrs. R. McCormick Shields, a
    neighbor in Darien. When Mr. Shields died, Groves had helped Mrs. Shields
    with the education of her two sons. Groves's private collection of books was
    donated to the library, as were a series of useful papers not available in the
    National Archives.

INDEX

Throughout this index, the initials *LRG* indicate references to Gen. Leslie R. Groves. Page numbers preceded by letters indicate photos in photo sections A, B, and C. Page numbers higher than 546 indicate material in Notes section.